Nanoscience for the Conservation of Works of Art

RSC Nanoscience & Nanotechnology

Series Editors:
Professor Paul O'Brien, *University of Manchester, UK*
Professor Sir Harry Kroto FRS, *University of Sussex, UK*
Professor Ralph Nuzzo, *University of Illinois at Urbana-Champaign, USA*

Titles in the Series:
1: Nanotubes and Nanowires
2: Fullerenes: Principles and Applications
3: Nanocharacterisation
4: Atom Resolved Surface Reactions: Nanocatalysis
5: Biomimetic Nanoceramics in Clinical Use: From Materials to Applications
6: Nanofluidics: Nanoscience and Nanotechnology
7: Bionanodesign: Following Nature's Touch
8: Nano-Society: Pushing the Boundaries of Technology
9: Polymer-based Nanostructures: Medical Applications
10: Metallic and Molecular Interactions in Nanometer Layers, Pores and Particles: New Findings at the Yoctolitre Level
11: Nanocasting: A Versatile Strategy for Creating Nanostructured Porous Materials
12: Titanate and Titania Nanotubes: Synthesis, Properties and Applications
13: Raman Spectroscopy, Fullerenes and Nanotechnology
14: Nanotechnologies in Food
15: Unravelling Single Cell Genomics: Micro and Nanotools
16: Polymer Nanocomposites by Emulsion and Suspension
17: Phage Nanobiotechnology
18: Nanotubes and Nanowires: 2nd Edition
19: Nanostructured Catalysts: Transition Metal Oxides
20: Fullerenes: Principles and Applications, 2nd Edition
21: Biological Interactions with Surface Charge Biomaterials
22: Nanoporous Gold: From an Ancient Technology to a High-Tech Material
23: Nanoparticles in Anti-Microbial Materials: Use and Characterisation
24: Manipulation of Nanoscale Materials: An Introduction to Nanoarchitectonics
25: Towards Efficient Designing of Safe Nanomaterials: Innovative Merge of Computational Approaches and Experimental Techniques
26: Polymer–Graphene Nanocomposites
27: Carbon Nanotube-Polymer Composites
28: Nanoscience for the Conservation of Works of Art

How to obtain future titles on publication:
A standing order plan is available for this series. A standing order will bring delivery of each new volume immediately on publication.

For further information please contact:
Book Sales Department, Royal Society of Chemistry, Thomas Graham House, Science Park, Milton Road, Cambridge, CB4 0WF, UK
Telephone: +44 (0)1223 420066, Fax: +44 (0)1223 420247
Email: booksales@rsc.org
Visit our website at www.rsc.org/books

Nanoscience for the Conservation of Works of Art

Edited by

Piero Baglioni and David Chelazzi
Department of Chemistry, University of Florence, Italy
Email: piero.baglioni@unifi.it; chelazzi@csgi.unifi.it

RSC Publishing

RSC Nanoscience & Nanotechnology No. 28

ISBN: 978-1-84973-566-7
ISSN: 1757-7136

A catalogue record for this book is available from the British Library

Published by The Royal Society of Chemistry,
Thomas Graham House, Science Park, Milton Road,
Cambridge CB4 0WF, UK

Registered Charity Number 207890

For further information see our web site at www.rsc.org

Printed in the United Kingdom by CPI Group (UK) Ltd, Croydon, CR0 4YY

To Laura, Tommaso and Cosimo,
who every day make me feel that life is worth living.

Preface

The last three decades have witnessed major developments in conservation science. Colloids and interface science, together with materials science, which belong to the realm of popular nanosciences, have contributed concepts, techniques, expertise and tools to increase understanding of the most common degradation processes of works of art. Likewise these disciplines have provided reliable methods for durable and, as far as possible, compatible conservation.

The idea of writing a book on the application of colloid science concepts to conservation traces back to a conversation with my friend Barry Ninham, a well known scientist in the field of colloids and soft matter. He was in Florence on sabbatical leave and we were discussing the "evolution" of scientific views: "a subject or discipline is defined by a freemasonry, the members of which are united by a common lore. The received beliefs of adepts are reaffirmed by recitation and repetition of the lore, and reinforced by the weight of authority embodied in the literature of the discipline. The more venerable the literature, the more conservative is the freemasonry. Challengers to the canon are at first dismissed as heretics or apostates, but if, later, their claims are conceded to have validity, they and their theories are eventually accepted into the fold. The discipline moves on".[1] What Barry was teaching me was what I had already experienced together with my mentor, the chemist Enzo Ferroni, when he devised the first scientific approach to the conservation of wall paintings, damaged by the Arno river flood in 1966 in Florence. His ideas gained great advantage by the strict cooperation with Dino Dini, the restorer in charge of restoration of the Beato Angelico's frescoes in San Marco Basilica. The rigorous approach of Ferroni's method was barely accepted among the restorers in Florence at that time, when most of the restoration interventions (and not only in Italy) relied on the use of polymers, which were later proven to be harmful for the paintings. I continued working, with Ferroni, on the development of the Ferroni method during the restoration of Beato Angelico in San Marco and on the

RSC Nanoscience & Nanotechnology No. 28
Nanoscience for the Conservation of Works of Art
Edited by Piero Baglioni and David Chelazzi

Published by the Royal Society of Chemistry, www.rsc.org

development of new methods for conservation; a few years later, Luigi Dei joined us. Restoration was indeed a fascinating field, but it was at the border of official science, so we worked on the subject in our spare time. It was a sort of hobby, exciting, but something that was not useful for an academic career or to improve our bibliometric indicators, not to mention funding. However, working in the historical center of Florence was an inexhaustible source of inspiration and motivated us to continue working in that field.

I was convinced that this research subject was not interesting to "real scientists". In 1998 Jean Marc di Meglio, a friend of mine working with Pierre Gilles de Gennes at College de France, asked me to give a speech at the meeting in Les Houches celebrating the 65th birthday of Pierre Gilles. That was the first speech I gave to scientists on conservation and, to my surprise, it attracted a certain interest in the audience. However the field was not yet mature; a few years later, Barry's idea of writing a book on "our" new methods for conservation failed miserably when a famous publisher considered the subject too interdisciplinary to be worthy of publication.

Since the Florence flood, conservation science has developed along two main lines: 1) the use of advanced diagnostic techniques for the study of artworks; 2) the production of innovative materials and methods for the cleaning, consolidation and protection of different artistic objects. The authors of this book, through their research work, have mainly contributed to the second area.

This book is intended to provide the reader with the basic concepts of colloid and interface science that have constituted our scientific background to deal properly with conservation issues, and builds on our research experience and active experimentation in colloids and surface science, fuelled by continuous exchanges and feedback with end-users belonging to the restoration community, such as freelance restorers, private and public conservation institutions, and museums.

There are several books dealing with conservation, but there is a lack of a unique treatise showing the potential, and in most cases real, impact of nanosystems to restoration. The main idea behind this book is indeed to fill the gap between the traditional restoration methodologies that are based on "classical chemistry", and the most recent methods that benefit from the cultural framework of colloids and material science. In this sense, this book is unique in providing detailed information on how to produce and to use advanced systems for conservation. In fact, in the last few years we have been solicited by several colleagues to write a theoretical and practical textbook on this subject. We are therefore grateful to RSC for giving us this possibility.

We have selected material from several Masters and PhD courses, held by the editors and authors of this book, in the Conservation program of the Faculty of Science of the University of Florence, to create and shape the chapters reported in this book, always taking into account the point of view of restorers and end-users.

Writing such a treatise was not a straightforward task. The most insidious risk was to emphasize the theoretical aspects of colloids science over the conservation applications, or *vice versa*, weakening the fundamental message that we wish to deliver, *i.e.* that the union of a strong expertise in colloids and material science with the practical knowledge derived from the restorers' needs is fundamental to developing innovative and effective methods that are compatible with the properties of the works of art to be preserved, to grant long-term preservation of cultural heritage.

Several sections of this book are indeed conceived as practical tools to guide conservators and restorers through the production and application of nanomaterials for the conservation of works of art. Real case studies are approached and discussed, where the use of nanoparticles and other systems belonging to the colloidal domain has proven fundamental and innovative with respect to traditional restoration techniques.

Besides conservators and restoration end-users, the audience for which this book is of interest includes both academic scientists who approach the subject of conservation science for the first time, and advanced undergraduate or postgraduate students. In this sense, this treatise can be used as a textbook for teaching or as a reference and inspiration for further research work in this challenging and exciting field. In particular, students will find the "Review and Questions" sections beneficial in order to ensure full comprehension of the treated topics and to stimulate discussion of the most crucial aspects. On the other hand, curiosity-driven scientists, who wish to get involved in this emerging field, will find a widely interdisciplinary scientific background, different from conventional conservation science, which exhibits a strong overlap between the different "Science" and "Humanities" ways of thinking.

This book is organized in chapters that provide the cultural background for the comprehension of the systems that are used in most of the interventions for the conservation of both movable and immovable works of art, and others that highlight the application of innovative systems to real conservation case studies.

Some recommendations are necessary for conservators that are particularly interested in the applicative and case study sections, which cover the most common conservation interventions. A warning message that we want to transfer is that most of the innovative formulations (especially those for cleaning) are based on self-assembling systems, meaning that any modification to their composition possibly leads to a dramatic change in the physico-chemical properties (and effectiveness) of these systems. If conservators really want to modify such systems the best approach, we suggest, is an interactive cooperation with scientists who are expert in the field of colloids and materials science in order to maximize the effectiveness of the adopted nanostructured systems. This would also produce a progressive refinement of the involved technology and the tuning of the proposed systems to address specific conservation issues. In one word, cooperation is the key.

As a final, but most important note, I would like to express my deep gratitude to David Chelazzi, who shares the editing of this book and has carried an impressive burden on his shoulders.

Together with David, I would like sincerely to thank all the authors with whom we are in debt for the realisation of this work. A special thanks to Pierandrea Lo Nostro for his invaluable criticism and constant support, through this intense writing experience. We also wish to thank all the conservators who indirectly contributed to the refinement of the conservation methods we developed. Finally, we wish to acknowledge both the European Union (which has financed the "Nanoforart" FP 7 project) and CSGI (Consorzio Interuniversitario per lo Sviluppo dei Sistemi a Grande Interfase - Center for Colloid and Surface Science), which is the main source of funding and the driving force of this "side activity", for financial support.

Piero Baglioni

Reference

1. B.W. Ninham and P. Lo Nostro, *Molecular Forces and Self Assembly In Colloid, Nano Sciences and Biology*, Cambridge University Press, Cambridge, 2010, p. *xi*.

Contents

RSC Nanoscience & Nanotechnology No. 28
Nanoscience for the Conservation of Works of Art
Edited by Piero Baglioni and David Chelazzi

Published by the Royal Society of Chemistry, www.rsc.org

CHAPTER 1

Cultural Heritage Artefacts and Conservation: Surfaces and Interfaces

G. CAMINATI

University of Florence, Italy
Email: gabriella.caminati@unifi.it

1.1 Thermodynamics of Interfaces: Surface Energy and Surface Tension

Our appreciation of any work of art belonging to our Cultural Heritage is definitely intertwined with the observation and interpretation of the surface of the object itself. As a matter of fact, whether we are observing a Renaissance fresco, a Maya painting, or the surface of a grotto painted by an unknown artist of our prehistory (see Figure 1.1), our attention is drawn entirely to the surface of the artefact.

An art object is devised so as to observe, read and experience its surface: the surface is the locus where the artist transferred their message and emotions, but it is also the place where different materials, with their own specific chemical composition and mechanical properties, coexist. Surface also plays a leading role in another unfortunately unavoidable process that affects our Cultural Heritage: all works of art deteriorate over time upon exposure to light, temperature stresses and relative humidity cycles, insects, or microorganisms, depending on the particular location and exposure to the environmental factors. The effects of deterioration may proceed deeper into the artefact but

RSC Nanoscience & Nanotechnology No. 28
Nanoscience for the Conservation of Works of Art
Edited by Piero Baglioni and David Chelazzi

Published by the Royal Society of Chemistry, www.rsc.org

Figure 1.1 (top left) Jatakas Cycle, 1st century BC to 1st century AD, Ajanta Grottos (India); (top right) Maya wall painting discovered in the archaeological site of Calakmul (Mexico); (bottom) Scene from the frescoes of Masaccio in the church of Santa Maria del Carmine, in Florence (Italy).

the first screen and the first point of attack is undoubtedly what the object exposes to the external surroundings: its surface, or better the interface with the world outside the art object. No matter whether the causes of deterioration are physical, chemical or biological, the surface will be the first frontier to be modified both in structure and composition. Exposure of the work of art to the atmosphere will result sooner or later in the growth of nano- or microlayers of different chemical composition readily adsorbed on the surface, as depicted in Figure 1.2. The mechanism and the kinetics of interactions of these compounds as well as their correct removal can be understood by proper application of the principles of physical chemistry of surfaces.

In fact, a proper conservative intervention should proceed only if the properties of the chemical media used are known and the mechanism with which these systems interact with the artefact's surface is understood.

It has long been recognized and described in several textbooks[1,2] and reference books[3] that surface science extends its branches in many realms of science and technology, but its importance in art restoration and conservation

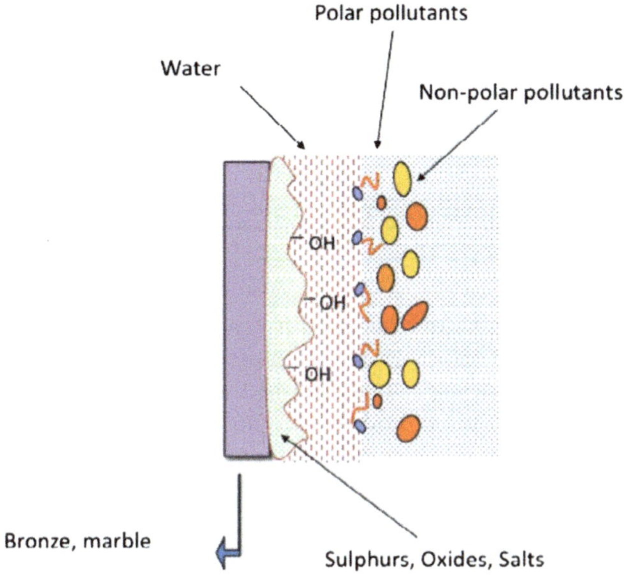

Figure 1.2 Cartoon depicting the hierarchical arrangement of adsorbed layers of different chemical composition on the surface of a work of art upon exposure to the atmosphere.

has often been undervalued and only in recent times has a schematic scientific framework been attempted.

Art conservators should therefore acquire a sound competence not only in the nature and behaviour of materials but they should also master the science of surface phenomena. In addition, surface phenomena play a leading role in dictating the behaviour of many of the nanosystems conceived for art restoration that will be described in this book. Nanosystems, including nanoparticle dispersions, micelles, micro- and nanoemulsions, and polymer gels, will be described and their applications to art conservation illustrated.

The subject of this chapter is the thermodynamics of surface and interfacial phenomena involved in many aspects of art restoration and conservation, including their interpretation in terms of basic physics and chemistry. Both the general thermodynamic principles and the theoretical approach for the determination of molecular properties are widely discussed in literature,[4,5] and will not be reported here.

The thermodynamic treatment will be presented in the case of a planar surface (as far as microscopic scale domains are concerned, curved surfaces found in statues or bas-reliefs are still considered planar). Extension to the important case of fluid curved surfaces will be described in section 1.4.2.

Interfacial properties are modified by changing the adjoining phases, so that the thermodynamic properties of these bulk phases must be mastered and understood in the first place. A complete description of classical thermodynamics is beyond the scope of this book but the interested reader may

Table 1.1 Application of thermodynamics of surfaces to the conservation of cultural heritage.

Thermodynamic variable	*Application fields*
Interfacial tension	Detergency, adhesion
Contact angle	Wetting
Capillarity	Capillary rise, porosity
Physi- and chemisorptions	Surface modification

refer to several textbooks on this subject reported in the "Further Suggested Reading" section.[4,5] This chapter will present the thermodynamic basis of surface and interface science, underlining those aspects that can be readily employed in art restoration and conservation. This chapter is directed to restorers searching for the scientific basis of their own work, as well as to students in applied chemistry and restoration techniques.

Table 1.1 reports a summary of some of the thermodynamic phenomena and variables that will be dealt with in the remainder of this chapter, indicating how they are correlated with the specific application in art conservation.

Readers unfamiliar with the physical chemistry of interfaces are certainly aware of many everyday life phenomena that are strictly correlated with the surface properties reported in Table 1.1: plants receive water and nutrients from the ground and many common insects such as water spiders may actually walk on the surface of a pond. The above phenomena share common roots in the physical chemistry of interfaces; in particular, they are examples of the larger domain of wetting phenomena, a domain that will be proven to be of relevance when conservation and restoration are involved.

1.1.1 Definition of Surfaces and Interfaces

This section deals with the discussion of a physico-chemical approach to surfaces and interfaces. The discussion on surfaces will lead to the definition of a very important physical entity, surface tension, which will be found repeatedly through this entire book. What we provide in this section is therefore a general and fundamental approach that underlies the following sections.

How to define and locate the surface of an object exactly is not a trivial question. The surface of a painted table or canvas may be approximated by a geometric plane, but when a section of the pictorial layer is enlarged we can easily prove that the surface of separation between the two phases is somewhat more extended and heterogeneous (see Figure 1.3).

Interfaces are boundaries between different phases, but the physical and chemical properties of the interface differ from those of the adjacent bulk phases. It is customary to use the symbols S, L and G to denote a solid, a liquid and gaseous phase, respectively. Using this terminology we can summarize different interfaces as SS, SL, SG, LL and LG interfaces as schematically described in Figure 1.4.

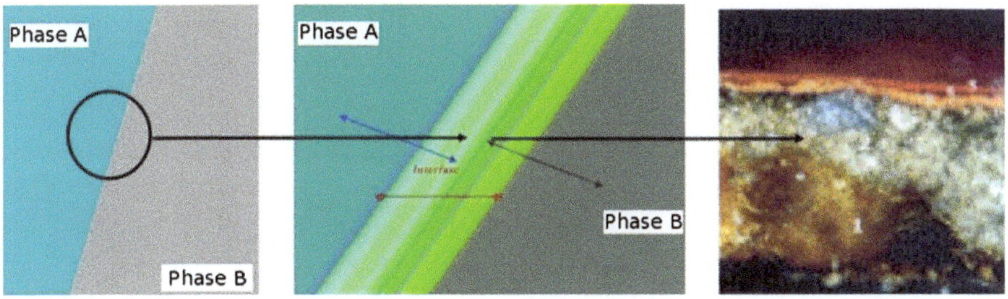

Figure 1.3 (left) Geometrical representation of the surface dividing two bulk phases; (centre, right) the two enlargements show the details of the hypothetical and the real interfacial region.

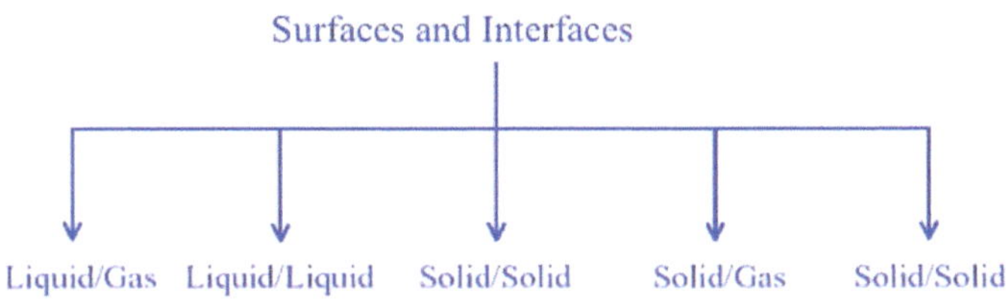

Figure 1.4 Summary of the possible interfaces between bulk systems.

Examples of such interfaces are ubiquitous in the real world. Specific cases where art conservation is involved will be dealt with separately in this book, including solid–liquid interfaces (colloidal particles for wall painting consolidation), liquid–liquid interfaces (oil–water: micro- and miniemulsion for the cleaning of paintings), solid–solid interfaces (glue–cement, glue–canvas: adhesives).

The notion of "interface" is indeed the most general one, whereas "surface" is more restrictive, for example it is a boundary between a gas phase and a condensed phasewhen gas–liquid or gas–solid boundaries are considered.

The term "surface" is often also used when referring to boundaries of a particle, no matter what its dimensions are, *i.e.* from macro to nanoparticles, and independently of what is around the particle. A surface has a different physico-chemical nature with respect to the associated bulk phases, but as we approach smaller dimensions surface effects become much more dominant. This concept is easily visualized in Figure 1.5: assuming that they have the same overall volume, a single large particle exhibits a total surface area smaller than the total surface area of a collection of smaller objects filling the same total volume. In other words, the higher the surface-to-volume ratio the larger the predominance of surface forces with respect to bulk forces.

The smaller the particles the greater the specific area defined as the surface area per unit weight. The same concept applies also to a sequence of crystalline solid phases: the more finely the material is divided, the larger the surface area. It is therefore apparent that the surface rules the behaviour of the entire system as far as nano-objects are concerned, *e.g.* $Ca(OH)_2$ (calcium hydroxide)

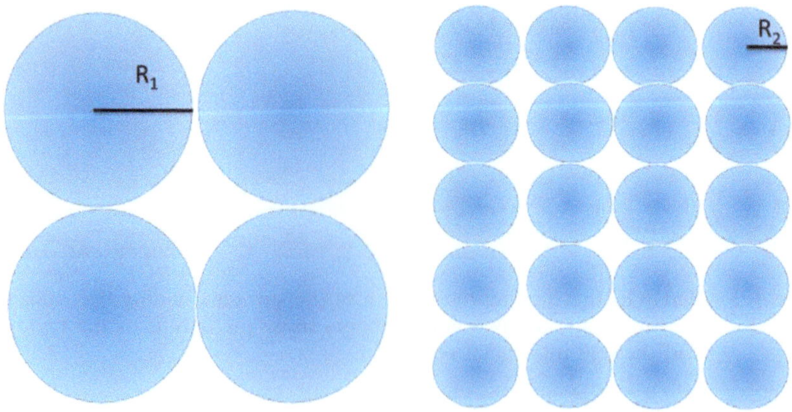

Figure 1.5 Surface-to-volume ratio for a sphere of radius R_1 (Volume = $4/3\pi R_1^3$) and for a collection of sphere of radius $R_2 < R_1$ filling the same total volume.

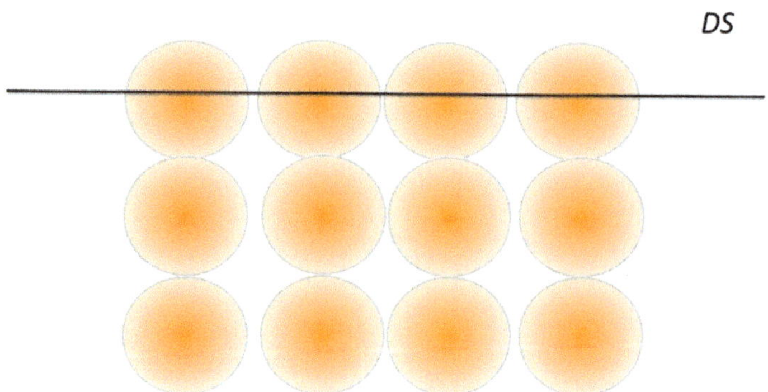

Figure 1.6 Dividing surface for the surface of a crystalline solid.

nanoparticles for the treatment of wood acidity or for the consolidation of wall paintings. Definition and modelling of interfaces are fundamental in order to describe the parameters that rule all the phenomena occurring between two phases (*e.g.* degradation reactions of artefacts).

The three-dimensional region of contact between two generic phases, α and β, is called the *interfacial region* or *interfacial layer*. In two-phase systems where one of the phases is crystalline it may be tempting to identify the division surface (DS) with the geometrical plane that intersects the centres of the atoms forming the first surface layer, as depicted in Figure 1.6. This simple case may be extended to any solid surface, as in a typical case of a marble, wood or bronze surface in contact with air.

When planar liquid phases are concerned, the location of the dividing surface is much more controversial: at the liquid–vapour and at some liquid–liquid interfaces the boundary layers extend over the dimension of few molecules, and more rarely several molecular layers can be involved. Although this boundary

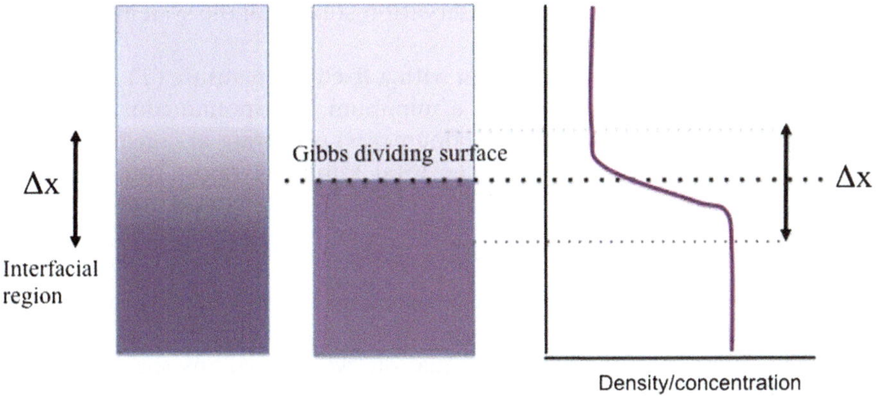

Figure 1.7 (left) Schematic representation of the interfacial transition layer between two condensed α and β phases: ΔX is the thickness of the transition layer. (right) Schematic representation of the interfacial transition layer between the α and β phases after the imaginary *Gibbs dividing surface* is located.

region may appear static, in real systems the interface is in a very turbulent state: for a liquid–vapour interface, the liquid is in equilibrium with its vapour, meaning that molecules from the vapour phase hit and condense on the surface while molecules from the liquid bulk phase escape from the surface and evaporate.

In many cases of interest for the treatment of works of art, liquid–liquid and solid–liquid interfaces will be involved. The interfacial region between two condensed phases is shown schematically in Figure 1.7. It is generally assumed that, in the absence of electrolytes, this region is a few molecules in thickness (approximately 1–2 nm) and only an extremely small fraction of the molecules in the system are present in the interfacial region owing to geometrical constraints.

If a system (for instance an artefact's surface) is in equilibrium with its surroundings, its macroscopic properties are fixed, and the system can be defined as a given *thermodynamic state*. Practically, a system is in *equilibrium* if no further spontaneous changes take place at constant surroundings. Out of equilibrium, a system is under stress, and tends to equilibrate to a fully relaxed state. Many degradation reactions occur at the interfaces of artefacts (metal oxidation, tarnishing, *etc.*) and such systems evolve to a stable state that may hinder the readability of the surface.

In order to inquire further into surface thermodynamics it is necessary to recall a fundamental state function,[1] *i.e.* the Gibbs free energy of the system, which is the maximum amount of work a system can do at constant pressure (isobaric changes). The importance of this function for the description of phenomena related to the chemistry of art conservation is immediately

[1]A *state function* describes univocally the thermodynamic state of a system. For an insight discussion of state functions, see the Further Suggested Reading section.

apparent once we notice that in conservation studies all the systems considered are usually at constant pressure.

For closed[2] systems at equilibrium with a fixed temperature (T) and pressure (P), the Gibbs free energy reaches a minimum. All spontaneous, irreversible processes (*e.g.* the degradation of monuments) occurring at constant T and P proceed in a direction such that the total Gibbs energy of the system still decreases:

$$dG_{T,P} \leq 0 \tag{1.1}$$

Thus the equilibrium state of a heterogeneous closed system is the state with the minimum total Gibbs energy attainable at the given T and P.

In the derivation of the previous equation, we ignored any special change at the dividing interface (boundary), or the effect of the variation of the interfacial area.

Gibbs[6] treated this thin layer as a *quasi-two-dimensional phase* having no volume (see Figure 1.7). The Gibbs dividing plane concept is a departure from the physical reality but it is consistent and allows us to apply thermodynamics to surface processes. Extensive thermodynamic quantities can thus be written as a contribution to the Gibbs free energy from the system bulk phases plus a surface term.

In the case of a system with surface area A:

$$dG_{TOT} = dG_{Bulk} + G_s dA \tag{1.2}$$

where G_s is the extra Gibbs free energy per unit area. Although the composition varies in the neighbourhood on the surface, according to Gibbs we consider the system as uniform up to this interface.

For reversible processes at a completely planar interface, the differential Gibbs energy per unit area, γ, can thus be considered a surface energy at constant temperature, pressure, and composition:

$$\gamma = \left(\frac{\partial G}{\partial A}\right)_{T,P,N} \tag{1.3}$$

The surface excess free energy term is correlated with the work done in generating an interfacial area increment (dA), which can be expressed as γdA. In other words, the surface free energy, γ, is the work that should be supplied to bring the molecules from the interior bulk phase to its surface to create a new surface of unit area ($1\ m^2$).[7]

The variable γ is of utmost importance in interfacial science and is called the interfacial, or surface, tension.

The dimension of γ is energy per unit area, $J\,m^{-2}$ in the SI system. However, these units are used exclusively for the case solid surfaces whereas for liquid interface the equivalent unit $N\,m^{-1}$ is adopted (force per unit length). In

---

[2]A *closed system* is one for which only energy transfer is permitted, but no transfer of mass takes place across the boundaries, and the total mass of the system is thus constant.

practical applications surface tension is reported in mN m^{-1}, equivalent to the obsolete dyn cm^{-1} units.

From the previous discussion it can be stated that in every interfacial region at constant pressure and temperature, there is a tendency for mobile surfaces to decrease spontaneously in area, in order to decrease the Gibbs free energy of the system, since interfaces are the seats of excess Gibbs energy.

If the interface is fluid, *i.e.* liquid–liquid or liquid–gas boundaries, the action of the interfacial tension determines thus the final shape of the interface. The smallest surface area for a given liquid volume is geometrically a sphere, without any external force acting on it. For example, as far as gravity is concerned large spherical drops tend to flatten, as in raindrops.

The formation of a new surface includes two separate steps: the first is the formation of two new surfaces leaving unaltered the arrangement of atoms and molecules in space. In the second step, atoms and molecules rearrange at the surface until a new equilibrium state with minimum energy is achieved. Molecular rearrangements in liquids are very fast and surface tension can generally be considered an equilibrium value. Theoretically, the surface tensions of real liquids should be strictly measured in liquid–vacuum conditions. However, since liquids will continually evaporate in a high (or complete) vacuum condition, it is physically impossible to measure their real surface tension. In practice, we can only measure the liquid–air interfacial tension instead, under room conditions.

The situation differs for solids, where the greatly reduced molecular mobility slows down the molecular rearrangement at the surface. Therefore, surface energy for solids strongly depends on the local crystalline structure and, owing to the slow kinetics of rearrangement, also on the specific history of surface formation. In the case of crystalline or polycrystalline solid phases the surface energy correlates with the atomic density and number of nearest neighbours on surface plane, therefore γ is a function of orientation of the surface plane and of the specific crystalline structure, *i.e.*, γ is not homogeneous.

Consequently, surface energies of solid surfaces are not as easily determined, and for solid–liquid and solid–gas interfaces the presence of the interfacial tension can only be established indirectly[8] (see also Section 1.4.1).

A rough estimate of the surface energy of a solid surface can be obtained from the example reported in Figure 1.8.

If we separate a rectangular crystalline material into two pieces, two new surfaces will be created. This process requires the breaking of bonds between two layers of molecules or atoms. Depending on the orientation of the slicing plane, a different atomic structure will be exposed at the newly created surface. In Figure 1.8, two typical examples of surfaces created from the rupture of a face-centred cubic crystal[3] are reported: the surfaces were obtained by cutting the same crystal with planes of different orientation. This translates into the

---

[3]Many metals (Co, Cu, Ag, Pt) adopt the *cubic close-packed* (also *called face-centred cubic*) structure. Others (Ti, Co, Zn) adopt the *hexagonal close-packed* or the slightly less efficiently packed *body-centred cubic* structure (*e.g.* Fe).

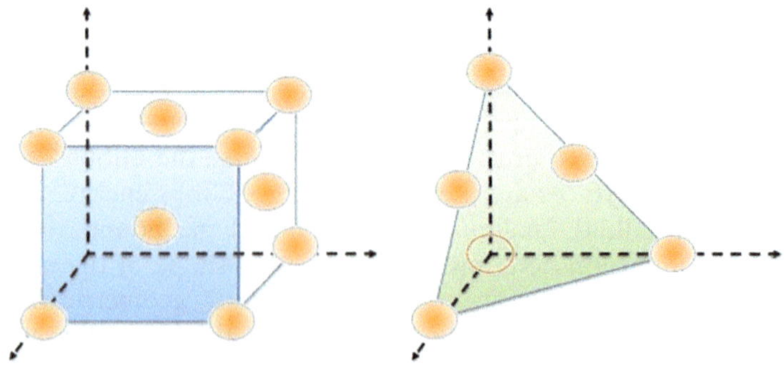

$$\gamma_{\{100\}} = \frac{1}{2}\frac{2}{a^2}\cdot 4\cdot \varepsilon = \frac{4\varepsilon}{a^2} \qquad \gamma_{\{111\}} = 2\sqrt{3}\cdot\frac{\varepsilon}{a^2}$$

Figure 1.8 Surface energy of a crystalline solid surface with specified orientation of the surface plane in the crystalline structure [100] and [111] faces of a face-centered cubic (fcc) crystal.

fact that each atom is located in an asymmetrical environment where the inward forces are not balanced by the bonds that have been broken to create the surface. The surface energy that we defined in eqn (1.3) can then be equated to the energy involved in bond breaking for unit area:

$$\gamma = \frac{1}{2}N_b\varepsilon\rho_a = \left(\frac{dG}{dA}\right)_{n_i,T,P} \tag{1.4}$$

where ρ_a is the surface atomic density, *i.e.* the number of atoms per unit area on the new surface, N_b is the number of broken bonds in each case and ε is half of the bond strength.

In the two cases depicted in Figure 1.8, surface energy will result in two significantly different values: the surface energy for the [111] plane is 0.87 times smaller than the surface energy for the [100] plane.

This rough approximation is only applicable to solids with rigid structure where no surface relaxation occurs; in all other cases where surface relaxation cannot be neglected owing to surface atoms moving inwardly or surface restructuring, the corresponding surface energy will be lower than that estimated.

1.2 Origin of Surface Tension: Intermolecular Forces

In the previous section, we discussed surface tension in terms of surface energy per unit area for a macroscopic ensemble of molecules; we should now scale down these concepts and connect the thermodynamic scenario to events occurring at the molecular scale.

In all states of matter, molecules interact with each other, in other words forces are exerted between them, and these interactions determine the dynamic and static properties of both bulk and interfacial systems. In reality "intermolecular forces" is a general term that includes a variety of distance-dependent interactions between atoms or molecules. In this book, the treatment of intermolecular forces is directed towards the understanding of the forces ruling the interactions between colloidal particles (such as alkaline nanoparticles dispersed in a medium, for the deacidification of paper) and molecular interactions at the interfaces, such as those occurring during degradation of artistic surfaces or their cleaning, as well as forces leading to the mixing of two different solvents. Therefore the derivation of intermolecular forces between macro-bodies from pair, or multibody, interactions is not treated explicitly and only referred to in the Further Suggested Reading section.[9]

Besides the former microscopic approach, the macroscopic counterpart that considers a more integral treatment has also developed.[10]

Although a detailed description of the intermolecular forces operating in nano and interfacial systems will be presented in Chapters 4, 5 and 6 of this book, here we introduce briefly the categories of fundamental forces:

1. van der Waals interactions, in particular London, or dispersion forces which are ubiquitous;
2. Electrostatic or double layer forces if the nano-object or the surface is charged;
3. Steric interactions exerted at short range for small molecules or occurring at larger scales for macromolecules.

Gravitational forces are also operative and are explicitly considered for particles of high specific density and when describing capillary phenomena. In specific cases, structural forces, originating from the modification in the liquid structure adjacent to surfaces, or magnetic forces, if magnetic nanomaterials are involved, may prevail and dominate the overall balance. For instance, magnetic forces can be used to ease the removal of magnetic gels used for the cleaning of artistic surfaces (*e.g.* wall paintings), avoiding any handling of the gel and minimizing the mechanical action onto the surface (see Section 11.4).

Interactions can be expressed in terms of forces, energies or potentials depending on the way in which the mechanical energy is conserved or dissipated and converted into heat during the process. If the process is path independent the force is conservative and can be related to the potential: van der Waals and gravitational interactions are examples of conservative forces. Examples in art conservation are numerous, for instance white spirit, a widely used cleaning solvent in art restoration, is mainly a mixture of hydrocarbons that carries out its action through van der Waals forces (see also Section 5.2).

The molecules near or at the surface experience intermolecular interactions that are different from those that involve molecules in the bulk phase: this translates into the fact that molecules that are situated at the interfaces behave differently from those in the bulk.

The physical mechanism underlying the occurrence of surface tension at the interface is necessarily to be correlated with the imbalance of forces acting in the outer surface layer. Underneath a water–air interface (GL), *e.g.* within a water droplet, the water molecules are surrounded by identical neighbouring molecules. Here, the molecules will interact with each other in a symmetrical way with interactions of comparable extent acting in every direction. On the other hand, in the vapour phase the interaction between molecules will be very weak. Thus, the water molecules on the GL interface will be attracted by the molecules that are sitting below and by their sides, but there will be no molecule in the vapour phase to interact with. As a consequence, the molecules at the interface will be under an asymmetrical force field, a stress that produces a resultant net force perpendicularly directed inside the liquid. The nearer the molecule to the surface, the greater is the magnitude of the force due to this asymmetry.

The simplistic view described in Figure 1.9 shows how the attraction among molecules of the liquid will be larger in the more condensed phase. The liquid composition is uniform from the interior bulk phase to the surface but the balance of the forces acting on the surface molecules will be different from that in the bulk liquid phase.

As a result, the surface molecules are continuously moving inwards more rapidly than interior molecules, which move upwards to take their places. This process decreases the number of molecules in the surface, and this diminishes the liquid surface area; this surface contraction continues until the interior accommodates the maximum possible number of molecules.

The inward attraction normal to the surface causes the surface to be under a state of lateral tension, thus for a planar surface the surface tension can be viewed as the force acting parallel to the surface. In other words, the cohesion among the molecules supplies a force tangential to the surface so that a fluid surface behaves like an elastic membrane that wraps and compresses the liquid below.

This concept is reflected in a common observation that, due to the surface tension, insects such as the marsh treaders and water striders exploit the surface tension to skate on the water without sinking; other examples include the floating of a metal coin (regardless of its density) on the surface of water.

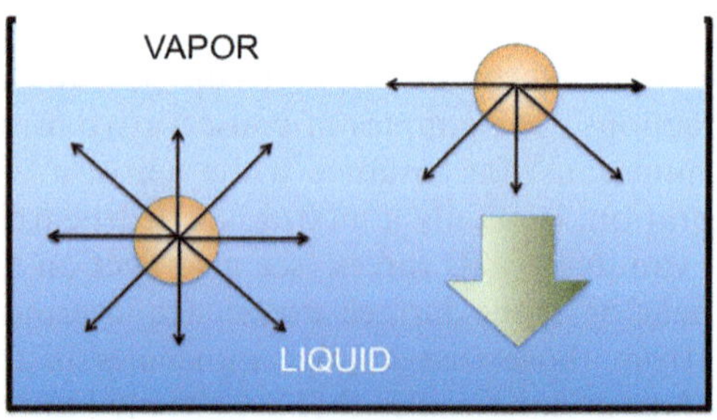

Figure 1.9 Schematic representation of a liquid molecule in the bulk liquid and at the surface. A downward attraction force is operative on the surface molecule owing to the unbalance of molecular interactions in the outer layer of the liquid phase in contact with a vapour phase.

A typical, dramatic effect of surface tension is found in archaeological and waterlogged wood conservation (*e.g.* historical shipwrecks): after the wreck is recovered, exposure to air starts the evaporation of water within the wood's pores. The resulting surface tension forces of the evaporating water cause the collapse of the cell walls in the wood. As a result, severe shrinkage and cracking may result (see Section 16.4). Instead, solvents whose surface tension is lower than that of water (see below, at the end of this section) can be used for the gradual dehydration of wood, before consolidation interventions.

The surface tension and the surface free energy of substances are dimensionally equivalent, and for pure liquids in equilibrium with their vapour, the two quantities are numerically equal. However these two terms are conceptually different and surface free energy can be regarded as the fundamental property in thermodynamic terms, while surface tension would be taken simply as its equivalent if there were no adsorption on a surface.

On the other hand, when we consider two immiscible phases and an interface between them, we should define the *interfacial tension*, γ_{12}, as the force that operates inwards from the boundaries of a surface perpendicular to each phase, tending to minimize the area of the interface. The *interfacial free energy* between liquids is dimensionally equivalent and numerically equal to their *interfacial tension.*

As a matter of fact, interfacial tension between two immiscible liquid phases is a central physical parameter that underlies the design of advanced systems for the cleaning of artistic objects of any kind (stone, wall and easel paintings, paper, wood *etc.*). Surfactants, in fact, act on this parameter, and assemble into complex systems such as microemulsions and micellar solutions, whose structure and effectiveness as cleaning tools will be widely discussed through several sections of this book (see Chapters 6–9).

Surface tension is currently measured with many different experimental set-ups that will be mentioned in Chapter 6.

A plethora of pure liquids have been studied and their surface tension is tabulated in many reference textbooks,[11] here we collect the results of some measurements run at 25 °C for solvents including water, alcohols, hydrocarbons and metals that are liquid at room temperature, such as mercury (see Table 1.2).

Acetone, ethanol and hexane (which is one of the main constituents of white spirit), are all standard solvents used in art restoration, for instance in cleaning interventions for wall and easel paintings (see Chapter 5).

Table 1.2 Typical values of surface tension, γ, for pure liquids at 20 °C.

Liquid	γ *(mN m^{-1})*	*Liquid*	γ *(mN m^{-1})*
Water	72.8	Ethanol	22.3
Benzene	28.9	Acetone	23.7
CCl_4	26.8	n-Hexane	18.4
Acetone	23.7	Mercury	472

1.3 Adhesion and Cohesion

Surface tension γ can be regarded as due to the net difference between intermolecular forces acting at the interface, therefore all processes occurring near any interface depend on the molecular orientations and intermolecular interactions. This leads to the introduction of forces of adhesion and cohesion that can be exerted between liquids or between a liquid and a solid, an aspect that finds major applications in surface and colloid science and that is particularly important for Cultural Heritage.

In fact, several processes can be explained in terms of adhesion and cohesion forces. For example we will see later that capillarity results from a combination of adhesion and cohesion involving liquid and solid phases, and the removal of solid or liquid deposit from the surface (dirt removal) may be ruled by these two phenomena.

A distinction is generally drawn between adhesive forces, which act to hold distinct molecules together (the working principle of several adhesives and glues, widely found in restoration) and cohesive forces, which act to hold together like molecules of a single phase. However both forces result from the same properties of matter.

In Figure 1.10 the cartoon explains schematically the process of cohesion and adhesion.

In liquids, the cohesion forces keep the molecules close to each other, while translational and rotational motion of molecules takes place within the liquid with considerable freedom.

The energy necessary to separate two identical interfaces from contact to a virtually infinite separation distance is called the energy of cohesion, and the work (per surface area unit) needed for this process, W^c, is given by:

$$W^C = \Delta G^C = 2\gamma_I \tag{1.5}$$

Where ΔG^C is the free energy of cohesion and γ the surface energy. W^C measures the attraction between the molecules of the two portions. The surface tension γ can be considered as half the work of cohesion; it measures the free energy change involved when molecules are brought from the bulk of the

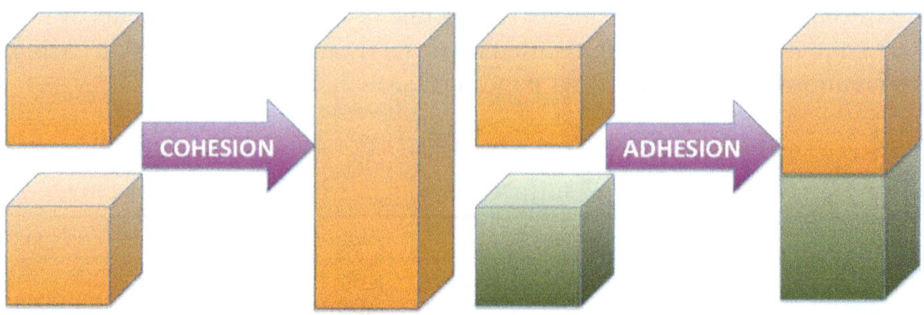

Figure 1.10 Description of the process of cohesion (left) and of the process of adhesion (right).

sample to the surface. Thus, two identical interfaces with high surface tension will cohere strongly.

Similarly, if two unlike surfaces are brought from contact to infinite separation distance, the energy required for the process is given by the difference between the final and initial states:

$$W_{\mathrm{ij}}^{\mathrm{A}} = \Delta G^{\mathrm{A}} = \gamma_{\mathrm{final}} - \gamma_{\mathrm{initial}} = \gamma_{\mathrm{j}} + \gamma_{\mathrm{i}} - \gamma_{\mathrm{ij}} \tag{1.6}$$

Where $W_{\mathrm{ij}}{}^{\mathrm{A}}$ is the work of adhesion and the various γ_j, γ_i and γ_{ij} are the corresponding surface and interfacial energies. This approach leads to the statement that two unlike interfaces with high surface tensions will cohere strongly. The adhesion energy is particularly important when solid surfaces are involved. The adhesion of colloidal particles to solid substrates is of fundamental and technological importance (*e.g.* pneumatic transport of powders, printing, filtration, detergency, air pollution, glues). Let us consider the case of a solid surface of any nature (frescoes, statues or bronze) with a dirt deposit on it as described in Figure 1.11.

The removal of the contaminating spot can be considered in terms of the surface energy involved. The work of adhesion between solid surfaces and dirt will be given by the equation

$$W_{\mathrm{SD}} = \gamma_{\mathrm{DW}} + \gamma_{\mathrm{SW}} - \gamma_{\mathrm{SD}} \tag{1.7}$$

The action of the detergent is to lower the surface tension of the dirt–water and solid–water interfaces, thus decreasing the work of adhesion W^{A} and enhancing the removal of the dirt by mechanical agitation. If the dirt is fluid (oil or grease), the addition of the detergent will lower the contact angle at the triple solid–oil–water boundary and its removal can be considered as a contact-angle phenomenon, which will be discussed later in this chapter.

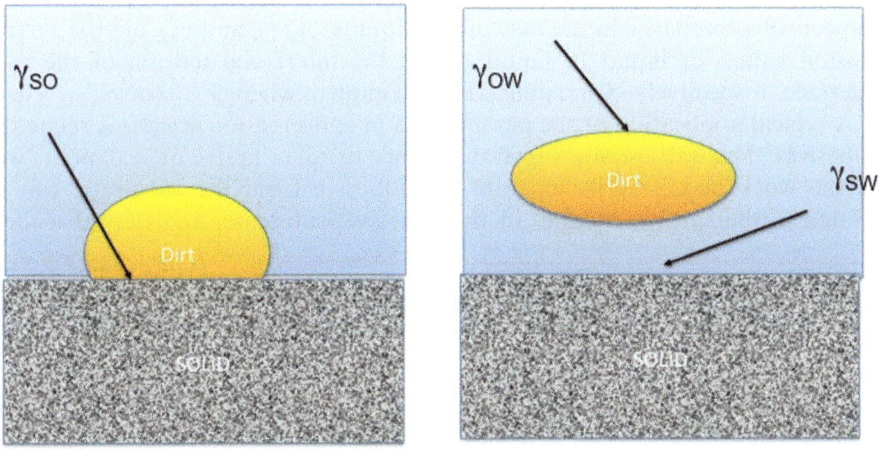

Figure 1.11 Cartoon depicting the removal of dirt from a solid surface.

In the case of liquid–solid interfaces it is useful to consider the tendency of the liquid to spread on the solid surface, that is the spreading coefficient, *S*.

In general, different limiting cases can occur depending on the nature of the solid and liquid phases and on the adhesion energy. When a drop of a non-volatile low density liquid (A) is placed on the surface of a high density solid sub-phase or on the surface of a liquid (B) which is practically immiscible with liquid (A), there are three possibilities:

1. Liquid (A) forms a non-spreading liquid lens with a defined edge, leaving the rest of the surface clean. The shape of the lens is constrained by the force of gravity;
2. Liquid (A) spreads as a monolayer on the surface;
3. If there is not enough space for all of the liquid (A) to spread fully, it spreads as a polylayer or a relatively thick film on the surface.

Figure 1.12 describes the possible scenarios.

If the liquid spreads over surface A, there is a contraction of the solid–vapour interface with an increment in the LV and SL interfaces. Spreading is spontaneous if the total surface energy decreases. Alternatively, we can describe this situation using the spreading coefficient *S*, *i.e.* the difference in the free energy (per unit area) between the solid surface A (or immiscible liquid surface) in contact with the vapour phase and the surface A fully covered with a thick layer of liquid B. Then we have:

$$S^{SLV} = \gamma^{SV} - \gamma^{SL} - \gamma^{LV} \quad \text{[liquid B on solid surface A]} \tag{1.8}$$

$$S_{A/B} = \gamma_B - \gamma_A - \gamma_{AB} \quad \text{[liquid B on liquid surface A]} \tag{1.9}$$

Where S is the spreading coefficient, and γ^{SV}, γ^{SL}, γ^{LV} are the interfacial tensions of the solid–vapour, solid–liquid and liquid–vapour interfaces, respectively. Similarly, in the case of two liquids, γ_B, γ_A and γ_{AB} are the surface tension values of liquid B, liquid A and the interfacial tension of the A–B interface, respectively. Spreading will be complete when S^{SLV} (or $S_{A/B}$) > 0.

A typical application of the parameter *S* in conservation science is related to adhesives. The wetting of a substrate surface by an adhesive or sealant, as well as the work necessary to separate the adhesive from the substrate, can be related to the surface energies of the adhesive, substrate, and the subsequent

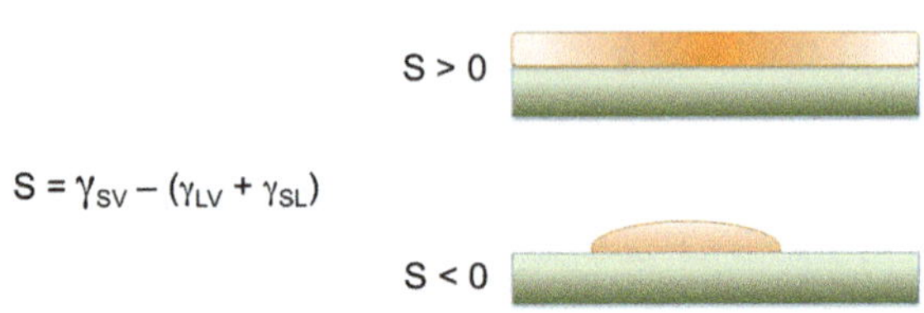

Figure 1.12 Schematic description of spreading and wetting processes.

interface. In an ideal situation for spreading or wetting, the surface of the substrate should always have a higher surface energy than that of the liquid adhesive.

1.4 Solid–Liquid Interfaces

Whereas liquid surfaces readily respond to perturbations in their equilibrium structures, converging in a new homogeneous interfacial composition, solid surfaces are generally in kinetic equilibrium in a highly non-homogeneous organization of atoms and molecules. It is therefore expected that the behaviour of solid surfaces will be highly dependent on their specific history. As a matter of fact, as far as works of art are concerned, the solid–liquid and solid–gas interfaces are by far the most important domains on which degradation and conservation operators act.

Conservative intervention requires in most cases the application of liquid systems on the solid surface of the artefact: this trivial action implies a deeper level of knowledge of why and how the liquid will be able to wet and spread on the surface. In this section we will examine how we can define, measure or predict the wetting behaviour of liquid phases on solid surfaces. We will also consider the common case in which the solid presents porosity of various types and sizes, examining the physical principles that rule capillary phenomena such as the capillary rise.

It is necessary to introduce some essential aspects of solid surfaces in order to understand the issues connected to high reactivity and large deviation from the ideal behaviour exhibited by solid surfaces. We will therefore examine some general aspects concerning the composition and structure of ideal solid surfaces.

Crystalline solids consist of periodically repeating arrays of atoms, ions or molecules. Unfortunately real surfaces are not that simple: not only are surface properties different depending on the facet exposed, but more generally the same surface will expose domains with different Miller indexes, as shown schematically in Figure 1.8. In addition, real surfaces are typically very imperfect and exhibit a heterogeneous mixture of surface sites including a variety of structural defects such as steps, vacancies and isolated adatoms, or sites associated with the terraces of the close-packed surfaces (see Figure 1.13).

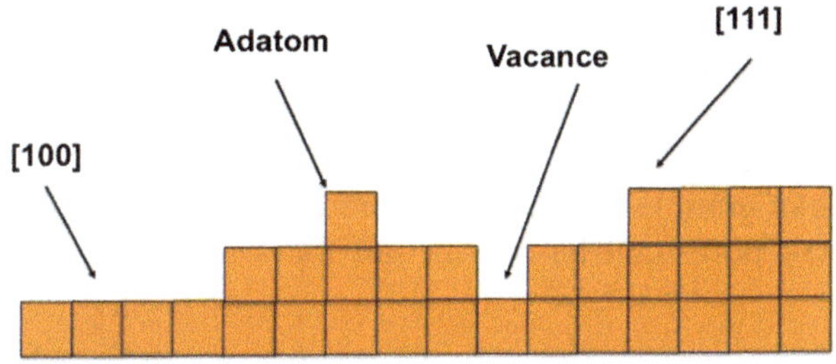

Figure 1.13 Examples of typical surface defects in solid structures.

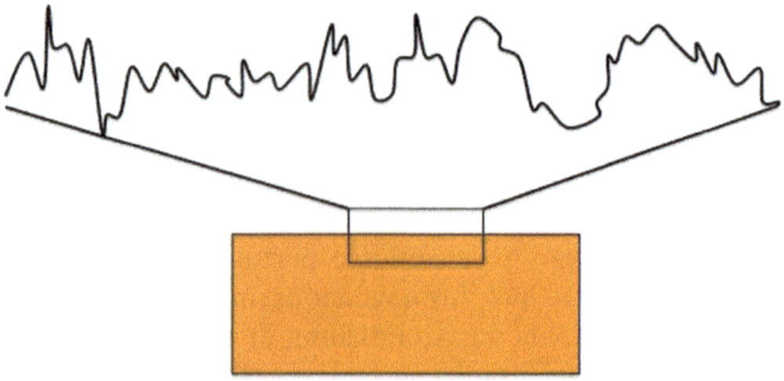

Figure 1.14 Cartoon describing an enlarged vision of a surface portion evidencing surface roughness.

The thermodynamic impetus *vs.* minimization of the surface free energy translates, in the case of solid surfaces with a fixed surface area, into a strong tendency to decrease ΔG. This can occur through a variety of mechanisms: (i) slow surface relaxation, the surface atoms or ions shift inwardly; (ii) solid state diffusion with *composition segregation* or *impurity enrichment* at the surface and surface adsorption of contaminants onto the surface. In fact, real surfaces are generally covered by a layer of chemisorbed and physisorbed atoms or molecules (see Figure 1.2) including salts, organic compounds, hybrid organic/inorganic layers, patinas, pollen and microorganisms.

Structural and morphological features of solid surfaces comprise thus a large-scale inhomogeneity at the nano- and micro- scales (see Figure 1.14), among the most common examples of which are surface roughness and porosity, two crucial factors when considering the degradation, cleaning and consolidation of artistic surfaces.

Generally, surface roughness is one of the factors that may influence greatly the process of wetting and adhesion at liquid–solid interface. Surface roughness, r, is defined as the ratio between the real surface extension and the geometrical area of the surface under examination.

Surface roughness can easily be visualized through modern techniques such as atomic force microscopy.[12] In Figure 1.15 we report two examples taken from the Cultural Heritage of Florence where surface roughness play a distinctive role in governing the state of conservation of these masterpieces: the Gates of the Baptistery and the David by Michelangelo.[13]

Real surfaces can also expose holes, or more generally pores, that extend also underneath the surface through the deeper layers. Porosity can in fact be of various types, but in all cases it has the effect of increasing the effective area of the exposed surface.[4]

The surfaces of granular materials and powders can be regarded as a typical porous material; pores can be of different types and sizes and are classified

[4]Porosity can be defined as the ratio V_V/V_T, where V_V is the volume of void space and V_T is the total volume of material. The ratio ranges from 0 to 1 (or from 0% to 100%).

Figure 1.15 (left) The Gates of Paradise in the Baptistery of Florence and the AFM images (4 μm×4 μm) of a bronze surface of the same composition used for the gates; (right) copy of David by Michelangelo with a detail of the left toe.

according to the average dimensions as: macroporous (pore size >50 nm), mesoporous (pore size 2–50 nm) and microporous (pore size < 2 nm). Marbles, with pores between 1 and 5 nm fall into the last category, whereas chemical gels for the cleaning of artistic surfaces exhibit pores with radii in the 5–10 μm range. Wood cell walls have a nanoscale porosity, with radii of even less than 10 nm.

It is worth recalling that, in the case of granular materials, porosity depends not only on the dimensions of the solid particles but also on their packing in space. The same number of spheres with the same radius may exhibit different porosity according to the way they are packed: for a cubic packing of spheres the theoretical porosity is 47.65%, whereas if the packing is rombohedrical the porosity decreases down to 25.95%.

Solid surfaces share many peculiar properties, among which are history-dependent surface properties, structural defects and high surface roughness, porosity, site-specific high reactivity, presence of chemical impurities and contaminants. For most works of art, the solid surface under consideration may exhibit many of these features and it is therefore of the utmost importance that a clear knowledge of the limits of the system is obtained before any intervention is planned.

1.4.1 Surface Wettability and Contact Angle

Surface wetting phenomena are ubiquitous and become essential in many applications and processes, such as transport in soil, development of biocompatible surfaces, control of biofouling, detergency, membranes, glues, protective coatings and more importantly art restoration and conservation issues. Wetting is a general term that includes a variety of wetting systems. Figure 1.16 shows a wetting system consisting of a liquid drop on a solid surface, immersed in a fluid that could be a gas or another liquid. A typical situation includes the deposition of a water microdroplet on the surface of a fresco painting or a bronze surface. There are many other possible configurations of solid–liquid–fluid systems, such as a liquid inside a porous medium, or a particle floating on a liquid–fluid interface.

Understanding the physical principles that rule wetting phenomena is therefore of crucial importance not only in understanding how and why a surface has been modified and degraded but also in order to envisage proper interventions to restore the original structure and appearance of the surface itself. Oxidation and corrosion, and formation of patinas, are only a few of the phenomena that inevitably descend from preliminary wetting phenomena.

On the other hand, wetting processes are present in everyday life: different types of fabric may adsorb water to different extents, or even repel it (Figure 1.17).

Restoration of fabric and tapestry cannot avoid taking this feature into proper consideration. Similarly, water collects in large drops on oily surfaces whereas aqueous thin films are formed on clean glass surfaces. A simplistic cartoon of various scenarios for a water drop in contact with a solid surface is sketched in Figure 1.18.

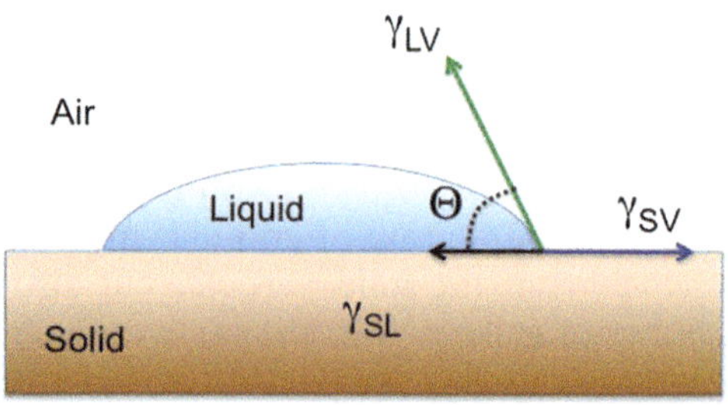

Figure 1.16 A liquid phase immersed in a fluid (gas or liquid) wets a solid surface, defining the contact angle.

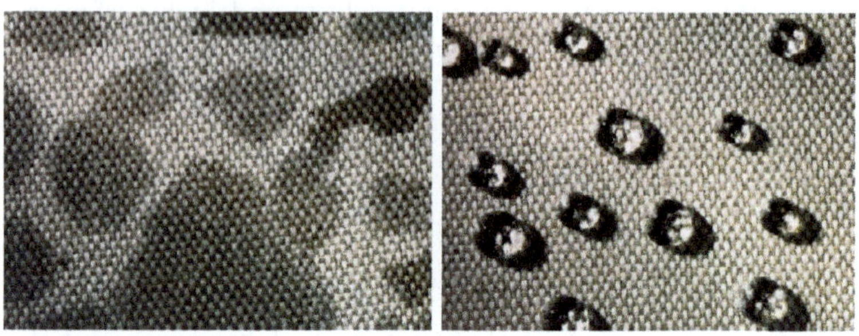

Figure 1.17 Water interacts with differently coated fabrics.

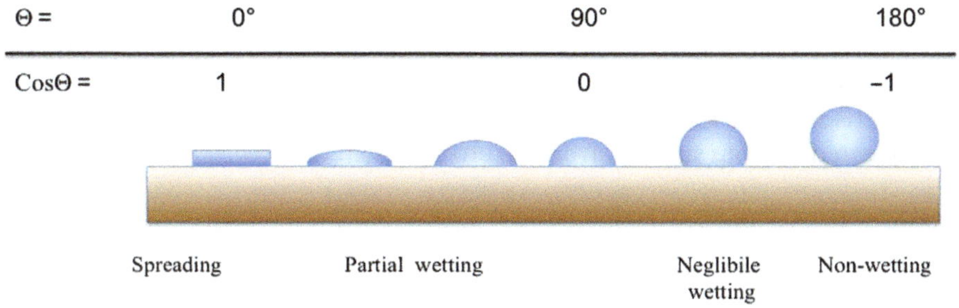

Figure 1.18 Wetting of water on different solid surfaces. The contact angle of water as a wetting index: a large contact angle implies poor wetting properties of the solid.

This behaviour reflects the fact that a liquid drop on a solid surface may adhere and spread differently depending on the nature of the solid surface and of the liquid.

We have already discussed the case of total spreading in Section 1.3, now we will explore the range of intermediate cases where wetting can be only partial.

In the above systems, the most important measurable parameter is the contact angle, θ, which is directly related to the wetting properties of the three-phase system under consideration. The contact angle is defined as the angle between the tangent to the liquid–fluid interface and the tangent to the solid interface at the contact line between the three phases (see Figure 1.16).

If we deposit a drop of water on a solid surface, the drop will probably start to spread. When spreading stops, each point of the perimeter of the drop will be at equilibrium. The equilibrium contact angle depends on the nature of the liquid and of the solid. Figure 1.18 shows how the contact angle of water on surfaces of different chemical nature changes, mirroring the wetting properties of the solid: water spreads on glass but does not wet wax surfaces.

The contact angle is a concept related to the adhesion and cohesion phenomena discussed in the previous section: liquid molecules are subject to cohesion forces that keep them close together but at the same time adhesion forces operate between the liquid molecules in contact with the surface and the molecules of the solid material. When the adhesion forces are larger than the cohesion forces the liquid wets the surface, whereas the liquid "refuses" the surface if cohesional attractions prevail. In the scenario depicted in Figure 1.18, a hydrophobic solid surface will be poorly wet by water and will exhibit a contact angle larger than 90°. Conversely, a water contact angle lower than 90° corresponds to a hydrophilic surface.

An applicative example is related to the treatment of hydrophilic artistic substrates, such as several stones (*e.g.* marble), with coatings so as to prevent water from wetting the surface and producing degradation effects (see Chapter 3). As an effect of the treatment, the contact angle will rise from values typically smaller than 30° (untreated stone) to values close to 80° (when the stone is coated). However, synthetic coatings often produce detrimental effects (both aesthetic and functional), leading to the need for their removal from the treated surface. In this case, the contact angle is a good parameter with which to assess the effectiveness of the cleaning intervention, and an inverted trend (roughly, for a treated marble surface, from 80° to 30°) will be expected upon cleaning.

1.4.1.1 Contact Angle for Ideal and Real Surfaces

When a drop of liquid is put in contact with a solid surface its perimeter will shift, contracting or expanding, until equilibrium is established for each point along the perimeter of the liquid drop, as represented in Figure 1.19 where γ^{SV} is the surface tension at the solid–vapour interface, γ^{LV} is the surface tension at the liquid–vapour interface and γ^{SL} is the solid–liquid interfacial tension; the three are in equilibrium.

Therefore, from the balance between the interfacial tensions along the parallel to the interface we may write eqn (1.10):

$$\gamma^{SV} = \gamma^{SL} + \gamma^{LV}\cos\theta_c \tag{1.10}$$

The pioneering correlation between the contact angle and the interfacial tensions is due to Thomas Young. The Young equation was developed for the case of an ideal solid surface, which is defined as smooth, rigid, chemically homogeneous, insoluble, and non-reactive.[14,15] Young's contact angle depends only on the physico-chemical nature of the three phases, and is independent of gravity. The latter may affect the shape of the liquid–fluid interface, but not θ_c. This corresponds to the state of the solid–liquid–fluid system that has the minimal Gibbs energy. Although it has been recognized that the three interfacial tensions may be influenced by each other at the contact line, for all practical purposes related to the measurement of contact angles of macroscopic drops, the effect of line tension on the contact angle is negligible.

It is worthwhile noting that the Gibbs energy *vs.* contact angle curve for an ideal solid–liquid–fluid system has only a single minimum at θ_c. In other words,

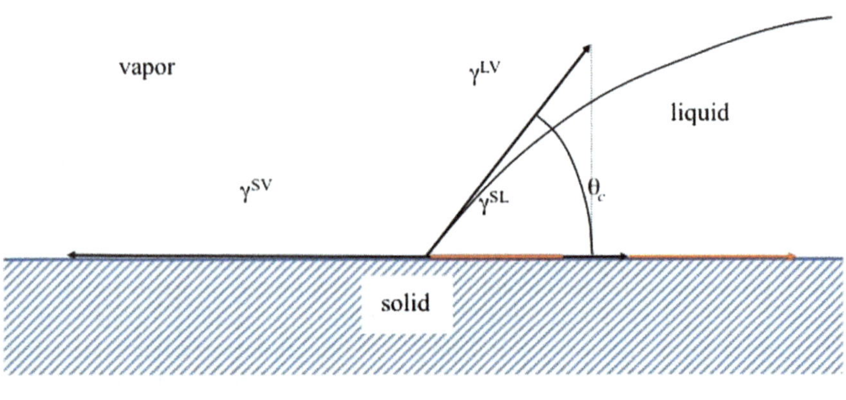

Figure 1.19 Definition of the Young contact angle.

Table 1.3 Contact angle for typical solid–liquid systems at T = 25 °C.

Solid	*Liquid*	*Contact angle*
Glass	Water	ca. 0°
	Mercury	148°
Wax	Water	106°
	Benzene	ca. 0°
Talc	Water	88°
PFTE	Water	108°
	Benzene	140°
	Mercury	86°
	Hexane	ca. 0°
Silicate rock	Water	ca. 0°

the ideal system is characterized by a single value of the contact angle, which is not always the case for real surfaces.

Since the Young equation defines $\cos\theta c = \frac{\gamma_{SV}-\gamma_{SL}}{\gamma_{LV}}$, a range of wetting behaviour $[-1 \leq \frac{\gamma_{SV}-\gamma_{SL}}{\gamma_{LV}} \leq 1]$ can be classified and predicted by measuring the contact angle, as described in Figure 1.18.

The Young equation is widely used to determine the contact angle of liquids on a variety of solids, including the surfaces involved in the conservation of Cultural Heritage items, both to monitor the degradation state of the works of art (such as marbles, frescoes, papers, tapestry, glass windows) or to control the surface modification imposed for protection purposes. In Table 1.3 we report the values of contact angle for typical solid–liquid couples. Such solid materials are often used in cleaning and restoration of works of art, for example wax is widely used for adhesives and coatings as well as a traditional cleaning tool. In this latter case a mixture of bleached beeswax, stearic acid, ammonia and deionized water (waxy emulsion or stearate emulsion) is used to treat the surfaces. This peculiar mixture is also known as "*Pappina Fiorentina*" (owing to its extensive usage by the conservators of the Opificio delle Pietre Dure, OPD, a public art restoration institute in Florence; see Chapter 9).

Polytetrafluoroethylene (PTFE), commonly known as teflon, is well known for its anti-adhesive properties and widely used for anti-stick coatings for cooking tools. In addition, SiO_2-based materials, together with carbonate rocks, are often encountered in Cultural Heritage, as in sandstone artefacts.

The apparent simplicity of eqn 1.10 when used to assess the wettability of solids is deceptive; first of all γ_{SV} and γ_{SL} cannot be measured directly, and more importantly the specific features of solid surfaces described above may induce large deviations of the measured contact angle from the ideal $\theta_{c.}$

The major causes of incorrect evaluation of the contact angle include surface roughness, chemical heterogeneity of the surface, and surface-active substances that can adsorb both at solid–liquid and at liquid–vapour interfaces and change the contact angle locally. All these effects introduce errors in the measurements, and the contour of a sessile drop is *very* irregular. This results in non-univocal values for the measurement of the contact angle and hysteresis of the contact angle that will be described in a following section.

1.4.1.2 Measurement of Static and Dynamic Contact Angles

Contact angle measurement is a low-cost method of analysis of surfaces that in most cases provides valuable information on the surface of the work of art for different purposes, *e.g.* to design a appropriate system for its cleaning or protection and to control the surface modification after the treatment. The method is non-invasive, as long as non-aggressive probe liquids are used and water-sensitive substrates are not involved, and can be performed easily with both water and non-polar organic solvents as probe liquids.

The techniques for measuring contact angles can be classified broadly into two categories.

1. Actual measurement of the angle by goniometric observation or some optical technique. This is the simplest and most direct method.
2. Methods that do not give θ_c directly but give $\gamma^{LV}\cos\theta_c$. Such methods usually involve a force measurement, or compensation of a capillary force, and can be carried out precisely and automatically.

Among the first class, direct visualization is the most convenient method, especially for samples of large area, but the results are often imprecise and scarcely reproducible. Automated systems with a computer-controlled dispensing systems are in most cases the best choice. From a technical point of view, the sessile drop method is most frequently used, because this is the most convenient method; however, other contact angle measurement methods for planar surfaces are used as well. Moreover, in art conservation, it is also possible to encounter other scenarios that require measurement of contact angle for more complex systems, for example in the case of tapestry and carpet restoration it may be necessary to asses the wetting properties of fabrics and, more often, fibres. In this case, contact angles are measured by suspending the fibre vertically from a microbalance and using the Wilhemy plate set-up.

The problem of contact angle measurement and interpretation is not yet completely solved; therefore it is necessary to understand the difficulties that are involved, in order to choose the best possible path for a specific application. The problems encountered in the measurement of the contact angle are due to the asymmetry of the drop and the relative dimensions of the drop compared with surface heterogeneity.

Solid surfaces are usually rough and chemically heterogeneous; a distinction is therefore necessary between the apparent contact angles, a quantity that can be measured, and the actual or intrinsic contact angle that depends only on the surface energy of the system interfaces. The actual contact angle is the angle between the tangent to the liquid–fluid interface and the actual, local surface of the solid. The apparent contact angle is the angle between the tangent to the liquid–fluid interface and the line that represents the nominal solid surface, as seen macroscopically (see Figure 1.20).

Whereas the Young equation predicts a single value for the static contact angle, it is found experimentally that a finite range of apparently stable contact

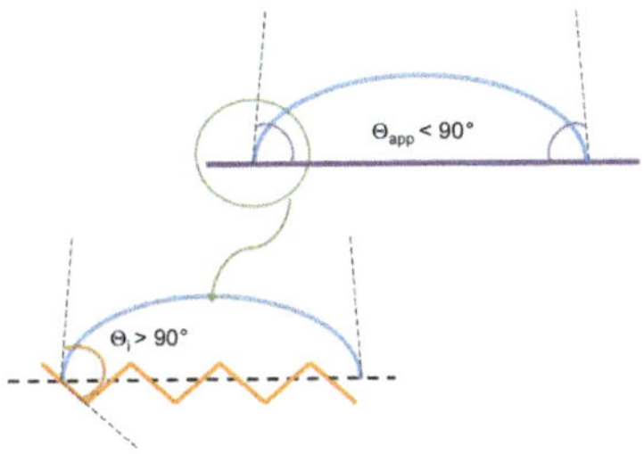

Figure 1.20 Schematic description of apparent Θ_a and intrinsic Θ_i contact angles for rough surfaces.

angles can be measured for real heterogeneous surfaces such as the ones involved in works of art.

Surface roughness may be very high, as in the case of fresco surfaces (for which also porosity acts to disguise the intrinsic contact angle of the material), but also for oil tempera on wood the presence of a siccative oil layer may alter the morphology of the surface. Moreover, most pictorial surfaces present a diversity of chemical composition in a small area, demanding a tailored theoretical treatment. The approach to rough surfaces has been treated by Wenzel,[16] whereas a model for contact angle interpretation of chemically heterogeneous surfaces has been proposed by Cassie and Baxter.[17] In 1936, Wenzel[9] developed in a rather intuitive way an equation for the apparent contact angle θ_{app} on a rough surface, averaging the fine details of the roughness. For smooth ideal surfaces we have already demonstrated that:

$$\gamma_{SV} - \gamma_{SL} = \gamma_{LV} \cos \theta_{true} \tag{1.11}$$

However, the surface roughness enlarges the surface area of contact between the surface and the liquid and the solid–vapour interface. Wenzel proposed that for rough surfaces the following equation holds:

$$r\,(\gamma_{SV} - \gamma_{SL}) = \gamma_{LV} \cos \theta_{app} \tag{1.12}$$

where r is the roughness ratio, defined as the ratio between the actual and projected solid surface area. With this definition, $r = 1$ for a smooth surface, and >1 for a rough surface. Therefore:

$$\cos \theta_{app} = r \cos \theta_{true} \tag{1.13}$$

The Wenzel equation is based on the assumption that the liquid completely penetrates into the roughness grooves, as shown in Figure 1.21. This wetting situation on rough surfaces is termed "homogeneous wetting".

Interestingly, as shown in Figure 1.21, the increase of surface roughness results in a decrease of the contact angle in the case of $\theta_{true} < 90°$, *i.e.* $\theta_{rough} < \theta_{true}$. On the contrary, if $\theta_{true} > 90°$, the apparent contact angle will be larger than the true one, $\theta_{rough} > \theta_{true}$. This means that extremely rough surfaces can exhibit extremely large contact angles and eventually leads to the super-hydrophobic effect (see below) that paves the way to new possibilities for

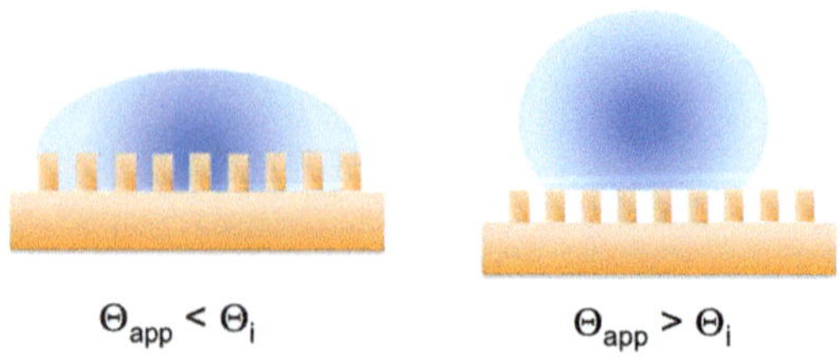

Figure 1.21 Contact angle variation due to surface roughness for (left) $\Theta_{true} < 90°$ and $\Theta_{app} < \Theta_{true}$ (Θ_i); (right) $\Theta_{true} > 90°$ and $\Theta_{app} > \Theta_{true}$ (Θ_i).

obtaining superhydrophobic protection layers for works of art located in outdoor or unprotected environments. For instance, organic–inorganic composite coating films are being developed to induce artificially an increased roughness of mineral substrates, resulting in water repellency (higher surface contact angle; see Section 14.3).

The Wenzel equation is an approximation that becomes better as the drop becomes larger in comparison with the scale of roughness; if the drop is larger than the roughness scale by two to three orders of magnitude, the Wenzel equation applies satisfactorily (for water, an average droplet has a size of a few millimetres, hence the scale of roughness that satisfies the Wenzel approximation is much smaller than 0.1 mm).

On chemically heterogeneous solid surfaces the surface tension varies from one location to another. Accordingly, the contact angle has a different value at each location also on smooth but inhomogeneous surfaces; this may be due to impurities, polycrystallinity or, in the case of works of art, on the intrinsic nature of the pigments used or on the aged deposits that contribute to the deterioration of the surface. In 1948, Cassie developed an equation for the apparent contact angle on a heterogeneous solid surface[17] and, for the case of a surface with only two different chemical species, the following equation applies:

$$\cos\theta_{composite} = f_1 \cos\theta_1 + f_2 \cos\theta_2 \quad (1.14)$$

In this equation, f is the area fraction for each species, and the subscripts 1 and 2 indicate the two different surface functionalities; θ_1 and θ_2 are the true contact angles that correspond to the area fractions f_1 and f_2. More complicated relationships have been proposed,[18] but the Cassie equation still seems to predict experimental results correctly.

This equation can be generalized to state that the cosine of the Cassie contact angle is the weighted average of the cosines of all the contact angles that characterize the surface. As in the case of the Wenzel equation, the Cassie equation is also an approximation that improves when the drop size becomes larger with respect to the scale of chemical heterogeneity.

Under some roughness conditions, air bubbles may be entrapped in the roughness grooves under the liquid, and both roughness and chemical composition will converge in the modification of the measured contact angle.

In this case, the solid surface may be considered chemically heterogeneous, and the Cassie equation 1.14 may be applied in the following form:

$$\cos \Theta_{CB} = f_1 r_f \cos \Theta_1 + f_2 r_f \cos \Theta_2 \tag{1.15}$$

Where Θ_{CB} is the CB (Cassie–Baxter) apparent contact angle, f_1 is the fraction of the area of the solid surface that is wet by the liquid, and r_f is the roughness ratio of the wet area. This equation was developed by Cassie and Baxter.[17] When $f = 1$, $r_f = r$ and the CB equation turns into the Wenzel equation 1.13. The transition from homogeneous wetting (Wenzel equation) to heterogeneous wetting (CB equation) has been recently analysed,[19] and it has been found to depend not only on the roughness ratio, but also on the specific geometry. Examples of surfaces where Cassie–Baxter treatment can be readily applied can be found among the superhydrophobic surfaces.

The interaction of surface roughness and capillary phenomena during wetting of a solid surface leads to a number of complex effects. Superhydrophobicity is the enhancement of hydrophobic properties due to roughness and it can be considered as a bio-inspired phenomenon. In fact, the best known example of superhydrophobic phenomenon is the Lotus effect, which involves superhydrophobicity and self-cleaning.[18] The Lotus effect is based on surface roughness caused by different nanostructures, together with the hydrophobic properties of the epicuticular wax, as shown in Figure 1.22.

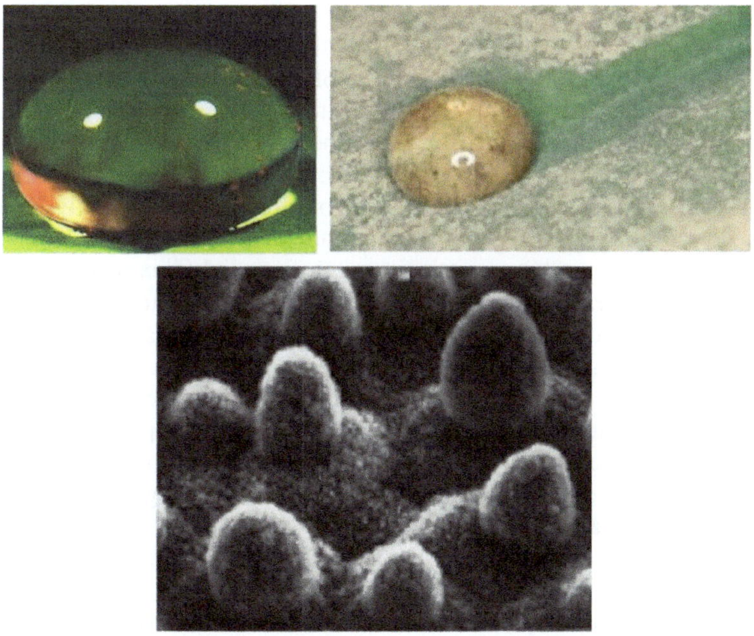

Figure 1.22 (top left) A water droplet on a lotus leaf with adhering dust particles. (top right) A water droplet removes dust as it rolls over a superhydrophobic surface. (bottom) Nanostructured islands or patches of wax on leaf surface.

The surface is covered with microscopic islands or patches of wax and such heterogeneity enhances the hydrophobicity of the leaf enormously. Small water droplets do not wet the surface at all; they remain spherical, and roll off the leaf upon minor disturbance, cleaning the surface of the contaminants in their way (see Figure 1.22). As discussed above, an initially slightly hydrophobic solid surface with a water contact angle $\theta_{true} > 90°$ becomes very hydrophobic after roughening, and it may have a θ_{app} approaching 180°. A roughness-induced superhydrophobic surface, according to the accepted definition, has $\theta_{app} > 150°$.[18] The effect of roughness-induced superhydrophobicity was theoretically predicted and experimentally observed in the 1930s, although the term "superhydrophobicity" was coined later, in the 1990s, when micro-patterning technology matured. It then became possible to build superhydrophobic surfaces with desired properties, and these systems started to be exploited as self-cleaning surfaces. Self-cleaning is the ability of many superhydrophobic surfaces to wash out contaminant particles with water drops running upon the surface, as opposed to conventional surfaces that have stronger adhesion to contaminants.

Many fields for application are possible (facades, paints, church windows and historical doors) where such surfaces are advantageous. For instance, it is worth mentioning another important area of application: underwater protection layers. An example of conservation where superhydrophobicity may play a major role could be the treatment of the underwater walls of Venetian buildings (Figure 1.23) treated with highly hydrophobic and nanostructured polymer coatings.[18]

Finally, we will report a case study of contact angle measurements for Cultural Heritage, concerning the degradation of bronze surfaces located on the doors of the Baptistery of Florence, one of the oldest buildings in the city, built between 1059 and 1128 (Figure 1.24).

Works of art located in outdoor environments are commonly affected by severe degradation induced by factors that are not as easily controlled or monitored as in indoor settings, such as light exposure, humidity, temperature, and rain and air pollution. The last two factors are difficult to monitor exhaustively and have often been overlooked, although their effect on the surface of a work of art can be quite devastating. Among the works mostly affected by the aggression of polluted urban areas stand doors and gates of monuments and historical buildings. In this respect the gates of the Baptistery of Florence represent a paradigmatic case study[18] that has been the object of a three-year-long multidisciplinary investigation that gathered many different characterization techniques to monitor the alteration as a function of time, together with daily analysis of temperature and air compositon.[20]

A variety of surface-specific techniques is accessible nowadays for conservation and restoration purposes, but contact angle measurement is particularly appealing because it is an unsophisticated experimental set-up that provides a variety of significant information on the state of alteration of metallic surfaces,[21] wood surfaces[22] as well as stone and marble.[23]

Figure 1.23 (left) Protection of an underwater wall along a Venetian Canal. (right) Detail of the Canal wall after water removal.

Figure 1.24 (left) Baptistery in Piazza San Giovanni in Florence; (right) the North Gate, where bronze replicas were located at different heights and orientations, to gain differential protection from water. Inset: one of the bronze replicas and the areas where contact angle was measured.

Blocks of multiple bronze replicas were prepared by fusion of the same alloys used in the original artefact[18] by Ghiberti and Pisano, and the sets of replicas were placed on the Gates of Pisano and Ghiberti that face either north towards the entrance of the Baptistery or to the south towards the pedestrian area. The replicas were located at different heights and orientations to gain different protection from water, as demonstrated in Figure 1.24. Continuous atmospheric monitoring was performed over a three-year period: single replicas from each set were collected every six months and characterized with a multitude of physico-chemical measurements, among which was contact angle determination.

Interfacial phenomena find here a major role in determining the state of conservation of these precious artefacts: the spontaneous growth of various layers on the surface occurs, each layer showing different contact-angle and wetting properties. The extent of liquid–vapour interfaces controls oxygen diffusion,[24] whereas solid–liquid interfaces have a direct influence on the rate of dissolution of corrosion products. The wetting properties of the surface play a key role in corrosion and in the adsorption of hydrophilic contaminants.

The effect of exposure to the external environment and to the cycling of the seasons was investigated by measuring the contact angle of all replica tiles in the selected positions. Measurements were run on 2×2 cm samples using ultrapure water as a liquid probe. Experiments were repeated in different spots of the sample to estimate the homogeneity and repeated after 24 hours to investigate in more detail the patina formed after the first exposure of the samples to water.

The data evidenced similar contact angles for all replica targets in the range 86°–98°, indicating a low degree of hydrophilicity and wettability of the bronze alloys, as expected for metal surfaces exposing copper and zinc.[25,26] When the measurements were repeated for replicas taken from the door location after six months of exposure, a dramatic change of contact angle was observed, as reported in Figure 1.25, indicating the formation of a surface layer with high hydrophilicity.

For all locations the average contact angle dropped down to small values; the decrease was paralleled by an increase in roughness of the surface, determined by means of atomic force microscopy. Comparison with results obtained for the same samples with analytical surface techniques demonstrated that such

Figure 1.25 Water contact angle measurements on the bronze surface before (left) and after (right) being exposed on the Baptistery doors.

behaviour is related to the formation of hydrophilic corrosion products such as oxides, sulfates or chlorides of copper and zinc.[27] Interestingly, the decrease in θ was larger for samples placed on the lower positions of the South Door, samples that are less available to rain events that would wash away the hydrophilic deposits. On the other hand, samples placed in water-protected zones showed higher contact angles owing to the presence of cuprite[28] or hydrophobic particles. Contact angle investigation was also correlated with the behaviour of temperature, rainfall and the concentrations of various pollutant gases and particles (NO_x, PM_{10}) in the atmosphere during the three years of study. The results revealed many interesting aspects related to these environmental parameters, as shown in Figure 1.26.

Hydrophilicity was found to increase up to summer 2006, followed by a stabilization or partial increase in θ, due to the formation of hydrophobic pollutants, after winter 2007. The increase in water contact angle was more important for unprotected replicas, due to the intense rain volume recorded in that period. A cartoon depicting the rationale suggested by the results obtained from contact angle measurements is reported in Figure 1.27.

The drastic decrease in θ observed in the first 18 months of exposure to the environment reveals the massive presence of hydrophilic corrosion products; interestingly such decrease occured during the summer season, and in particular in periods where extremely high concentrations of ozone were recorded in the central urban area of Florence. This supports a recently proposed hypothesis[29] that high ozone concentration may act synergistically with other pollutants to induce corrosion effects that exceed the threshold value established for buildings included in the Cultural Heritage designation.

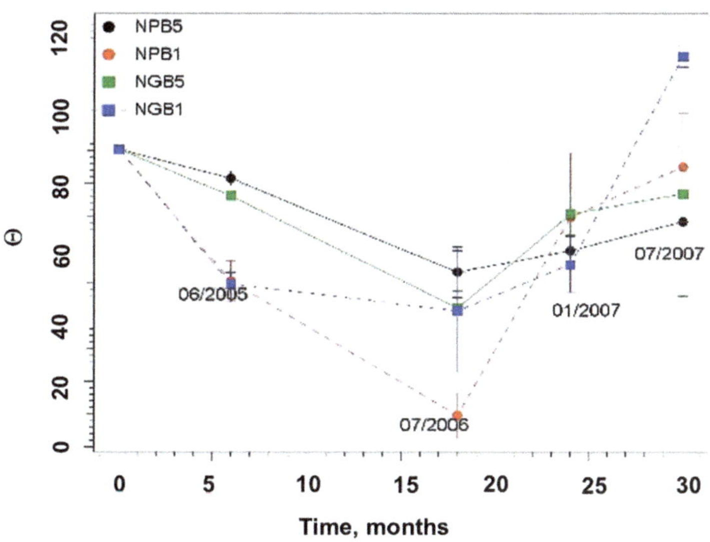

Figure 1.26 Change in water contact angle as a function of time for bronze replicas placed on the South Door (circles) or on the North Door (squares), either protected (solid line) or exposed (dashed line) to rain events.

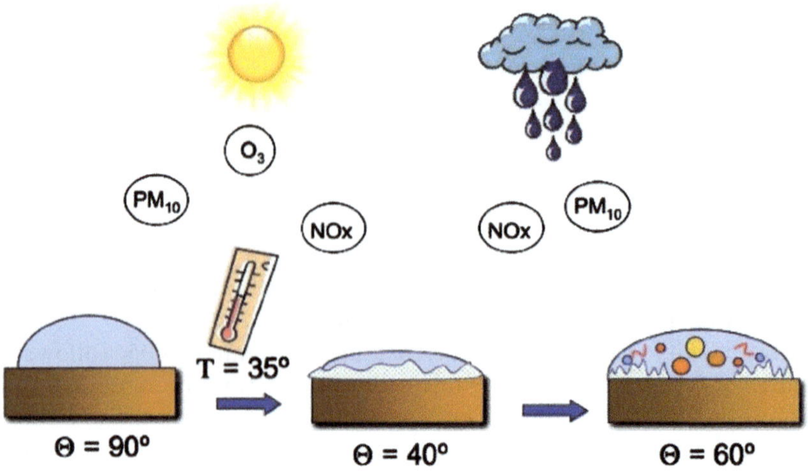

Figure 1.27 Effect of seasonal cycling and pollutants on the state of alteration of the bronze surfaces.

1.4.2 Curved Surfaces and Capillarity

The concept of surface or interfacial tension described in Section 1.2 implies the presence of a planar interface that separates two adjacent bulk phases.

In smaller-scale systems such as nano- or microdroplets, bubbles, gas cavities in fluids, or liquid in pores, this description may fail because the system is highly heterogeneous.[30] These examples are often encountered when nanoscience is applied to art conservation, and in the following chapters a comprehensive description of such heterogeneous nanosystems will be provided. Here we sketch the basic concepts underlying the physical behaviour of such interfaces. A more detailed discussion of thermodynamics is made available by consulting the Further Suggested Reading section.

We have already stated that γ is affected by temperature but also by interfacial curvature (see Section 1.1). This peculiar role of curved surfaces is easily recognized in phenomena such as the difference in vapour pressure of a liquid with curved interfaces, or in the rise of liquids in capillary tubes or complex porosities.

The first observations of liquid rise in thin tubes can be traced at least to medieval times: the phenomenon initially escaped explanation and was described by the Latin word *capillus*, meaning hair. It became clearly understood only during recent centuries that many phenomena share a unifying feature that involves the interface between two materials situated adjacent to each other. The following chapters will describe how the effect of capillary forces affects the process of deterioration of the Cultural Heritage in many complex systems, spanning from walls, frescoes, woodwork to fabric

Figure 1.28 Example of capillary rise of coloured water in a small tube. The inset shows the curvature of the interface.

and paper artefacts (see also Section 2.4.1). But no matter how important its effect will be in complex systems, capillarity is a phenomenon that can be described simply as the rise of a liquid in tubes of small diameter, as depicted in Figure 1.28.

The rise of the liquid above the level of zero pressure is due to a net upward force produced by the attraction of the liquid molecules, *i.e.* water molecules, to a solid surface, *e.g.* glass or soil. When the adhesion of the liquid to the solid wall is greater than the cohesion of the liquid to itself, the quantity of the liquid that rises in the tube increases until equilibrium with the weight of the liquid column is reached. For the same liquid, the height of the column increases with a decrement in the tube diameter. Moreover, if liquids with different surface tension are used or surface active agents are added to water, the height reached by the liquid in the same tube will vary. These empirical observations lead to the general relationship between H, the height in the tube, R, the tube radius and the surface tension γ:

$$H = C\gamma/R \tag{1.16}$$

Here C is a proportionality constant. Furthermore, changing the thickness of the walls has no effect on the surface, thus suggesting that the forces giving rise to the phenomenon can be significant only at extremely small distances (a more precise analysis indicates a large portion of these forces to be at most molecular in range).

Molecules being pulled toward the walls will force other molecules aside in all directions, resulting in a spread along the walls that is only partly compensated by gravity. Liquid is forced upward along the walls, and cohesive forces carry the remaining liquid column with it.

Equation 1.17 predicts the height reached by a liquid with surface tension γ when rising inside a tube of diameter r, provided the contact angle between the liquid and solid walls is zero:

$$h = \frac{2\gamma}{\rho g r} \tag{1.17}$$

where ρ is the density of the liquid and $g = 9.8\,\mathrm{m\,s^{-2}}$.

In the case of water ($\gamma = 72.8\,\mathrm{mN\,m^{-1}}$, $\rho = 1000\,\mathrm{kg\,m^{-3}}$), the column of the liquid will reach different heights depending on the radius of the capillary tube, as reported in Table 1.4.

Moreover, the equation also predicts that the lower the surface tension, the lower the height of the column in the capillary, meaning that the presence of contaminants or "surface active" substance (*e.g.* surfactants) will affect the magnitude of h.

If the radius of the capillary is larger than 0.5 mm (see Figure 1.29, below) it is not correct to assume the meniscus to be hemispherical and the capillary equation needs to be corrected by adding explicitly a factor correlated to the contact angle θ that the liquid forms on the wall of the tube. The equation will become:

$$\rho g h = 2\gamma(\cos\theta)/r \tag{1.18}$$

Table 1.4 Height of liquid rise in capillary tubes of different radii.

h (cm)	*r (cm)*
0.0015	100
0.15	1
14	0.01

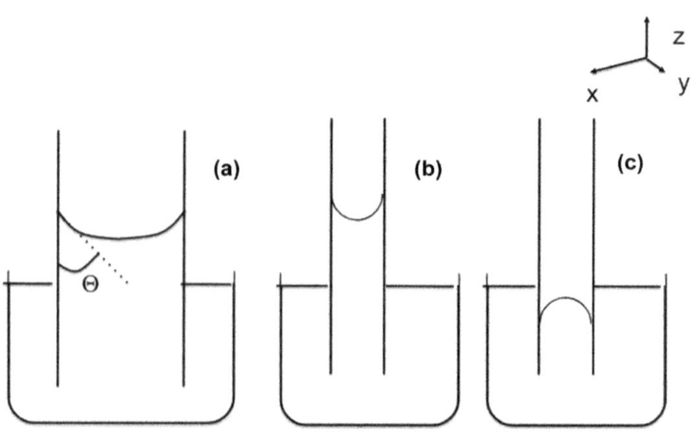

Figure 1.29 Schematic description of the meniscus formed in capillary tubes. (a) General case. (b) Hemispherical meniscus with contact angle $\Theta = 0°$ and $r_{\text{meniscus surface}} = r_{\text{capillary tube}}$. (c) Capillary depression for liquid that does not wet the tube walls, *i.e.* large contact angles.

with r being the actual radius of curvature of the meniscus. Tables for such corrections are published in the existing literature (see Further Suggested Reading section). From eqn (1.18), it is clear that the rise or fall of a liquid in a tube will be governed by the sign of $\cos\theta$.

For instance, in the case of liquid mercury on glass, the contact angle at hydrophilic solid interfaces is very high (180° in the case of hemispherical meniscus), therefore $\cos(180°) = -1$. In this case we say that the liquid is non-wetting. A *capillary depression* is observed, as described in Figure 1.29 (c): the picture depicts a situation where $r < 0$, that is to say the liquid meniscus will be inverted and the liquid recedes along the tube walls. In this case, the adhesive forces are weaker than the cohesive forces among the molecules of the liquid, and the fluid is drawn away from the walls of the tube, causing the fluid to sink slightly. The hydrostatic equilibrium is reached by a lowering of the liquid level.

Capillary forces play an important role in all systems where liquids are present in a porous environment, although real-world capillaries or pores are not always circular in shape, as in the case of porous or granular materials encountered in artistic substrates (*e.g.* paintings on wood and paper, or frescoes).

In any system in which the fluid flows through porous material, it would be expected that capillary forces will be one of the most dominant factors, no matter the shape or extension of the pores, and the rise in capillaries of other shapes, such as rectangular or triangular, can be measured.[30] For example, for a porous solid such as paper (or compacted granular media), the height and the time of rise of the liquid follow the well known Washburn equation that allows measurement of the contact angle or surface tension of a porous medium that is regarded as an ensemble of tiny capillary tubes:

$$L^2 = (2\gamma rt\cos\theta)/4\eta = (\gamma rt)/2\eta \tag{1.19}$$

where L is the height of rise, γ is the liquid surface tension, r is the pore diameter, t is the time of rise, θ is the contact angle and η is the liquid viscosity.

Conversely, for a liquid of known properties the height of rise is affected by the pore radius and by the contact angle. This rather simple measurement has important effects on the characterization of porous media, allowing for the estimation of the average porosity of powders, and of granular and porous materials.

Capillary phenomena in porous media also affect other processes that may cause important degradation of pictorial surfaces, especially in settings where the capillary rise is important, *e.g.* church and crypts walls, as in capillary condensation described in the next section.

1.4.2.1 Capillarity in Action

As stated above, many industrial and natural processes rely on capillary phenomena, ranging from blood flow in the veins to oil recovery in a reservoir. The sponge absorbs water or other fluids as a result of capillary forces, a process known as *wicking* (as in candlewicks). Properties of fabrics and paper are also governed by capillary forces (wetting *etc.*) as is the rise of moisture from the soil in the case of plants and trees.

It is common to witness the ability of liquids derived from the soil to penetrate into cracks and pores in building walls. Moreover, along with water, a variety of water-soluble elements may rise and diffuse in the porous media, giving rise to efflorescence, growth of microorganisms and eventually the detachment of layers of the surface. In the case of Cultural Heritage, several materials are affected by these phenomena, *e.g.* marble and stone artefacts, ceramics, frescoes and decorated walls, painted wood or paper. The presence of water inside the porosities of these materials is detected by the appearance of a whitish border, often formed by the crystallization of salts coming from the soil and migrating to the surface of the artefacts, as shown in the example of Figure 1.30.

In the case of walls, the water may come from groundwater or from accidentally dispersed water as in liquid spills from badly collected rainwater and aqueducts or, with more devastating effects, in floods. The height from the ground at which the diffuse border appears may vary and it depends on the type of porosity involved including the diameter of the pores and type of porosity, *i.e.* closed, connected or open; connected porosity is by far the most prone to massive degradation phenomena.

It should be mentioned that the extent of the water rise is also dependent on the thickness of the wall base in contact with the ground, and on its ratio with

Figure 1.30 (a) Example of typical capillary rise of water from the ground in church walls. (b) Detail of the salt crystallization.

the surface exposed to water evaporation: maximum rise will be observed for internal walls but it will be less important for isolated pillars. The quantity of capillary water may be significant and reach 30% in volume for construction materials such as mortar and cement; water quantities as high as 300 kg can be adsorbed per cubic metre of wall. In the case of frescoes, or external location of other works of art, the aqueous solution trapped in the porosity of the material may also freeze in severe and uncontrollable climatic conditions; in turn the increase in volume inside the pores will induce a mechanical stress in the material, resulting in cracks and disruption of the object, which is often observed for stone artefacts placed outdoors.

1.5 Solid–Gas Interfaces

In the previous section we introduced the distinctive features of the solid surface and demonstrated how the solid surface, and hence also the solid–air interface, differs from the surface of a liquid. We recall that the shape and surface structure of a solid surface is generally under kinetic equilibrium and strongly depends on the "history" of its formation, whereas the large mobility of molecules in the liquid phase induces rapid rearrangements of the surface, allowing for fast re-equilibration of the interface.

Moreover, solid surfaces present sites with different properties and surface energies together with many intrinsic defects; this translates into a strong reactivity of the solid surface, which is therefore prone to spontaneous adsorption of any substance that may lower the solid surface energy.

This applies also to the solid–air interface where impurities and contaminants are almost unavoidably adsorbed at the surface. A cartoon of this behaviour was presented in Figure 1.2: these properties of the solid–air interface render any surface exposed to the environment subject to the aggression of gases, particulates or bio-contaminants. With regard to Cultural Heritage, it is evident that, in the case of buildings, facades, church gates or statues exposed in an outdoor environment, the quantity and quality of the air contaminants cannot be controlled and will cause massive adsorption processes that lead to deterioration of the exposed surface. In indoor collections (frescoes, paintings on wood or canvas, books, tapestries and carpets), the concentration of contaminants in the air phase in contact with the solid surface may be partially controlled but only in museum settings does such an approach constitute proper protection of the surface.

The contaminants at the interface may interact following two different routes. In the first case, they may adsorb weakly at the interface and, in many favourable cases, they can be removed by appropriate cleaning procedures. Unfortunately, adsorption may also lead to a chemical reaction between the contaminants and the surface that will irreversibly and severely damage the artefact surface and its readability. These two processes follow different physico-chemical paths that will be briefly described and compared. Classical examples of adsorption-induced damage will be further discussed in this book, including oxidation and corrosion phenomena, formation of patinas, tarnishing upon moisture and sulfur adsorption. A careful investigation of the adsorption processes may lead not only to an understanding of the phenomena that induced the degradation of the artefact, but it will eventually suggest ways and strategies to protect, by adsorption of appropriate substances, the surface from further damage. We will focus in this section on adsorption at the solid–air interface, although the same principle applies also to solid–liquid and liquid–air interfaces.

The thermodynamic driving force for adsorption is the quest for a reduction of surface tension that will result in a minimization of the surface energy. Complex adsorption phenomena (see Adamson (1997), Further Suggested Reading section) are the result of different types of physical interactions at the interface that we can broadly group into the following.

1. *Physisorption*: a process prompted by van der Waals interactions between the molecules at the interface and the surface. This phenomenon is therefore associated with weak adsorption energies and it is the prevailing process at low temperatures. The molecules coming from the vapour phase collide with the surface but they are free to diffuse at the solid–air interface in search of energetically favourable surface sites.
2. *Chemisorption*: as the compound approaches the surface, it reacts to form a chemical bond with the atoms of the solid surface; this phenomenon involves higher energies and is favoured by an increase in temperature.

In the case of ionic substances, electrostatic interactions between the adsorbate and the surface also contribute to the adsorption process, and the charge of the compound plays a major role, *e.g.* trivalent ions such as Al^{3+} are attracted by an OH^- group more strongly than an Na^+ ion.

Two popular theoretical models have been developed to describe adsorption: the Langmuir model and the Brunauer–Emmet–Teller (BET) approach. They apply to monolayer and multilayer formation at the interface, respectively.

In the case of chemisorption and of strongly localized physisorption, Irving Langmuir proposed in 1918 to consider the surface as "a checkerboard with finite number of equivalent adsorption sites".

In the equilibrium between the gas phase and the adsorbed phase at constant temperature, an increase in pressure will increase the number of molecules adsorbed on the surface; on the other hand the number of molecules adsorbed at any pressure p will also depend on the strength of the adsorbate–surface interaction. At any surface pressure, the amount of adsorption is expressed by the parameter Θ, which assumes a fractional value whose maximum limit is 1, as shown in Figure 1.31.

At the equilibrium, the following equation applies:

$$\text{Rate of adsorption} = k_{ads}\, p\,(1 - \Theta) = \text{Rate of desorption} = k_{des}\Theta \quad (1.20)$$

Where k_{ads} and k_{des} are the rate constants for adsorption and desorption, respectively.

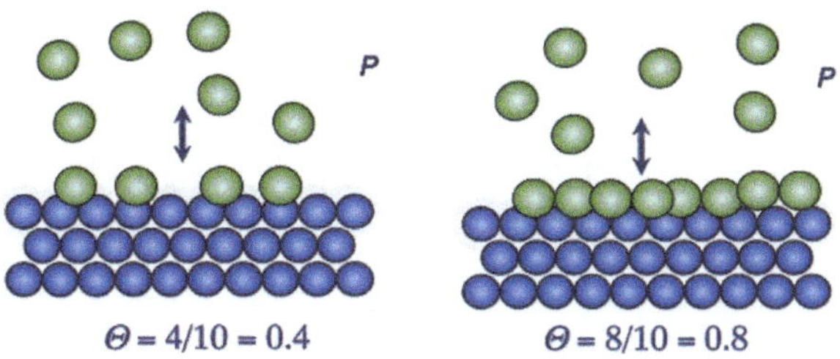

Figure 1.31 Cartoon of adsorption process at different pressures.

If we define a parameter b as $b = k_{ads}/k_{des}$ we obtain a function that relates the fractional coverage to the pressure, expressed by eqn (1.21):

$$\Theta = bP/(1 + bP) \tag{1.21}$$

which is an equation popularly known as the Langmuir isotherm. The parameter b is the analogue of the equilibrium constant K_{ads} for the adsorption process, thus reflecting the strength of the adsorbate–surface interaction. The Langmuir model is limited to perfectly equivalent adsorption sites and it is not valid for heterogeneous surfaces exhibiting sites with different atomic ordering and defects, or cases where lateral interactions between adsorbates in the monolayer cannot be neglected. Nevertheless, the Langmuir model is a useful approximation in many instances, although other complex mathematical models[31] may fit the experimental data better. The Langmuir model was later extended by Brunauer, Emmet and Teller in 1938,[32] to include the formation of multilayer at the solid–air interface for strong adsorbate–adsorbate interactions.

The properties of the adsorbing surface also affect the adsorption process: roughness, wettability and porosity of the solid surface will drive both the kinetics and the mechanism of adsorption.

The quantity of adsorbed substance, or surface coverage, as a function of the equilibrium concentration of the adsorbate at constant temperature describes what is called the adsorption isotherm. The form of the isotherm is specific for each adsorbate–interface couple.

In Figure 1.32 we report typical International Union of Pure and Applied Chemistry (IUPAC) isotherms that are tabulated in many textbooks (see Further Suggested Reading section). We can extract specific information on the interaction parameters between adsorbate and adsorbent, between adsorbate and adsorbate molecules, and on the porosity of the adsorbing solid by examining the shape of the isotherms.

The first part of the isotherm provides information on the adsorbate–adsorbent interaction and on the adsorption energy: the higher the slope of the curve in this region, the larger the interactions between the molecules and the surface. In the second part of the isotherm, information on adsorbate–adsorbate interaction at the solid–air interface can be extracted: the presence of only one plateau region (type I) means that only one monolayer of adsorbed molecule can be formed and hence that adsorbate–adsorbate

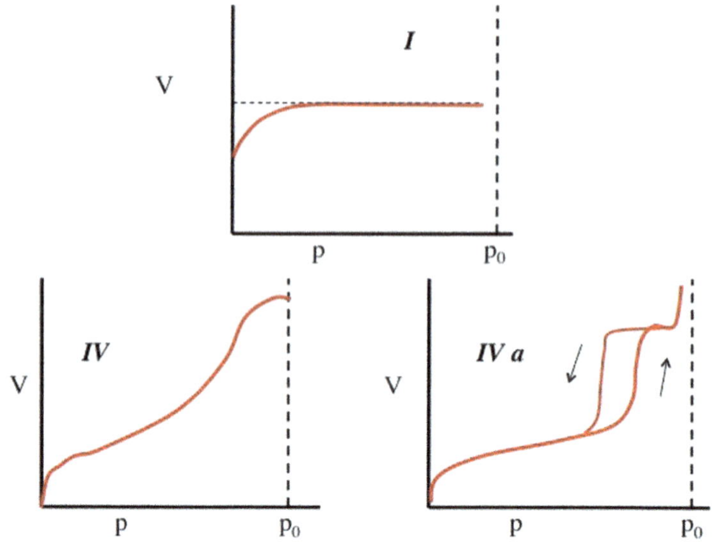

Figure 1.32 IUPAC classification of some typical adsorption isotherms of types I, IV and IVa.

interactions are not established. Conversely, if more than one linear arrest is observed, as in type IV, the growth of two or more layers of adsorbate molecules at the interface can be inferred, indicating the presence of significant interaction energies between two adjacent layers of adsorbed molecules.

Furthermore, information on the structure of the adsorbent, and more specifically on its porosity, is described in the last part of the curve: in the case of multilayer adsorption, adsorption and desorption isotherms may not be identical and will show what is called a hysteresis phenomenon.

Such behaviour is often found for porous solids, and is due to a phase transition of the adsorbed gas to the liquid state so that desorption from the pores is hindered. Capillary condensation is due to an increased number of interactions between vapour phase molecules inside the confined space of a capillary. A capillary does not necessarily have to be a tubular, closed shape, but can be any confined space with respect to its surroundings, as for the case of granular materials used in frescoes.

Capillary condensation is an important factor in both naturally occurring and synthetic porous structures. The study of this phenomenon, by means of experimental adsorption isotherm measurements, provides access to the determination of specific area values and pore sizes for a variety of materials.[33]

1.6 Liquid–Liquid Interfaces

Liquid molecules rearrange quickly after a new surface is created, reaching a new thermodynamic equilibrium, whereas solid surfaces are only kinetically stable and may take a long time to relax at their final equilibrium state.

The differences between interfacial and bulk molecular interaction energies are due mainly to the magnitude of the interactions, to the two-dimensional geometry of the surface and also to differences in interfacial structure. In principle, it would be possible to calculate the energy of cohesion between molecules within a single phase and the energy of adhesion for this interface, if the potential energy functions and the geometrical deformation at an interface are known.

However, in practice we are far away from such success. First, the potential energy functions are not well enough known; second, the structures of liquids and the interfacial structures between two liquids are not completely understood. Consequently, at present, it is necessary to apply semi-empirical approaches to interfacial free energy problems, defining the surface free energy and surface tension concepts exclusively in thermodynamics terms.

In practice, the need for interfacial investigation and measurements arises whenever immiscible liquids – those incapable of mixing – reside within the same vessel. It is common experience to see the lighter liquid material rise to the top and the heavier liquid material settling at the bottom of the container. Many processes of everyday life are controlled by the interactions occurring at the interface between two immiscible liquids: the stability of emulsions mainly depends on the interaction of surfactants at oil–water interfaces; solvent extraction and phase transfer catalysis rely on the optimization of reactions at the boundary between two liquids.

Liquid–liquid interfaces are most commonly found in the diverse separation processes that are essential to many sectors, including oil recovery and, more importantly for our purposes, in nanotechnology for the cleaning of works of art with multi-phase systems that will be described in the following chapters (see Chapters 6–9), *e.g.* emulsions and microemulsions.

Broadly speaking the interface may be complex: an emulsion layer with narrow and precise boundaries (a rag layer) may form between the two liquids, and more frequently a broader gradient of mixed liquid phases can be observed.[34] On the other hand, when a drop of an insoluble oil is placed on a clean water surface, it may behave in different ways: either it does not spread and it remains as a lens as shown in Figure 1.33(b) or it spreads as a monolayer, leaving excess oil as lenses in equilibrium.[35] A third situation may also occur: the liquid spreads as a thin uniform duplex film, a film which is thick enough for two interfaces to establish with characteristic surface tensions, as in Figure 1.33(a).

We may exemplify this statement introducing, in the case of oil on water, what Harkins[36] defined as the *initial spreading coefficient*, $S_{O/W}$:

$$S_{O/W} = \gamma_{WA} - (\gamma_{OA} + \gamma_{OW}) \tag{1.22}$$

Where γ_{WA} is the interfacial tension of the water–air interface and γ_{OA} and γ_{OW} the interfacial tension of the oil–air and oil–water interface, respectively. The various interfacial tensions are measured before mutual saturation of the liquids in question has occurred.

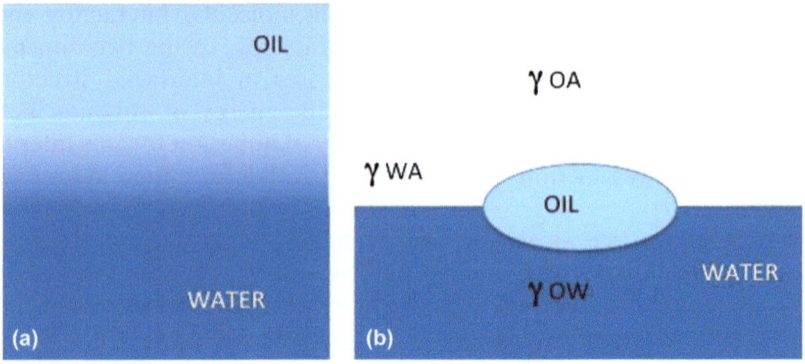

Figure 1.33 (a) Cartoon describing a planar interface between two immiscible liquids, and (b) a drop of non-spreading oil on a water surface. The corresponding interfacial tensions, γ, are also shown for each interface.

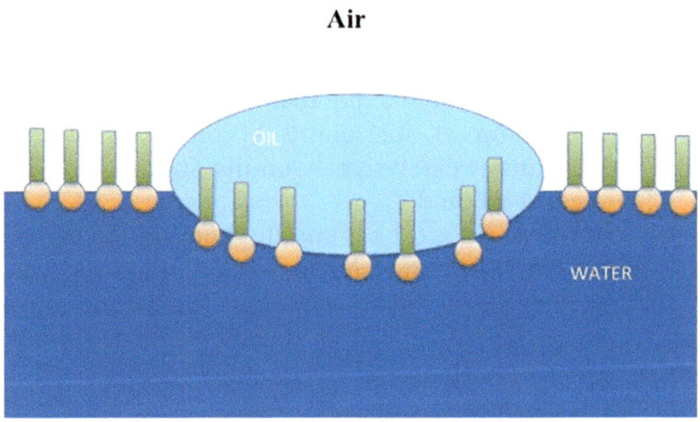

Figure 1.34 Spreading of *n*-hexanol on a water surface.

The condition for initial spreading is therefore that $S_{O/W} \geq 0$, which means that the oil adheres to the water more strongly than it coheres to itself. Impurities in the oil phase (*e.g.* oleic acid in hexadecane) can significantly reduce γ_{OW} to make $S_{O/W}$ positive.

Impurities in the aqueous phase normally reduce $S_{O/W}$ because γ_{WA} is lowered more than γ_{OW} by the dissolved substance, especially for low γ_{OW}. Hence, short-chain hydrocarbons such as octane will spread on a clean water surface but not on a contaminated surface. For example, when hexanol is spread on water, the initial spreading coefficient at 20 °C will be:

$$S_{H/W} = \gamma_{WA} - (\gamma_{HA} - \gamma_{HW}) = 72.8 - (24.8 + 6.8) = +41.2 \text{ mN m}^{-1}$$

so that hexanol is expected to spread on water.

However, if we consider the mutual saturation of hexanol and water (as sketched in Figure 1.34) the water–air surface tension, $\gamma_{WA,}$ is reduced owing to the presence of adsorption of hexanol molecules at the interface, and $S_{H/W}$ decreases to:

$$S_{H/W} = \gamma_{WA} - (\gamma_{HA} - \gamma_{HW}) = 28.5 - (24.7 + 6.8) = -3.0 \text{ mN m}^{-1}.$$

The final state of the interface is now just unfavourable for spreading. This causes the initial spreading to be stopped, and can even result in the film retracting slightly to form very flat lenses, the rest of the water surface being covered by a monolayer of hexanol.

Liquid–liquid interfaces, and in particular water–hydrocarbon interfaces, are very common systems where cleaning of artefacts surfaces is concerned; for example, the removal of waxes from wall-paintings by means of oil-in-water microemulsions (see Section 9.5) relies on the behaviour of water–hydrocarbon interfaces stabilized by surfactants, which has to be properly evaluated.[36,37]

A comprehensive list of spreading coefficient estimation based on interfacial tension values can be found in the pertinent literature[38,39] and in physical chemistry handbooks.[40]

The interfacial tension at the oil–water interface (γ_{OW}) used to determine S_{OW} as discussed above, can be measured experimentally by some of the methods mentioned in Chapter 4. Moreover, γ_{ow} is a crucial parameter that underlies the formulation of oil-in-water (o/w) systems. Change of the interfacial forces at the oil–water boundary can be achieved by adding small quantities of substances that, migrating and adsorbing at the interface, may induce a lowering in the contribution to surface tension. In this way, one can indeed *disperse* oil in water, or *vice versa* (see Chapters 6 and 8).

For instance, solvents with low γ_{ow} can be used to favour the formation of o/w systems formed by water and solvents with high γ_{ow}. Such systems can be very useful for the cleaning of undesired coatings from a variety of works of art. Some representative oil–water surface tension values are reported in Table 1.5.

Some of the solvents reported in Table 1.5 (hexane, ethyl ether, ethyl acetate) are used in restoration practice (see Chapter 5), while pentanol has been used for the formulation of o/w systems such as micellar solutions and microemulsions for the removal of natural and synthetic coatings from wall-paintings (see Sections 7.5 and 9.5).

Table 1.5 Interfacial tension at water–organic liquid inteface at 20 °C.

Water/oil interface	γ_{ow}, *mN m*$^{-1}$	*Water/oil interface*	γ_{ow}, *mN m*$^{-1}$
n-Hexane	50.8	n-Decanol	10
n-Octane	51.6	n-Octanol	8.5
CS_2	48.0	n-Hexanol	6.8
CCl_4	45.1	Aniline	5.9
$Br–C_6H_5$	38.1	n-Pentanol	4.4
C_6H_6	35.0	Ethyl acetate	2.9
$NO_2–C_6H_5$	26.0	Isobutanol	2.1
Ethyl ether	10.7	n-Butanol	1.6

Further Suggested Reading

We report here a comprehensive list of fundamental books on physical chemistry and colloid science, to support the content of this chapter and provide the reader with further insight:

W. Adamson, *Physical Chemistry of Surfaces*, 5th edn, Wiley, New York, 1990.

A. W. Adamson and A. P. Gast, *Physical Chemistry of Surfaces*, 6th edn, Wiley Interscience, New York, 1997.

P. Atkins and J. de Paula, *Atkins' Physical Chemistry*, 9th edn, Oxford University Press, 2009.

R. S. Berry, S. A. Rice, J. Ross, *Physical Chemistry*, 2nd edn, Oxford University Press, 2000.

K. S. Birdi, Surface and Colloid Chemistry, Principles and Applications, CRC Press, London, 2010.

H. Y. Erbil, *Surface Chemistry of Solid and Liquid Interfaces*, Blackwell Publishing Ltd, Oxford, 2006.

P. Hiemenez, *Principles of Colloid and Surface Chemistry*, 3rd edn, Marcel Dekker, New York, 1997.

R. J. Hunter, *Foundations of Colloid Science*, 2nd edn, Oxford University Press, Oxford, 2001.

G. Kumar and K. N. Prabhu, Review of non-reactive and reactive wetting of liquids on surfaces, *Advances in Colloid and Interface Science*, 2007, **133**, 61–89.

J. Lyklema, *Fundamentals of Interface and Colloid Science*, vols I–V, Academic Press, London, 2001.

D. A. McQuarrie and J. D. Simon, *Physical Chemistry: A Molecular Approach*, University Science Books, 1997.

J. van Oss, *Interfacial Forces in Aqueous Media*, Marcel Dekker, Ins., New York, 1994.

Review and Questions

This chapter introduces the concept of an interface, outlining the background that supports the following parts of this book. Theoretical parts are integrated with practical examples, to cover a wide range of aspects and implications.

The main points include:

- The definition of surfaces, interfaces and surface tension.
- The concepts of adhesion and cohesion, and their theoretical implications in cleaning of works of art.
- The concepts of surface wettability and contact angle, and a case study concerning the conservation of the doors of the Florence Baptistery.
- Capillarity and its implication in conservation issues.

- A theoretical approach to chemisorption and physisorption of gases onto solid surfaces.
- The definition of the initial spreading coefficient and the interfacial tension of several solvents with water.

Questions

1) The surface energy of solids may differ greatly according to the environmental history of the sample itself. Would the surface energy of a clean and polished crystal surface be larger or smaller than the real surface of the same material? What would be the effect of defects roughness of the surface?
2) Would the adsorption of atmospheric gases and related contaminants be higher or lower on high energy surfaces? Consider three exposed clean surfaces of different materials: copper ($\gamma = 2000\,\mathrm{mN\,m^{-2}}$), gold ($\gamma = 1500\,\mathrm{mN\,m^{-2}}$) and iron ($\gamma = 2400\,\mathrm{mN\,m^{-2}}$). Assuming defect-free surfaces, which of the three surfaces will undergo most likely surface contamination through gas adsorption and further damaging reactions?
3) Measurements of adsorption isotherms of a gas on a solid surface present a significant hysteresis phenomenon (see Figure 1.32). Would we expect the appearance of capillary phenomena for that particular material?
4) The apparent contact angle of water on a surface is $\theta = 52°$, while the intrinsic contact angle determined for the ideal surface of the same component is 61°. Applying the Wenzel approach, what would the estimated roughness of the test surface be?
5) If the measured capillary rise of water (surface tension $\gamma = 72.5\,\mathrm{mN\,m^{-1}}$) is 52 mm, what is the average diameter of a hydrophilic capillary tube? What would be the rise if the diameter were 0.5 mm? Assuming $\theta = 0°$, at what height is water expected to rise on a church wall if the average pore diameter of the material is 0.03 cm?

Answers

1) Real crystal surfaces may expose domains with different organization of atoms in space; this translates in significantly different surface energies for each exposed face. For example: surface energy is defined in Section 1.1.1 as $\gamma = (1/2)N_b\varepsilon\rho_a$ with ε related to the bond strength, N_b the number of broken bonds and ρ_a the surface density of the atoms in that specific face. In the case of a face-centered-cubic crystal:

$$\gamma(100) = 4\varepsilon/a^2$$
$$\gamma(111) = 2\sqrt{3}\,\varepsilon/a^2 = 3.46\,\varepsilon/a^2$$

Surface defects result generally in sites with large surface energy.

2) High surface energy sites are prone to adsorption of any contaminant that would minimize the total surface energy. Therefore, gold surfaces are less vulnerable to surface adsorption and degradation, as expected from everyday experience.
3) The shape of the desorption isotherm is shifted compared with the curve corresponding to the adsorption cycle. This reflects the porous nature of the solid under investigation; in fact, provided the pore dimensions are small enough, the vapour phase condenses inside the pores and desorption is hindered. Therefore, for such a solid important capillary rise phenomena are expected.
4) The Wenzel equation for rough surfaces relates the roughness factor, r, which is given by the ratio between real and geometrical surface area, to the apparent and intrinsic contact angle by the following equation: $\cos\theta_{\text{app}} = r\cos\theta_{true}$.
   Therefore, $r = \frac{\cos\theta_{app}}{\cos\theta_{true}} = \frac{\cos 52°}{\cos 61°} = \frac{0.61}{0.48} = 1.27$. The surface is moderately rough.
5) The height of capillary rise, h, is given by the following equation [eqn 1.18]:

$$h = \frac{2\,\gamma\cos\theta}{\rho\, g\, R}$$

where γ is the surface tension of the liquid θ is the contact angle between the liquid and the capillary tube, ρ is the density ($\rho = 1000\ \text{kg m}^{-3}$) and R is the radius of the liquid meniscus. Assuming null contact angle ($\cos\theta = 1$), the radius of the capillary tube equals R and proper substitution yields: $h = \frac{2\times 72.8\times 10^{-3}\cdot 1}{10^3\times 9.8\times R} = \frac{14.84\times 10^{-6}}{R} = 52\times 10^{-3}\text{m}$ and $R \approx 0.28$ mm. If the diameter is 0.5 mm, the radius is $0.25\ \text{mm} = 0.25\times 10^{-3}$ m and $h = \frac{14.84\times 10^{-6}}{R} = 59$ mm. The expected water rise along a church wall if the average pore diameter of the material is 0.03 mm ($R = 0.015\times 10^{-3}$ m) is 0.989 m.

References

1. J. Lyklema, *Fundamentals of Interface and Colloid Science*, Academic Press, New York, 2001, vol. I.
2. K. W. Kolasinski, *Surface Science. Foundations of Catalysis and Nanoscience*, John Wiley & Sons, New York, 2002; H. Ohshima, *Biophysical Chemistry of Biointerfaces*, John Wiley and Sons, New York, 2010.
3. K. Holmberg, *Handbook of Applied Surface and Colloid Chemistry*, John Wiley and Sons, New York, 2002; L. L. Schramm, *The Language of Colloid & Interface Science*, ACS Professional Reference Book, ACS, Washington, DC, 1993.
4. P. W. Atkins, *The Second Law*, W. H. Freeman, New York, 1984.

5. T. L. Hill, *An Introduction to Statistical Thermodynamics*, Dover Publications, New York, 1987.
6. D. K. Chattoraj and K. S. Birdi, *Adsorption and the Gibbs Surface Excess*, Plenum Press, London, 1984.
7. G. Job and F. Herrmann, *Eur. J. Phys.*, 2006, **27**, 353.
8. D. Sands, *Introduction to Crystallography*, Dover Books on Chemistry, 1980.
9. J. Lyklema, *Fundamentals of Interface and Colloid Science*, Academic Press, London, 2001, vol. IV.
10. G. C. Maitland, M. Rigby, E. B. Smith and W. A. Wakeman, *Intermolecular Forces: Their Origin and Determination*, Clarendon Press, Oxford, 1987.
11. M. J. Jaycock and G. D. Parfitt, *Chemistry of Interfaces*, Ellis Horwood, Chichester, 1986.
12. P. Eaton and P. West, *Atomic Force Microscopy*, Oxford University Press, Oxford, 2010.
13. R. Fontana, M. C. Gambino, M. Greco, L. Marras, M. Materazzi, E. Pampaloni and L. Pezzati, *Optical Metrology for Arts and Multimedia*, ed. Renzo Salimbeni, Proceedings of The International Society for Optics and Photonics, Vol. 5146, 2003.
14. H. Y. Erbil, *Surface Chemistry of Solid and Liquid Interfaces*, Blackwell Publishing Ltd, Oxford, 2006.
15. A. W. Adamson, *Physical Chemistry of Surfaces*, 5th edn, Wiley, New York, 1990.
16. G. Wolanskya and A. Marmu, *Colloids Surface A*, 1999, **156**, 381; P. S. Swain and R. Lipowsky, *Langmuir*, 1998, **14**, 6772.
17. A. B. D. Cassie and S. Baxter, *Trans. Faraday Soc.*, 1944, **40**, 546.
18. S. Sethy, L. Ge, L. Ci, P. M. Ajayan and A. Dhinojwala, *Nanolett.*, 2008, **8**, 822; M. Nosonovsky and B. Bhushan, *Current Opinion Coll. Interf. Sci*, 2009, **14**, 270; P. Manoudis, S. Papadopouolu, I.Karapanaagiotis, A. Tsakalof, I. Zuburtikudis and C. Panayiotou, *J. Phys. Conf. Series*, 2007, **61**, 1361.
19. W. Choi, A. Tutela, J. M. Mabry, R. E. Cohen and G. H. McKinley, *J. Colloid Interf. Sci.*, 2009, **339**, 208; P. Roach, N. J. Shirtcliffe and M. I. Newton, *Soft Matter*, 2008, **4**, 224.
20. G. P. Bernardini, M. C. Squarcialupi, R. Trosti-Ferroni, M. Matteini, C. G. Lalli, G. Lanterna, M. Rizzi and I. Tosini, in *Protection and Conservation of the Cultural Heritage of the Mediterranean Cities*, ed. Galan and Zezza, Sevilla, 2002, pp. 43.
21. G. Caminati, F. Gambinossi and L. Dei, *Progetto Battistero: studio dell-'impatto ambientale sulle Porte dlBattistero di Firenze*, ed. M. Matteini and R. Rimaboschi, Ed. Opera del Duomo, Firenze, 2011.
22. G. Kumar and K. N. Prabhu, *Adv. Colloid Interfac.*, 2007, **133**, 61.
23. J. Van den Bulcke, J. Van Acker and M. Stevens, *Int. Biodeter. Biodegr.*, 2007, **59**, 25.
24. M. Brugnara, E. Degasperi, C. Della Volpe, D. Maniglio, A. Penati, S. Siboni, L. Toniolo, T. Poli, S. Invernizzi and V. Castelvetro, *Colloid Surface A*, 2004, **241**, 299.

25. T. H. Muster, A. K. Neufeld and I. S. Cole, *Corrosion Sci.*, 2004, **46**, 2337.
26. S. Takahashi and O. Kuboi, *J. Mater. Sci.*, 1996, **34**, 1797.
27. G. Li, B. Wang, Y. Liu, T. Tan, X. Song, E. Li and H. Yan, *Sci. Technol. Adv. Mater.*, 2008, **9**, 025006.
28. E. Bernardi, C. Chiavari, B. Lenza, C. Martini, L. Morselli, F. Ospitali and L. Robiola, *Corrosion Sci.*, 2009, **51**, 159.
29. D. de la Fuente, J. Simancas and M. Morbillo, *Corrosion Sci.*, 2008, **50**, 268.
30. S. M. Thompson, K. E. Gubbins, J. P. R. B. Walton, R. A. R. Chantry and J. S. Rowlinson, *J. Chem. Phys.*, 1984, **81**, 530.
31. J. L. M. Poiseuille, *Phys. Rev.*, 1921, **18**, 273.
32. Adsorption: Theory, Modeling and Analysis, Ed. J. Toth, *Surfactant Sci. Series*, vol. 107, M. Dekker, New York, 2002.
33. F. Rouqerol, J. Rouquerol and K. Sing, *Adsorption by Powders and Porous Solids: Principles, Methodology and Applications*, Academic Press, London, 1999.
34. I. Benjamin, *Ann. Rev. Phys. Chem.*, 1997, **48**, 407.
35. I. Tsuyumoto, N. Noguchi, T. Kitamori and T. Sawada, *J. Phys. Chem. B*, 1998, **102**, 2684.
36. W. D. Harkins, *J. Chem. Phys.*, 1941, **9**, 552; D. Bon, J. Eggers, J. Indekeu, J. Meunier and E. Rolley, *Rev. Mod. Phys.*, 2009, **81**, 739.
37. G. Caminati, D. Senatra and G. Gabrielli, *Langmuir*, 1991, **7**, 1969.
38. G. Caminati, G. Gabrielli and E. Ferroni, *Removal of hydrophobic impurities from pictorial surfaces*, Intnl. Pat. 91105096.1.
39. V. Shah and D. Broseta, *Langmuir*, 2007, **23**, 12598.
40. H. Dobbes and D. Bonn, *Langmuir*, 2001, **17**, 4674.

CHAPTER 2

The Degradation of Works of Art Materials

L. DEI* AND R. GIORGI

University of Florence, Italy
*Email: luigi.dei@unifi.it; giorgi@csgi.unifi.it

2.1 Works of Art Materials

When speaking about works of art we often refer to paintings, even if the term "art" includes many other creative activities, such as the production of statues, mosaics, glasses, ceramics, *etc.* Given that it is impossible, for obvious reasons, to treat all the artistic masterpieces in a single chapter, we chose to discuss here only paintings, and in particular mural and easel paintings, because they cover a large part of both classic and contemporary Cultural Heritage.

Dealing with paintings means, for chemists, the study of pictorial surfaces: indeed, the artistic message is materially confined within a few microns of coloured matter. Pictorial surfaces can be considered from a physico-chemical point of view as a solid–gas interface where the solid phase is intrinsically micro-heterogeneous.

Both easel and mural paintings are made up of a macroscopic support coated by a thin layer of a painting material, including the binder that provides adhesion to the support. The painting layer is thus the true interface between the work of art (solid phase) and the environment, which in most cases is the atmosphere (gas phase). It is important to note that the pictorial surface, defined in physico-chemical terms, is the region where the thermodynamic system "work of art" exchanges matter and energy with the surrounding environment.[1] In other words, the pictorial surfaces are the preferential site of

RSC Nanoscience & Nanotechnology No. 28
Nanoscience for the Conservation of Works of Art
Edited by Piero Baglioni and David Chelazzi

Published by the Royal Society of Chemistry, www.rsc.org

degradation due to matter and energy exchanges; this is particularly important if we consider that the pictorial surface can be depicted, from an artistic point of view, as the form imprinted on the matter by the artist in a way that makes the particular artefact a unique masterpiece. As outlined in Chapter 1, surface science should fulfil the task of understanding all the physico-chemical mechanisms that are responsible for matter/energy exchanges. In turn, these determine decay. Moreover surface science is responsible for developing methods to prevent any modifications of this crucial interface. Conservation aims at applying the results of surface science to inhibit or retard all the interfacial processes that modify the work of art.

Given that the pictorial surfaces of easel and mural (or wall) paintings differ greatly in both their chemical and physico-structural properties, we prefer to discuss their physico-chemical behaviour in two separate sections.

2.1.1 Wall Paintings and Limestone

Mural paintings are very peculiar works of art, because they do not exist as single isolated masterpieces, but always belong to buildings such as churches, monasteries, castles, palaces, tombs, and so on. Indeed, the wall painting is the actual painted surface – the external "skin", a few microns thick – of the architectural work: the first important physico-chemical aspect is related to the adhesion of this painted surface onto the wall support.

In order to understand this point we briefly recall the main features of these paintings, which are often called "frescoes".[2] This term refers to the technique used to execute the artefact: the Italian expression "*a fresco*" literally means "in wet conditions", *i.e.* the coloured pigments are put onto the wall when the aerial plaster that covers its surface is still wet.[3,4] Therefore, although the external layer (the pictorial surface) can be observed on the surface of a mural painting, the cross-section of the artefact is much more complex. In fact, as a function of the depth,[5] there are many superimposed layers, starting from the inner rough wall up to the visible painting layers. This stratigraphic nature of wall paintings reflects the sequence of the operations carried out by the painter. First of all we recall that the plaster on which wall paintings are made is aerial, a mixture of lime [calcium hydroxide $Ca(OH)_2$], sand and water. The first stratum (*arriccio*) of this plaster is distributed on the wall (stone or brick wall), and is made up of a sand–lime mixture (about 3 : 1 by volume) with enough water to make the mixture pasty. It is interesting to note that the sand is constituted of coarse particles (whose size is of the order of hundreds of microns) and that the surface of this layer is made very rough using a trowel when the mortar is still wet. The specific area of this curled surface (*arriccio* is the Italian word for "curl") is thus much greater than that of a perfectly smoothed surface. This high specific area has important consequences for the application of the next layer, the so-called *intonaco* (*intonaco* is the Italian word for "lime plaster"). As a matter of fact, the adhesion of the second stratum on the first one is strongly improved by the high specific area of the adhesion surface, according to the basic principles of surface chemistry.

After the deposition of the *arriccio*, the artists usually wait several days before putting on the *intonaco* layer, in order to ensure the complete setting of the lime plaster constituting the *arriccio*. In fact, the setting of an aerial plaster consists of two steps: (i) dehydration of the pasty mixture of sand and lime, and (ii) chemical transformation of the $Ca(OH)_2$ (lime) into binding calcite, $CaCO_3$, by reaction with atmospheric CO_2.[2–6] The first process exhibits a kinetics rate of the order of hours, while the carbonation stage requires several days and depends strictly on the plaster thickness. The carbonation proceeds until CO_2 succeeds in reaching the inner portions of the plaster; as the setting advances, CO_2 penetration becomes hindered by the compact external layers that have already set. The carbonation generates a complex network of calcite crystals, which nucleate and grow at a very slow rate, determining a kind of macroscopic cage constituted of sub-micrometric and micrometric units. This network entraps the sand particles and produces a compact material that has the physico-structural properties typical of stones.

Therefore, the second layer (*intonaco*) is put on the *arriccio* when the latter is dry and perfectly compact (full setting). To prepare the *intonaco* the artist classically used a 2 : 1 (or 1 : 1) sand : lime volume ratio, and chose very fine sand particles (a few μm in size). The reason for this is twofold: (i) the surface of the *intonaco* must be as smooth as possible (very fine sand particles) because the artist has to paint on it, and (ii) the plaster needs to exhibit a high binder [$Ca(OH)_2$] content, as the binder is meant to fix the coloured pigments and to provide perfect adhesion and stability of the paint layer on the wall support.

The third layer comprises the coloured pigments that are applied by brush. Typically, the pigments were finely ground and suspended in water at different concentrations, depending on the chromatic and relief effect the artist wanted to achieve.[2–4] The most important aspect is that the painting layer must be laid on the *intonaco* when the latter is still wet, before the carbonation of the *intonaco* has begun. The treatises recommended painting over a wet *intonaco*, hence the origin of the English word *fresco*. In fact, if the pigment suspension is spread onto a wet surface, the following carbonation step succeeds in entrapping not only the sand particles, but also the pigment grains among them. The reactions that take place are those reported in Scheme 2.1 below:

It is interesting to observe that these reactions occur at the interface between the $Ca(OH)_2$ saturated solution that impregnates the solid part of the plaster and the gaseous CO_2 coming from the surrounding atmosphere. The reaction

$$Ca(OH)_2 \leftrightarrows Ca^{2+} + 2OH^-$$

$$CO_2 + H_2O \leftrightarrows HCO_3^- + H^+$$

$$HCO_3^- \leftrightarrows CO_3^{2-} + H^+$$

$$Ca^{2+} + CO_3^{2-} \leftrightarrows CaCO_3\downarrow$$

Scheme 2.1

itself generates two interfaces: (i) between the growing $CaCO_3$ crystals and the remaining excess $Ca(OH)_2$ aqueous solution, and (ii) between the same $CaCO_3$ crystals and the solid fraction of the plaster (sand and calcium hydroxide), which includes the pigment particles. The result is a perfect adhesion among the pigments in the plane of the pictorial surface, and between the paint layer and the *intonaco* underneath.

The nature of the painting layer is quite complex, because it consists of a number of pigment sub-layers laid one after the other to obtain the desired artistic effects.[2–4,7,8] It is worthwhile to recall that not all the inorganic pigments can be used,[2,9] according to the fresco technique. In fact, pigments that are not stable at the high pH imparted by the $Ca(OH)_2$-rich environment cannot be distributed on the wet *intonaco*. For these pigments it is necessary to wait for the complete drying and setting of the surface; they are then applied by means of an appropriate binder. For these pigments the painting technique is called "*a secco*" (which means in dry conditions), as opposed to the above-discussed "*a fresco*".[2] Generally, common binders include organic substances such as egg, casein, animal glue and certain vegetable gums.[6] More rarely is it possible to find lipid substances such as linseed and poppy-seed oil used as binders.

According to the *a fresco* technique, mural paintings contain regions painted on different days: each morning the artists had to decide what portion of the wall was to be painted during that day, and consequently they laid on the corresponding portion of the *arriccio* a well-controlled amount of *intonaco*, onto which the painting had to be safely concluded before the evening. A single day can be considered an appropriate time to avoid *intonaco* carbonation that would compromise any further fixing of the pigments on the surface. The day after, or some days after, the artists painted another portion of the *arriccio* for another day's work. The borderline between two neighbouring regions, corresponding to different working days, can be easily detected. As a matter of fact a map of the whole *fresco* painting can be traced. Since the Italian word for "days" is "*giornate*", each wall painting is characterized by the map of the "*giornate*" the artist took to create the work of art.[2] As an example, the wall painting "*Madonna del Parto*" ("Madonna of the Childbirth") by Piero della Francesca was done very quickly: in only one single week of work.[10]

For limestone, we must note that these materials refer to stones used in architecture and in the manufacture of statues. With the word limestone we refer to sedimentary rocks mainly constituted of $CaCO_3$. Depending on the petrographic, mineralogical and physico-structural properties, they suffer from an acidic environment as wall paintings do, because they share the same chemical nature.

2.1.2 Easel Paintings

Unlike wall paintings, easel paintings are movable works of art whose support is usually a piece of wood or canvas made of textiles. The support can be also made of metal, paper, cardboard, parchment, ivory, glass or stone, but in the present review we will consider only the most commonly used materials: wood and canvas. These works of art exhibit complex structures from a physico-chemical point of view: a stratigraphic nature, as already observed in mural paintings, is also found in this case. The main layers, from bottom to top, are: the support (wood or canvas), a ground layer, the painting layers, and the final thin varnish layer.[11] As found for mural paintings, the various interfaces between different layers play a crucial role in determining deterioration: in fact, these surfaces are mostly affected by mechanical and chemical alterations. Given that the most important properties of wood and canvas supports mostly concern the mechanical and biological aspects, we will focus our attention on the other three layers (ground, paint layers and varnish) that are more interesting from a physico-chemical point of view.

The main function of the ground layer is to form a homogeneous link between the support and the paint layers; owing to this function, it must adhere perfectly to the support while allowing the binding between the paint layers and the support itself.[11,12] The ground layer provides a uniform surface to the support, contributes to the final texture and colour, and decreases the porosity at the interface of the paint layers. In this sense it constitutes a homogeneous and continuous absorbent layer for the pigments and the binding media forming the paint layers. The main components of the ground are the media and the fillers: the main physico-chemical properties of the ground are determined by the type of medium and by the filler–medium proportion. The fillers decrease and smooth the porosity of the support surface, and determine the colour of the surface on which the pigments will be laid. The media act as binders that hold the filler grains together strongly while granting their adhesion onto the support surface. Common fillers used by the artists are chalk or powdered limestone ($CaCO_3$), calcined gypsum, called plaster of Paris ($CaSO_4.1/2H_2O$), and lead white ($2PbCO_3.Pb(OH)_2$). In the case of coloured grounds, red, yellow and brown ochres, bone charcoal, smalt (from 17th century), barytes, zinc white and titanium white (from the 19th–20th centuries) were added. Sometimes the ground comprises more than one coloured layer: this occurs, for example, in some of the 17th century Dutch paintings.

At the beginning of this section we stated that one of the most important properties of the ground is its absorbent power; this feature should not be very strong, since this would result in rapid penetration of the paint layers into the support, resulting in flattening of the pictorial effect. If the ground was too absorbent, many artists used to add another layer between the ground and the paint layers, consisting of sizes or resins. This layer is named primer or *imprimatura* and acts as a barrier to prevent the painting mixture penetrating into the support. The most common medium is animal glue, a mixture of protein substances applied in their *sol* form which produce a *gel* phase[13] that leads to the formation of a compact network to entrap the filler grains and adhere to the rough surface of the support. Sometimes drying oils, such as linseed oil, were added to provide the ground layer with both hydrophilic and hydrophobic substances.

While in *fresco* paintings the paint layers are stable and fixed to the support by an inorganic binder ($CaCO_3$ from lime carbonation), in the case of easel paintings the binders in the paint layers are organic compounds.[14,15] The pigment grains are suspended in aqueous or oily phases containing an appropriate binding medium, and this mixture is put on the prepared ground by means of a brush. Therefore, rather than the pigments, the most relevant difference between *fresco* and easel paintings is in the binder; furthermore, in *fresco* paintings the binder is unique, whereas in easel paintings many different binding media have been used.[14–20] Until the 20th century only natural substances were used as binding media: egg (an oil-in-water emulsion of egg yolk), egg white, casein, animal glue, natural gums such as Arabic gum, drying oils (linseed, walnut, poppy-seed oils) and resins (Venice turpentine, mastic and copal). Since the beginning of the 20th century, synthetic polymers have also been employed as binding media. According to the hydrophilic or hydrophobic nature of the binding media two main classes are defined: oil-in-water emulsions, called "*tempera*", and pure oily media, called "oil paints". In the following sections we will describe in detail the properties of both pigments and binders.[14,15,20]

The final difference between frescoes and easel paintings is the external surface: in the case of *fresco* paintings this is formed by the last paint layers, the so-called "final drawing, highlights and shadows", while in the easel paintings the outer surface is the varnish layer. The varnish provides the paint layers with depth, enhancing colour effects and protecting the paint layers underneath.[11,21–25] Varnishes typically consisted of drying oils, alone or in a mixture with natural resins (before the 17th century), that were applied after heating owing to their high viscosity at low temperatures. Subsequently, natural resins (dammar, mastic) dissolved in organic solvents such as turpentine were employed, and sometimes egg white varnishes were used.[6,11] Nowadays, solutions of polymeric materials in organic solvents are used as picture varnish.

2.2 Natural Weathering of Materials

Weathering of works of art materials depends on interactions with the environment. In thermodynamics, this means that these artefacts exchange energy and matter with the surroundings. Obviously, the mechanisms of degradation depend on the specific properties of the materials used to create the artefacts; this means that stone, glass, ceramic, wood, paper, textiles and bronze will all follow different mechanisms and reaction pathways. However, all these processes are affected in similar ways by the same environmental parameters: temperature, relative humidity, light and pollution.

It is important to underline the meaning of the term "natural weathering". Obviously it is not fully correct, because pollution, artificial lighting and thermal control of buildings are all anthropic factors. "Natural weathering" consists thus of all the processes works of art are exposed to, under conditions that cannot be completely eliminated (*i.e.* pollution) or fully controlled (*e.g.* illumination and heating of buildings where the objects are exposed).

Nonetheless, many efforts can be made in order to optimize the best practice for preventive conservation. In the following sections, some of the main degradation mechanisms will be described with special emphasis on what is usually done to inhibit the effects of weathering by controlling the environmental parameters.

A detailed description of the action of light on different substrates is beyond the scope of this text. Some deeper insights, *e.g.* about relationships between irradiation and chemical changes of original materials, will be given each time this topic can be related to the most recent advancements provided by nanotechnology to fix specific conservation issues.

In this section, the degradation mechanisms of stones and wall paintings (and more briefly easel paintings) will be presented. Most of this knowledge can be easily transferred to other kinds of materials and degradation processes. In later chapters, more detailed information about the interaction of environmental conditions with paper and wood artefacts (see Chapters 15 and 16) will be provided.

2.2.1 The Role of Temperature, Humidity, and Light

In this section the influence of thermo-hygrometric fluctuations (temperature and relative humidity) on the degradation of materials will be discussed. In particular, it will be shown that these parameters are the key to understanding the degradation processes that depend on the presence of saline solutions within stone (or walls). To illustrate better the behaviour of salt solutions confined in porous media, it is useful to understand fully the definition of relative humidity of a solution at equilibrium for a given temperature. For that reason, it is necessary to recall some definitions.

- *Vapour pressure of water*. This is the pressure exerted by the water molecules in the vapour phase. It depends on the temperature and on the water content in the "volume" of atmosphere that is considered.
- *Saturated vapour pressure*. A given volume of atmosphere, at a fixed temperature, contains a finite amount of water molecules in the vapour phase. If the air is in equilibrium with liquid water, the vapour reaches the saturation point (*i.e.* the amount of vapour remains fixed unless the ambient temperature and/or pressure change), and the corresponding pressure is called "saturated vapour pressure". At constant temperature and pressure, further addition of water vapour would cause the formation of liquid droplets by condensation.
- *Absolute humidity*. This is defined as the mass of water vapour for a certain volume of air.
- *Relative Humidity (RH)* is defined by the equation:

$$RH = 100(P / P_S) \tag{2.1}$$

where P is the vapour pressure of water and P_S is the saturated vapour pressure at the same temperature.

The vapour pressure of any substance increases non-linearly with temperature according to the Clausius–Clapeyron equation. This means that closed rooms with *RH* about 65%, where exchange of the air is impossible (constant air composition), may reach a high humidity level up to saturation as a result of a temperature drop of 10–12 °C. These variations usually occur inside large historic buildings (churches, palaces, *etc.*), and in museums where the heating system operates depending on the tourist fluxes. The main consequence is that cyclic fluctuations of thermohygrometric parameters may have a daily frequency. An important parameter is the dew-point, defined as the temperature below which the water vapour in a volume of humid air at constant pressure will condense into liquid water ("dew" indicates the condensed water formed on a solid surface). Table 2.1 shows the water vapour content in saturated air at different temperatures.

A saline solution (for the moment we consider a solution of only one component) has a saturated vapour pressure (or tension) lower than that of pure water at the same temperature. In other words, the relative humidity (*RH*) of the air in equilibrium with the solution is less than 100%. This is due to the fact that the water molecules sitting at the surface of the liquid are "diluted" by the presence of the salt ions, as expressed by Raoult's law (see also Section 4.1.7). The strong water–ions interactions decrease the "tendency" of water to evaporate. The greater the concentration of the salt, the lower the relative humidity will be. Since there is a limit to the amount of salt that can be dissolved in water (see Section 4.1.2), correspondingly, the relative humidity decrement is limited, and in the case of a saturated salt solution in water, *RH* will reach a minimum value, at constant temperature. This lower limit is called the "relative humidity of equilibrium", and represents the *RH* of the air in equilibrium with a saturated solution of a given salt. Saline species are characterized by different values of relative humidity of equilibrium. In Table 2.2, a short list of some common salts is reported.

If a dilute salt solution, which typically impregnates a porous wall or stone, is exposed to a gradual decrease of the relative humidity of the environment, water will evaporate and the concentration of the solution will increase. When the environmental *RH* reaches the equilibrium value, the saline solution will be saturated. Any further reduction of *RH* will lead to gradual evaporation of the solvent and incipient crystallization of the salt. On the other hand, if humidity

Table 2.1 Water vapour content in saturated air. (Source: Lange, *Handbook of Chemistry*, New York, McGraw Hill Book Company, 1967, p. 1424.)

Temperature (°C)	*Water amount ($g\,m^{-3}$)*
10	9.5
20	17
25	23
30	30
35	40

Table 2.2 Values of equilibrium relative humidity (*RH*; %) of certain salts, measured at 20 °C.

Salt	*Equilibrium RH (%)*
Sodium sulfate	93
Potassium sulfate	98
Calcium sulfate	99.9
Magnesium sulfate	90
Ammonium sulfate	81
Sodium nitrate	75
Potassium nitrate	94
Calcium nitrate	56
Magnesium nitrate	53
Ammonium nitrate	66
Sodium chloride	75
Potassium chloride	85
Calcium chloride	33
Magnesium chloride	34
Ammonium chloride	80

increases, salt crystals tend to adsorb water from the air, dissolve in the water and form diluted solutions.

It follows that a porous medium contaminated with a saline solution will suffer mechanical stresses within the pores (due to the repeated crystallization of the salt) whenever the relative humidity fluctuates above or below the equilibrium *RH* value of the salt. Every time *RH* drops below the equilibrium *RH* value, crystallization takes place. Dissolution follows when *RH* exceeds the equilibrium value.

In order to minimize these harmful processes, the environmental *RH* should never coincide with the values of equilibrium. If *RH* is always kept above or below the equilibrium value for the saline species, these will remain either in solution, or in the solid state. The values reported in Table 2.2, therefore, should be sufficient to define the best conditions for storing the works of art in a museum. However, it is very rare that works of art materials are contaminated by a single salt species. Analysis usually reveals the presence of several ions simultaneously (*e.g.* sodium, magnesium, potassium, calcium, sulfate, nitrate, chloride).

Which environmental conditions should then be maintained in these cases in order to minimize the effects of this process?

The simplest hypothesis is to assume that the behaviour of the different salts is independent. In such a case, it would be enough to avoid the typical values of equilibrium *RH* for each species. This is, unfortunately, very complicated, because this assumption is acceptable only for very dilute solutions. On the contrary, crystallization is a process that originates in conditions close to saturation and therefore the assumptions do not apply in the real situation. It follows that it is not absolutely reliable to define the optimal conditions for inhibiting the processes of cyclic crystallisation–dissolution. The only viable solution is to maintain conditions of constant *RH*, *i.e.* to avoid or minimize as much as possible *RH* fluctuations.

It is important to remark upon the role of salts whose equilibrium *RH* is low (typically below 50%); these salts are defined as deliquescent (or hygroscopic). A classical example is represented by calcium chloride, which has a value of equilibrium *RH* of 33% at 20 °C. Typically, deliquescent salts may maintain porous materials in a constantly wet condition, because of the continuous adsorption of water from the air. This is not acceptable for good conservation because several side effects have been observed. The most relevant is represented by biological attack.

Finally, the action of light on works of art materials can be briefly described in terms of oxidation capability. This depends on the formation of radicals, which are due to the catalytic activity of light. The energy provided by light is sufficient, especially under conditions of air pollution (as will be described in the following sections), to promote homolysis of water (–OH bonds) and many other organic pollutants usually present in urban areas. In principle, photo-oxidation of works of art materials may involve a large number of different organic substrates. Organic biopolymers (*i.e.* cellulose- and protein-based materials), colorants, dyes, inks, natural and synthetic varnishes or adhesives are all under potential oxidation attack, if not properly stored or exhibited.

Photo-oxidation is mainly influenced by high relative humidity conditions, which favour this reaction. Therefore, good preventive conservation practice must be conceived by using an integrated approach, where temperature, relative humidity, lighting and (possibly) pollution are strictly interconnected.

2.2.2 Salt Crystallization

Without any doubt, the presence of water within a porous matrix is thus the major cause of deterioration of a work of art. Directly or indirectly, humidity is involved in many processes of alteration of both stone and plaster.

The increase in volume that water undergoes upon liquid–solid phase-transition below 0 °C may induce significant stresses, depending on the properties of the stone. This phenomenon, known as gelivity, *i.e.* freezing in the pores, is quite common in places characterized by a very rigid winter climate. On the other hand, hot temperatures may contribute to stone deterioration, because most materials dilate with increasing temperature (anisotropically for most crystalline species), and the differential thermal expansion can lead to the formation of microcracks. For example, the anisotropy of expansion of calcite crystals upon heating can lead to the formation of small cracks in marble, which is composed largely of iso-orientered calcite crystals.

The moisture content can have different origins. Water can impregnate a porous matrix by adsorption from the subsoil (upward motion), by the action of rainfall, from condensation of water vapour in the atmosphere at the temperature of the dew-point, or by adsorption promoted by deliquescent substances that may be present into the wall.

Pure water solubilizes large amounts of salts, and is rich in dissolved gases. During its migration, water carries the dissolved salts, which are distributed and deposited anywhere within the material, with important consequences that will be described in detail. The presence of water is also of paramount importance in all the processes of chemical corrosion, biological attack (fungi, algae, lichens) and physical alterations, such as freezing.

The paint layer is the "interface" between the wall and the surrounding environment and is strongly susceptible to degradation due to the mechanical stresses following salt crystallization, usually occurring at the surface. Saline solutions impregnate the porous network of the wall and move through a capillarity mechanism (for capillarity see Section 1.4.2). Crystallization takes place when the salt concentration exceeds saturation, often due to water evaporation at the interface. This process is generally associated with volume expansion. Temperature and humidity fluctuations favour cyclic salt dissolution–crystallization that can be repeated several times. The consequence is that the pores of the walls periodically withstand mechanical stresses due to the nucleation and growth of crystalline phases. The repeated action over time leads to a heavy mechanical stress, which manifests itself in the weakest points in a porous matrix, *i.e.* the pores. As a matter of fact, the formation of new crystals inside the pores and/or at the interface between the paint layer and the plaster generates mechanical stress and results either in lifting and detachment of the paint layer, or in cracks and fissuring of the plaster.[4]

When the internal humidity of the porous medium is not in equilibrium with the ambient relative humidity, surface phenomena of evaporation or condensation of water will take place, and this occurs periodically over time, owing to fluctuations of temperature and humidity conditions (temperature and *RH*). If the saline solution becomes supersaturated by evaporation of the solvent, the dissolved salts begin to crystallize, resulting in "efflorescence".

The conditions of supersaturation of a saline solution that impregnate a porous material are highly dependent on the type of salt or salt mixture, and on the specific properties of the substrate. For instance, sulfates are commonly found as contaminants of wall paintings, since they result from the chemical degradation of calcium carbonate caused by acid pollutants (SO_2 and SO_3).

Within the porous matrix, the percentage of empty spaces with respect to the total volume may vary within very wide limits, depending on the type of material and its genetic process. It is also worth noting that different materials may present the same total open porosity, but different pore distribution, size and shape (see also Section 1.4). Obviously, all these parameters may determine very complex phenomena, depending on the type of stone and the conditions of temperature and humidity.

In the range of stability of unsaturated solutions crystallization does not occur. If, at constant temperature, the concentration of the salt increases, *e.g.* as a result of evaporation of the solvent, the salt remains in solution up to the saturation point (*i.e.* its solubility), then it crystallizes. As their size increases, crystals exert pressures on the capillary walls.

The same type of salt may crystallize with different degrees of hydration. Given that each hydration status corresponds to a different specific volume, the internal walls of the porous matrix can be repeatedly exposed to high pressures. According to the literature,[26] the change from a state of hydration to the anhydrous form (*e.g.* $NaSO_4{\cdot}10H_2O$ mirabilite and $NaSO_4$ thenardite) does not take place in the solid phase, but through the solution phase; in fact, the metastable compound is more soluble than the corresponding stable species, so that it partially dissolves. This process results in a supersaturated solution, and the simultaneous re-precipitation of the two phases may occur.

Although most of the relevant degradation processes are connected to the action of salts in solution, they are not the only ones to be considered.

2.2.3 Effects of Light on Easel Paintings

The interaction of light with artefacts materials has been widely debated. Light provides matter with energy, favouring chemical reactions that cannot be easily controlled or inhibited. This process is particularly relevant for some pictorial techniques and supports. Typically, watercolours, gouache paintings (glue-medium paintings), pastel drawings, varnishes applied as final protective and coating on easel paintings, and modern synthetic inks are all subject to molecular changes because of light adsorption. These changes are due to oxidation mechanisms, which produce fading or darkening effects, depending on the materials involved and the supports. It is beyond the scope of this paragraph to examine in detail all these processes. Some aspects will be focused on here in order to define what kind of problems can now be approached by using nanotechnologies. The removal of degraded materials (*i.e.* varnishes), which hinder the readability of the painting, is one of these. The darkening of varnishes upon photo-oxidation is a typical case that requires the intervention of conservators, who must remove as much as possible of the aged layer in order to revive the original colours. Unfortunately, the oxidation of varnish favours the cross-linking of the polymer chain through radical mechanisms. As a consequence, the varnish layer may no longer be soluble in classical solvents (*e.g.* the same solvents used to spread the varnish onto the paint layer).

2.3 Environmental Pollution

The mere exposure of architectural surfaces, monuments and mural paintings to the atmosphere leads to heavy deterioration, which has been particularly accelerated in recent years due to pollution.

An air pollutant is gas or particulate matter, which, in high concentrations, can be harmful to human health and/or materials. Pollution can be of natural origin (particulate matter created by erosion, tropospheric ozone caused by lightning, volcanic emissions) or anthropogenic (combustion, industrial processes). Primary pollutants are carbon monoxide, hydrocarbons (HCs), nitrogen oxides, sulfur oxides, volatile organic compounds (VOC) and particulate matter. Secondary pollutants are aldehydes, peroxides and ozone. The latter are formed from the former through reactions that usually involve light and oxygen. The pollutants are removed from the atmosphere through dry or wet deposition, which may cause adverse effects on receptors (humans, flora, fauna and materials).

Air pollution is sensitive to seasonal variation, and is more relevant during the winter months, because of both domestic heating and the climatic conditions characterized by stagnant air. Climatic conditions also strongly affect the formation of so-called *smog* (a term that originated from combination of the words "smoke" and "fog"), which usually indicates polluted air.

In the following section, we will examine the most important pollutants first, and then the main chemical reactions that cause surface corrosion will be briefly discussed.

2.3.1 Outdoor Pollution

Several pollutants contribute to the alteration of works of art materials. Obviously, their effects are more evident over surfaces exposed to outdoor environmental conditions. In fact, the action of pollutants strongly depends on thermo-hygrometric fluctuations, which exhibit a greater variation outdoors than indoors. However, in principle, the degradation mechanisms which will be described in this section also affect the surfaces of artefacts that are stored under (more protected) indoor conditions. The main pollutants directly involved in the degradation mechanisms are: sulfur compounds, nitrogen oxides, hydrochloric acid and atmospheric suspended particulates. The sulfur compounds may be in the gaseous (hydrogen sulfide, sulfur dioxide), liquid (sulfuric acid solutions), or solid state (calcium sulfate, ammonium sulfate). With the exception of hydrogen sulfide, whose formation is mainly due to biological reactions, all the compounds mentioned are primarily due to the combustion of solid and liquid fossil fuels (oil, coal, *etc.*), which contain appreciable amounts of sulfur. The combustion of coal containing up to 1–3% sulfur produces from 17–51 kg of SO_2 per ton of coal, which is emitted into the atmosphere together with carbon dioxide (from coal combustion). The sulfur dioxide in the atmosphere remains as such for a limited period ranging from several hours to several days. It is oxidized to sulfur trioxide (SO_3) following different reaction mechanisms (homogeneous phase reaction, heterogeneous phase reaction, photochemical catalysed oxidation). Sulfur trioxide is strongly hygroscopic, and is quickly transformed into sulfuric acid (H_2SO_4), which can remain suspended in the atmosphere as micro-droplets (aerosols) of acidic solution.

Nitrogen oxides (generically indicated with the formula NO_x) are the main emissions from vehicle engines as exhaust gases, and are typical of an urban atmosphere characterized by intense motor traffic and high solar irradiation. Under ultraviolet (UV) light, NO_x react with oxygen and VOC, always present in a polluted atmosphere, to form ozone and organic radicals, which are the main components of the so-called "photochemical smog", particularly harmful owing to its oxidizing properties. Moreover, nitrogen oxides, following a complex series of reactions, are transformed into nitric acid and in this form they contribute to the corrosion of surfaces (especially architectonic).

In addition to being dispersed in the atmosphere in the gaseous state, the acidic pollutants may also be found in atmospheric particles, together with soluble salts, carbonaceous material and oxides of heavy metals. The aggressive action of the particles is carried out through a range of phenomena due to their different composition: mechanical erosion, change in the surfaces' water permeability, acid etching, deposition of hygroscopic material and chemical reactions in the heterogeneous phase catalysed by carbonaceous material and/or by metal oxides, which may lead, for example, to the direct oxidation of SO_2 on the surface of the material.

To complete this picture a few comments can be added about the gaseous pollutant that is mostly involved in the chemical corrosion of stone and wall paintings: sulfur dioxide.

This compound, almost absent from the atmosphere in past centuries, is today massively released into the environment. It is formed during the combustion of materials containing sulfides and mineral oils. It has been shown that fine particles of coal, suspended in air and formed through the incomplete combustion of fossil coal, may catalyse the oxidation of SO_2 to SO_3, which in water is transformed into sulfuric acid. The latter corrodes carbonate-based stone surfaces. This phenomenon, already observed for limestone and white marble, is also very concerning for painted surfaces. The amount of gypsum (selenite), $CaSO_4{\cdot}2H_2O$, which is formed after chemical attack of carbonate binder, is often very large. The crystalline unit cell of gypsum is larger than that of carbonate. Therefore the transformation occurs with an increase in volume. Gypsum is also a sparingly soluble salt (it crystallizes with two molecules of water), thus it can diffuse within the porous stones and walls, undergoing crystallization–dissolution cycles under ordinary climatic regimes.

However, it is worth noting that "sulfation" of stone materials and paintings usually occurs in locations far away from urban areas, which suggests that air pollution is not the only cause of the formation of gypsum in porous materials. This observation encouraged further research and analysis of other possible contributing factors, such as the activity of bacteria (sulfobacteria) that settle in the ground. In the case of mural paintings it is possible that the action of sulfobacteria could be favoured by the presence of "eggnog-like" (ancient conservation materials made with egg yolk and vinegar) used to "revive" the colours for "conservative" purposes. Obviously, relevant amounts of soluble sulfates may also come from the ground through capillary rise sorption. Eventually, the level of contamination of stones and wall paintings by sulfate can be very high, and it is not rare to find concentrations from 5–10% w/w. This value is usually indicated as the "sulfation" effect, but it would be more correct to use this term to indicate only the reaction product (gypsum) of the acidic attack on carbonate.

Sulfation involves thus these concomitant reactions (Scheme 2.2):

$$SO_2 + H_2O + 1/2O_2 \rightarrow H_2SO_4$$

$$H_2SO_4 + CaCO_3 + H_2O \rightarrow CaSO_4{\cdot}2H_2O + CO_2$$

Scheme 2.2

Or Scheme 2.3:

$$SO_2 + H_2O \rightarrow H_2SO_3$$

$$H_2SO_3 + 1/2O_2 \rightarrow H_2SO_4$$

$$H_2SO_4 + CaCO_3 + H_2O \rightarrow CaSO_4 \cdot 2H_2O + CO_2$$

Scheme 2.3

Sulfation appears as black crusts over architectural stone surfaces, which are typically exposed to outdoor conditions. This phenomenon is a well-known consequence of atmospheric pollution in urban areas and is due to the inclusion of air particulates (carbon being the major constituent) during the formation of calcium sulfate upon corrosion of calcium carbonate. The black patina also includes other materials such as fine grains of powder (originated by wind erosion of rocks) and other salts (suspended in the air as a marine aerosol).

Typically, cities with cold and rainy winters, with consistent pollution due to total suspended particulate (TSP), exhibit sulfation over stones with porosity not exceeding 5–8%. The affected areas are covered by black crusts, if protected from rainwater run-off, and by white deposits if exposed to rain, with severe signs of corrosion. When the stone porosity is higher, the blackening is more widespread but less intense, and the crusts are more easily detached. Black crusts can be considered a very clear signal of the damage caused by air pollution, and a careful analysis provides useful information for an understanding of the phenomena involved in deterioration.

Once formed, the black crusts remain for a certain period in a situation of apparent stability and then fall off after the occurrence of micro-cracks and lifting of the flaps.

It is worth noting that black patinas adsorb light and this phenomenon consistently increases their temperature. Carbonate-based materials, mainly constituted by calcite that is anisotropic to thermal variations, are thus subjected to mechanical stresses along the directions of thermal dilation (the thermal expansion of gypsum is five times higher than that of calcite). Eventually, this process causes the irretrievable detachment of patina together with the outer layer of stones (peeling effect).

While there is a clear correlation between the presence of SO_2 in the air and the formation of calcium sulfate in calcareous materials, a similar match is not found between the amount of nitrogen oxides in air and the formation of nitrates on stone surfaces. A possible explanation may be found by comparing the Gibbs free energy change (ΔG) for the two reactions, which is -69 kcal mol^{-1} in the first case and only -26 kcal mol^{-1} in the latter. The formation of sulfate is thus thermodynamically more favoured. Kinetics studies have also shown a higher rate for the formation of calcium sulfate compared with nitrate.

Nitrates are sometimes present as salt contaminants on mural paintings. Deposits are often found where an extensive decomposition of organic material has occurred. This is the case, for instance, in churches that have been largely used as burial areas. In this case, the mechanism of formation of nitrates is probably biogenic. The ammonia formed by decomposition of organic material is oxidized to nitrous acid by nitrifying bacteria (*Nitrosobacter*), and then the acid is oxidized to nitric acid by nitrating bacteria (*Nitrobacter*).

2.3.2 Indoor Pollution

Works of art that are housed in buildings cannot be considered totally protected and immune from deterioration. The presence of pollutants and reactive species that are able to interact with works of art materials is, in most cases, unavoidable. Several research programmes have been recently established to understand the formation mechanisms and the degradation pathways that materials undergo during storage in museums, archives and libraries. Many efforts have been made in order to determine the critical levels of pollutants in indoor conditions.

It is worth noting that most of the modern air-purification systems (heating, ventilation and air conditioning, HVAC) are not able to separate the indoor from the outdoor system completely. Besides, some pollutants may originate from different sources and, eventually, dangerous outdoor pollutants are also detrimental in indoor conditions. This is the case for TSP. In fact, studies performed in museums have detected a higher concentration of particles inside than outside the building, especially in the summer and during the day. This phenomenon is associated with the fact that the major influx of visitors (more relevant during the summer) involves transport phenomena such as thermophoresis.

The use of passive monitoring sampling is nowadays quite widespread. Several factors have been investigated in order to define the best practice to preserve the displayed works of art: the deposition rates of pollutants, the emission from display materials, the chemical transformation routes of some pollutants under indoor conditions have all received particular attention. The critical levels of pollution are still debated; however, general guidelines for effective preventive conservation are agreed.

Obviously, some critical points are still in need of better solutions. This is partly due to the nature of the buildings. The interior of a church or historical building, in fact, cannot be considered exactly as an indoor system because the interactions with outdoor conditions are very intense; on the other hand, typical indoor conditions may be found inside a museum or a library, but in this case, further pollution sources must be considered when the influx of visitors is large. In fact, it is quite common in these cases to find heaters, which are sources of combustion pollutants, photocopying machines and laser printers, which are a source of ozone, display materials made of wood or plastic, which emit organic acids and other VOC, and public areas (restaurants, cafes and study rooms), which contribute to pollution by cooking and smoking. These factors complicate the estimation and prediction of the long-term effects of indoor exposition on works of art.

2.4 Physical, Chemical, and Mechanical Stresses

The deterioration of works of art materials, apart from the biological aspects that are not the topic of this book, deals with physical, chemical and mechanical stresses. Physical phenomena are associated mainly with temperature, electromagnetic radiation and frost formation mechanisms. Chemical alterations are due to the interaction between the artefacts materials and substances – mostly gaseous – spread in the environment. Finally, mechanical stresses can result from both physical and chemical processes where changes in volume arise. As an example, sulfation causes mechanical stress due to the drastic increase of the elementary cell passing from calcite to selenite ($CaSO_4{\cdot}2H_2O$; the molar volume of selenite is twice that of calcite). Sections 2.2 and 2.3 dealt with the description of natural weathering and the effects of environmental pollution on chemical alterations. In the following sections the chemical aspects of materials deterioration will not be discussed; instead, the discussion will focus on phenomena that are only indirectly associated with chemical decay, such as the general effects of water on porous matter. Therefore, water condensation, wetting and capillary suction will be illustrated to discuss the critical parameters involved in these processes, which must be taken into account for real and concrete preventive conservation. In particular, condensation at the dew point and the absorption of water by hygroscopic and deliquescent salts will be treated in detail. One of the most crucial problems associated with water absorption within porous materials of artistic or architectonic interest is the interaction between the absorbed liquid water and soluble salts that are possibly present inside the material structure or on its surface. Thus, the subject of salt efflorescence, which is strictly connected with water absorption or condensation, will be discussed, with particular emphasis on the reciprocal role of the crystal type and crystallization matrix on determining the entity of the deterioration effects. On the other hand, the mechanical stresses originated by the crystallization induced by water absorption will be discussed, showing that the physico-chemical properties of the porous stone material are crucial for determining the damage.

Concerning the physical causes of the degradation of wall paintings and stones, great importance will be given to the ageing of organic materials (such as binders or coatings on artistic and architectonic surfaces) due to the combined action of oxygen in the air and light (see Section 2.4.2).

2.4.1 Water Condensation, Wetting, and Capillary Suction

Humidity within the wall is one of the most dramatic causes of deterioration in wall paintings, because moisture can determine many physico-chemical processes that result in the degradation of the pictorial surface. We recall that the maximum amount of water that a wall can contain without presenting any damage is 3–5%; above this limit both the painted layers and the plaster underneath lose their mechanical properties and become subjected to rapid deterioration. The main sources of moisture in walls are:[2] (i) infiltration,

(ii) capillarity, (iii) condensation, and (iv) rain and wind causing excessive cooling of the wall, resulting in condensation on the inner side of the wall. From a physico-chemical point of view the most interesting causes of moisture are capillary suction, condensation from a wet atmosphere on the pictorial surfaces, and wetting of the pictorial surfaces by condensed liquid water. We will briefly examine these three phenomena in relationship to the type of degradation they produce on the wall painting surfaces. A more theoretical treatment was provided in Chapter 1 (Sections 1.4.1. and 1.4.2).

Water condensation on pictorial surfaces can occur according to two different mechanisms: (i) by reaching the dew point, and (ii) through absorption from the wet atmosphere owing to hygroscopic and deliquescent salts that are sometimes present in the wall.[27,28] From a thermodynamic point of view, water vapour condensation on a surface from a wet atmosphere with a relative humidity RH^* ($0 \leq RH^* \leq 100$) occurs when the temperature of the surface $T_s \leq T_{sat}$, where T_{sat} is the temperature at which the vapour pressure relative to RH^* reaches the value of saturation. In these conditions the stable phase for water on the surface is the liquid, and according to nucleation and growth mechanisms droplets of liquid water are formed over the painted surface. Liquid water penetrates the surface layers by capillary absorption and the inner wall very soon becomes highly wet. The mechanism through which the wall becomes wet is driven by the wetting properties of the porous surface with respect to the liquid water. In other words the three interfaces, water droplet–porous solid substrate, water droplet–air, and porous solid substrate–air, all play a fundamental role in determining whether the liquid will penetrate into the wall or spread over the pictorial surface. We know, in fact, that for porous materials three different regimes of behaviour can be defined in relationship to the spreading *vs.* depletion processes.[29,30] For the porous materials we are considering – *i.e.* aerial plasters – absorption generally prevails over spreading, with a constant contact line, a contact angle decrease, and an increase in the droplet radius of curvature.[31] Nevertheless, we have to consider that the aerial plasters are painted, so that pigments present on the surface can strongly modify this behaviour and may possibly reverse the regime so that spreading prevails over depletion. This can be rationalized in terms of different contact angles of liquid water on surfaces that differ because of different pigments.

We can conclude that the type of pigment alters the interaction between the condensed water and the pictorial surface. Furthermore, regions painted with the "*a secco*" technique behave in a different way, owing to the presence of organic materials – *i.e.* the binder – in the surface paint layers. These organic materials modify the wetting properties of the pictorial surfaces, determining differentiated degradation.

The second source of moisture in walls is associated with the presence of hygroscopic or deliquescent salts.[27,28] Hygroscopic and deliquescent salts may produce water condensation from moist air at temperatures well above T_{sat}. This means that, in the presence of these salts, even if the temperature of the pictorial surface is far from the dew point, condensation occurs, and the hygroscopic or deliquescent salts form saturated solutions that migrate in the entire wall. Some studies[27,28] have shown that a deliquescent salt commonly found in wall paintings, $Ca(NO_3)_2{\cdot}4H_2O$, produces water condensation at temperatures typical of an interior (church or building) for $RH \geq 56\%$.

The source of moisture in walls other than water condensation is capillary rise from the ground (see also Section 1.4.2.3). This is the most common source of moisture, and is intrinsic in wall painting substrate (in churches, buildings, *etc.*). This means that no preventive conservation measure can be adopted to remove humidity coming from this source. We remind the reader that capillary rise from the ground can reach a height up to 6 m from the stamping plane. Indeed, the rising moisture comprises aqueous solutions of salts such as nitrates, sulfates, and sometimes chlorides. When these rising solutions reach the wall painting surface, salt crystallization occurs, generating salt efflorescence (*vide infra*). It is important to note that water rising from the ground not only dissolves salts and allows them to migrate and diffuse into the wall structure, but can also dissolve gases such as CO_2, NO_x and SO_2, behaving thus as a carrier for polluting agents.[32] The wall properties that mainly drive the mechanism of capillary rise are porosity, which depends on the pore size distribution.

The wall porosity can be very different from that of the pictorial surface layers, owing to both the procedure used to spread the *intonaco* and to the presence of pigments that can form more or less compact paint layers depending on the degree of grinding. In any case the pictorial surfaces of wall paintings are highly permeable to both liquid and gaseous water: this is a very important characteristic that imparts high durability to mural paintings. In fact, wall painting surfaces must continuously exchange water (liquid or vapour) with the environment in order to remain in thermodynamic equilibrium in certain environmental conditions, which may vary over a wide range of values.

The processes of alteration promoted by moisture are, essentially: (i) cycles of crystallization–dissolution of salts, (ii) degradation of pigments that are sensitive to humidity, (iii) biological degradation, (iv) dissolution or diffusion/spreading of polluting agents such as gases or aerosols, and (v) frost action.[2] Among these processes, the most significant and important is the first.

As illustrated in Section 2.2.2, one of the most dramatic causes of surface deterioration of wall paintings is associated with the crystallization/dissolution cycles of salts, both inside the wall and on the pictorial surface.[27,28,33] These cycles, depending on the environmental conditions (mainly temperature and *RH*), initiate physical and chemical processes that are potentially dangerous for both the paint layers and the support. We have already shown that the damage is strictly related to the hygroscopicity and deliquescence of the salts involved. Many studies have shown that the most harmful salts are those that are less hygroscopic, because the damage they cause is principally due to the crystallization process, which produces crystallization pressures inside the porous fresco matrix.[33] Concerning this aspect, a study[31] pointed out the role of KNO_3, a poorly hygroscopic salt, in the mechanism of degradation of porous artistic/architectonic materials (wall paintings, sandstones, marble). Figure 2.1 shows the effects of the growth of crystal phases of KNO_3 on a pictorial surface of a famous fresco.

Figure 2.1 Effects of the growth of nitrate crystalline phases in the porous structure of '*Ultima Cena*' ('St. John's face'), a wall painting by Taddeo Gaddi (14th century) in the Refectory of Santa Croce in Florence, after the Arno river flood in 1966. Top: the damage shown before the restoration; middle: a detail of the nitrate efflorescences; bottom: the painting after restoration. (Reprinted with permission from L. Dei, M. Mauro, P. Baglioni, C. Manganelli Del Fà and F. Fratini, *Langmuir*, 1999, **15**, 8915; Copyright American Chemical Society.)

It is markedly visible that nitrate efflorescence produces both modifications in the appearance of the pictorial surface and severe damage to the porous material underneath the paint layers. The "explosion" of a salt efflorescence on a pictorial surface causes a considerable loss of pigment, which lifts from the *intonaco*, losing its adherence to the support. Interestingly, the porous substrate where crystallization occurs has a very important role in determining the morphology and the *habitus* of the salt efflorescence, maintaining all the other parameters constant.

In particular, KNO_3 forms lamellar crystals on "*Pietra di Firenzuola*" and "*Pietraforte*", as shown in Figure 2.2(a) and (b). This morphology can be linked to the "layered" structure of sandstones, the KNO_3 crystals growing parallel to the original stone layers. *Travertine*, though very compact, presents a

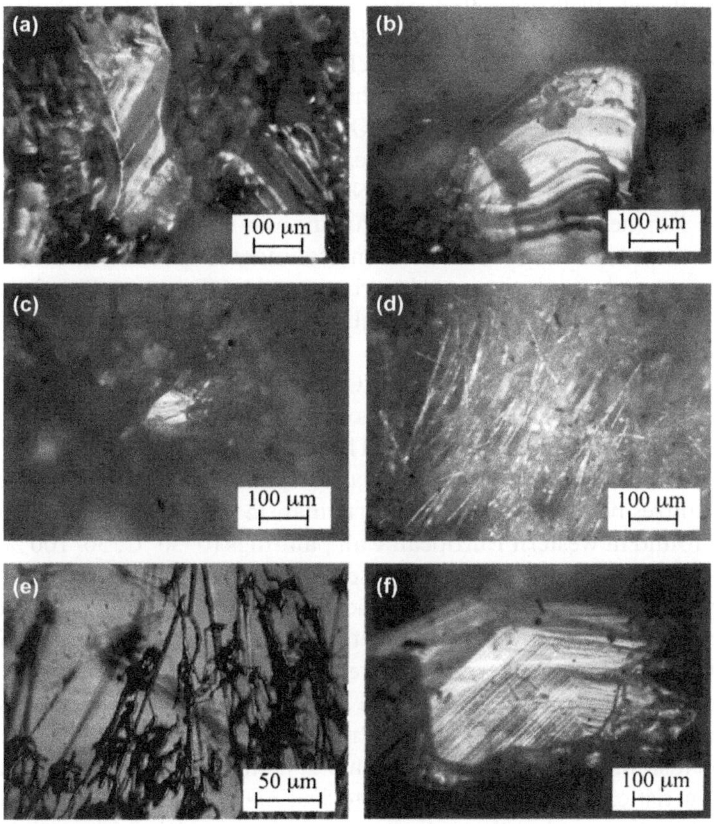

Figure 2.2 Scanning electron micrographs of the KNO_3 crystalline phases grown on different porous surfaces: (a) *Pietra di Firenzuola*, (b) *Pietraforte*, (c) *Travertine*, (d) *Carrara* marble, (e) *Carrara* marble, (f) aged marble. (Reprinted with permission from L. Dei, M. Mauro, P. Baglioni, C. Manganelli Del Fà and F. Fratini, *Langmuir*, 1999, **15**, 8915; Copyright American Chemical Society.)

number of millimetre-sized macrocavities on the external surface, where crystals are able to grow and accumulate in a random way. These agglomerates of disordered crystals can be clearly observed by scanning electron microscopy [Figure 2.2(c)]. On the "*Carrara*" marble KNO_3 forms needle-shaped "whiskers"[34] with partial alignment in one direction [Figure 2.2(d) and (e)]. It should be noted that the width of the "whiskers" is very small; this confirms what usually occurs with "dense" materials. This needle-like habitus disappears when crystallization occurs on surfaces of aged marble with a high percentage of sulfation and high loss of cohesion among the marble grains. In the latter case the morphology is again lamellar [Figure 2.2(f)]. The result suggests that the needle-shaped habitus depends strictly on the crystallinity of the "*Carrara*" marble's calcite. These drastic morphological changes induced by the stone surfaces can be very important in determining the damage caused by salt crystallization to many masterpieces of art and architecture.

Therefore, in principle, hygroscopic and deliquescent salts are not so important in the conservation of wall paintings and stones; indeed, it has been recognized that hygroscopic and deliquescent salts can produce wetting solutions containing ionic species that can initiate crystallization of less hygroscopic salts. For example, $Ca(NO_3)_2{\cdot}4H_2O$ is deliquescent and can act as a carrier for NO_3^- anions, which can react with K^+ or Na^+ cations to produce the less hygroscopic KNO_3 and $NaNO_3$ that result in persistent salt efflorescence. Furthermore, we must bear in mind that each hygroscopic or deliquescent salt, under well-defined environmental conditions, absorbs water to form a saturated solution, or desorbs water to give crystallization and salt efflorescence, either on the pictorial surfaces or within the wall layers (crypto-efflorescence).

As a matter of fact, $Ca(NO_3)_2{\cdot}4H_2O$, among the various hygroscopic and deliquescent salts that can be present as contamination agents in wall paintings, has received particular interest.[27,28] This salt can act as a "surface wetter" because it absorbs water when the relative humidity is greater than 53–54%, and crystallizes below this value. In particular, in the environmental conditions typically found in western European wall paintings (0–30 °C, 30–100% *RH*) the threshold of 53–54% is very often crossed in both directions (above and below), meaning that when this salt is present, several dissolution–crystallization processes occur during a year. The kinetic and thermodynamic properties of the system $Ca(NO_3)_2{\cdot}H_2O$ have been extensively studied, but one of the most interesting features not yet understood is the kinetics of water absorption as a function of the substrate where the nitrate is present.[27]

In conclusion, the growth of inorganic crystalline phases within the porous media of artistic, architectonic and historical materials originates efflorescence and crypto-efflorescence that affect many physico-chemical properties (*e.g.* total porosity and pore size distribution), causing potential damage to the work of art or architecture. Furthermore, if the salts formed are hygroscopic or deliquescent, the dew-point temperature should be considered as a misleading parameter in the environmental control regime designed to prevent wetting of the pictorial surface.

2.4.2 Photo-oxidation of Materials

In the conservation of easel paintings, one of the critical points is the ageing of the perishable organic substances present as original materials or superimposed by conservators in previous conservation treatments. Indeed, the most crucial mechanism by which ageing occurs is the photochemical oxidation that takes place as a result of the combined action of light and air. Varnishes and drying oils are the main compounds that are subjected to this type of decay.[35–37]

In a recent paper,[38] the ageing and yellowing of triterpenoid varnishes such as mastic and dammar resins were studied. The authors addressed the influence of resin composition and ageing conditions.

One of the factors that needs to be further investigated is the role played by light *vs.* darkness in relationship to oxidation and the kinetics of the process. Indeed, although many authors have already studied this topic,[39–42] several aspects remain to be unravelled. Dieteman *et al.*[38] demonstrated that the role played by light is not so critical in determining the oxidation kinetics of mastic and dammar resin. In fact, the study shows that the ageing mechanism is driven by auto-oxidation, mostly independent of the presence of light. The only difference between light and darkness conditions is the slower kinetics of self-formation of radicals in dark conditions. It is interesting to note that the presence of polymeric materials in mastic resin is found to inhibit oxidation but simultaneously increase the yellowing. This can be ascribed to the ability of polymer molecules to act as radical stabilizers, slowing down the oxidation kinetics and enhancing the intrinsic yellowing tendency of mastic. One of the goals of conservation of easel paintings should be to find possible routes to enhance natural resin stability against photo-oxidation. In spite of the studies carried out involving photostabilizers,[43] more work has to be done in this sense, above all to assess their possible use as new conservation materials.

Drying oils constitute another vast class of substances subjected to photo-oxidation processes. These materials have been commonly used for centuries as binders in the painting techniques. Even if the kinetics of photo-oxidation of drying oils is slower than that of varnishes, nevertheless yellowing phenomena associated with these processes often occur. Several studies[44–47] have been carried out to study both photo-oxidation and thermo-oxidation of these compounds. It is worthwhile recalling that photo-oxidation is the mechanism through which drying oils dry during the setting process. The photo- and thermo-oxidation we are treating in this chapter refer to processes involving dried films, *i.e.* the ageing of the set oil binders. Unlike varnishes, in this case light plays a very crucial role: indeed, photo-oxidation of dried linseed oil films is critically important, while thermo-oxidation does not cause specific damage to the dried films. Some authors[48] have found that the degradation of the dried oil network under photo-oxidation at wavelengths above 300 nm is consistent and results in the breaking of the chain. On the other hand, dried films are rather stable against thermo-oxidation. The photo-instability is attributed to the presence of crosslinks that are sensitive to radical attack because of the weakness of the hydrogen atom on the tertiary carbon atoms. Another very interesting result reported by the same authors[48] is that the yellowing of the cured samples rapidly decreases as a result of irradiation, because the oil contaminants that are mainly responsible for the yellowness are photo-oxidized. Instead, yellowing slowly but continuously increases during thermo-oxidation at 100 °C. In conclusion, photo-oxidation processes are very important for the works of art where drying oils and/or varnishes have been used both as original materials and as materials superimposed in the past. Much work has to be done to solve the critical problems associated with these ageing effects, starting from the well-known yellowing of easel painting surfaces.

Further Suggested Reading

On works of art materials:

B. Berrie (ed.), *Artists' Pigments: A Handbook of Their History and Characteristics*, National Gallery of Art, Washington DC and Archetype Publications, Washington DC, vol. 4, 2007.
M. Doerner, *The Materials of the Artists*, revised edition, Hart-Davis, MacGibbon, London, 1969.
R. L. Feller, A. Roy, E. Fitzhugh and B. Berrie (ed.), *Artists' Pigments: A Handbook of Their History and Characteristics*, National Gallery of Art, Washington DC and Cambridge University Press, Washington DC and Cambridge, vol. 1, 1986.
J. S. Mills and R. White, *The Organic Chemistry of Museum Objects*, Butterworth-Heinemann Series in Conservation & Museology, London, 1999.
A. Roy (ed.), *Artists' Pigments: A Handbook of Their History and Characteristics*, National Gallery of Art, Washington DC and Oxford University Press, Washington DC and Oxford, vol. 2, 1993.
E. West Fitzhugh (ed.), *Artists' Pigments: A Handbook of Their History and Characteristics*, National Gallery of Art, Washington DC and Oxford University Press, Washington DC and Oxford, vol. 3, 1997.

On environmental and urban pollution:

D. Camuffo and A. Bernardi, *Sci. Total Environ.*, 1996, **189/190**, 235–245.
M. Laurenzi Tabasso and M. Marabelli, *Il degrado dei monumenti in Roma in rapporto all'inquinamento atmosferico*, Betagamma, Viterbo, 1992.

On physical, chemical, and mechanical stresses:

B. M. Feilden, *Conservation of Historic Buildings*, Butterworth-Heinemann, Oxford, 1994, p. 89.
A. Goudie and H. Viles, *Salt Weathering Hazards*, Wiley, Chichester, 1997.
G. W. Scherer, *Cement Concr. Res.*, 2004, **34**, 1613.
G. Scott, *Atmospheric Oxidation and Antioxidants*, Elsevier Science Publishers B.V., Amsterdam, 1993.

C. Selwitz and E. Doehne, *J. Cult. Herit.*, 2002, **3**, 205.
N. Stolow, in *On Picture Varnishes and Their Solvents*, ed. R. L. Feller, N. Stolow and E. H. Jones, National Gallery of Art, Washington DC, 1985, p. 47.
H. W. S. Chan, in *Autoxidation of Unsaturated Lipids*, ed. H. W. S. Chan, Academic Press, London, 1987, p. 1.

Review and Questions

This chapter deals with the main causes that lead works of art to be subjected to degradation during their ageing.

- The main physico-chemical properties of wall paintings, limestone, and easel paintings have been illustrated with the aim of focusing the attention on those aspects that are critical for determining the various mechanisms and phenomena driving deterioration by ageing.
- The natural weathering phenomena are discussed, pointing out the role of temperature, relative humidity, light, and crystallization of salts in governing the kinetics of decay.
- All the aspects dealing with environmental pollution, both outdoor and indoor, are treated in detail, in order to sketch out the best conditions for the preventive conservation of works of art.
- We focused our attention on capillary phenomena (water condensation, wetting and capillary suction) in order to illustrate the effects of these physico-chemical processes on works of art, referring to the theoretical basis discussed in detail in Chapter 1.
- Salt efflorescence, already described in this chapter as salt crystallization, has been reviewed in recalling the role played by deliquescent and hygroscopic salts, especially for monitoring the environmental conditions for the best preventive conservation.
- Some remarks have been made concerning the combined effects of light and oxygen from air on organic materials that may be present in the works of art, either as original or superimposed materials in past conservation treatments.

Questions

1) What is the main chemistry involved in the setting process of lime, $Ca(OH)_2$?
2) Ground is a fundamental layer in easel paintings: what is its composition and what are its main functions?
3) Why are black crusts so harmful for stones?
4) What are the best hydrothermal conditions (temperature, *RH*) to preserve works of art in a museum?
5) Why are deliquescent salts within a porous matrix so important?
6) What are the substances most sensitive to photochemical oxidation, and why are these substances so sensitive to this kind of deterioration?

Answers

1) The main chemistry involved in the setting process of lime, *i.e.* $Ca(OH)_2$, is the reaction between the carbonate anion, CO_3^{2-}, originated by the dissolution of atmospheric CO_2 in the strong alkaline environment of the fresh prepared aerial plaster, and the Ca^{2+} cations from the saturated solution of calcium hydroxide which surrounds both sand and pigment particles. The nucleation and growth of calcite, $CaCO_3$, crystals is slow, owing to the low concentration of CO_2 in the atmosphere (around 390 ppm).
2) The ground layer consists of a filler or inert and a binder or medium. In the classical Italian ground the filler is gypsum, $CaSO_4.2H_2O$, or lead carbonate (*biacca*), and the binder is animal glue (from rabbits). Indeed, other fillers can be suitable, such as kaolin, $CaCO_3$, and in more recent times ZnO, $BaSO_4$ and TiO_2; from the 20th century natural glues have been substituted by synthetic polymer resins. The main functions are the following: (i) perfect homogenization of the surface onto which one has to paint, (ii) reducing to the optimum level the porosity of the support in order to minimize penetration of the painting material inside the porous structure of wood or canvas.
3) Black crusts absorb light and cause temperature increase on artistic surfaces. Carbonate-based materials are thus subjected to mechanical stresses along the directions of thermal dilation. Eventually, this process causes the irretrievable detachment of patinas together with the outer layer of stones (peeling effect). Stone surfaces apparently unaffected by patina may indeed have been exposed to rainfall, with the consequent detachment of the peeling layers.
4) The optimal hydrothermal conditions to preserve works of art in a museum require the minimum fluctuations of both *RH* and temperature. Temperatures of about 18–21 °C represent the best compromise to prevent condensation while favouring a decrease in the rate of all the degradation reactions. The *RH* should be around 50–55% in order to avoid dehydration of materials (which occurs below 50%) and to discourage bio-deterioration and all the degradation processes favoured by the presence of water.
5) The presence of possible deliquescent salts inside a porous matrix is a fundamental topic because the *RH* conditions that determine water condensation are greatly different with respect to those in the absence of the salts, the latter deriving uniquely from the dew-point conditions. Instead, the presence of deliquescent salts can lower the relative humidity threshold for water absorption, even by some tens of percentage points.
6) The substances most sensitive to photochemical oxidation are organic chemicals, and particularly those with unsaturated bonds that possess suitable sites for oxygen attack catalysed by the UV component of light. In this sense, varnishes, owing to their chemical structure, are among the most sensitive substances to this type of deterioration process.

References

1. E. Ferroni, in *Proc. Internat. Workshop 'Ecological Physical Chemistry'*, ed. C. Rossi and E. Tiezzi, Elsevier, Amsterdam, 1991, p. 345.
2. P. Mora, L. Mora and P. Philippot, *Conservation of Wall Paintings*, Butterworths, London, 1984.
3. A. C. Cennini, *Il libro dell'arte, Manuscript of 1437*, ed. G. Milanesi and C. Milanesi, Le Monnier, Florence, 1859; new edition by F. Brunello and L. Magagnato, Neri Pozza, Vicenza, 1971.
4. U. Procacci and L. Guarnieri, *Come nasce un affresco*, Bonechi, Florence, 1975.
5. P. Baglioni, L. Dei, F. Piqué, G. Sarti and E. Ferroni, *Stud. Conserv.*, 1997, **42**, 43.
6. M. Matteini and A. Moles, *La chimica nel restauro. I materiali dell'arte pittorica*, Nardini, Florence, 2007.
7. Dionysius of Fourna, *Ermeneia*, from the French edition by M. Didron, Manuel d'Iconographie Chrétienne, Grecque et Latin, Paris, 1865.
8. Theophilus Presbyter, *Schedula diversarum artium*, from the French edition of Comte Charles de l'Escalopier, Paris-Leipzig, 1843.
9. H. Tintelnot, *Die barocke Freskomalerei in Deutschland*, F. Bruckmann Publisher, Munich, 1951.
10. A. M. Maetzke, in *Piero della Francesca - La Madonna del Parto. Restauro e iconografia*, Marsilio, Venice, 1993, p. 23.
11. H. Kühn, *Conservation and Restoration of Works of Art and Antiquities*, Translation by A. Trone, Butterworths, London, 1986, vol. 1, p. 3.
12. P. Hendy and A. Lucas, *Museum XXI*, 1968, **4**, 245.
13. J.-M. Guenet, *Thermoreversible Gelation of Polymers and Biopolymers*, Academic Press, New York, 1992.
14. X. DeLanglais, *La technique de la peinture à l'huile*, Flammarion, Paris, 1959.
15. D. V. Thompson, *The Practice of Tempera Painting*, Dover Publishers, New York, 1962.
16. A. P. Laurie, *The Painter's Methods and Materials*, Dover Publishers, New York, 1968.
17. F. Margival, *Peintures, pigments, vernis*, 1965, **41**, 390.
18. P. Duell and R. J. Gettens, *Tech. Stud. Field Fine Arts*, 1940–41, **9**, 75.
19. A. H. Church, *The Chemistry of Paints and Painting*, Seeley & Co., London, 1890.
20. G. L. Stout, *Tech. Stud. Field Fine Arts*, 1938, **7**, 59.
21. M. Doerner, *The Materials of the Artists*, Hart-Davis, MacGibbon, London, revised edn, 1969.
22. R. J. Gettens and G. Stout, *Painting Materials: a Short Encyclopaedia*, Dover Publishers, New York, 1966.
23. K. Wehlte, *Werkstoffe und Techniken der Malerei*, Ravensburg, 1967.
24. R. L. Feller, in *Conservation of Paintings and the Graphic Art - Lisbon Conference*, IIC Publishers, London, 1972, p. 201.

25. R. Feller, N. Stolow and E. H. Jones, *On Picture Varnishes and their Solvents*, National Gallery of Art and AIC Publishers, Washington, 3rd revised edn, 1985.
26. A. E. Charola and J. Weber, in *Proceedings of the 7th International Congress on Deterioration and Conservation of Stone, Lisbon, June 15–18*, ed. J. Delgado Rodriguez, F. Enriquez and F. Telmo Jeremias, Laboratorio nacional de engenharia civil, Lisbon, vol. 2, 1992, p. 581.
27. F. Piqué, L. Dei and E. Ferroni, *Stud. Conserv.*, 1992, **37**, 217.
28. F. Piqué, P. Baglioni, L. Dei and E. Ferroni, *Sci. Tech. Cultur. Herit.*, 1994, **3**, 155.
29. M. Denesuk, B. J. J. Zelinski, N. J. Kreidl and D. R. Uhlmann, *J. Colloid Interface Sci.*, 1994, **168**, 142.
30. M. Denesuk, G. L. Smith, B. J. J. Zelinski, N. J. Kreidl and D. R. Uhlmann, *J. Colloid Interface Sci.*, 1993, **158**, 114.
31. L. Dei, P. Baglioni, M. Mauro, C. Manganelli Del Fà and F. Fratini, *Langmuir*, 1999, **15**, 8915.
32. A. Arnold and K. Zehnder, in *The Conservation of Monuments in the Mediterranean Basin*, ed. F. Zezza, Grafo, Brescia, 1990, p. 31.
33. A. Arnold and K. Zehnder, *J. Cryst. Growth*, 1989, **97**, 513.
34. A. Arnold and K. Zehnder, in *Proceedings of the 5th International Congress on Deterioration and Conservation of Stone*, Lausanne, 1985, p. 269.
35. E. R. De la Rie, *Anal. Chem.*, 1989, **61**, 1228.
36. A. C. Elm, *Ind. Eng. Chem.*, 1949, **41**, 319.
37. S. B. Crecelius, R. E. Kagarise and A. L. Alexander, *Ind. Eng. Chem.*, 1955, **47**, 1643.
38. P. Dietemann, C. Higgitt, M. Kälin, M. J. Edelmann, R. Knochenmuss and R. Zenobi, *J. Cult. Herit.*, 2009, **10**, 30.
39. D. Scalarone, J. Van der Horst, J. J. Boon and O. Chiantore, *J. Mass Spectrom.*, 2003, **38**, 607.
40. R. L. Feller, *Accelerated Aging, Photochemical and Thermal Aspects, Research in Conservation*, The Getty Conservation Institute, Los Angeles, 1994.
41. E. R. De La Rie, *Stud. Conservat.*, 1988, **33**, 53.
42. E. R. De La Rie and C. W. McGlinchey, *Stud. Conservat.*, 1989, **34**, 137.
43. E. R. De La Rie and C. W. McGlinchey, in *Cleaning, Retouching and Coatings. Technology and Practice for Easel Paintings and Polychrome Sculpture, Contributions to the IIC Brussels Congress, 3–7 September 1990*, ed. J. S. Mills and P. Smith, IIC Publisher, London, 1990, p. 160.
44. E. N. Frankel, *J. Am. Oil Chem. Soc.*, 1984, **61**, 1908.
45. E. N. Frankel, W. E. Neff and E. Selke, *Lipids*, 1981, **16**, 279.
46. K. Miyashita, N. Hara, K. Fujimoto and T. Kaneda, *Lipids*, 1985, **20**, 578.
47. W. J. Muizebelt, J. C. Hubert and R. A. M. Venderbosch, *Prog. Org. Coat.*, 1994, **24**, 263.
48. J. Mallégol, J.-L. Gardette and J. Lemaire, *J. Am. Oil Chem. Soc.*, 2000, **77**, 257.

CHAPTER 3

Conservation Treatments: Cleaning, Consolidation and Protection

L. DEI

University of Florence, Italy
Email: luigi.dei@unifi.it

3.1 Introduction

The preservation, conservation and transmission of our cultural heritage to future generations are some of the main duties of humanity. Unfortunately natural, archaeological, historical and artistic materials are constantly subject to the action of many detrimental actions due to environmental pollutants, wrong handling practices, natural disasters, accidental damage, or simply to neglect. Thus, in order to ensure the durability of the whole human cultural heritage, minimizing any further decay, and to set up the best conservation approaches and procedures, all the so-called hard sciences and technologies should play a primary role.

From the thermodynamic point of view, every tangible work of art can be considered as an open system continuously exchanging both matter and energy with the external environment. From this perspective, a degradation phenomenon includes all the spontaneous and irreversible biological, physical and chemical processes that, through this exchange of matter and energy, cause the chemical and/or the physico-structural alteration of the materials constituting the work of art with respect to their original status. As a

RSC Nanoscience & Nanotechnology No. 28
Nanoscience for the Conservation of Works of Art
Edited by Piero Baglioni and David Chelazzi

Published by the Royal Society of Chemistry, www.rsc.org

consequence, the thermodynamic function of an "ideal" conservation treatment should be to stop or at least to slow down the kinetics of these processes, with the final aim of increasing both the stability of the "object" and its lifetime in order safely to optimize the virtuous promotion of the cultural, economic and social value of the artefact. Paul Philippot wrote that "conservators should remedy, as far as they can, to the fatal divorce that time – but often men – cause between the ageing matter and the form it embodies".

In order to achieve this scope, the responsibility of the management of the cultural heritage safeguard should be handled by people that have, each with their specific function, a global view of the main problems related to all the phases of the conservation process. The method for the complete understanding and the correct management of the system's – *i.e.* cultural heritage – complexity can be achieved through: (i) identification of the nature and role of the materials constituting the artefacts, and (ii) evaluation of their molecular properties, their physico-chemical history and evolution, and the dynamics of their surface and bulk interactions with the surrounding environment. Once these data are obtained, the critical and comparative knowledge of all the available conservation and restoration technologies and materials has to be carried out to plan the best conservation strategy.

This methodological approach is mandatory for both correct interpretation and prediction of the dynamics of the degradation processes (including the development of theoretical models), and the set up of appropriate conservation treatments. In the present chapter some fundamental ideas about the three main stages of every conservation intervention – cleaning, consolidation, and protection – will be provided and discussed in a dialectical manner.

3.2 Restoration and Conservation: Meanings, Purposes, and Strategies

The first aspect to consider when dealing with conservation and restoration is to illustrate and clarify the difference between these two terms. Indeed, in many countries the term restoration is considered as one of the several actions involved in the complex activity called conservation. However in some countries, especially (but not only) Italy, such distinction of terms does exist, thus it is meaningful to start this paragraph with some net definitions. We can start from the definition of restoration as all the actions involving the direct – often manual – interaction between humans and the work of art as mediated by the matter constituting the masterpiece. In this sense cleaning, consolidation and protection are typical operations that can be related to the term restoration. On the other hand, conservation is a broader term, because it refers to the many different measures adopted for the correct safeguard of the artefact, and not only to those that are strictly linked to direct intervention on the artefact. For instance climate checks, environmental control, the choice of the best protocols for the intervention, and definition of the best practices, are all aspects to be included in the meaning of the term conservation. According to this idea, conservation also includes restoration. Therefore, when discussing the topic of this paragraph, we will refer to conservation, considering in particular the three operations that constitute the main target of this book: cleaning, consolidation and protection.

If we consider the first of these operations, cleaning, we can immediately state that it deals with a strictly irreversible process, in the sense that once the "foreign" material has been removed, it would be absolutely impossible to go back to reconstitute partially or completely what was removed.[1] In this sense, the rule of reversibility cannot be adopted.

Instead, the correct terms that should be indicated for cleaning are *controllability* and *selectivity*. The main topic focuses thus on possible operations that may be beyond control: for instance, in the past the use of excessively aggressive solvents led to the removal of original materials from paintings.[2] Both studies and the daily practice of conservators have shown that traditional solvents exhibit two main disadvantages: insufficient specificity or selectivity, and poor controllability of both solubilization power and penetration within the layers of the works of art. This conclusion cannot be considered as a kind of universal condemnation of the use of organic solvents, pure or in mixture: in fact, even traditional solvents in the hands of very expert conservators succeed in obtaining controllable and selective cleaning.[3]

Nonetheless, in the last 25 years many attempts have been made to develop new materials and systems for more controllable and selective cleaning. It is sufficient here to recall the studies by Wolbers,[4] who identified gels as potential systems for improving both selectivity and controllability (advanced gel systems will be discussed in Chapter 11). On the other hand, if gels allow a controlled and slow release of the cleaning solvents, they also suffer the limitation of being formulated with non-volatile substances, because the gelators are mostly polymeric. The controllability and selectivity issue is thus replaced by the problematic removal of non-volatile compounds, such as thickening agents and/or surfactants.[5] Another critical question concerning cleaning is the possible reaction among the solvents used in blends. A classical example of this phenomenon is reported in two papers and deals with the mixture named ***4A*** (*Acetone*, 1,2-dimethylketone; *Ammoniaca*, NH_3; *Alcol*, ethanol; and *Acqua*, water) and the mixture between ammonia and ammonium carbonate solutions. Dizabo and Pepe[6] have shown that the ***4A*** mixture has strong reactivity, leading to the formation of many different coloured compounds. Studying the ammonia–ammonium carbonate binary system, it was demonstrated that highly volatile solvents can remain entrapped within the porous structure, where they can keep reacting even after a long period of time.[7] In summary, cleaning is a very delicate and complex operation: research in the field of new materials, including nanoscience, can open up new perspectives with the aim of developing cleaning systems that are completely controllable and selective, according to the well-known idea of smart materials.[8]

Consolidation and protection of works of art can be misleading if one does not immediately clarify the difference between these two approaches.

Consolidation is strictly connected to the recovery of the artefact's mechanical properties. It can refer to some superficial parts of a masterpiece, or alternatively it may concern the bulk of a material object. On the other hand, protection simply refers to a kind of surface treatment aimed at slowing down the kinetics of surface alteration or deterioration. Consolidation of the external layers can thus be considered simultaneously as consolidation and protection, owing to the location of the intervention. On the other hand, bulk consolidation is a very complex operation, and the penetration of the consolidating material has been the subject of many publications in the last 20 years.[9–15] Generally, most consolidating materials have a polymeric nature and they are often chosen from the vast class of industrial adhesives. The process of adapting such materials to artefacts consolidation may thus be complex. Given that most of the consolidating agents are polymeric, the critical feature is the penetration of large molecules, dispersed in suitable solvents, within the porous structure of the work of art, reaching all the internal pores where the original mechanical properties have been lost. The situation is completely different for different types of support (stone, mortar, wood, paper, stucco, wall paintings, easel paintings, textiles, *etc.*) and for each case a specific and fully dedicated polymeric product must be designed. In any case, when the deepest layers are to be consolidated, it is almost impossible to achieve good penetration.

To overcome this problem, a possible solution is to use smaller molecules, *e.g.* monomers that can undergo a polymerization *in situ*.[16–18] The results obtained so far are quite interesting but an extensive application is still unlikely.

On the other hand the so-called inorganic approach for consolidating limestone, mural paintings, stuccoes and historical plasters does not suffer from this limitation because inorganic substances are monomeric in solution, or consist of very small particles (colloids, nanomaterials) in dispersed systems (see Chapters 12 and 13). Nevertheless, their penetration can be difficult and sometimes not very controllable (depending on the substrate topology), and hydroxide particles need atmospheric CO_2 to penetrate through the layers for complete setting to occur. Moreover, the adhesive properties of inorganic consolidants are very poor; however, they are highly compatible with many artistic substrates from the physico-chemical point of view.

In the field of consolidation much work remains to be done for different reasons. Perhaps hybrid inorganic–organic nanocomposites might combine the features of the two different approaches and minimize the disadvantages related to the separate materials (see Chapter 14).

Finally, protection is very often the last intervention of conservation to be performed, and much debate has been carried out in the past about its role. For outdoor monuments or decorated architectural surfaces the term protection is synonymous to hydrorepellency,[19–21] which can be achieved through polymeric coatings, mainly using siloxanes and fluoro-derivatives. Nevertheless, there are interesting examples of protective coatings that involve inorganic materials, such as the barium hydroxide technique, oxalate coatings, and recently phosphate crystallization in the first layers of stone or mortar supports.[22–25] The protection of easel paintings is normally achieved with natural and synthetic varnishes. Natural varnishes usually include dammar and mastic,[26–28] whereas synthetic treatments are based on many different kinds of materials, mainly acrylic, vinyl, alkyd and aliphatic.[29]

Several topics related to the strategies and purposes of the conservation and restoration of wall paintings will be discussed in various parts of this book. But given that easel paintings will not be extensively treated in other chapters, it is worthwhile to include some words on their preservation. Restoration of easel paintings is highly complex and involves many complicated steps that often must be modified *in itinere*, depending on the specific situation. Indeed, another important issue deals with the fact that the procedures adopted for the ground, the paint layers and the varnishes can be greatly different. We choose here to omit any consideration of the support (wood or canvas), because the object of the present book is more dedicated to the degradation and restoration of pictorial surfaces, generally considered to be formed by the varnish layer, the paint layers and the first layer of the ground. Generally, restoration is carried out using different techniques for the ground, the paint layers and the varnish layer, because the degradation mechanisms for these three components are quite different, leading to varied deterioration morphologies. In any case, however, we will have to take into account the same steps, *i.e.* cleaning, consolidation and protection. For each stage the meanings, strategies and purposes can be summarized according to the rule of the minimum intervention.

In very unique situations the degradation involves simultaneously the ground, the paint layers and the varnishes. This is what occurred in the case illustrated by Burmester and co-workers,[31] who developed a methodology based on ion-exchange resins that solved a highly complex situation. However, in almost all issues the approach for the ground, the painting layers and the varnishes is absolutely different and therefore the various aspects dealing with these three parts will be discussed separately.

The principal degradation phenomena of the ground can be associated with the loss of adhesion between the support (canvas or wood) and the ground itself. These processes can initiate blisters or localized detachments between the support and the ground. During the conservation treatment these "voids" must be filled in, and adhesion between the support and the ground or between the ground and the paint layers must be restored: in one word consolidation has to be carried out. Molten wax–resin mixtures are used to fill in the "voids", whereas water-soluble adhesives are mainly employed to re-adhere detached regions. Great care must be taken when using wax and resins, because they can produce translucent effects and glossy portions: in these cases the exclusive use of water-soluble adhesives is recommended. Indeed, water-soluble adhesives (animal glue or synthetic resins) are used as emulsions of the hydrophobic active components in water, and the adhesive is in the form of micro-droplets dispersed in an aqueous medium. Before consolidating a panel painting it is often necessary to keep the painting in a climatic box at relative humidity of *ca.* 65%, in order to allow the support to expand, minimizing the stresses that initiated the detachments and/or the blisters.[32,33] If degradation of the ground has produced powdering, it may be difficult to restore the correct cohesion by the above-described methodology, owing to the impossibility of performing a fair impregnation of the ground with the selected adhesives. In this case, the only solution is the removal of the support and the ground from the paint layers and the creation of a new ground to be applied from behind, in a way similar to the detachment and re-positioning of wall paintings on a suitable support.

The restoration of paint layers is a very delicate step. Very often, in the past, unsuitable treatments have caused damage and irreversible modifications. In particular, great care must be taken in the cleaning of the paint layers, because over-cleaning with an excessive amount of solvent or very aggressive cleaning mixtures can result in the drastic loss of pigments and/or binding media. Generally, restoration of paint layers consists in three sequential steps: (i) cleaning, (ii) filling the 'voids', *i.e.* consolidating, and (iii) applying a protective coating. Cleaning of the paint layers is critical only in cases where varnish is not present; otherwise conservators and restorers do not clean so deep as to reach the paint layers. In unvarnished paintings the first pictorial surface consists of the paint layers and the cleaning procedure can be performed either by means of a solvent, or by dry methods using dirt-absorbing solid media. Both solvents and dirt-absorbing putties must be used with great care in order to avoid severe undesired side effects.[34] Conservators should use very pure solvents, and avoid the use of mixtures that could induce ageing effects and chemical reactions. After the cleaning, the paint layers have to be consolidated by filling the "voids".[35] This can be attained by using appropriate fillers such as mixtures of animal glue and chalk, or molten wax–resin mixtures. While the animal glue–chalk filler is applied as a paste at room temperature, the thermoplastic wax–resin mixtures must be applied with a pre-heated spatula. Sometimes, in the presence of coloured ground, the fillers are charged by fine powdery pigments in order to avoid colour discontinuities in the ground. In certain circumstances the cleaning has to be stronger, if later over-paintings are discovered. Indeed, X-radiography and infra red (IR)-reflectography are useful tools to identify the over-paintings on a pictorial surface.[32] These two techniques also allow control of the removal of the over-paintings, and allow one to check the successful, selective and progressive removal of the over-paintings without touching the original paint layers. The choice of a suitable solvent for cleaning, and of the correct application times, is crucial during the restoration of pictorial surfaces; it has been shown that the effects of solvents on the mechanical and surface properties of paint films can be very important.[36]

Finally, varnish layers are the first sites where degradation occurs, because they are located at the work of art–environment interface. Their restoration is thus very critical, owing to the risk of drastic alteration of these layers upon cleaning. In fact, varnish layers had for the artist a very important role in determining the aspect of the painting. Varnishes are subject to many natural ageing phenomena, as described in Chapter 2, so it is impossible for a restoration or conservation treatment to restore the varnish layers to their original aspect. Cleaning this pictorial surface irreversibly brings about the loss of material and leads to an almost matt surface. Nonetheless, it is always necessary to clean the varnish layers and to restore the loss of varnish. Either dry or wet methods are used. In the case of the wet method the most critical step is the choice of the solvents to be used (see also Chapter 5); in fact solvents, besides cleaning the varnish layers, can remove them and penetrate into the layers underneath, causing undesired effects of swelling and infiltration.[37] In order to avoid solvent penetration into the layers below the varnish, conservators have recently started to use gels that succeed in enhancing the retention of the solvent by the colloidal system (*vide infra*). One of the most debated questions concerns the removal of the aged varnish layers. The main problem is the selectivity of the solvent used to remove the varnish.[38] The best solvent should solubilize the varnish without altering the paint layers underneath. It is a very difficult task to predict whether the paint layers are going to be damaged, even when all the chemical components of the varnish and paint layers are well known. In fact, ageing can produce important changes in the solubility properties of varnishes and binders of the paint layers. It is practically impossible to classify the solvents as harmless or risky: a solvent can be completely harmless in the removal of varnish from a specific painting, while being highly dangerous on the paint layers of another painting.

In any case, restoration of the varnish layer consists of two separate steps: (i) cleaning, with partial or total removal of the varnish, and (ii) application of new varnish layers onto the restored easel painting. As described above, cleaning is performed by means of solvents and/or gels, while the application of new varnish layers is a very controversial subject. Some prefer to use synthetic resins rather than natural resins such as dammar and mastic. Synthetic resins exhibit less yellowing, but often they are not aesthetically satisfactory, while natural resins are more compatible with the original materials and are nowadays preferred by restorers.

In the presence of micro-cracks that produce multiple scattering effects, a particular procedure is adopted. Alcohol vapours succeed in restoring the original transparency because they react with the varnish at the micro-interfaces of the cracks, operating a micro-sealing between the two rims of a micro-crack. As a consequence, the scattering effect is minimized and the transparency of the varnish film is increased.[32]

3.3 Reversibility of the Treatment

For several years during the second half of the 20th century, reversibility of treatments was considered as a kind of golden rule that had to be strictly respected every time a new material was applied to a work of art. The main idea behind this rule was the respectable and important principle that, when a new material is cast on a work of art, despite the deepest investigations in the laboratory, there is no warranty regarding the behaviour and possible undesired effects in the short, medium and long term. Thus, reversibility should allow at every moment the possibility of removing any material that could be a potential damaging agent. Typical examples are polymers applied as consolidating or protective agents in solution in suitable organic solvents. These materials have long been considered to be reversible, based on the assumption that the same solvent could be used for their removal. Therefore they had great success in many conservation applications. At the beginning of the 1980s, some studies showed that the combined action of light, temperature, oxygen and relative humidity could lead to undesired effects such as the formation of an opaque layer, yellowing, formation of micro-fractures, and solubility loss. The last effect captured the attention of many scientists, who started to carry out systematic studies aimed at quantifying the loss of solubility. It was discovered that the same solvents that had been used to apply the consolidating and/or the protective polymeric resins often became almost ineffective, even after only a few years of exposition. Therefore, reversibility started to be critically discussed and in 1999 an important International Congress organized by the British Museum had the intriguing title "Reversibility: does it exist?".[30] Many scholars, scientists and conservators joined this conference and discussed the question vibrantly for three days, eventually providing a revolutionary answer. The conclusion was that, for many scientific reasons, reversibility does not exist! It is interesting to recall some of the most important contributions to this Conference, because they can be considered relevant even today.

One of the most topical questions about reversibility is its link with the second law of thermodynamics. It has been clearly shown[39,40] that reversibility, rather than a golden rule for conservation and restoration, is to be considered as a ghost,[39] or an illusion.[40] As a matter of fact, the second law of thermodynamics states that time has no symmetry, as seen in mechanics, because time is characterized, at least in real systems, by an arrow strictly directed towards the future. In this sense the authors state that reversibility is an illusion, since it is absolutely impossible completely to reverse any conservation or restoration treatment. Therefore, it is important to draw some fundamental conclusions about the reversibility of treatment. For reasons of clarity we adopt the concluding words by Andrew Oddy[41] in his contribution to that Conference.[30] "A generation ago, reversibility was taken to mean that any material added to an object should be easily removable. Now it includes all conservation processes applied to antiquities, and conservators must consider not only the shape of the object, but also potential changes to its chemical composition and to its micro-structure. The result is to realize that very few processes are, in fact, truly reversible, so the decision on whether and how to conserve becomes not only a practical one, but also an ethical one. As Smith pointed out, "the time has come for professional judgment to be substituted for the principle of reversibility,"[42] and Appelbaum stated that, 'the notion of re-treatability is one that is often more helpful ... than the idea of reversibility itself."[43] We can add to the words of Applebaum that re-treatability implies the concept of physico-chemical compatibility and durability, which will be treated in the next section.

3.4 Durability and Physico-chemical Stability

The topic of reversibility, discussed in the previous section, cannot be separated from the equally important and strategic concepts of durability and physico-chemical stability. The rule of reversibility is generated by the idea that some materials undergo deterioration even in quite short timeframes, and can thus be removed because they are, indeed, reversible. In a certain sense reversibility was born from the awareness that some of the new materials synthesized during the golden age of polymers could have extraordinary performance as freshly applied, but be very risky in the short and medium term.

While polymers were having great success in many fields of conservation and restoration, the thought arose that physico-chemical compatibility and stability leading to durability could be an analogous golden rule to that of reversibility. In conservation of wall paintings this dualism, reversibility *vs.* compatibility, found a clear representation during two memorable workshops: on the Beato Angelico wall paintings cycle in the San Marco Convent in Florence,[44–46] and those by Piero della Francesca ('*The Legend of the True Cross*') in the San Francesco church in Arezzo.[47] While restorer Leonetto Tintori in Arezzo was fascinated by the amazing progress of the chemical industry that allowed conservators to test many different polymers for the purpose of consolidation, adhesion and protection, Dino Dini admitted to several perplexities, being afraid of possible undesired effects in the future due to the poor physico-chemical compatibility between the organic – synthetic – materials and the mostly inorganic nature of wall paintings (apart from the regions where the fresco technique was not used).

After 30 years of natural ageing we can conclude that Dini was right and that compatibility in many circumstances is to be preferred to reversibility. Ferroni, during those years, developed a method based on successive applications of two inorganic salts: ammonium carbonate, $(NH_4)_2CO_3$, and barium hydroxide, $Ba(OH)_2$ (see also Section 13.4). A new question entered the world of conservation and restoration: can physico-chemical compatibility and durability be considered as a new golden rule to compare constantly with reversibility? This question implied a conceptual dualism (organic–reversible *vs.* inorganic–irreversible–compatible) and a balance between the positive and negative aspects of the two approaches.

Initially, the inorganic approach remained confined in Tuscany and more specifically in Florence, so that the method changed its appellation from the Ferroni to the Florentine technique, or the barium method. Slowly, the method started to be applied in many workshops in Italy, elsewhere in Europe, and across the world. The slow acceptance of the Ferroni method was paralleled by the gradual perception, during the following years, of the problems caused by some polymers that had been professed to be reversible.

According to this idea of durability and physico-chemical compatibility, other materials and techniques have been developed, as previously described in Section 3.2.[22–25]

3.5 Modern Criteria in Conservation and Restoration

Conservation and restoration have been for a long time empirical rather than scientific practices. This does not imply that conservation and restoration in the past did not achieve important and very positive results. Simply, conservators and restorers had to face conservation issues by relying only on their skill, intuition and creativity, using a "trial and error" approach. Therefore, they did not take advantage of the incredible progress that science and technology made in the second half of the 20th century.

The watershed between the merely empirical and the scientific approach to conservation and restoration is marked by the events following the Florence flood in 1966, when scientists were compelled to offer their competence in order to solve a large number of very delicate and complex problems concerning highly damaged works of art.

Since then, research groups have been born worldwide that are specifically aimed at providing insights and contributions to all the various fields of cultural heritage conservation. We can state that conservation developed similarly to medicine, by focusing on two fundamental steps: diagnosis by means of all the available technology, and then the set up of the correct therapy. Concerning the latter, materials science surely plays a fundamental role, as this book tries to demonstrate by dealing with nanoscience applied to the conservation of works of art.

Thus, the time is ripe to reconsider the role of scientists and technologists in cultural heritage conservation and their relationship with conservators and restorers. It is interesting to reconsider the needs of the conservation community, and in this sense a redefinition of the profession of conservation scientist is a current question in the debate about modern criteria in conservation and restoration. Undoubtedly, an important feature in modern conservation and restoration is the continuous exchange between conservators–restorers and scientists–technologists.

In particular, there is the perception that scientists may need several years of practice to achieve skills in specific conservation applications, following their classical training. Moreover, traditional scientists might have difficulty in illustrating their results to art historians, architects and archaeologists.

In this frame, at the beginning of the 21st century some Italian university *curricula* were set up to educate a new profile of scholar, *i.e.* a technologist or scientist with competences in all the various branches of natural sciences, strictly applied to cultural heritage conservation, and with a satisfying knowledge in humanities. This professional should be able to overcome the limits of the traditional scientists–conservators. Two types of course, at first and second university level, were established in 2001 in several universities in Italy, on the occasion of the Reformation of the University Courses according to the Bologna Protocol. These courses include all scientific disciplines, considering both the basic subjects and the applications in cultural heritage conservation. The topics are multidisciplinary and include credits of art, architecture and law. The technologist coming from this educational path is neither a chemist, nor a physicist, geologist, mineralogist, or biologist: skills in all the natural sciences and also a partial knowledge of the human sciences, all aimed at meeting the needs of the conservation world. The technologist should succeed in carrying out diagnostic analysis on all the materials in a cultural heritage object, and they should also be able to communicate successfully with other professionals, including non-scientific personnel working on a conservation or restoration project.

Of course, after only three years of university training, the graduating professionals cannot face complex problems dealing with materials science in general. In order to educate first-level graduates at a higher level, a two-year course, called Science for Cultural Heritage Conservation, was developed. Indeed, some universities preferred to have a true specialization in one single topic, while others chose to continue with a multidisciplinary education. Florence, for instance, followed the latter approach. The final thesis period covers almost a semester in the second-level course, and consists of a true scientific research project, with publication of the results in scientific peer reviewed journals.

As an example of this new and modern approach to science and technology applied to conservation and restoration, we mention below the Florentine experience, considering how the preparation of the students by the five-year course allowed them to contribute both to applicative and research work in conservation science. As a final remark, stressing the comparison with medicine, it is important to highlight here the dual role of the work provided by conservation scientists: diagnostics and therapy.

Scientific instrumental diagnostic work is necessary before any therapeutic intervention on a cultural heritage object. Of course, diagnostics must be correctly addressed to identify the appropriate conservation procedure. In this sense, the scientific and technological community should reflect and even exert self-criticism, because often scientists perform analyses to achieve published results, lacking true reference to the pressing and practical questions of the professionals involved in the conservation workshops. It is thus fundamental to stress that scientific diagnostics must be strictly considered as the first step for the conservation of cultural heritage.

On the other hand, the development of new materials and techniques for conservation and restoration is of paramount importance, exploiting all the potentialities of the tools offered by nanosciences, and in particular colloid and material science.

Further Suggested Reading

On restoration and conservation: meanings, purposes, and strategies:

P. Mora, L. Mora and P. Philippot, *The Conservation of Wall Paintings*, Butterworths, London, 1984.

P. Philippot, *Readings*, 1996, **21**, 216.

M. Petzet, *Restauro*, 1995, **131–132**, 80.

On reversibility of the treatment:

P. W. Atkins, *The Second Law: Energy, Chaos, and Form*, Scientific American Library (W.H. Freeman & Co.), New York, revised edn, 1994.
C. Brandi, *Teoria del restauro*, Einaudi, Turin, 6th edn, 1977.

On durability and physico-chemical stability:

The Getty Conservation Institute, *Permanence and Application of Conservation Materials* (1983–1996), http://www.getty.edu/conservation/our_projects/science/permanence/index.html.
K. L. Gauri and J. K. Bandyopadhyay, *Carbonate Stone: Chemical Behaviour, Durability and Conservation*, John Wiley & Sons, New York, 1999.

Review and Questions

This chapter is an introduction to the various aspects of cleaning, consolidation and protection that are the main topics of the book, which is aimed to show how nanoscience and nanotechnology might be a valid support for developing new materials for conservation and intervention procedures.

- The difference between conservation and restoration has been stated, and the main purposes and strategies have been discussed.
- Two fundamental rules of scientific conservation and restoration have been introduced: reversibility and physico-chemical compatibility.
- The importance of the physico-chemical compatibility of new conservation materials with the original materials constituting the work of art has been stressed. The case of nanolime, widely discussed in several parts of this book (*e.g.* Chapter 13), is a typical example.
- The chapter concludes by stating some general criteria for modern conservation and restoration, individuating materials and colloids science as two research fields with strong potential for the development of new art-compatible materials.

Questions

1) What is the main difference between the terms "restoration" and "conservation"?
2) How can science and technology help in setting up the most appropriate conservation treatment?
3) What does it mean that the golden rule of reversibility has recently become a utopian objective rather than a real and absolutely reachable scope?
4) What is the exact meaning of "physico-chemical compatibility" in cultural heritage conservation?
5) Is it sufficient to state that modern criteria in conservation and restoration simply mean that scientific diagnostics is mandatory before each intervention?

Answers

1) The term restoration is strictly associated with the operations that are carried out to solve all the problems dealing with the correct and preventive conservation of a work of art. In other words restoration is confined to the interaction between the human and the work of art, mediated by the matter constituting the masterpiece. On the contrary, conservation refers to all the measures one adopts for the best safeguard of the work, including climate checks, environmental control, the choice of the best protocols for the intervention, and the definition of the best practices. Conservation includes restoration and this is the reason why in many countries there is no such difference between these two terms, conservation being the only one used, with the implicit assumption that restoration is a part of the conservation practice.
2) Science and technology are fundamental for setting up the most appropriate conservation procedure in every workshop. Indeed, science and technology help in determining the status of conservation and the causes of deterioration, the elucidation of materials properties, and the determination of the most suitable conservation environment. Thanks to science and technology, conservation and restoration are no longer empirically approached and become truly scientific.
3) After the Conference at the British Museum in 1998, entitled "Reversibility: does it exist?", the scientific conservation and restoration community has agreed on the idea that, although reversibility is an important rule and a valuable principle, we must take into account that it is almost impossible for a procedure to be considered as completely reversible. For instance, cleaning is irreversible by definition! Therefore, reversibility cannot be considered as an absolute rule, but rather a fundamental guideline to be constantly related to the other very meaningful behaviour rule, named 'compatibility'.
4) Very often conservation and restoration involve the use of materials to be applied on the work of art. Thus, the chemical nature of the substrate – the work of art – should be the driving factor in selecting and formulating new materials. It is well known that engineered materials in medical sciences have to respond to the rule of biocompatibility, and green chemistry imposes eco-compatible materials for sustainability. Similarly, art-compatible materials need to match the properties of the materials constituting the works of art.
5) Scientific diagnostics is fundamental, being a necessary condition for a modern approach in conservation and restoration, but it is not sufficient. Indeed, diagnostics helps in understanding the causes of deterioration, the ageing mechanisms, the nature and properties of materials, and so on. However, science and technology should also help by creating new materials to overcome conservation issues. Materials science, and in particular nanoscience, can make a fundamental contribution to this task. In this context, we hope that this book will help in opening new horizons.

References

1. P. Ackroyd and D. Bomford, in *Reversibility: does it exist?*, British Museum Occasional Papers Nr. 135, ed. A. Oddy and S. Carroll, The British Museum, London, 1999, p. 53.
2. D. Erhardt and J.-S. Tsang, in *Cleaning, Retouching and Coatings*, ed. J. S. Mills and P. Smith, IIC, London, 1990, p. 93.
3. R. White and A. Roy, *Stud. Conserv.*, 1998, **43**, 159.
4. R. Wolbers, *Cleaning Paintings. Aqueous Methods*, Archetype Publications, London, 2000.
5. A. Burnstock and R. White, in *Cleaning, Retouching and Coatings*, ed. J. S. Mills and P. Smith, IIC, London, 1990, p. 150.
6. P. Dizabo and C. Pepe, *Sci. Technol. Conserv. Restaur. Oeuvres d'Art Patrim*, 1988, **1**, 57.
7. L. Dei, P. Baglioni, G. Sarti and E. Ferroni, *Stud. Conserv.*, 1996, **41**, 9.
8. P. Ball, *Smart Stuff*, A talk delivered at the Department of Materials, University of Oxford http://www.philipball.co.uk/index.php?option=com_content&view=article&id=80:smart-materials&catid=18:materials&Itemid=20, website consulted on May 30th, 2012.
9. E. Garrod, *Stone Consolidation – Halts Decay and Prolongs Life*, http://www.buildingconservation.com/articles/stoneconsol/stoneconsol.htm, website consulted on June 5th, 2012.
10. M. J. Mosquera, D. M. de los Santos, A. Montes and L. Valdez-Castro, *Langmuir*, 2008, **24**, 2772.
11. C. Miliani, M. L. Velo-Simpson and G. W. Scherer, *J. Cultural Heritage*, 2007, **8**, 1.
12. S. Vicini, A. Mariani, E. Princi, S. Bidali, S. Pincin, S. Fiori, E. Stefano, Pedemonte and A. Brunetti, *Polym. Adv. Technol.*, 2005, **16**, 293.
13. A. P. Ferreira-Pinto and J. Delgado Rodrigues, *J. Cultural Heritage*, 2008, **9**, 38.
14. P. Maravelaki-Kalaitzaki, N. Kallithrakas-Kontos, D. Korakaki, Z. Agioutantis and S. Maurigiannakis, *Progr. Organ. Coat.*, 2006, **57**, 140.
15. A. P. Ferreira-Pinto and J. Delgado Rodrigues, *J. Cultural Heritage*, 2012, **13**, 154.
16. N. Proietti, D. Capitani, S. Cozzolino, M. Valentini, E. Pedemonte, E. Princi, S. Vicini and A. L. Segre, *J. Phys. Chem. B*, 2006, **110**, 23719.
17. L. Toniolo, F. Casadio and F. Cariati, *Ann. Chim.*, 2001, **91**, 823.
18. S. Fiori, A. Mariani, L. Ricco and S. Russo, *Macromolecules*, 2003, **36**, 2674.
19. C. Tonelli, T. Trombetta, M. Scicchitano and G. Castiglioni, *J. Appl. Polym. Sci.*, 1995, **57**, 1031.
20. C. Botteghi, U. Matteoli, S. Paganelli, R. Arbizzani, F. Ciardelli, M. Aglietto, E. Taburoni and V. Fassina, *Sci. Technol. Cultur. Herit.*, 1992, **1**, 111.
21. A. L. Logothetis, *Progr. Polym. Sci.*, 1898, **14**, 251.
22. E. Doehne and C. B. Price, *Stone Conservation – An Overview of Current Research*, the Getty Conservation Institute, Los Angeles, 2nd edn, 2010.

23. B. Doherty, M. Pamplona, R. Selvaggi, C. Miliani, M. Matteini, A. Sgamellotti and B. Brunetti, *Appl. Surf.*, 2007, **10**, 4477.
24. L. Ramazzi, A. Andreotti, I. Bonaduce, M. Colombini, C. Colombo and L. Toniolo, *Talanta*, 2004, **967–977**, 63.
25. M. Matteini, S. Rescic, F. Fratini and G. Botticelli, *Internat. J. Architect. Herit.*, 2011, **5**, 717.
26. A. Nevin, D. Comelli, I. Osticioli, L. Toniolo, G. Valentini and R. Cubeddu, *Anal. Bioanal. Chem.*, 2009, **395**, 2139.
27. C. Theodorakopoulos, V. Zafiropulos, J. J. Boon and S. C. Boyatzis, *Appl. Spectrosc.*, 2007, **61**, 1045.
28. R. L. Feller, *Stud. Conserv.*, 1958, **3**, 172.
29. C. Blackman, *Choosing Varnishes. In between the concept and the reality falls the practicing conservator...*, http://www.e-conservationline.com/content/view/568/145/, website consulted on June 5th, 2012.
30. Various authors, *Reversibility: does it exist?*, British Museum Occasional Papers Nr. 135, ed. A. Oddy and S. Carroll, The British Museum, London, 1999.
31. A. Burmester, J. Koller and H. Kawinski, in *Cleaning, Retouching and Coating. Preprints of the IIC Brussels Congress*, IIC Publishers, London, 1990, 177.
32. H. Kühn, *Conservation and Restoration of Works of Art and Antiquities*, Translation by A. Trone, Butterworths, London, vol. 1, 1986.
33. F. du Pont Cornelius, *Stud. Conserv.*, 1972, **17**, 1.
34. N. Stolow, in *Conservation and Restoration of Pictorial Art*, ed. N. Bromelle and P. Smith, London, 1976, p. 153.
35. Various authors, in *Proceedings of the 6th Triennal Meeting of the ICOM Committee for Conservation*, Ottawa, 1981.
36. G. A. Hedley, M. Odlyha, A. Burnstock, J. Tillinghast and C. Husband, in *Cleaning, Retouching and Coating. Preprints of the IIC Brussels Congress*, IIC, London, 1990, p. 98.
37. S. Michalski, in *Cleaning, Retouching and Coating. Preprints of the IIC Brussels Congress*, IIC, London, 1990, p. 85.
38. R. Feller, N. Stolow and E. H. Jones, *On Picture Varnishes and their Solvents*, National Gallery of Art and AIC Publishers, Washington, 3rd revised edn, 1985.
39. S. Palazzi, in *Reversibility: does it exist?*, British Museum Occasional Papers Nr. 135, ed. A. Oddy and S. Carroll, The British Museum, London, 1999, p. 175.
40. N. G. Seeley, in *Reversibility: does it exist?*, British Museum Occasional Papers Nr. 135, ed. A. Oddy and S. Carroll, The British Museum, London, 1999, p. 161.
41. A. Oddy, in *Reversibility: does it exist?*, British Museum Occasional Papers Nr. 135, ed. A. Oddy and S. Carroll, The British Museum, London, 1999, p. 1.
42. R. D. Smith, *Restaurator*, 1988, **9**, 199.
43. B. Appelbaum, *J. Amer. Inst. Conserv.*, 1987, **26**, 65.

44. D. Dini, *Gli affreschi del Beato Angelico nel Convento di San Marco a Firenze. Rilettura di un capolavoro attraverso un memorabile restauro*, U. Allemandi & Co. Publishers, Turin, 1996.
45. E. Ferroni, V. Malaguzzi-Valerj and G. Rovida, in *Actes de l'8^{em} Colloque sur l'analyse de la matière*, Florence, 1969, Philips S.p.A., Milan, 1970, p. 7.
46. E. Ferroni, in *Proc. Internat. Workshop 'Ecological Physical Chemistry'*, ed. C. Rossi and E. Tiezzi, Elsevier, Amsterdam, 1991, p. 345.
47. E. Ferroni, in *Piero della Francesca ad Arezzo, Atti del Convegno Internazionale di Studi*, Arezzo (Italy) 7–10 marzo 1990, ed. G.A. Centauro and M. Moriondo Lorenzini, Marsilio, 1990, p. 219.

CHAPTER 4

Cleaning I: Solvents and Solutions

D. BERTI* AND P. LO NOSTRO

University of Florence, Italy
*Email: berti@csgi.unifi.it; pln@csgi.unifi.it

4.1 Solution Thermodynamics

Solubility is one of the most important properties of materials. It has wide applications in different scientific and industrial fields. The knowledge of the solubility of a material is essential for different reasons: to purify a product through recrystallization, to separate two or more compounds by exploiting their different solubilities in a common solvent, to control the concentration of a solute in a given medium, or to apply the chemical to a specific system (*e.g.* the deposition of a polymer onto a surface), and more. In cultural heritage conservation science, solubility finds fundamental applications in the cleaning of artistic substrates with solvents (see Chapter 5), and in the preparation of cleaning tools, such as microemulsions and micellar solutions (which will be thoroughly discussed in Chapters 6–9).

Solubility is characteristic of a given substance in a specific solvent at a given temperature, for example lactose is very soluble in water, soluble in ethanol, and insoluble in diethyl ether and chloroform.

In this chapter we will review the most important concepts, definitions and theories that describe the process of solubilization, the factors that affect the solubility of a solute in a solvent, and the interactions that are involved in this process. These concepts will serve as an essential theoretical basis to support the contents and discussion in the following chapters.

RSC Nanoscience & Nanotechnology No. 28
Nanoscience for the Conservation of Works of Art
Edited by Piero Baglioni and David Chelazzi

Published by the Royal Society of Chemistry, www.rsc.org

4.1.1 Definition of Solvent

A *solution* is a molecularly homogeneous mixture that always comprises a *solvent* and a *solute*. The solvent is usually the major component, a liquid, solid or gaseous material that dissolves the solid, liquid or gaseous solute. But there are cases (such as syrups) in which the major component is the dissolved species (sugar). In a solution, as opposed to a simple mixture, the solvent and the solute molecules interact with each other, with a more or less compact structure of the solvent molecules around the solute entities.

Gases always mix, regardless of their chemical composition and concentration. However, when the two compounds are not soluble or are only partially soluble, a precipitate is formed (*e.g.* from calcium carbonate in water), or a liquid–liquid interface appears (*e.g.* in oil–water systems). In a real solution no residue sets at the bottom of the container.

As we will see in the following sections, the intermolecular interactions between the solute and the solvent constitute one of two major factors that determine whether a solution is going to be formed or not. What we can anticipate is that, while in the gas state, the two components in their pure state experience quite low intermolecular interactions, and upon mixing the set of intermolecular interactions does not significantly change. On the other hand, when a solid is dissolved in a liquid, or when two liquids are mixed, the final mixed state is usually characterized by a large change in the intermolecular interactions, as depicted in Figure 4.1. For example, when NaCl is dissolved in

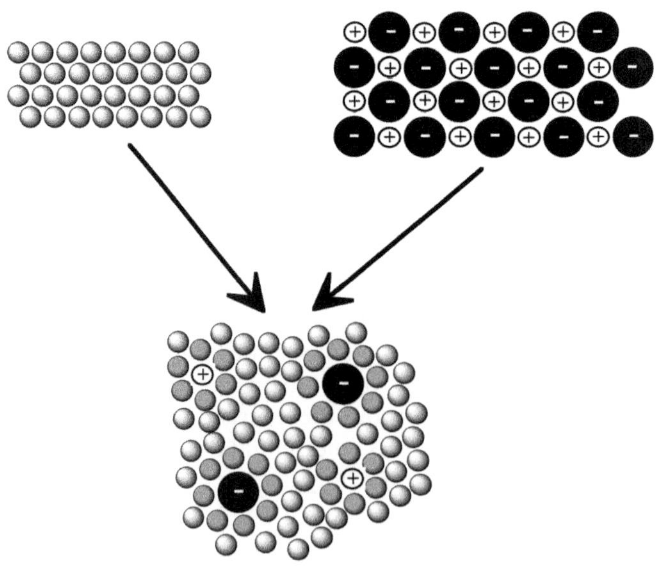

Figure 4.1 Mixing of a solid solute in a liquid solvent. The solvent molecules disrupt the crystal lattice, and the ions get separated. Some solvent molecules (dark circles in the bottom image) solvate the cation and the anions.

water, the original compact crystalline structure of the salt vanishes in the presence of H_2O, and the ions are separated and solvated by the solvent molecules. At the same time, some of the water molecules will change their normal bulk state and distribution and will be oriented by the electric field of the ions in the solvation shells, with a remarkable change in their dielectric constant and polarization. A similar situation occurs when two liquids (for example ethanol and acetone) are mixed.

In the solubilization process both the intermolecular interactions and the distribution of molecules are modified, and therefore a change in the enthalpy (heat at constant pressure, ΔH) and in the entropy (ΔS) will take place.

Generally, organic solvents are used in dry cleaning (*e.g.* tetrachloroethylene), as paint thinners (*e.g.* toluene, turpentine), as polish removers and glue solvents (acetone, methyl acetate, ethyl acetate), in spot removers (*e.g.* hexane, petrol ether), in detergents, in perfumes and other cosmetic products, and particularly in chemical synthesis. The use of inorganic solvents (other than water) is typically limited to research chemistry and some technological processes. The more recent introduction of ionic liquids has expanded the concepts and uses of this class of materials.

4.1.2 Solubility: Definition and Units

The solubility, *i.e.* the amount of solute dispersed in the solvent matrix, depends on the nature of the two mixing chemicals, on pressure and – especially – on temperature. In particular, the solubility of a solid (*e.g.* NaCl or urea) is defined as the concentration of the compound in a solution that is in equilibrium with the solid phase at the specified temperature and pressure conditions. Some liquids are totally miscible with water (for example acetone or ethanol), others are partially miscible. In the latter case – depending on temperature and composition – the liquid mixture will phase separate and form two coexisting phases in equilibrium, one richer in water and the other richer in the solute. In these cases the solubility is the concentration of the specified compound in the water-rich phase at equilibrium. In the case of gases in water, the solubility is the concentration in the water phase when the partial pressure of the compound above the solution is 1 atm.

The solubility of a solute can be expressed in different ways, depending on the kind of solute (liquid, solid, gas) and on some conventional rules that are related to the final use of the solution:

$$S = 100\,\frac{m_s}{m_s + m_w} \tag{4.1}$$

where m_s are the grams of the solute (for example urea) dissolved in m_w grams of solvent (for example water).

Sometimes the solubility is reported as grams of solute (m_s) per litre of final solution, which is particularly useful for certain cultural heritage applications and technical procedures, *e.g.* in the case of calcium sulfate, calcium nitrate and barium sulfate.

When the solute is a liquid, a convenient way to express the solution concentration is based on the volume. However, in this case we have to pay attention to the fact that usually volumes are not additive, therefore it is better to express the concentration as volume of solute per 1 L of final solution.

Other units require the molecular mass of the solute (M_s) and of the solvent (M_w):

- molality, *m*: number of moles of solute dissolved in 1 kg of solvent
- molarity, *M*: number of moles of solute dissolved in 1 L of solution
- mole fraction, *x*: moles of solute divided by the total number of moles in the solution.

For solutes that are poorly soluble it is convenient to use the concept of the solubility product, K_{sp}, which can be determined through direct measurement of the concentration in solution, or it can be calculated if the standard free energy of formation for the different species is known. For example, calcium carbonate is only sparingly soluble in pure water. Incidentally, this is the reason why works of art made up of marble – whose main component is calcite – are not dissolved by water but by acidic rain. Its aqueous saturated solution will present some $CaCO_3$ crystals at the bottom of the vessel, in dynamic equilibrium with hydrated free ions, $Ca^{2+}{}_{(aq)}$ and $CO_3{}^{2-}{}_{(aq)}$ according to Scheme 4.1.

For a general salt M_mX_n the solubility product is defined as the equilibrium constant for the dissolution process:

$$K_{sp} = [M^{p+}]^m \cdot [M^{q-}]^n \tag{4.2}$$

For example, for aluminum sulfate, $Al_2(SO_4)_3$, we have: $K_{sp} = [Al^{3+}]^2 \cdot [SO_4{}^{2-}]^3$.

Of course if at least one of the dissociated ions is pH-sensitive (*e.g.* $CO_3{}^{2-}$, S^{2-}) then the solubility of the salt will be greatly affected by pH changes. For example, insoluble carbonates and sulfides are much more soluble in acidic solutions because of the hydrolysis reactions they originate. This explains why acidic rain corrodes marble statues and other masterpieces (see also Section 2.3.1). The K_{sp} depends on temperature and on the composition of the solvent. In fact the addition of soluble inert electrolytes to water may increase the solubility of a poorly soluble salt. The opposite effect is known as the "common ion effect", and leads to a lower solubility. For example, AgCl is already poorly soluble in water ($K_{sp} = 1.8 \cdot 10^{-10}$). The addition of NaCl to a solution of silver chloride will enhance the precipitation of AgCl, since according to equation 4.2, $[Ag^+] = K_{sp}/[Cl^-]$ and any increment in $[Cl^-]$ will reduce $[Ag^+]$. Table 4.1 lists the K_{sp} at 25 °C for some inorganic salts.

$$CaCO_{3(c)} \leftrightarrows Ca^{2+}{}_{(aq)} + CO_3{}^{2-}{}_{(aq)}$$

Scheme 4.1

Table 4.1 K_{sp} at 25 °C for some inorganic salts.

$BaCO_3$	2.6×10^{-9}	$Ca(OH)_2$	5×10^{-6}
$CaCO_3$	3.4×10^{-9}	$Mg(OH)_2$	5.6×10^{-12}
$SrCO_3$	5.6×10^{-10}	$Fe(OH)_3$	3×10^{-39}
$MgCO_3$	6.8×10^{-6}	$Zn(OH)_2$	3×10^{-17}
$CaSO_4$	4.9×10^{-5}	PbS	3×10^{-7}

The list in Table 4.1 includes some artefact materials ($CaCO_3$ in marble), artefact degradation compounds ($CaSO_4$ – a typical compound that produces efflorescence on wall paintings, especially in the dihydrated form $CaSO_4 \cdot 2H_2O$ or gypsum, and PbS, a black salt resulting from the degradation of white lead pigment due to H_2S), and restoration materials [$Ca(OH)_2$ and $Mg(OH)_2$, used for the consolidation of wall paintings and for the pH adjustment of artistic substrates such as paper or wood]. If salt solutions wet a porous surface (such as in wall paintings), their solubility is of crucial importance, because soluble salts can migrate and can produce more serious damage than insoluble salts (see Section 2.2).

In order to calculate the solubility of $CaCO_3$ in $\mathrm{mol\,L^{-1}}$ from its K_{sp} value, we simply observe that at saturation the concentration of Ca^{2+} and CO_3^{2-} ions will be the same, therefore if S represents the saturation concentration in molar units, we will have: $K_{sp} = 3.4 \times 10^{-9} = S^2$, from which $S = (3.4 \times 10^{-9})^{1/2} = 5.8 \cdot 10^{-5}\,\mathrm{mol\,L^{-1}}$.

Useful solubility charts can be found in the literature,[1] in several textbooks and online.[2]

4.1.3 Physico-chemical Properties and Classification of Solvents

The solvent capacity of a given liquid strictly depends on its nature, and its physico-chemical properties. One of the most important features is the polarity, which is due to an asymmetric distribution of the charge. We recall here that the electronegativity χ reflects the capacity of a given atom in a molecule to attract the bond electrons. The most electronegative elements are placed in the uppermost right corner of the Periodic Table, for example fluoride, oxygen, nitrogen and chloride. Water is a very polar molecule owing to the fact that the electronegativity of the oxygen atom is much larger (about 3.5 on the Pauling scale) than that of the hydrogen (2.1), and therefore a fraction of the electronic cloud is displaced on the oxygen atom, resulting in a separation of partial positive and partial negative charges within the same molecule, and therefore in a net dipole moment (μ), as illustrated in Figure 4.2.

For a simple biatomic molecule A–B, where A and B possess different electronegativity, the dipole moment is a vector defined as $\mu = (\delta q)d$ where δq is the charge displacement between the two atoms, and d the internuclear distance (the vector magnitude μ is measured in Debye units) (see Figure 4.2). The dielectric constant (ε_r) of the solvent provides a rough measure of its polarity. A large ε_r indicates that the molecules will be strongly influenced by an electric

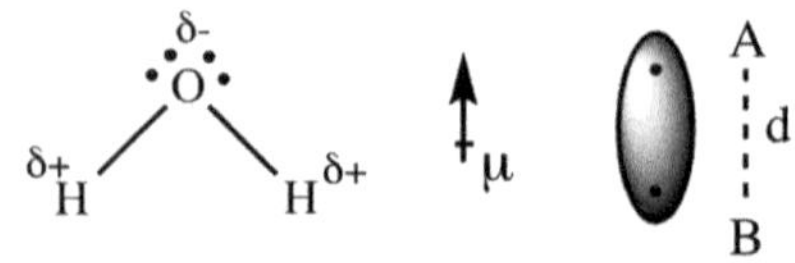

Figure 4.2 Structure of a water molecule. Owing to the partial charge separation, the molecule acquires a permanent dipole moment μ, indicated by the arrow.

field and will be oriented along its direction. For example, water has a permanent dipole moment of 1.85 D and a relative dielectric constant of about 80 at 20 °C. In general, a solvent with $\varepsilon_r < 15$ is considered to be non-polar, for example hexane–white spirit.

The polarity, dipole moment, polarizability (see Section 4.1.4) and hydrogen bonding (see Section 4.1.4) of a solvent all determine what type of compounds a solvent will be able to dissolve, and its miscibility with other solvents. As a rule of thumb, polar solvents dissolve polar compounds best and non-polar solvents dissolve non-polar compounds best: "like dissolves like". Strongly polar compounds such as sugars (*e.g.* sucrose) or ionic compounds, such as inorganic salts (*e.g.* the salts in Table 4.1), dissolve only in very polar solvents such as water, while strongly non-polar compounds such as oils or waxes dissolve only in very non-polar organic solvents such as hexane. Similarly, water and hexane (or vinegar and vegetable oil) are not miscible with each other and will quickly separate into two layers, even after being shaken vigorously.

Solvents with a dielectric constant greater than 15 can be further divided into protic and aprotic. A protic solvent contains hydrogen atoms bound to either oxygen or nitrogen atoms, and can release H^+ ions upon dissociation. Conversely, aprotic solvents cannot donate H^+ ions. Water and alcohols are protic solvents. Protic solvents solvate anions (negatively charged solutes) strongly *via* hydrogen bonding (see Section 4.1.4). Aprotic solvents with large dipole moments, such as acetone or dichloromethane, solvate positively charged species *via* their negative dipole.

The density of any material is defined as the ratio between its mass and its volume: $\rho = m/V$ and is measured in kg m^{-3} or in g mL^{-1}. Generally, organic solvents possess a lower density than water, meaning that if they do not mix with water they will form an upper layer on top of the aqueous phase. Most of the halogenated solvents ($CHCl_3$, CH_2Cl_2, CCl_4, *etc.*) are heavier than water. In some cases the specific gravity is given instead of the density. Specific gravity (SG) is defined as the density of the solvent (ρ_{solv}) divided by the density of water (ρ_{water}) at the same temperature. As such, specific gravity is a unitless value. It readily communicates whether a water-insoluble solvent will float (SG < 1) or sink (SG > 1) when mixed with water.

Another important property of solvents is their boiling point (bp), which is defined as the temperature at which the vapour pressure of the liquid equals the external pressure. Normal bp are measured at 1 atm (1.013 bar), and standard bp are defined at the pressure of 1 bar (0.987 atm). The value of the bp also determines the evaporation rate of a solvent: small amounts of low-boiling

Table 4.2 Parameters for ethylene carbonate at 40 °C, the others at 25 °C. (Reprinted with permission from Ref. 3. Copyright 2006 American Chemical Society.)

Solvent	ε_r	ρ
Ethylene carbonate	90.5	0.626
Propylene carbonate	65.5	0.751
Butylene carbonate	56.1	0.659
Glycerin carbonate	109.7	0.779

solvents such as diethyl ether, dichloromethane or acetone will evaporate in seconds at room temperature, while high-boiling solvents such as water or dimethyl sulfoxide need higher temperatures, an air flow or the application of vacuum for fast evaporation. In general a solvent is defined as a low-boiling liquid if its bp is lower than 100 °C, as a medium-boiling liquid if its bp is between 100° and 150 °C, and as a high boiling liquid when its bp is greater than 150 °C, at 1 atm. In conservation applications such as cleaning, the right choice of the solvent must take into account its bp: volatile solvents (low bp) in fact limit the penetration of dissolved materials through porous substrates.

Table 4.2 lists the values of the dipole moment μ and of the dielectric constant ε_r for some cyclic carbonates, including some solvents commonly used in conservation applications, which will be mentioned also in other chapters.

Finally we recall that a solvent often needs to be cleared of impurities or water before use. The specific procedure that one needs to adopt depends on the nature of the solvent and on its final use.[4]

4.1.4 Solute–Solvent Interactions

In order to understand the factors that drive the dissolution of a solute in a solvent we need to refer to the intermolecular forces that involve the molecules in the original pure state, that is the solute and the solvent before they are mixed, and those that occur in the final solution state. Besides the solvent, a solution can also comprise an electrolyte (and therefore a source of cations and anions), a neutral small or large molecule (for example sucrose or cellulose), or another more or less polar solvent. The set of interactions that take place in the solution will depend on the composition and properties of the intervening species. In general, apolar solutes dissolve in apolar solvents, and ionic (salts) or polar molecules dissolve in polar solvents ("like dissolves like").[5]

Depending on the nature of the species present, there are different interactions that involve a solute and a solvent, or two solvents in a liquid mixture. The main factors that control these interactions are the charge, the dipole moment, the polarizability and the ionization potential. In this section we will give a brief outline of the main factors that affect solute–solvent interactions (in terms of interaction potentials) and that explain some phenomena that will be fully discussed in other parts of this book.

4.1.4.1 Interactions between Ions and Dipoles

According to the Coulomb law, two charges z_1e and z_2e (where z_1 and z_2 indicate the valency of the ion, and "*e*" is the unit charge) interact with a potential energy given by:

$$V(r) = -\frac{|z_1 z_2| e^2}{4\pi\varepsilon_0\varepsilon_r r} \quad (4.3)$$

where ε_0, ε_r and r are the vacuum permittivity, the dielectric constant of the solvent, and the distance between the two interacting charges. If the solvent has a large dielectric constant (about 80 for water at room temperature), it will effectively dissolve electrolytes (*e.g.* sodium chloride, NaCl). In fact when the salt is in contact with water, the Coulomb interactions between the cations and the anions that hold the crystalline lattice are largely reduced and the solid structure collapses, releasing free ions into the solution. These are then solvated by water molecules and produce hydrated ions. At a more sophisticated level we have to consider that the presence of the other ions in solution screens the ion–ion interaction by a factor that depends on the concentration of the electrolyte and on the temperature.

Another way for two ions to interact is through the polarizability. This parameter (α) reflects the softness of the electron cloud that surrounds the nucleus. Small and/or polyvalent ions such as fluoride (F^-) or calcium (Ca^{2+}) possess a relatively low polarizability, while large univalent ions such as iodide (I^-) or cesium (Cs^+) have quite large values of α. In very dilute salt solutions, Coulomb interactions dominate and determine the behaviour of the solution; at moderate or high concentration (say above 10 mM) non-electrostatic specific interactions prevail.

Next, once the crystalline lattice of the salt has collapsed owing to the presence of water, the ions are separated and hydrated (or more generally solvated). Now the strong local electric field of the ion will produce a remarkable orientation of the water dipoles around it. Averaging over the entire range of angles q that the dipole can assume, and including the randomizing effect of thermal agitation (k_BT) we obtain expression 4.4:

$$\langle V_{ion-dipole}(r) \rangle = -\frac{q^2\mu^2}{6k_BT(4\pi\varepsilon_0\varepsilon_r r^2)^2} \quad (4.4)$$

Note that in this formula the temperature appears at the denominator, indicating that the ion–dipole interaction decreases at higher temperature. The ion–dipole interaction is stronger in solvents with a large dipole moment, such as water or ethylene carbonate or acetamide. Figure 4.3A illustrates the charge q of the ion and the permanent dipole μ that experiences the effect of the ionic electric field. Figure 4.3B represents the first solvation shell (dotted line) containing the water molecules that are strongly oriented by the ion's electric field.

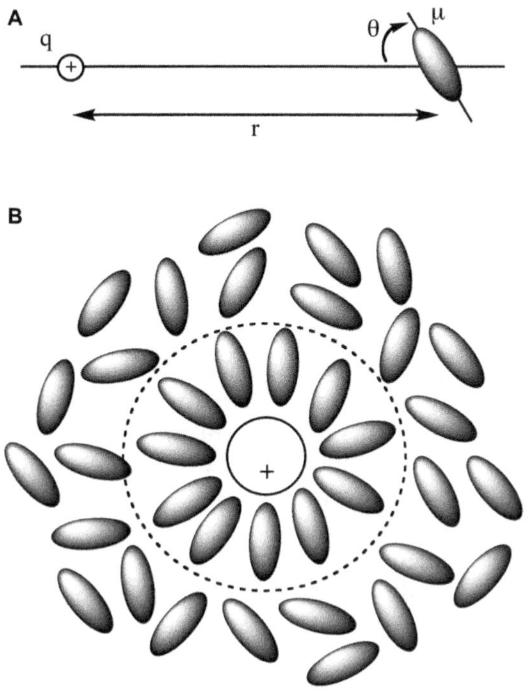

Figure 4.3 A: Interaction between a point charge (q) and a permanent dipole μ; r is the centre-to-centre distance, and θ is the angle formed between the dipole axis and the charge–dipole direction. B: Solvation shell around a central cation. The solvent molecules comprised inside the dotted circle are strongly oriented by the ion's electric field, compared with the bulk solvent molecules outside.

4.1.4.2 Dipole–Dipole Interactions

We consider now the forces that take place between neutral molecules. These depend on the permanent dipole moment if present, on the polarizability, and on the ionization potential of the intervening species.[5]

The dipole–dipole orientation or Keesom interactions need the presence of two permament dipoles, μ_1 and μ_2 (for example two molecules of water, or one molecule of methanol and a molecule of water). The minimum energy state is when the two dipoles are perfectly aligned. Taking into account the randomizing effect of thermal motion, and averaging over all possible angles, we obtain equation 4.5 for the interaction energy:

$$\langle V_{orientation}(r)\rangle = -\frac{2}{3k_BTr^6}\left(\frac{\mu_1\mu_2}{4\pi\varepsilon_0\varepsilon_r}\right)^2 \tag{4.5}$$

Note that in this case the dependence on the distance is expressed as r^{-6}, meaning that this is a short-range interaction and its effects are relevant only when the two polar molecules are close. Again, the larger the dipole moments μ, the stronger the interaction will be.

The presence of a permanent dipole in a molecule A can lead to the formation of an induced dipole in another molecule B, because of the distortion of the electron cloud in B. A must be polar (for example acetone or diethyl ether), B can be polar or apolar (for example chloroform or a hydrocarbon). The dipole-induced dipole or Debye interaction therefore represents the polarizing effect of a permanent dipole (μ) on an apolar molecule (with polarizability α), and its energy is proportional to the square magnitude of the dipole, to the polarizability (α) of the induced dipole, and to r^{-6}.

The induced dipole–induced dipole interactions, also called dispersion or London forces, are of quantomechanical origin, and occur in all cases, regardless of the nature and composition of the ions and molecules. They depend only on the polarizability (α) and ionization potential (I) of the involved species. Their derivation is quite complicated, but we can still use a semi-classical approach. Suppose we have a sample of argon gas, constituted by spherical atoms. Suppose also we have a special digital camera with which we can ideally take instantaneous pictures of single atoms. Each picture will show a different distribution of the nuclei and of their associated electron clouds. On the average the distribution will be randomly uniform with zero dipole moment, but at every instant one single electron distribution in one atom will induce an instantaneous dipole moment in another nearby atom, and the average is not zero. In this case the energy is given by:

$$V_{disp}(r) = -\frac{3\alpha_1\alpha_2 I_1 I_2}{2(4\pi\varepsilon_0\varepsilon_r)^2 r^6 (I_1 + I_2)} \tag{4.6}$$

The dispersion term is always important, but with polar molecules such as water the orientation contribution dominates.

Another very important kind of intermolecular interaction is *hydrogen bonding* (HB; see also Section 5.2). It occurs in molecules that possess at least one hydrogen atom bound to a more electronegative element, for example N, O, F and Cl. In water, hydrogen bonding determines an entire set of peculiar (and sometimes anomalous) properties.

Figure 4.4 illustrates the HB in gaseous acetic acid and in liquid water. Note the strong directionality of HB, as the two interacting molecules must be in the correct position in order to originate the hydrogen bonds.

The picture shows that the H atoms (empty circles) of each water molecule interact with the non-bonding electron pairs of a water molecule in the vicinity. The average characteristic length and the energy of HB depend on the groups involved. For water the length is about 0.18 nm, and the energy is approximately 20–30 kJ mol^{-1}. Although HB are much weaker than common covalent bonds (in H_2O an O–H bond has an energy of about 463 kJ mol^{-1}), they are stronger than other kinds of intermolecular interaction. The HB is highly directional, and imparts a significant order to the water molecules (see Figure 4.5). Owing to the characteristics of HB, and to its strong tridimensional

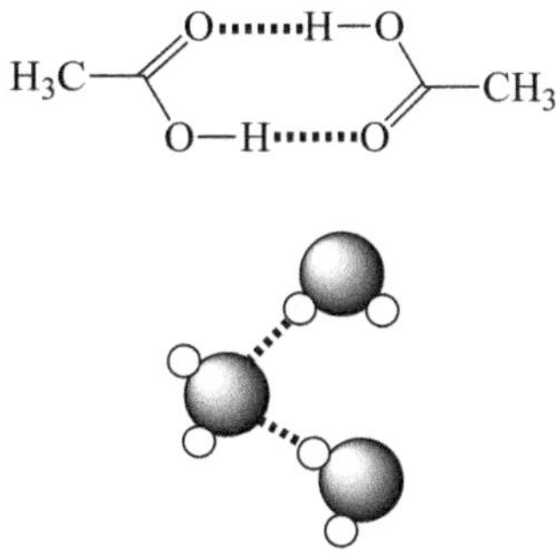

Figure 4.4 Hydrogen bonding (dotted lines) in acetic acid (top) and in water (bottom).

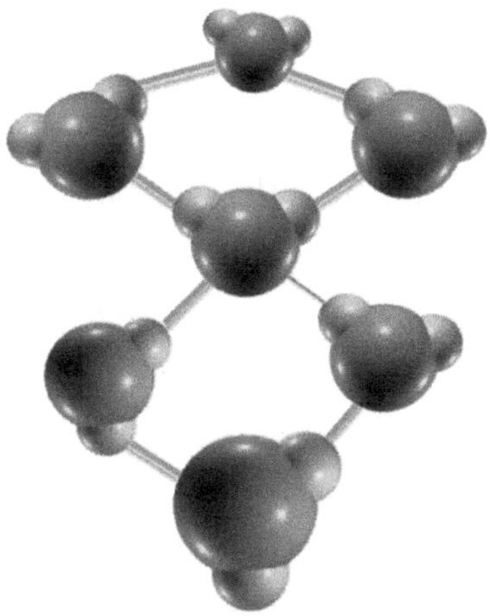

Figure 4.5 Tridimensional order of hydrogen bonding in liquid water.

network in liquid water, water has quite high melting and boiling points, a large heat capacity and enthalpy of vaporization, and large values of surface tension and dielectric constant.

4.1.5 How to Control the Solubility

Besides the nature of the solute and of the solvent, which determine the intermolecular interactions, there are other factors that can affect the solubility of a chemical in a certain solvent. These are temperature, pressure, presence of electrolytes (see Sections 4.1.2 and 4.1.4.1), the physical state of the solute, formation of complexes, and other chemical reactions. By appropriate control of these parameters we may enhance or reduce the solubility of a given product.

4.1.5.1 Temperature

The solubility of a solute in a certain solvent, *i.e.* the maximum amount that can be dissolved, usually depends on temperature. In many cases the solubility increases with temperature, unless the temperature is so close to the critical temperature that the properties and structure of the solvent are severely modified.

Some inorganic salts such as $Ba(NO_3)_2$ become very soluble at higher temperatures, for others like NaCl the solubility is almost unchanged by temperature variations, and in a few other cases the salt is less soluble in water as the temperature raises. There are also examples (Na_2SO_4) in which the solubility is not a monotonic function of temperature. In general, sulfates and nitrates of sodium, potassium, calcium and magnesium are formed on the surface of wall paintings because of the presence of SO_2 and NO_x gases in polluted air (see Sections 2.2.2 and 2.3.1). We recall here that barium salts can be formed on wall paintings after treatment with the Ferroni method. In particular, barium nitrate is soluble in water (90 g L^{-1} at 20 °C) but less soluble than calcium nitrate (1210 g L^{-1} at 20 °C), and its formation does not imply further damage to the work of art. This is an important aspect if the Ferroni method is to be used when calcium nitrate and sulfate are present.

4.1.5.2 Pressure

Usually pressure (P) does not affect the solubility of a solid or of a liquid, and its effect can be neglected. However, an important example of pressure-dependent solubility is that of $CaSO_4$, which becomes less soluble as P is decreased. The pressure dependence of solubility does occasionally have practical significance. For example, precipitation fouling of oil fields and wells by calcium sulfate (which decreases its solubility with decreasing pressure) can result in decreased productivity with time.

4.1.5.3 Physical State of the Solute

Given that solubilization is essentially a process that occurs at the water–solid interface, the extension of the solute's surface area, which is inversely proportional to its size, can have an effect on its solubility. In particular, the solubility product K_{sp} depends on the surface tension and on the molar specific area.

The specific crystalline structure of a solid is another important factor. For example, the dissolution of calcium carbonate depends greatly on whether it is in the form of calcite or aragonite (see also the following paragraph).

4.1.5.4 Formation of Complexes

The formation of complexes can efficiently alter the solubility of a solute. For example, alumina is very insoluble in water, with a solubility product of $[Al^{3+}]\cdot[OH^-]^3 \approx 3 \times 10^{-34}$ at 25 °C in pure water at pH = 7. In principle, according to the common ion effect, a pH increment should lower the solubility in a significant way. Instead, alumina is more soluble because of the formation of hydroxo-complexes such as $[Al(OH)_4]^-$. An interesting example for cultural heritage conservation studies is the solubilization of insoluble copper salts with ammonia, due to the formation of a very stable amino-complex (Scheme 4.2).

$$CuO_{(c)} + 4\,NH_{3(aq)} + H_2O \rightleftarrows [Cu(NH_3)_4]^{2+}{}_{(aq)} + 2\,OH^-{}_{(aq)}$$

Scheme 4.2

Therefore special care must be taken during treatment with chemicals that produce ammonia, such as ammonium carbonate $(NH_4)_2CO_3$ in the Ferroni method, when copper-based pigments (*e.g.* azurite, malachite) are present in wall paintings.

Another example is the formation of complexes between Mg^{2+} or Ca^{2+} (that cause the hardness of water) and ligands such as ethylenediamine tetraacetic acid (EDTA), sodium citrate, or phosphates. These ligands inhibit the formation of insoluble salts and are used, for example, in laundry detergent formulations. The same ligands are used in the removal of carbonate crusts and soils from works of art, and thus the treatment of carbonate-based substrates with these chemicals must be done with special care because they are not selective and may remove or harm the artefact.

4.1.6 Solubility Parameters

The need to predict the miscibility of two or more solvents, and the solubility of a given polymer in a certain liquid, has prompted the adoption of the solubility parameters (δ).

Hildebrand originally developed the concept of describing the thermodynamics of mixtures of non-electrolytes, and then applied it to the solubility of gases and solids in solvents, the miscibility of liquids, metallic solutions, polymer solutions, interfacial phenomena, and critical properties. The parameters can be calculated from the enthalpy of vaporization and the molar volume of a liquid.[6]

The Hildebrand solubility parameters δ can be used for a qualitative evaluation of the solubility of a material in a liquid: a good match between the δ values of two different compounds indicates miscibility. The δ parameters of several compounds can be found in the literature.[6]

Another approach is due to Hansen,[7] in which the solubility parameter is considered as the sum of three contributions: a dispersion term (δ_d), a polar term (δ_p) and a hydrogen bonding term (δ_h) that contribute to a total solubility parameter δ_t, more or less equivalent to δ:

$$\delta_t^2 = \delta_d^2 + \delta_p^2 + \delta_h^2 \tag{4.7}$$

Table 4.3 shows the different contributions to the Hansen parameters for water, ethanol and acetone. Other values can be found in the literature.[6]

Once the Hansen parameters δ_t for two solvents or for a polymer and a solvent have been calculated, we can predict their mutual solubility as for the Hildebrand's parameter. In general polar solvents have large δ_p values, and protic solvents possess large δ_h terms. For example, acetone is more polar than ethanol but shows less hydrogen bonding.

For cost, environmental, or other technical reasons, it is sometimes not convenient to use a single solvent. In order to illustrate how we work with

Table 4.3 Solubility parameter contributions for water, ethanol and acetone. (Reprinted with permission from Ref. 7. Copyright 2001 American Chemical Society.)

Solvent	δ_D	δ_P	δ_H
Water	15.6	16.0	42.3
Ethanol	15.8	8.8	19.4
Acetone	15.5	10.4	7.0

Hansen parameters in such cases, suppose we want to replace chloroform as a solvent – for example because of environmental concerns – with a mixture of solvents that possesses the same solvent capacity but does not include halogenated liquids. Let us first determine the δ_t value for $CHCl_3$, which is 18.9. We can easily calculate, with a trial-and-error procedure or with appropriate software, that a 50% vol : vol mixture of toluene ($\delta_t = 18.2$) and dioxane ($\delta_t = 19.8$) is a valid substitute. Here we use the simple equation for an average solubility parameter $\overline{\delta}$:

$$\overline{\delta} = \phi_1\delta_1 + \phi_2\delta_2 \tag{4.8}$$

where ϕ and δ are the volume fraction and solubility parameter of each liquid. If the molar volumes are similar, $V_1 \approx V_2$, then:

$$\overline{\delta} = x_1\delta_1 + x_2\delta_2 \tag{4.9}$$

where x is the mole fraction of each component in the mixture.

For the 50% vol/vol mixture of toluene and dioxane we obtain $\overline{\delta} = 19.0$, very close to the δ_t of choroform. If we then need to substitute toluene for safety reasons, we can operate in a similar way and find another suitable mixture of less harmful solvents.

The "M numbers" are considered when a simple estimate of the miscibility between two organic solvents is required. In this framework, 31 standard solvents are ordered in a table, depending on their lipophilicity *i.e.* affinity for an oil-like substance.[8] The table can be used to estimate the miscibility of two solvents A and B in the following way.

1. If the difference in the M numbers between A and B is less than 16, they are miscible.
2. If the difference is 16, the mixture separates between 25 °C and 75 °C.
3. If the difference is 17 or more, the two liquids do not mix at all, or they mix above 75 °C.

For some solvents, the literature reports two numbers, meaning that these solvents are immiscible with very lipophilic and very lipophobic solvents, *i.e.* at both ends of the lipophilicity scale. The larger number sets the miscibility limit with solvents that possess a low M number. For example, glycol acetate

(M = 12,19) is miscible with solvents having M of 19 − 15 = 4 and 12 + 15 = 27. The data were taken from Ref. 8.

The Teas parameters, largely used in cultural heritage restoration, will be discussed in Chapter 5.

4.1.7 Ideal Solutions and Raoult's Law

In this section we deal with mixtures of liquids, for example acetone and methanol, or toluene and chloroform. Raoult found that the vapour pressure of component "i" (p_i) is related to its mole fraction in the liquid mixture (x_i), according to the equation that bears his name, Raoult's law:

$$p_i = p_i^0 x_i \tag{4.10}$$

in the formula, p^0_i is the vapour pressure of pure "i".

A solution that follows Raoult's law in the whole composition range between the two pure components is called an *ideal solution*. This situation is usually met when the two components have similar shapes and establish similar intermolecular interactions. This occurs for example with hexane + heptane, or in benzene + toluene mixtures.

For a bicomponent liquid mixture we obtain:

$$x_2 = \frac{p_1^0 - p_1}{p_1^0} \tag{4.11}$$

therefore the mole fraction of component 2 (the solute) in the liquid mixture is measured by the relative decrement in vapour pressure. This is one of the colligative properties, because it depends directly on the number of particles that are dissolved in the solution. We will briefly come back to this subject in Section 4.1.9.

4.1.8 Regular Solution Theory

In order to explain the deviation from ideality of liquid mixtures, Hildebrand elaborated the regular solution theory based on the behaviour of solutions of molecular solids (such as iodine) in apolar and polar solvents. Hildebrand proposed that, in a regular solution, the entropy change of mixing is the same as in an ideal solution. Basically, the theory calculates the activity coefficient of one component in the liquid mixture (γ_1) in terms of the composition (expressed by the mole fraction x) and of an interaction parameter ω_{12}, *i.e.* the potential energy change when one molecule of component 1, surrounded by other molecules of 1, is exchanged with a molecule of 2, surrounded by other molecules of 2.

For regular solutions, γ depends on temperature, which means that we should detect a thermal effect (enthalpy) but no volume change upon mixing two liquids. Regular solutions can show either negative (*e.g.* a mixture of $CHCl_3$ and CH_3COCH_3) or positive [*e.g.* a mixture of CS_2 and $CH_2(OCH_3)_2$] deviations, as illustrated in Figure 4.6.

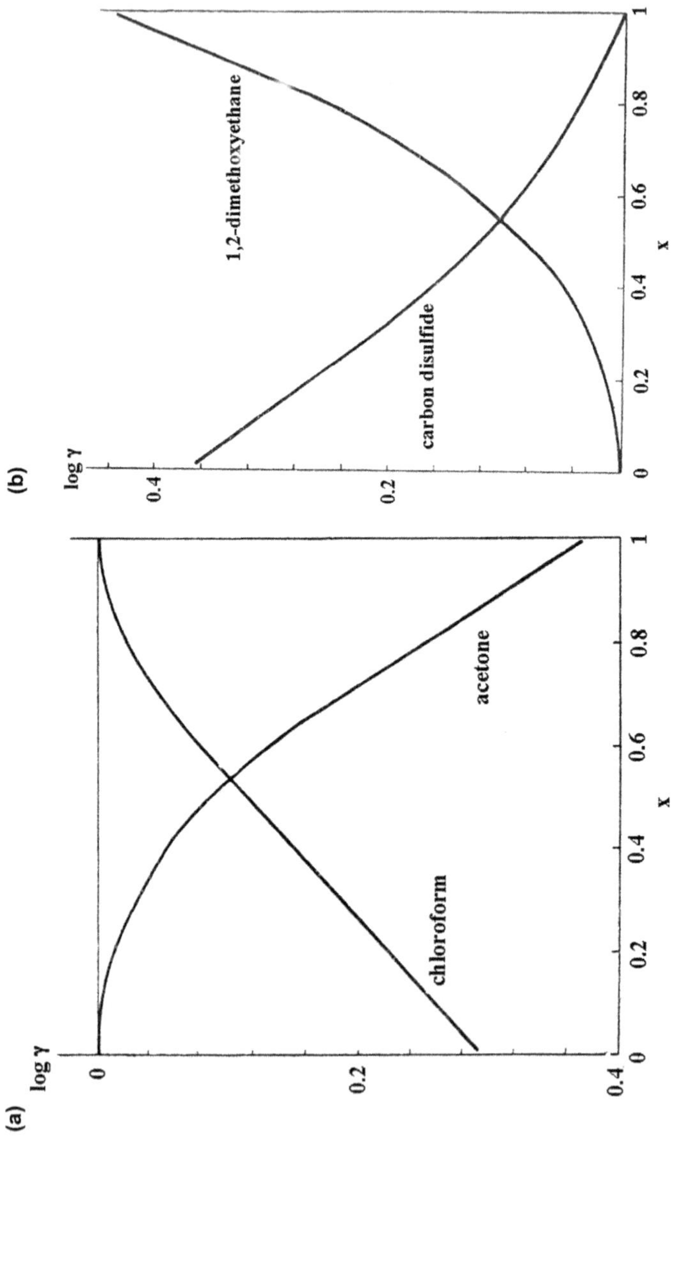

Figure 4.6 Activity coefficient (as log γ) *vs.* the mole fraction for chloroform + acetone (a) and for carbon disulfide + 1,2-dimethoxyethane (b) mixtures.

When $\omega_{12} = 0$ we have a perfect solution, and Raoult's law for ideal solutions holds. For large values of the interaction parameter, and in particular when $|\omega_{12}| > 2k_BT$, the mixture phase separates at a certain critical temperatures. Figures 4.7–4.9 illustrate the miscibility curves for a system that shows an upper critical temperature and a lower critical temperature. The presence of an upper critical solution temperature (UCST) indicates that for higher temperatures the system is monophasic and the two components are fully miscible, but below the UCST the mixture phase separates into two coexisting phases. This occurs for example in methanol + *n*-heptane or in propylene carbonate + water mixtures (see Figures 4.7 and 4.8).

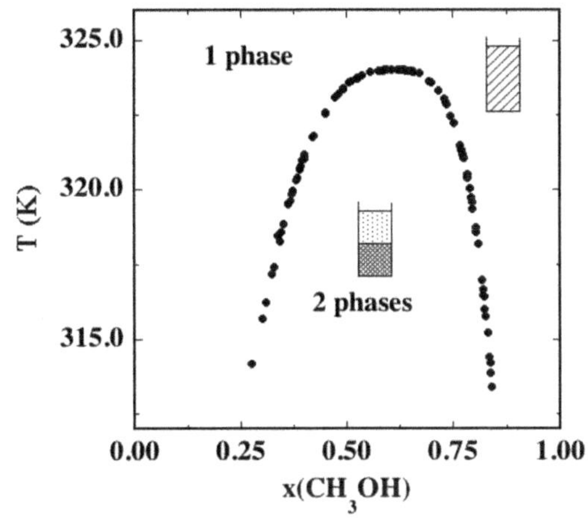

Figure 4.7 Phase diagram for methanol + n-heptane mixture, with a UCST. (Reprinted with permission from Ref. 9. Copyright 1961 American Chemical Society.)

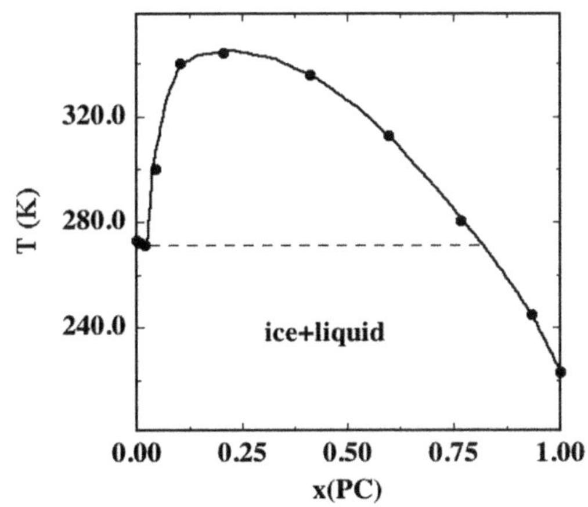

Figure 4.8 Phase diagram for propylene carbonate + water mixture, with a UCST; x(PC) indicates the mole fraction of propylene carbonate in the mixture. (Reprinted with permission from Ref. 10. Copyright 1982 American Chemical Society.)

On the other hand, a lower critical solution temperature (LCST) indicates that complete miscibility is reached at lower temperatures, as for example with triethylamine + water (LCST = + 18.5 °C, see Figure 4.9). Liquid–liquid systems that exhibit LCST behaviour are less common. A few systems, such as water + ethylene glycol butyl ether, possess both an UCST and an LCST, resulting in a miscibility gap in the middle of the phase diagram (see Figure 4.10).

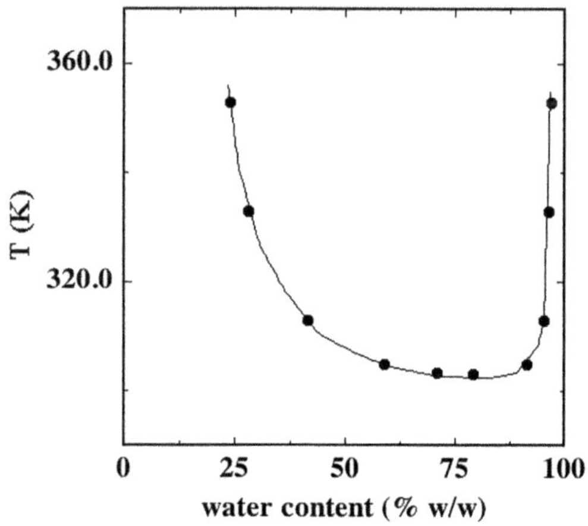

Figure 4.9 Phase diagram of the triethylamine + water mixture showing an LCST. (Adapted with permission from Ref. 11. Copyright 1966 American Chemical Society.)

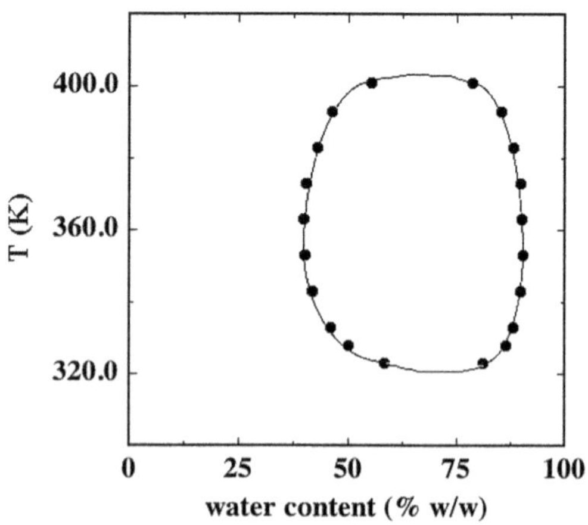

Figure 4.10 Phase diagram of the water + ethylene glycol butyl ether mixture, showing a miscibility gap.
(Reprinted with permission from Ref. 12. Copyright 2005 American Chemical Society.)

4.1.9 Colligative Properties

Besides the lowering in the vapour pressure, there are three other colligative properties that concern solutions: the freezing point decrement, the boiling point elevation, and the osmotic pressure. The name "colligative properties" refers to the fact that they do not depend on the nature and composition of the dissolved solute, but only on its concentration, or rather on the number of particles that are dispersed in the solution. Classical examples of colligative properties are the spreading of salt on iced roads, which lowers the freezing point of water, and the swelling or shrinking of cells upon dilution or concentration of the extracellular environment. The interested reader can find a detailed presentation of colligative properties in any physical chemistry textbook (see a short list at the end of this chapter, in the Further Suggested Reading section).

4.1.10 Solubility of Gases in Liquids and Henry's Law

We make use of a simple system to introduce this section. Take a bottle of soda water. We know that when the cap is removed and the bottle is opened, a significant amount of gas will come out, and if the temperature of the bottle is relatively high, such as in the summertime, this gas development will be so abundant that some water will spill out of the bottle. This occurs because carbon dioxide, as a gas, is added to mineral water, and the amount of the gas that can be contained in the liquid is larger at lower temperature. The added carbon dioxide remains in the water as long as the system is closed in the capped bottle, but when the cap is removed and the water is exposed to atmospheric pressure a significant amount of CO_2 will re-enter into the atmosphere.

The importance of gas solubility in liquids may be witnessed in some examples. All aquatic species (*e.g.* algae, invertebrates, fish) make use of gas that is dissolved in seawater or fresh water. They take up O_2 and release CO_2 through the water medium. An iron nail in a waterlogged shipwreck will become oxidized and will rust under water because of the oxygen that is dissolved in the liquid (see also Chapter 16 in this book).

The law that describes the phenomenon of gas solubility in a liquid is known as Henry's law. Henry's law applies to ideal dilute solutions of one or more gases that are dissolved in water, and it states that:

$$p_i = K_{H,i} x_i \tag{4.12}$$

where p_i is the partial pressure of gas "i" in the gaseous phase above the solution, x_i is its concentration in the liquid (expressed in mole fraction units), and $K_{H,i}$ is the Henry's constant for that component. $K_{H,i}$ depends on the specific gas, on the liquid in which the gas is dissolved, and on temperature, and has the dimension of pressure (Pa, atm, mmHg, *etc.*).

The gas concentration in the liquid depends on temperature, according to this equation:

$$\ln x_i = A + \frac{100B}{T} + C\ln\left(\frac{T}{100}\right) \tag{4.13}$$

The reference system is the pure gas with a partial pressure of 1.013×10^5 Pa (or 1 atm).

For example, O_2 has a solubility in water of about 2.5×10^{-5} at 20 °C and of 2.1×10^{-5} at 30 °C. In this case, $A = -66.7$, $B = 87.5$, and $C = 24.5$. CO_2 is more than 10 times more soluble than O_2: 7.1×10^{-4} at 20 °C and 5.4×10^{-4} at 30 °C in water.

4.2 Polymers in Solutions

A polymer is a macromolecule, *i.e.* a molecule with a high molecular weight, formed by repetition of simpler units, called monomers, connected by covalent links.

Polymers, like surfactants, are ubiquitous in commercial formulations of a multitude of products, from paints to household products to personal care, and they are usually added to stabilize dispersion or to confer particular rheological properties. They are often employed as components in formulations used as consolidants or protective coatings in the restoration of works of art; nowadays their removal is a central challenge in modern conservation science, as it will be made clear in Chapter 7.

Polymer solutions differ from ordinary molecular solutions, because of their high molecular weight. Their behaviour can be modelled by starting from the chemical properties of the monomers, but the covalent connectivity between them and their high molecular weight have dramatic implications for their phase and solution behaviour.

In view of the above considerations, after having introduced solvents and some basic solution thermodynamics, the last part of this chapter will be devoted to an introduction to the relevant structural and thermodynamic parameters that govern polymer solutions.

4.2.1 Structural Parameters and Conformation

The term "polymer" encompasses a broad class of chemical substances, whose properties are determined not only by the chemical nature of the monomers, but also by their molecular weight, determined by the total number of monomers, *i.e.* by the degree of polymerization, and by their structural architecture; a sketch of the possible branching modes around the polymer chain is shown in Figure 4.11. In the commonly accepted nomenclature the term **poly** precedes the name of the monomer. The simplest form is the linear one. Several examples among synthetic polymers fall into this class: polyethylene, polyacrylic acid (Figure 4.12) and so on.

Polymers are often encountered in living systems; polysaccharides, proteins, gelatine and DNA are probably the most famous examples. Natural polymers can have more than 10^6 monomers in a chain and all the chains can have the same length. Synthetic polymers commonly have a lower number of monomers but there is always a distribution of chain lengths. If a typical monomer length is approximately 1 nm, then 10^6 monomers would have a contour length

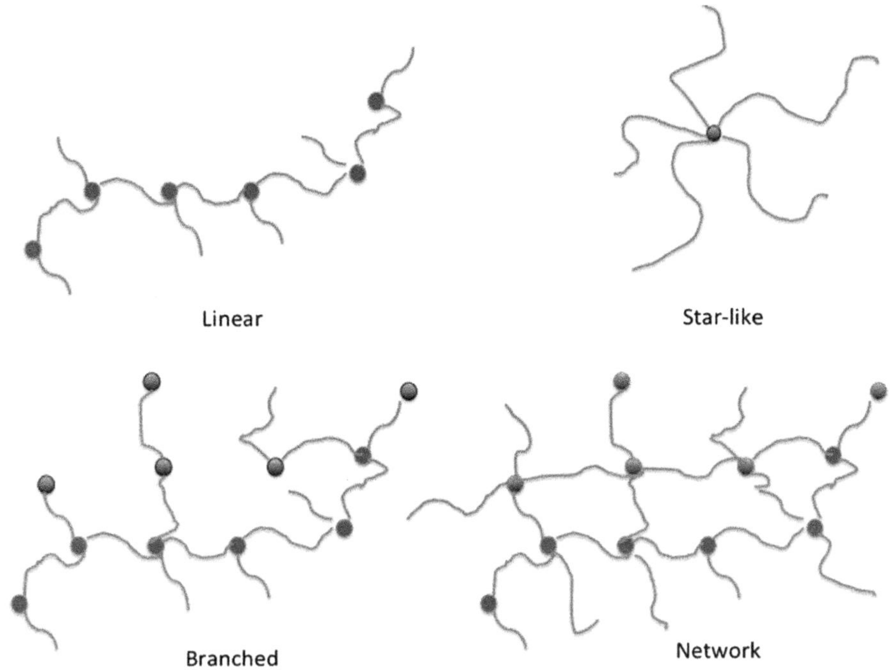

Figure 4.11 Architecture of polymers with different branching patterns.

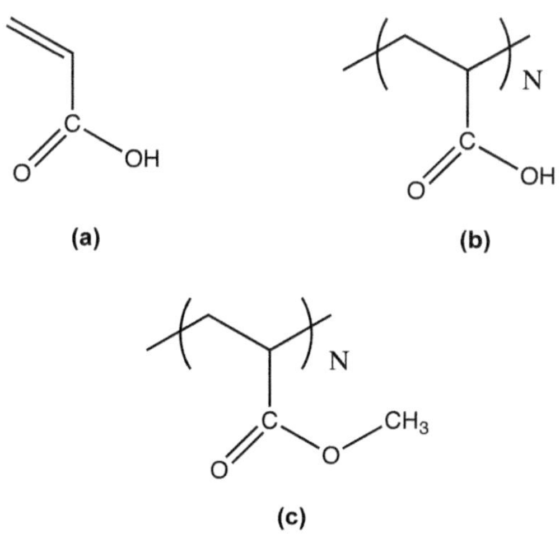

Figure 4.12 Chemical structure of acrylic acid (a), poly(acrylic) acid (b) and poly(methyl acrylate) (c). The number *n* is an integer that indicates the degree of polymerization, and the repeating unit is shown.

of 1 mm. Polymers are therefore "giant" molecules, with a pronounced asymmetry between the size of their cross-section and the chain length (the so-called contour length). The chains can be also crosslinked, to form a polymer network; this particular situation will be dealt with in Chapter 10, devoted to gels.

As Figure 4.12 shows for a famous class of polymers (poly acrylates), the architecture of linear polymers can be visualized as a chain with side groups that confer the overall properties, such as solubility. For instance polyacrylic acid (4.12b), derived from the polymerization of acrylic acid, is water-soluble because the side carboxyl groups are sufficiently polar and can be charged, depending on the pH of the solution. Conversely poly(methyl acrylate) (4.12c), which originates from the polymerization of methyl acrylate, is more soluble in organic solvents; in the same way, polystyrene shows affinity for aromatic solvents, owing to the presence of a phenyl ring in each monomer.

If N is the degree of polymerization and m is the monomer mass, the molecular weight of the polymer is given by:

$$M_\mathrm{n} = Nm \tag{4.14}$$

Typical degrees of polymerization are 10^2–10^4. Practically, for synthetic polymers it is very difficult to produce a single molecular weight in a polymerization reaction and we usually refer to a molecular weight distribution. Given that the properties of polymers are dictated by their degree of polymerization, it is easily understandable that they will also depend on the dispersion of molecular weights around an average value, *i.e.* on the polydispersity.

This quantity can be obtained with the help of several experimental methods, but the reader should be aware that different techniques measure different properties, whose functional dependence on molecular weight might vary. Table 4.4 collects the definition of the different averages for molecular weight, and the experimental accessibility to each of these quantities.

Some simple arithmetic, following from the definitions in Table 4.4, allows one to infer that:

$$M_\mathrm{n} < M_\mathrm{w} < M_\mathrm{z} \tag{4.15}$$

except in the case of monodispersity, *i.e.* the presence of a single molecular weight, when these three quantities coincide.

Therefore when considering the molecular weight of a synthetic polymer, it is very important to define the technique used to measure it. Another

Table 4.4 Molecular weight averages and techniques to measure them.

Kind of Average	*Definition*	*Method*
Number Average M_n	$M_n = \sum_i \frac{M_i n_i}{n_i}$	Osmometry
Weight Average M_w	$M_w = \sum_i \frac{M_i^2 n_i}{M_i n_i}$	Light scattering
Z-Average M_z	$M_z = \sum_i \frac{M_i^3 n_i}{M_i^2 n_i}$	Sedimentation (ultracentrifugation)

fundamental piece of information is the width of this distribution, which can be expressed with the so-called polydispersity index *P*, defined as:

$$P_i = M_w / M_n \tag{4.16}$$

From what has been stated above, it follows that this number is >1, or 1 for monodisperse polymers, and that its deviation from unity is directly connected to the width of the distribution. The value of P_i is usually indicated in the technical reference sheet for commercial polymers, together with the kind and value of average molecular weight. Polymers with a $P_i = 1.1$ are used as standards to calibrate instrumentation, while the components of commercial formulations usually have a P_i around 3.

Before introducing the thermodynamics of polymer solutions, it is worthwhile to present a qualitative picture of polymer conformation in solution.

We have said that a polymer results from the repetition of a fundamental building block, *i.e.* the monomer. If we now imagine each monomer as a segment freely jointed to the adjacent ones, and whose relative orientation with respect to them can vary, it is easy to imagine that, as the degree of polymerization grows, the number of conformations with the same overall energy that can be adopted will increase accordingly.

This means that if we have *N* polymeric chains in solution, and imagine following the contour length of the primary chain, each macromolecule will assume a different conformation with respect to the others, while maintaining the same overall conformational energy. As *N* gets larger, the polymer will adopt a so-called "random-coil" conformation, shown in Figure 4.13, with the dots indicating the starting and the end points of the chain.

Figure 4.13 Random coil conformation of a polymer. The dots are the starting and the end points of the chain.

Therefore, the structural parameters should be defined in a statistical fashion, even for a strictly monodispersed polymer, because many different values should be possible. Moreover, for a given chain, the same structural parameters vary with time.

The careful reader will have noticed that we have for the moment not explicitly mentioned the other key character of the story, which is of course the solvent. Remember that, in contrast to the case of simple solutes, which are completely surrounded by solvent molecules, in this case the connectivity with adjacent units imposes more constraints, but also implies a possible partial solvation, because contacts with the other segments are also an option. The solvation of each unit will depend on the affinity for the solvent, as we will see in the next paragraph.

An intuitive conformational analysis will be useful in understanding what will be explained in mathematical terms. Three possible extreme scenarios are possible:

- a fully extended conformation, when the affinity with the solvent is very high and repulsion between the segments is prevailing;
- a compact globular conformation, when the opposite situation occurs;
- a random coil conformation, where the interactions are balanced.

This is summarized pictorially in the Haug triangle, shown in Figure 4.14. Depending on the energetics of interaction, several intermediate situations can be possible.

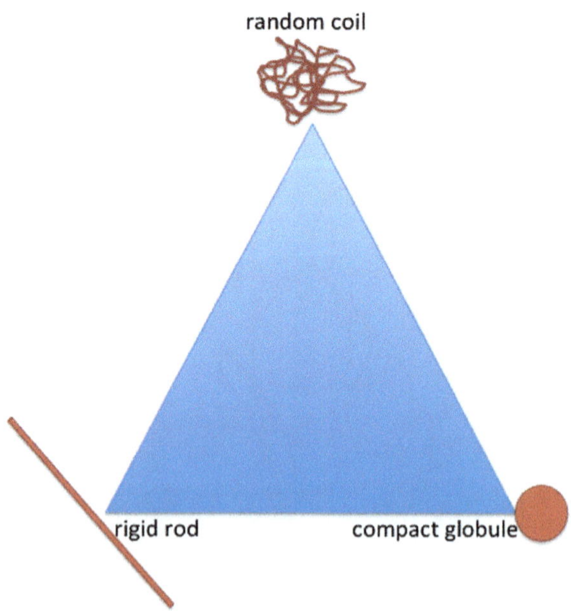

Figure 4.14 Haug triangle sketching the possible conformations of polymers in solution.

4.2.2 Flory–Huggins Theory

The need to predict the solubility and the conformation of polymers in a certain solvent has stimulated the development of several models. We will here review briefly the most important one, based on the Flory–Huggins theory, which leads to the introduction of the Flory–Huggins parameter. The conformation of polymers in solution depends on the balance of the interactions between the solvent and the polymer segments. The Flory–Huggins parameter describes these interactions on more quantitative grounds and categorizes the solvents in terms of solubility for a given polymer.

The solution is described as a three-dimensional cubic lattice, composed of sites of equal volume, each occupied either by a polymer segment or by a solvent molecule, as represented in Figure 4.15(a). The main assumption here is that the polymer segments and the solvent have equal volume. Each lattice site is surrounded by z neighbouring sites, or, in other words, z is the coordination number.

The polymer segments are indicated in red and the constraint of connectivity is indicated by the joining segments.

Let us now define the volume fractions occupied by the polymer φ_p and by the solvent φ_s:

$$\varphi_s = \frac{n_s}{n_s + Nn_p} \tag{4.17}$$

$$\varphi_p = \frac{Nn_p}{n_s + Nn_p} \tag{4.18}$$

where N is the polymerization degree (assuming a monodispersed macromolecule), n_s is the number of solvent molecules and n_p is the number of monomer molecules.

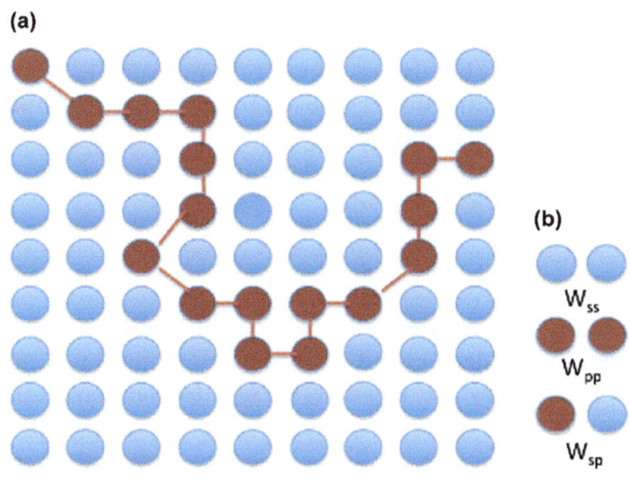

Figure 4.15 (a) A slice of the cubic network of sites, the starting point of the Flory–Huggins model; (b) possible pair contacts occurring in a binary solution.

We recall that the free energy of mixing is given by

$$\Delta G_{mix} = \Delta H_{mix} - T\Delta S_{mix} \tag{4.19}$$

where ΔH_{mix} and ΔS_{mix} are the enthalpic and entropic variations of the solution with respect to the separated components. The entropic term is connected to the number of possible configurations of the polymer chain in the lattice, while the enthalpic variation is due to the energetic balance of interactions between the different components.

The entropic term is thus connected to the number of arrangements; we will not go through the details of this derivation, for which the reader is referred to more specialized textbooks:

$$\Delta S_{mix} = -k_B\left(n_s \ln \varphi_s + n_p \ln \varphi_p\right) \tag{4.20}$$

This contribution recalls the mixing entropy of two simple liquids, but note that we are using *volume* fractions and not mole fractions. The entropic term is always positive, *i.e.* there is a gain in configurational freedom upon dissolving a polymer in solution. In deriving the enthalpic term, we introduce the balance of interactions between segment–solvent, segment–segment and solvent as indicated in Figure 4.15b.

If we assign an energy of interaction w_{ss} to a pair of solvent molecules, w_{pp} to two chain segments and w_{sp} to the couple solvent–polymer segments, and consider the volume fractions for each component, we get:

$$\Delta H_{mix} = \frac{1}{2} z n_s \varphi_p \omega_{12} \tag{4.21}$$

where z is the coordination number and ω_{12} is defined as follows:

$$\omega_{12} = 2w_{sp} - w_{ss} - w_{pp} \tag{4.22}$$

It is important to keep in mind that we are considering only intermolecular interactions: the covalent bonds that connect the macromolecular segments are not considered in this treatment, because of the restriction of connectivity of the lattice sites occupied by the polymer segments.

At this point, the Flory–Huggins parameter, indicated by χ, is usually introduced as:

$$\chi = \frac{1}{2}\frac{z\omega_{12}}{k_B T} \tag{4.23}$$

i.e. a measure of the energetic balance in thermal energy units (k_BT). The overall sign of ω_{12} depends on which pair potential predominates. If w's are attractive terms, they are negative by sign convention. For instance, in the case of polyelectrolytes in water, w_{pp} will contain electrostatic repulsive interactions between charges of the same sign (see Figure 4.13b), while the interaction with water molecules will be dominated by charge–dipole attractive interactions. As a result the ω_{12} term will have a relatively large negative value.

The final free energy of mixing is thus given by:

$$\Delta G_{mix} = k_B T(n_s \ln \varphi_s + n_p \ln \varphi_p + n_s \varphi_p \chi) \quad (4.24)$$

According to the balance of interactions and temperature, the dimensionless χ parameter can assume different values:

a) if $\chi < 0.5$: the solvent is a good one for the polymer (repulsion between the polymer segments prevails);
b) if $\chi \approx 0.5$: the solvent is called a theta-solvent, and the interactions with the solvent are balanced in terms of thermal energy;
c) if $\chi > 0.5$: the solvent is a poor one (attraction between the polymer segments prevails).

In the last case, the solubility is limited, even if the solvent can still swell the polymer. Swelling is an important phenomenon related to the cleaning of works of art, as will be discussed in Chapter 7.

For a given solvent–polymer pair, the theta conditions are satisfied at a given temperature, because the value of the Flory–Huggins parameter depends inversely on this parameter. It is therefore important to know the theta temperature for a given polymer in a solvent, because this will act as a good one only at temperatures above this value. For instance PMMA [poly(methyl methacrylate)], commercialized by Rohm and Haas under the trademark Plexiglas, has a theta temperature of 25.0 °C in MEK [(methyl-ethyl-ketone)/isopropanol (50 : 50)]; this means that, above this temperature, the mixture of solvents acts as a good one.

More difficult is the physico-chemical definition of copolymers, such as Paraloid B-72, a common adhesive agent in conservation, formed by random copolymerization of ethylmethacrylate and methylacrylate. Usually, acetone is used as the solvent to coat Paraloid films on surfaces, but the final mechanical properties of the coating depend on the casting solvent mixture, and different proportions of toluene, acetone and ethanol are used to modulate the final performance. The solvent properties of mixtures of liquids towards copolymers are usually tackled through phenomenological approaches, as presented in Chapter 5 when introducing the Teas graph. For an extended set of theta temperatures, please refer to Ref. 13.

4.2.3 Polymer Solution and Phase Behaviour

The value of the χ parameter categorizes solvents in terms of affinity for a polymer chain; it is therefore obvious that it also strongly affects the conformation of polymers in solution. In water an uncharged polymer chain is generally found in a random conformation in solution, while for a polyelectrolyte the charge repulsion will cause the chain to adopt a more rigid and expanded conformation. If the ionic strength, *i.e.* the quantity of background electrolyte, is high, the charges will be screened and the polyelectrolyte chain will collapse to a more random conformation (essentially identical to a neutral chain).

Therefore, coming back to the Haug triangle, reported in Figure 14.4, it is clear that each vertex portrays different solvent conditions. Temperature, ionic strength and pH can shift the conformation.

We would like to stress, at this stage, that polymer conformation also has a dramatic effect on the characteristics of the dry polymer film, when used as a coating (paint, varnish) or as an adhesive. The adhesion efficiency and the gas permeability of the film, as an example, depend on the conformation of the polymer in the solvent used for casting: the more uncoiled the polymer (*i.e.* the better the solvent quality) the higher the contact area with the substrate, *i.e.* the higher the adhesion.

In terms of a phase diagram, the theory predicts an upper critical solution temperature (UCST), just like those seen for "ordinary" solutes, and this has been found experimentally for many non-aqueous polymer solutions. However, many aqueous polymer solutions also show a lower critical point (LCST). This is very often the case for poly(ethylene oxides) and has the same chemical foundation as the hydrophilic–lipophilic balance (HLB) temperature dependence of CiEj amphiphiles, which will be treated in Chapter 8. As the temperature is increased the ethylene oxide groups become less and less hydrated and the solvent quality decreases dramatically, eventually leading to phase separation. This peculiar behaviour should be borne in mind when dealing with polymers in good solvents. We remind the reader that CiEj amphiphiles are used in the formulation of nano-structured systems for the cleaning of works of art (see also Section 9.5: Triton X), and thus their stability in solvents (*e.g.* water) is of utmost importance.

4.2.4 Phenomenological Approaches to Predict Solubility

The Flory–Huggins theory is specific for a polymer solution; as we have seen earlier in this chapter (Section 4.1.6), a set of phenomenological tools are needed to predict solubility in a given solvent; the same concepts apply to polymer solutions.[13] Here the Hansen parameters are referred to the monomer. You will see these concepts at work in the next chapter.

As a final remark, we would like to mention that the Flory–Huggins model and the Hansen formalism are two approaches to the same problem; therefore the relevant quantities that they introduce must be intimately related. Indeed χ is experimentally accessible from osmotic pressure or static light scattering measurements; however if the solubility parameters are known, they can be used to evaluate the Flory–Huggins interaction parameter:

$$\chi = 0.34 + \frac{V_s}{RT}\left(\delta_s - \delta_p\right) \qquad (4.25)$$

where V_s is the solvent molar volume, $\delta_{s,p}$ are the Hansen solubility parameters for the solvent and the polymer, respectively, and 0.34 is a correction factor that accounts for entropic contributions.

Further Suggested Reading

The interested reader can find further information and develop the concepts and applications addressed in this chapter in several books. For example:

P. Atkins and J. de Paula, *Atkins' Physical Chemistry*, Oxford University Press, Oxford, 9th edn, 2010.

J. Goodwin, *Colloids and Interfaces with Surfactants and Polymers*, John Wiley and Sons, Chichester (UK), 2009.

P. C. Hiemenz and R. Rajagopalan, *Principles of Colloid and Surface Chemistry*, Marcel Dekker, New York, 1997.

C. V. Horie, *Materials for Conservation*, Elsevier Butterworth-Heinemann, Burlington (MA), 1987.

I. R. Levine, *Physical Chemistry*, McGraw-Hill, New York, 6th edn, 2009.

D. A. McQuarrie and J. D. Simon, *Physical Chemistry: A Molecular Approach*, University Science Books, Sausalito (CA), 1997.

R. Wolbers, *Cleaning Painted Surfaces*, Archetype Publications Ltd, London, 2000.

A tutorial on solubility parameters can be found at following url: http://aic.stanford.edu/sg/bpg/annual/v03/bp03-04.html (accessed on 25 May, 2012);

Review and Questions

In this chapter we have discussed the main properties of solvents, solutions and polymer solutions. In particular we introduced these main concepts:

- Solubility (definition and units).
- The most important physico-chemical properties of solvents.
- The solute–solvent interactions that depend on their chemical composition.
- The ways we can control the solubility of a solute in a liquid.
- The Hildebrand solubility parameters as a guideline for prediction of the miscibility.
- The meaning and properties of an ideal solution.
- The colligative properties that depend only on the concentration of the solution.
- The solubility of gases in liquids.
- Polymers: some basic definitions.
- Molecular weight, polydispersity, structural parameters of a polymer.
- Solubility of a polymer: the Flory–Huggins theory.
- The χ parameter: the goodness of a solvent.

Questions

1) Calculate the solubility in water of magnesium hydroxide $Mg(OH)_2$ from its solubility product, $K_{sp} = 5.6 \times 10^{-12}$.

2) Discuss the solubilization of calcium hydroxide and magnesium hydroxide in water from their K_{sp} values (5×10^{-6} and 5.6×10^{-12}, respectively).
3) Based on the *M* numbers listed below assess the mutual miscibility of the following pairs of solvents: acetone + ethanol, propylene carbonate + 1-pentanol, and p-xylene + 1-pentanol. Propylene carbonate: 9 or 17; 1-pentanol: 17; acetone: 15 or 17; ethanol: 14; p-xylene: 24.
4) The monomer of PMMA [poly(methyl methacrylate)], also known as Plexiglas®, has the formula ($C_5H_8O_2$). The density of the polymer is $1.18\,g\,cm^{-3}$. The polymer is dissolved in acetone at a concentration of 50 mg per 10 mL solution. Evaluate the molar concentration of the polymer if the degree of polymerization is a) 10^2; b) 10^3. c) Determine the volume fraction of the polymer in the two cases. (Acetone density = 0.791 $g\,cm^{-3}$.)

Answers

1) When $Mg(OH)_2$ dissolves in water, for each Mg^{2+} ion two OH^- ions will be released. Therefore, indicating with *S* the solubility of the hydroxide, $K_{sp} = [Mg^{2+}]{\cdot}[OH^-]^2 = 5.6 \times 10^{-12} = S{\cdot}(2S)^2 = 4S^3$, from which $S = (5.6 \times 10^{-12}/4)^{1/3} = 1.1 \times 10^{-4}\ mol\,L^{-1}$. This value corresponds to about 6.4 mg of $Mg(OH)_2$ per litre of solution.
2) In the previous question we have calculated that the solubility of $Mg(OH)_2$ is about $6.4\,mg\,L^{-1}$. Repeating the same calculation for $Ca(OH)_2$ we obtain about $0.8\,g\,L^{-1}$. This result suggests that the solubilities of these materials are so low that it is more convenient to work with colloidal dispersions rather than with simple aqueous solutions.
3) All pairs produce perfect mixtures, because in any case the difference between the *M* numbers is lower than 15. In particular, for propylene carbonate the two values (9,17) suggest that this liquid will mix with any other solvent that possesses an *M* number spanning between $9 + 15 = 24$ and $17 - 15 = 2$, and 1-pentanol has an *M* number of 17.
4) The molar mass of the monomer is $100.12\,g\,mol^{-1}$, as calculated from the chemical formula. In case a) the molar concentration of the polymer is $5 \times 10^{-4}\,mol\,L^{-1}$, while in case b) it is 10 times lower, because the molar mass is 10 times as large. When dealing with volume fractions, the concentration is identical in the two cases. In 10 mL of solutions we have $0.04\,cm^3 = 0.04\,mL$ ($= 0.05\,g/1.18\,g\,cm^{-3}$) occupied by the polymer chains. Therefore $\varphi_p = 0.04\,mL/10\,mL = 0.004 = 0.4\%$

References

1. D. R. Lide (ed.), *CRC Handbook of Chemistry and Physics*, Taylor and Francis, Boca Raton (FL), 87th edn, 2007.
2. http://en.wikipedia.org/wiki/Solubility_table (accessed on 22 May, 2012).
3. Y. Chernyak, *J. Chem. Eng. Data*, 2006, **51**, 416.

4. D. D. Perrin and W. L. F. Armarego, *Purification of Laboratory Chemicals*, Pergamon Press, Oxford, 3rd edn, 1988.
5. B. Ninham and P. Lo Nostro, *Molecular Forces and Self Assembly in Colloid, Nano Sciences and Biology*, Cambridge University Press, Cambridge, 2010.
6. H. Mollet and A. Grubenmann, *Formulation Technology*, Wiley-VCH, Weinheim, 2001.
7. C. M. Hansen and L. Just, *Ind. Eng. Chem. Res.*, 2001, **40**, 21.
8. N. B. Godfrey, *Chemtech*, 1972, 359.
9. R. W. Kiser, G. D. Johnson and M. D. Shetlar, *J. Chem. Eng. Data*, 1961, **6**, 338.
10. C. S. Hong, R. Wakslak, H. Finston and V. Fried, *J. Chem. Eng. Data*, 1982, **27**, 146.
11. B. J. Hales, G. L. Bertrand and L. G. Hepler, *J. Phys. Chem.*, 1966, **70**, 3970.
12. S. P. Christensen, F. A. Donate, T. C. Frank, R. J. LaTulip and L. C. Wilson, *J. Chem. Eng. Data*, 2005, **50**, 869.
13. P. R. Sundararajan, in *Physical Properties of Polymers Handbook*, John Wiley and Sons, Chichester (UK), 2009, Part III, p. 259.

CHAPTER 5

Cleaning I: Application

E. CARRETTI* AND L. DEI

University of Florence, Italy
*Email: carretti@csgi.unifi.it; luigi.dei@unifi.it

5.1 Introduction

In the previous chapter the basic concepts of liquid solvents and solutions were described. In the present chapter we will discuss the application of the properties of pure and mixed liquids to the cleaning of works of art surfaces. Indeed, among the various types of cleaning techniques – mechanical, physical, chemical and optical methods – the approach based on the use of pure or mixed solvents has played and continues to play a very important role. Therefore, here we will illustrate the use of pure or mixed solvents for cleaning, referring to the mechanism of solubilization of the coating materials to be removed from an artistic substrate.

The solubilization process will be presented by referring to the ancient Latin motto "*similia similibus solvuntur*" (like dissolves like) and we will discuss the meaning of this empirical rule of thumb in terms of intermolecular forces. We will introduce the reduced solubility parameters that are fundamental to conservators in selecting the appropriate solvent or mixtures of solvents for an appropriate cleaning situation. The parameters will be discussed in detail in relationship to the Teas diagrams, very powerful tools for determining the correct cleaning liquid mixtures. A paragraph dealing with the safety of liquid solvents will follow to make conservators aware of all the problems associated with the usage of such chemicals. Finally, details of the triangle of solubility software tool developed at the *Istituto Superiore Centrale di Restauro* (Institute for the Conservation and Restoration, ISCR) in Rome will be given.

RSC Nanoscience & Nanotechnology No. 28
Nanoscience for the Conservation of Works of Art
Edited by Piero Baglioni and David Chelazzi

Published by the Royal Society of Chemistry, www.rsc.org

The whole chapter is thus conceived as a powerful but compact introduction to the chemical concepts behind the application of liquid solvents as cleaning agents for works of art conservation. It provides a fundamental discussion on the state-of-the-art knowledge on chemical cleaning, which constitutes a necessary preliminary training and review step for the following chapters on innovative cleaning methodologies based on nanotechnology (Chapters 7 and 9).

5.2 Extraction of Materials through Solubilization

The solubilizing power of a liquid solvent represents its capability to dissolve a substance to form a solution. When dealing with the cleaning of artistic or architectonic surfaces we are often in a situation where solid "foreign" substances have to be removed from the artefact, by using liquid solvents that can solubilize the "foreign" molecules from their solid aggregation state without affecting the original materials. This is the basis for the term "selective cleaning". The mechanism of solubilization of a solid substance in a liquid solvent – or a mixture of liquid solvents – can be schematically represented as indicated in Figure 5.1.

Here M refers to the solid substance, and the attractive M–M intermolecular interactions are responsible for the stability of the crystal lattice or glassy amorphous structure. In a similar manner the solvent molecules S are brought together by attractive forces. When the solubilization of M by S occurs, the M–M and the S–S interactions are broken and new attractive S–M interactions

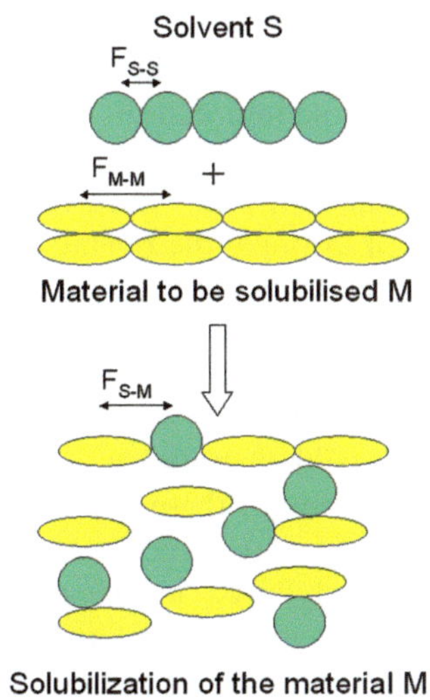

Figure 5.1 Representation of the solubilization mechanism of a solid substance M by a solvent S. The solubilization process is governed by the balance of the potential energies associated with the attractive forces F_{M-M}, F_{S-S} and F_{S-M}.

take place. Solubilization is governed by the balance between the three potential energies associated with these three types of attractive force.

From the thermodynamic point of view, the solubilization of M occurs spontaneously when the variation of the Gibbs free energy of the process (ΔG_{solub}) is negative:

$$\Delta G_{solub} = G_{SM} - G_{SS} - G_{MM} < 0 \quad (5.1)$$

where G_{SM} is the Gibbs free energy of the solution, and G_{SS} and G_{MM} are the Gibbs free energies of the pure solvent and of the pure solute, respectively. A rigorous approach to predict whether a solubilization process is thermodynamically favoured or not will be discussed in Sections 5.3 and 5.4.

Practically, the choice of the appropriate solvent for the cleaning of a surface is based on the rule of thumb "like dissolves like". At the molecular level, this implies that solubilization is granted when the intermolecular solute–solvent interactions are comparable to those occurring between the molecules of the pure species M and S.

According to the theory of intermolecular interactions,[1] the attractive forces typically acting between neutral molecules can be due to three different contributions, whose relative strength depends on the structure of the interacting molecules (see Sections 1.2, 4.1.3 and 4.1.4).

The first contribution is due to hydrogen bonds (see Section 4.1.4.2). Hydrogen bonding is typical of molecules containing hydrogen atoms (hydrogen bond donors) bound to electronegative atoms (hydrogen bond acceptors, usually oxygen, fluorine or nitrogen).[2]

The second contribution to intermolecular attractive forces is due to dipolar interactions, typical of polar molecules or polar groups. Dipolar interactions (see Section 4.1.4.2) can occur between two permanent dipoles (Keesom interactions) or between a permanent dipole and a polarizable molecule (permanent dipole–induced dipole interactions). The latter are called induction or Debye interactions.[3,4]

The value of the dipole moment (μ) may have strong consequences on macroscopic physical properties. For example the boiling points of water ($\mu = 1.847$ D) and ammonia ($\mu = 1.47$ D) are 100 °C and –33 °C respectively. The enthalpy of vaporization is 40 kJ mol^{-1} for water and 23.35 kJ mol^{-1} for ammonia.

The third contribution to the attractive intermolecular forces derives from London or dispersion forces (see Section 4.1.4.2).[5] This term is ubiquitous for all substances, independently of their polarity.

As an example of the application of these theoretical concepts to the practice of cleaning in conservation, we refer to three different solvents, ethanol, acetone and n-hexane, which are largely used in the field of conservation for the cleaning of surfaces of historical, artistic and architectonic interest (the structures of the solvents are reported in Figure 5.2).

Ethanol is a highly polar molecule characterized by the presence of the alcohol –OH group. Owing to its chemical structure, attractive interactions will be established between this solvent and the solid material, mainly hydrogen bonding and Debye–Keesom interactions.

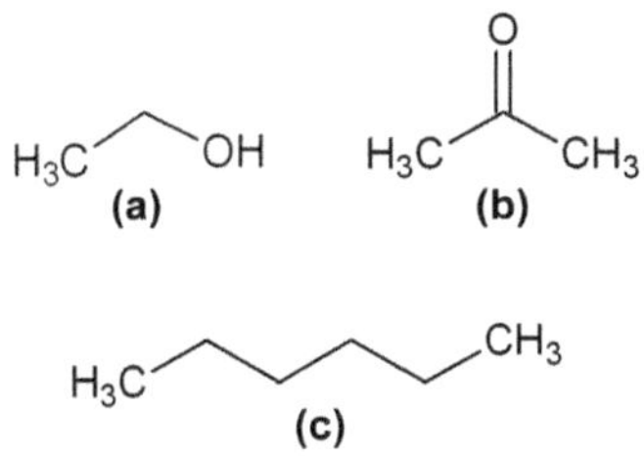

Figure 5.2 Chemical structures of ethanol (a), acetone (b) and n-hexane (c).

Acetone (dimethylketone) is a polar molecule that carries a C=O group. Such residues can interact through Debye–Keesom and London forces. Hydrogen bonding can be formed with a solute if this contains hydrogen atoms bound to electronegative atoms (F, O, N, Cl).

Finally, n-hexane presents only C–C and C–H covalent bonds (the difference in electronegativity between C and H is very low, 0.4) and therefore we expect it to establish interactions only with other apolar molecules *via* London forces.

It is important to recall that the materials that are to be solubilized and removed from artefact surfaces are mostly macromolecular (polymeric) substances. These materials are often characterized by a low physico-chemical stability that results in the alteration of their original structure (simultaneous reactions of depolymerization and crosslinking between different polymeric chains), mainly due to oxidation reactions induced by the atmospheric oxygen and catalysed by light.[6–8] One of the main consequences of this spontaneous process is a decrease of solubility in almost every solvent commonly used by conservators.[9]

In these cases the cleaning mechanism is a rather complex process involving two different steps. Once the solvent is in contact with the material to be removed, it penetrates inside the polymer layer and produces a swelling effect (see Section 7.3); the solvent absorption by the polymer increases together with the degree of swelling. In the following phase the polymer molecules lose adhesion with the layers underneath, allowing cleaning with a soft mechanical action.

5.3 Polarity of Solvents: f_d, f_p, f_h Parameters

In order to predict the behaviour of a solid substance (M) in the presence of a liquid solvent (S), the most commonly used approach is the one proposed by Hansen.[10,11] In this model the interaction forces between M and S molecules are supposed to be additive.

Hansen's theory defines the cohesive energy density δ_t^2 (see Section 1.3 for the theoretical description of the cohesion phenomena) for a solvent as a function of three partial solubility parameters δ_d, δ_p, δ_h (see Section 4.1.6). The values of the Hansen parameters δ_t can be calculated following physico-chemical models as indicated in the literature,[12] and they are usually reported in physical chemistry handbooks.

From the Hansen's parameters the reduced solubility parameters f_d, f_p and f_h can be calculated as:

$$f_d = \frac{100\delta_d}{\delta_d + \delta_p + \delta_h} \tag{5.2}$$

$$f_p = \frac{100\delta_p}{\delta_d + \delta_p + \delta_h} \tag{5.3}$$

$$f_h = \frac{100\delta_h}{\delta_d + \delta_p + \delta_h} \tag{5.4}$$

where:

f_h, f_p and f_d are the reduced solubility parameters due to hydrogen bonds, to polar forces, and to dispersion forces, respectively. These parameters are calculated as the percentage of the different forces that contribute to the overall interaction between the solvent molecules. They are defined in such a way that: $f_d + f_p + f_h = 100$.

In this way, each solvent can be univocally identified by a group of numerical parameters (see Table 5.1), whose values take into account the relative strength of hydrogen bonding, polar and dispersion interactions among the solvent molecules.[13]

The data reported in Table 5.1 show that, for non-polar solvents such as n-alkanes, the interactions among the molecules are dominated by dispersion forces (for n-hexane, f_d is equal to 100 while both f_p and f_h are zero).

Given two different substances, M and S, in order to predict their reciprocal solubility their respective f_d, f_p and f_h parameters have to be calculated and compared. The more similar these parameters, the more the substances are mutually soluble.

Table 5.1 Reduced solubility parameter for several solvents.

Solvent	f_d	f_p	f_h
Methanol	30	22	48
Ethanol	36	18	46
n-Propanol	40	16	44
n-Butanol	43	15	42
Benzyl alcohol	48	16	36
Dimethylsulfoxide	41	36	23
Acetone	47	32	21
Ethyl acetate	51	18	31
Chloroform	67	12	21
Xylenes	83	5	12
Cyclohexane	96	1	3
Mineral spirits	90	4	6
n-Hexane	100	0	0
Water	18	28	54
Formamide	28	42	30

5.4 Choice of Solvents and Teas Diagrams

The reduced solubility parameters for each solvent can be graphically represented in a triangular diagram (Teas chart) as indicated in Figure 5.3. On each side is reported the value of one parameter, spanning between 0 and 100.[14] Conventionally, the horizontal axis, the base of the triangle, represents f_d, the right side f_p and the left side f_h. Thus, every solvent is univocally identified in the triangle by a point defined by the values of the three coordinates, as sketched in Figure 5.3 for acetone ($f_d = 47, f_p = 32, f_h = 21$). From the coordinate on each axis, a line is drawn parallel to the nearby clockwise axis. The three lines cross at one point and locate the particular solvent on the Teas chart.

Solvents having similar structure are usually mutually miscible, possess similar solvating properties and are in close proximity on the Teas chart. Hydrocarbons, both aliphatic and aromatic, are located in the lower right corner of the diagram. Being apolar substances, they are characterized by low contributions of the polar and hydrogen bonding terms. On the other hand, the interactions among polar solvents (*e.g.* ketones) have a strong polar contribution and their position is shifted towards the left side of the diagram (Figure 5.4).

In restoration practice, the use of blends of solvents is very common, so that a discussion of the application of the Teas chart to mixtures of solvents is useful for the choice of the optimal cleaning tool, in terms of both efficacy of the cleaning action and toxicity of the liquid mixture used.

For example, a mixture of acetone and ethanol (60 : 40 w/w) is one of the classic blends used for many cleaning purposes. The values of f_d, f_p and f_h for

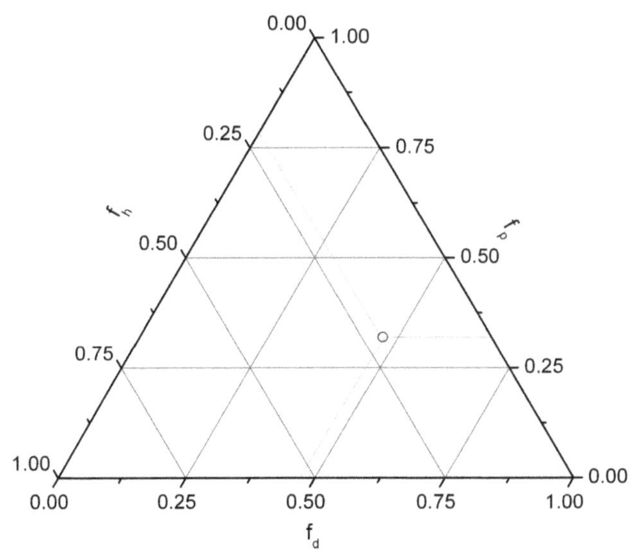

Figure 5.3 Position of acetone in the Teas diagram ($f_d = 47, f_p = 32, f_h = 21$).

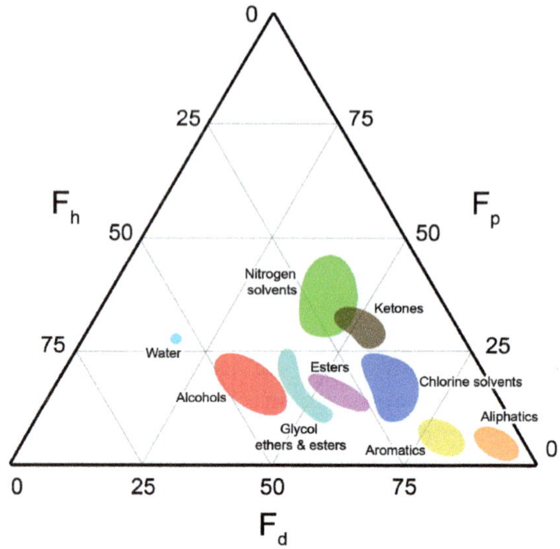

Figure 5.4 Teas solubility diagram indicating the solubility parameters of common families of solvents having similar properties (image by Michele Baglioni).

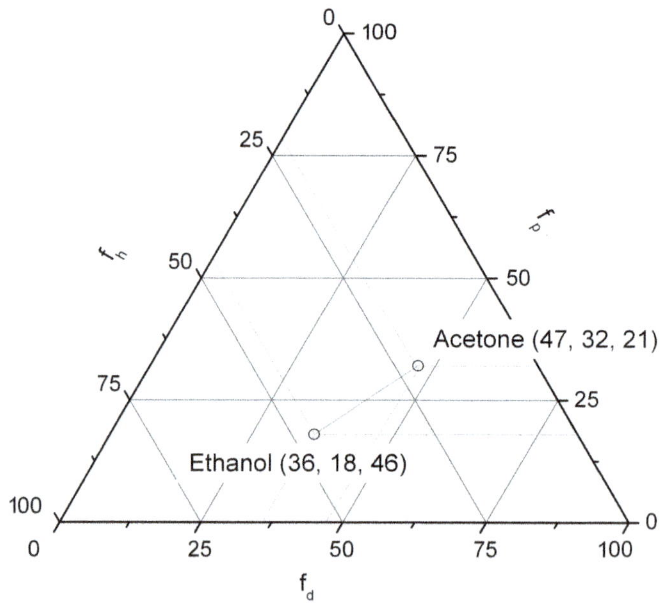

Figure 5.5 Position of acetone ($f_d = 47, f_p = 32, f_h = 21$) and ethanol ($f_d = 36, f_p = 18, f_h = 46$) in the Teas diagram.

the two solvents are indicated in Table 5.1. First of all, the two points corresponding to the solubility parameters of each solvent are identified as described above. Every mixture of ethanol and acetone will be represented by a mark on the segment that links the two pure solvents (Figure 5.5). The exact position of the mark will depend on the relative amount of the two solvents in the blend.

In order to find the specific position of the acetone–ethanol 60 : 40 mixture in the graph, we divide the segment into 10 equidistant points; the point corresponding to the mixture, in this case, will be the fourth starting from the acetone side of the segment (Figure 5.6). To obtain the coordinates of this point, we draw the lines parallel to the f_d, f_p and f_h axes, as described before, and read the corresponding intersection values (see Figure 5.6).

As an alternative to the graphical method, it is possible to calculate the value of the solubility parameters of the mixture by multiplying the values of the single solvents for the percentage of each component, and summing them to obtain the mixture parameters. The calculations for the acetone–ethanol 60 : 40 mixture are reported in Table 5.2.

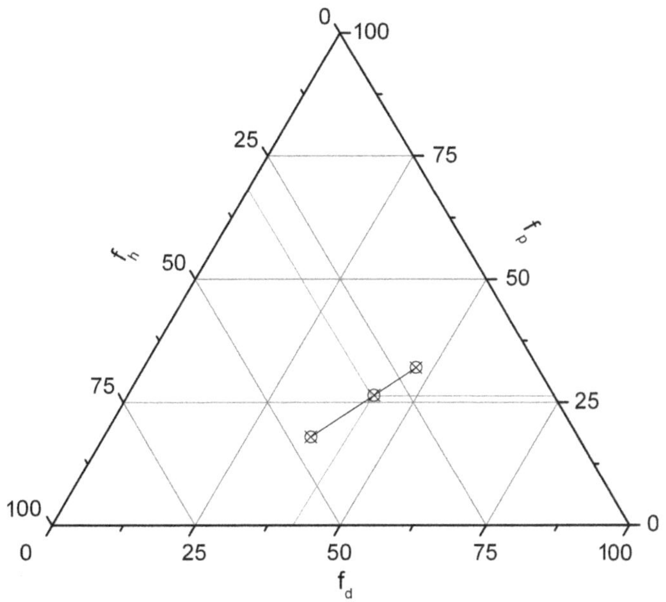

Figure 5.6 Graphical calculation of the values of the solubility parameters for a mixture acetone–ethanol 60 : 40.

Table 5.2 Calculations of the f_d, f_p and f_d parameters for the acetone–ethanol 60 : 40 mixture.

	f_d	f_p	f_h	*Contribution to the solubility parameters of the mixture*		
Ethanol	36	18	46	(36 18 46)×40/100	(14.4 7.2 18.4)	+
Acetone	47	32	21	(47 32 21)×60/100	(28.2 19.2 12.6)	=
Mixture					(42.6 26.4 31)	

Table 5.3 Calculations of the f_d, f_p and f_d parameters for a mixture of ethanol, n-butylacetate and acetone.

	f_d	f_p	f_h	*Contribution to the solubility parameters of the mixture*		
Ethanol	36	18	46	(36 18 46)×30/100	(10.8 5.4 13.8)	+
n-Butylacetate	60	13	27	(60 13 27)×30/100	(18 3.9 8.1)	
Acetone	47	32	21	(47 32 21)×40/100	(18.8 12.8 8.4)	=
Mixture					(47.6 22.1 30.3)	

For multicomponent mixtures the approach is exactly the same: the solubility parameters of a mixture of ethanol, n-butylacetate and acetone are calculated in Table 5.3.

The Teas charts also allow one to obtain approximate solubility diagrams for solid substances. By testing the solubility of a given solute in various solvents, it is possible to identify an approximate solubility area for the material by connecting all the points representing the solvents in which the studied material is soluble. All the solvents and mixtures of solvents lying within that region are effective in the solubilization of the material considered.[9]

For example, in the case of Paraloid B72 [poly(EMA/MA) 70 : 30], a copolymer largely used in conservation as a coating, fixative or consolidant, its solubility in many solvents of different polarity is considered. The results indicate that the copolymer is soluble in several solvents, including toluene, xylene, dichloromethane, terachloromethane, 2-butoxyethanol, ethyl acetate, n-butyl acetate, acetone, butanol, n-methyl-2pyrrolidone and *N*,*N*-dimethyl formamide. All these solvents define a solubility region in a triangular diagram, as indicated by the open dots in Figure 5.7.

When the solubility region is available for a given product, it is possible to predict whether a solvent or a mixture of solvents will be able to dissolve it or not, simply by checking the solvent position with respect to the solubility region.

A further advantage given by this approach is the possibility of avoiding the use of toxic solvents, simply by using a liquid (either a pure solvent or a mixture) whose solubility parameters are as similar as possible to the toxic solubilizing agent. For example if we have a product that is soluble in chloroform – a highly toxic solvent, carcinogenic and harmful for the environment – we can replace this solvent ($f_d = 67$; $f_p = 12$; $f_h = 21$) with a similar liquid or with a mixture that possesses similar solubility parameters (for example a mixture composed of isopropyl alcohol, 38 v/v%, n-octane, 46 v/v% and methyl ethyl ketone 16 v/v%).

As a final note on the application of solubility parameters and Teas charts, it is worth recalling the Feller test, a useful method to find the correct solvent or solvent blend for the cleaning of a painted surface (see also the Further Suggested Reading section).

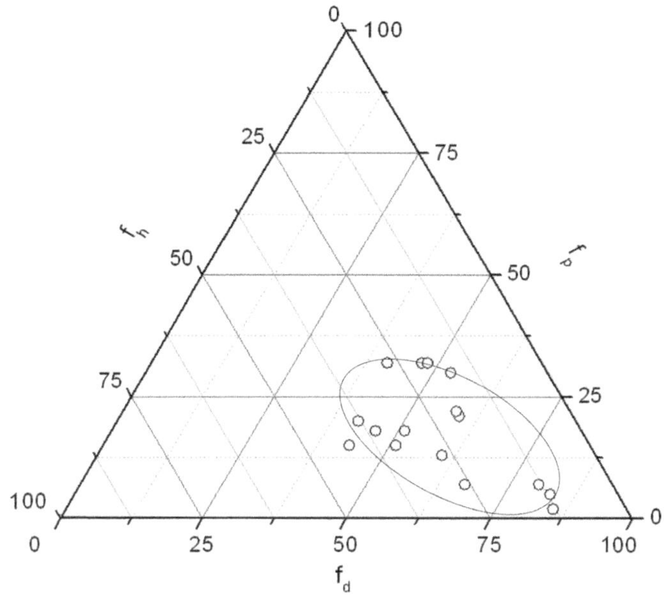

Figure 5.7 Solubility diagram of Paraloid B72. The open circles indicate the solvents in which the copolmer is soluble.

The test can be outlined as follows:

1. Preparation of 13 mixtures composed of different amounts of cyclohexane, toluene and acetone, and thus characterized by different polarities (quantitatively indicated by the f_d values);
2. Application of the mixtures (by means of a cotton swab roll) onto 13 small areas of the painted surface, in a region representative of the surface conservation status;
3. Identification of the mixture that provides the best performance by means of visual analysis (a mobile optical microscope can also be used);
4. Choice of a solvent or solvent blend having an f_d value equal to the one of the best mixture. A dedicated software program, similar to the one described in Section 5.6, can be used for this purpose. The main criteria that drive the choice of the solvent or blend are low toxicity and high volatility.

5.5 Use and Safety of Liquid Solvents

Given that solvents can solubilize various materials, they are widely used as cleaning tools for painted surfaces. It is not possible to classify a solvent as "safe" in an absolute way, and even if some substances are less hazardous than others, the manipulation of every solvent is characterized by specific risks that have to be considered. For example, it is possible that a solvent is "safe" if used

Table 5.4 Flash point (F_p) values for some of the solvents commonly used in conservation.

Solvent	*Flash points, F_p (°C)*
Isopropanol	11.7
Methyl ethyl ketone (MEK)	–9.9
n-Methyl-pyrrolidone (NMP)	91
Acetone	–17 °C

in low amounts or in the presence of correct ventilation for short periods of time, but otherwise it becomes harmful (*e.g.* alcohols). Furthermore, some people may be allergic to specific substances although the same products are completely harmless to other persons.

The most important hazards typical of solvents are mainly related to three different factors. The first is their flammability and/or the possibility of explosion. The flammability of a solvent is quantitatively indicated by the flash point (F_p),[15] which is the lowest temperature at which a volatile liquid can evaporate in air to form a mixture that can ignite in the presence of an ignition source. In Table 5.4 the F_p values for some of the solvents commonly used in conservation, for the cleaning of paintings, are reported.

The second important factor that defines a solvent hazard is toxicity. The word "toxic" is related to the biological effects that exposure to a solvent can cause on a human body. The mechanisms of exposure to the action of a solvent can be mainly due to the inhalation of vapours, to direct contact (eyes or skin) or to accidental ingestion. From the clinical point of view, the effects of the exposure to solvents can be categorized into:

- Acute, short-term effects such as irritation of the eyes, nausea, headache;
- Chronic, long-term effects such as liver problems, cancer and/or problems affecting the reproductive organs.

The concentrations of the harmful substances, and the duration of the exposure, are important parameters to be taken into account. Their balance is a crucial factor for the effects of the contact between the solvents and the operator.

For every substance it is possible to identify a specific "threshold level", *i.e.* the maximum level of exposure below which no toxic effects are observed. In the United States the exposure limits (threshold limit values, TLV, available online at http://www.acgih.org) are rigorously expressed by the American Conference of Governmental Industrial Hygienists (ACGIH) as the average concentration (in ppm) of a substance in the air to which an average worker can be exposed for 8 hours per day or 40 hours per week. For highly toxic solvents the TLV are also calculated for very short periods (*i.e.* 15 minutes). The US TLV values are acknowledged also by the European Union through two different guidelines: 200/39/CE and 2006/15/CE (available online at http://eur-lex.europa.eu).

From the physico-chemical point of view, one of the most important parameters in determining the toxicity of a solvent is its vapour density, V_D (the density of the vapour of a solvent with respect to that of the uncontaminated air). If a solvent is characterized by a V_D value higher than that of air, it will accumulate at the bottom of the room. Specific V_D values can be calculated by means of the Antoine vapour–pressure equation, as indicated in the literature.[16]

The third important hazard factor is given by the possibility of violent reactions occurring between the solvents and other chemicals, particularly acids, causing fire or explosions. In the presence of an ignition source, the flames can propagate causing a fire as high and dangerous as the amount of flammable solvents reachable by the fire itself.

In order to minimize all the risks connected with their use, all solvents have to be stored in special cabinets, far away from any ignition source. Small amounts of liquid (the minimum quantity needed in 24 hours) can be left out of the cabinets without any risk for the operators only in the presence of a functioning ventilator hood, and respecting all the safety recommendations indicated on the labels on the solvent bottles. Furthermore, all liquid substances that react violently with each other (*e.g.* acids and bases, or oxidizing and reducing chemicals) should be stored separately, in order to avoid any possible contact.

As indicated in Figure 5.8, all chemical bottles are characterized by labels that indicate the most important physico-chemical properties of the substance (molecular weight, boiling temperature, density, purity, *etc.*) and the main recommendations regarding storage, hazards and safety. In Europe the correct procedure for the classification, labelling and packaging of substances and mixtures is defined by the CLP Regulation (which aligns the EU law to the US Globally Harmonized System of Classification and Labelling of Chemicals (GHS), substituting for the European Union Directive 67/548/EC – both these documents are available online at http://eur-lex.europa.eu/). Thus in order to store each product correctly, all the information must be taken into account every time a chemical product is manipulated. Figure 5.8 reports the image of an example of a label of a chemical with the explanation of the meaning of all the symbols and all the information reported on it.

The different sections that appear on a chemical's label are (see Figure 5.8):

1. Product name.
2. Volume or mass of the package.
3. Purity (sometimes the degree of hydrolysis is also specified).
4. Catalogue number (for any given chemical, each company has a specific catalogue number).
5. CAS registry number. This is a numerical code defined by the Chemical Abstract Service and assigned to every substance. It allows the unequivocal identification of a substance.
6. Elemental formula.
7. Fundamental physico-chemical information: chemical formula, formula mass, density (in $g\,L^{-1}$ or $kg\,dm^{-3}$), boiling point, melting point and flash point. All temperatures are expressed in degrees Celsius (°C).

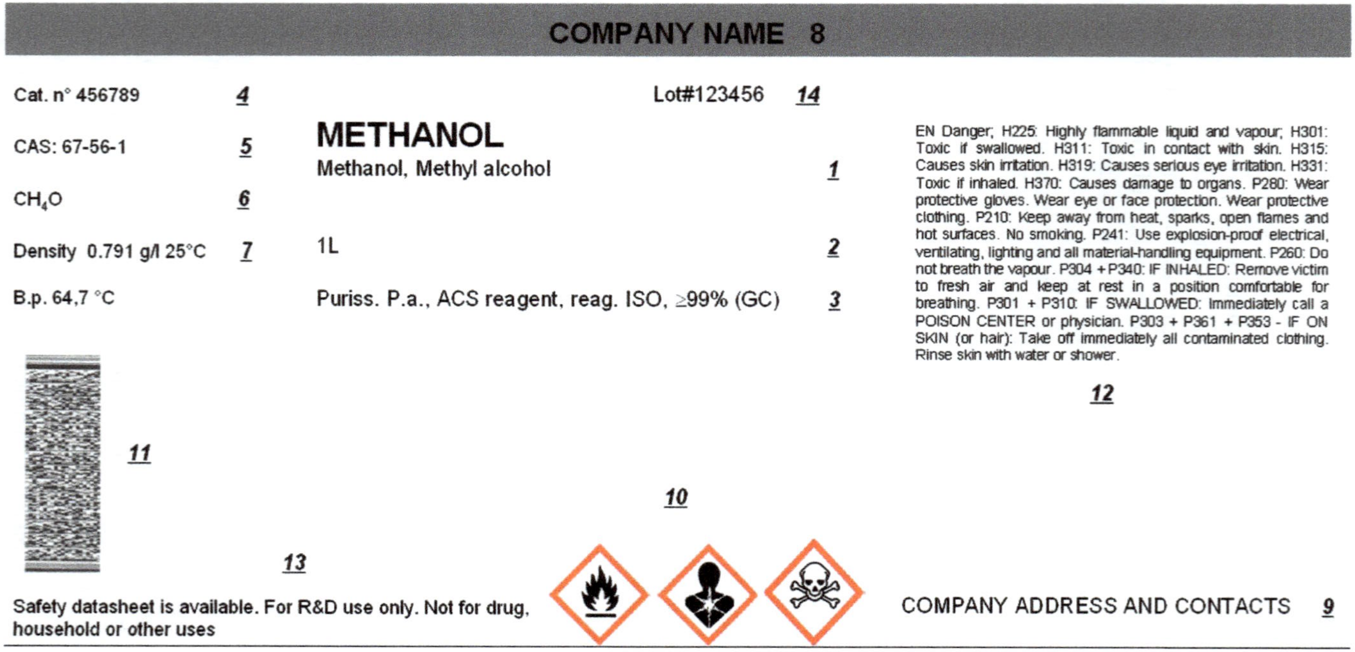

Figure 5.8 Example of a label on the bottle of a chemical (in this case methanol).

8. Name of the company that sells the chemical product.
9. Address and contact details of the company that sells the chemical product.
10. Hazard pictograms, *i.e.* images forming part of the Globally Harmonized System of Classification and Labelling of Chemicals (GHS) that graphically indicate the most important hazards typical of each substance. They are shown in Table 5.5.
11. Bar code.
12. Hazard and Precautionary Statements.
13. Supplementary information.
14. Lot number (indicates the lot of production, which is useful for the exact identification of every bottle of the chemical).

Other indications for the correct manipulation and storage of chemicals are the Hazard and Precautionary Statements (H/P statements, or H/P phrases); these codes are used to classify chemicals depending on the risk related to their manipulation (there are 69 Hazard phrases, of which 20 are present only in the CLP Regulation but not in the GHS system, which are known as H phrases and are available online at http://eur-lex.europa.eu/) and on the safety precautions to be adopted in the manipulation of chemicals (136 Precautionary phrases, called P, available online at http://eur-lex.europa.eu/). For every compound it is possible to define an H/P statement that consists of a hazard part (H) and a precautionary part (P), followed by a combination of numbers (each number indicates a specific phrase). The complete list of H and P phrases is usually reported in all the catalogues of chemicals produced by the most important international companies (*e.g.* Sigma Aldrich, Fluka, but they are also available online at http://www.hse.gov.uk). Examples are given in Tables 5.6 and 5.7.

The actual H/P phrases replace the old Risk and Safety Statements (R/S statements, or R/S phrases) reported in the Annex III of the European Union Directive 67/548/EEC (available online at http://eur-lex.europa.eu/): the R and the S phrases have the same function as the Hazard statements and the Precautionary statements respectively.

In general the precautions to be adopted for the correct handling of a solvent can be indicated schematically as follows:

1. In the presence of volatile solvents and/or other harmful chemicals, always work in a fume hood, or with local exhaust ventilation (LEV), or in a well-ventilated area.
2. Keep all containers tightly closed, and dispose of the empty cans properly.
3. Never use open flames near flammable solvents, use electrical heating instead. In case of extremely volatile and flammable solvents (diethyl ether, carbon sulfide, *etc.*) use only a warm water bath if it is necessary to increase the temperature of the solution, to avoid the sparks originated by electrical heating devices.

Table 5.5 Hazard pictograms.

Description of the pictogram	*Pictogram*	*Hazard class*
Exploding bomb		– Explosives, self-reactive substances and mixtures – Organic peroxides
Flame		– Flammable gases, aerosols, liquids, solids – Self-reactive substances and mixtures – Pyrophoric liquids and solids – Self-heating substances and mixtures – Substances and mixtures which, in contact with water, emit flammable gases – Organic peroxides
Flame over circle		Oxidizing gases and liquids
Gas cylinder		Gases under pressure: – Compressed gases – Liquefied gases – Refrigerated liquefied gases – Dissolved gases
Corrosion		Corrosive to metals, skin corrosion, eye damage
Skull and crossbones		Acute toxicity (oral, dermal, inhalation)
Exclamation mark		– Toxicity (oral, dermal, inhalation) – Skin and eye irritation – Skin sensitization – Specific Target Organ Toxicity for a single exposure
Health hazard		– Respiratory sensitization – Mutagenicity – Carcinogenicity – Reproductive toxicity – Specific Target Organ Toxicity for a single or a multiple exposure – Aspiration hazard
Environment		Hazardous to the aquatic environment

Table 5.6 Hazard and Precautionary Statements for acetone.

Acetone	
Hazard pictograms	
Hazard statements	H225-H319-H336
Precautionary statements	P210-P261-P305 + P351 + P338
Recommended personal protective equipment	Eyeshields, faceshields, full-face respirator (US), gloves, multi-purpose combination respirator cartridge (US)
*Risk statements	R:11-36-66-67
*Safety statements	F:9-16-26

*Both for risk and safety statements, dashes separate different phrases; these statements have been replaced by Hazard and Precautionary statements.

Table 5.7 Hazard and Precautionary Statements for methanol.

Methanol	
Hazard pictograms	
Hazard statements	H225-H301-H311-H331-H370
Precautionary statements	P210-P260-P280-P301 + P310-P307 + P311
Recommended personal protective equipment	Eyeshields, faceshields, full-face respirator (US), gloves, multi-purpose combination respirator cartridge (US)
*Risk statements	R: 11-23/24/25-39/23/24/25
*Safety statements	S: 7-16-36/37-45

*Both for risk and safety statements, dashes separate different phrases; these statements have been replaced by Hazard and Precautionary statements.

4. Never flush flammable solvents down the drain.
5. Avoid the inhalation of solvent vapours.
6. Avoid contact of the solvent with the skin – many solvents are easily absorbed through the skin. They also tend to dry the skin and may cause sores and wounds. Always use goggles and gloves.
7. Make sure you know the safety procedures and the location of fire extinguishers, eye wash station, safety showers, spill cleanup materials, first aid kit, fire alarm and emergency phone numbers. Never work alone in a laboratory. If necessary, notify your presence in the laboratory to other people, especially during weekends and holidays.
8. Read the material safety data sheets (MSDS) for all information.

5.6 The Triangle of Solubility at ISCR, Rome

In order to make the use of the Teas-triangle of solubility easier for restorers, a visual software has been developed at the *Istituto Superiore per la*

Conservazione ed il Restauro (ISCR) in Rome, by Maurizio Coladonato and Paolo Scarpitti. It is a multi-language, open-source tool available online.[12,17]

This free software enables the user to improve the formulation of solvent mixtures and to replace harmful solvents with less harmful liquids. Thus, it allows the evaluation of solvents according to needs in conservation and restoration, taking into account occupational health and safety regulations. In the following images, an overview of the main functions of the program is presented.

In the three fields labelled A, B and C (right top, red box) the three different main classes of solvents most commonly used for the cleaning of paintings are shown: alcohols, hydrocarbons and ketones. For each solvent the values of the solubility parameters f_d, f_p and f_h are reported. Once the solvent has been selected (in the example reported in Figure 5.9, isopropyl alcohol, belonging to group A), it will be identified in the Teas-triangle on the left side of the page by the letter corresponding to the appropriate group (in this case group A, red box on left side of Figure 5.9).

The solubility regions of the most diffused classes of materials, which may be spread on the surface of paintings, are also reported in the Teas-triangle, in

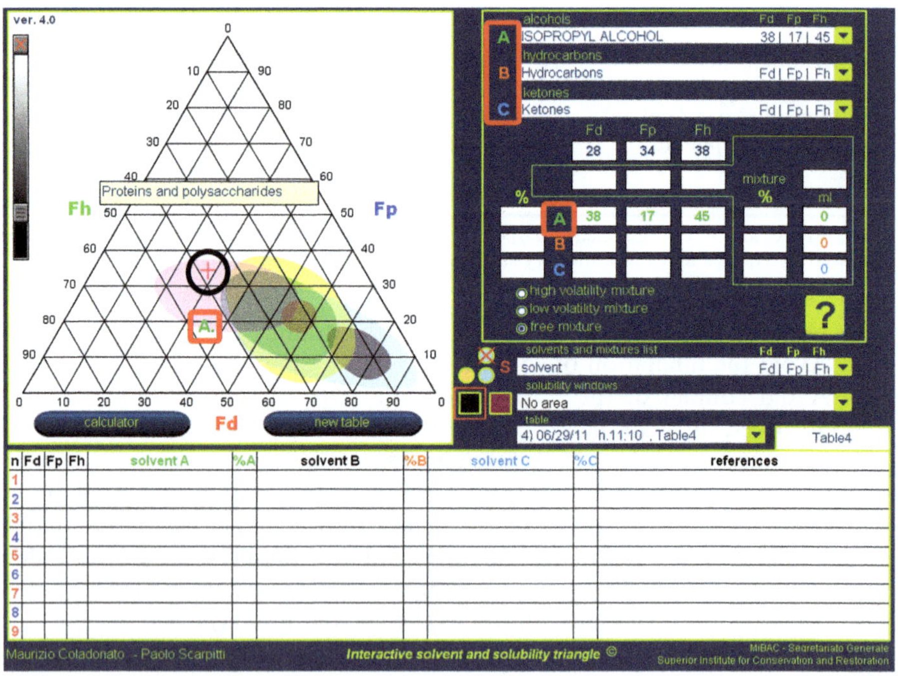

Figure 5.9 Use of the visual software developed at the *Istituto Superiore per la Conservazione ed il Restauro* (ISCR) in Rome by Maurizio Coladonato and Paolo Scarpitti. Step 1.
Image taken from the website http://iscr.beniculturali.it/flash/progetti/TriSolv/TriSolv.html and readapted by Emiliano Carretti.

different colours (the pink area corresponds to proteins and polysaccharides, the yellow one to synthetic resins, the light blue to oils, *etc.*). The identification of the materials is done simply by means of the computer mouse: as soon as the cursor passes onto a region (black circle in Figure 5.9), a label with the name of the material appears.

In Figure 5.10 the example of a blend composed of three different solvents is shown: ethanol (A; $f_d = 36, f_p = 18, f_h = 46$), isooctane (B; $f_d = 98, f_p = 1, f_h = 1$) and acetone (C; $f_d = 47, f_p = 32, f_h = 21$). The red box highlights a table in which the composition of the mixture (in volume %) can be specified.

When several formulations are chosen, the properties of each formulation (volume % and the f_d, f_p, f_h values calculated following the procedure described in Section 5.3) are reported in the table on the bottom of the page (black box). By clicking with the computer mouse inside the triangular area defined by the points corresponding to the three solvents (triangular yellow box), it is possible to identify new mixtures (see the mixtures labelled with the numbers between 1 and 6 inside the black box of the triangular graph), as a function of the f_d, f_p, f_h values.

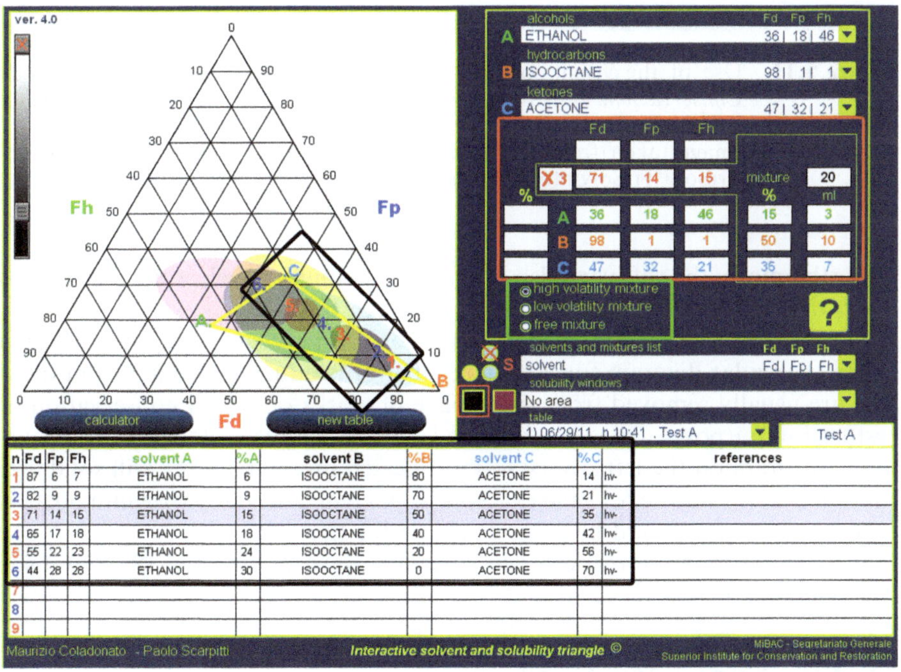

n	Fd	Fp	Fh	solvent A	%A	solvent B	%B	solvent C	%C	
1	87	6	7	ETHANOL	6	ISOOCTANE	80	ACETONE	14	hv-
2	82	9	9	ETHANOL	9	ISOOCTANE	70	ACETONE	21	hv-
3	71	14	15	ETHANOL	15	ISOOCTANE	50	ACETONE	35	hv-
4	65	17	18	ETHANOL	18	ISOOCTANE	40	ACETONE	42	hv-
5	55	22	23	ETHANOL	24	ISOOCTANE	20	ACETONE	56	hv-
6	44	28	28	ETHANOL	30	ISOOCTANE	0	ACETONE	70	hv-
7										
8										
9										

Figure 5.10 Use of the visual software developed at the *Istituto Superiore per la Conservazione ed il Restauro* (ISCR) in Rome by Maurizio Coladonato and Paolo Scarpitti. Step 2.
Image taken from the website http://iscr.beniculturali.it/flash/progetti/TriSolv/TriSolv.html and readapted by Emiliano Carretti.

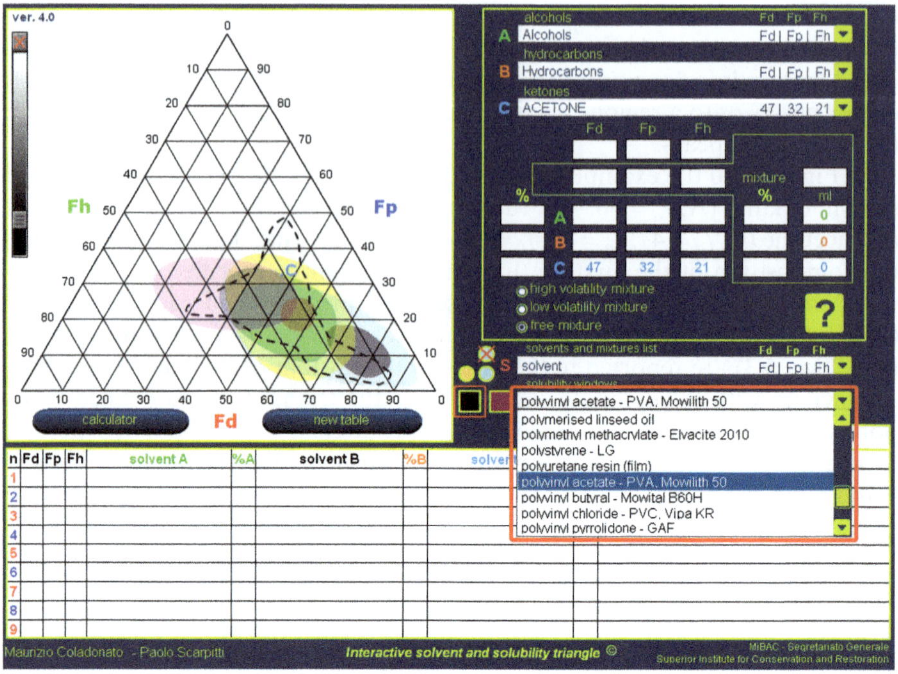

Figure 5.11 Use of the visual software developed at the *Istituto Superiore per la Conservazione ed il Restauro* (ISCR) in Rome by Maurizio Coladonato and Paolo Scarpitti. Step 3.
Image taken from the website http://iscr.beniculturali.it/flash/progetti/TriSolv/TriSolv.html and readapted by Emiliano Carretti.

Furthermore, the software also offers the possibility to select the mixtures as a function of their volatility (Figure 5.10, green box).

Finally, as indicated in Figure 5.11, it is possible to identify the solubility area in the triangular diagram for a wide number of products, selecting from those that are usually removed from the surface of works of art for conservation or restoration purposes. Once a substance is selected, a black dotted line appears in the diagram on the left. Every solvent located inside that area will solubilize the material (in Figure 5.11 the example of polyvinyl acetate, Mowilith 50®, is reported; acetone should be a good solvent for this product).

We suggest that the reader gains further insight visiting the ISCR website (see reference 17).

Further Suggested Reading

On the polarity parameters of solvents:

C. Reichardt, Empirical parameters of the polarity of solvents, *Angewandte Chemie Int Edn*, 1965, **4**, 29.

On the Feller test:

R. L. Feller, N. Stolow and E. H. Jones, *On Picture Varnishes and their Solvents*, National Gallery of Art (Washington), 1985.

On safety rules:

http://www.acgih.org/
http://europa.eu/geninfo/atoz/en/index_1_en.htm
http://www.cdc.gov/niosh/ipcs/icstart.html

Review and Questions

This chapter deals with the solubilization of substances by solvents, referring to practical criteria that can be used in conservation practice. The following points have been considered:

- Given two different substances, a solvent S and a solute M, the solubilization process is driven by the intermolecular interactions. The solubilization process is favoured when the structure (and the polarities) of the interacting molecules are similar, as stated by the empirical ancient rule *"similia similibus solvuntur"* (like dissolves like).
- Quantitatively, in order to obtain the solubilization of M in S, the intermolecular interactions occurring between the solvent and the solute molecules should be similar to those typical of the pure species M and S.
- Every pure solvent can be univocally identified through a specific set of parameters (f_h, f_p, f_d) that respectively define the percentage contribution of hydrogen bonds, polar interactions and dispersion forces to the overall intermolecular interaction energy.
- The f_h, f_p, f_d values can be reported in a ternary diagram (Teas chart) where each pure solvent or solvent mixture is identified by a single point. In a Teas chart it is possible to identify solubilization areas for a specific material or class of materials. For easy and correct use of the Teas chart there is free-source software available online.
- All the chemicals should be handled and stored following the indications reported on the labels and the technical sheets, particularly taking into account the security information (mainly R and S phrases and the hazard pictograms).

Questions

1) What is the practical function of the solubility parameters f_d, f_p, f_h?
2) Try to describe the hydrogen bond by using a maximum of 200 words. Is it correct to mention hydrogen bonding for ammonia? And for hexane?
3) Calculate the values of the f_d, f_p, f_h parameters for a binary mixture of acetone–2-propanol 30 : 70 and for a ternary mixture composed of acetone–2-propanol–isoctane 20 : 50 : 30.

4) What is the practical relevance of the phrase "*similia similibus solvuntur*"? How can this phrase be explained in terms of the solubility parameters f_d, f_p, f_h?
5) By using the ISCR free of charge application "Interactive triangle of solvents and solubility©" try to identify the best solvent and/or the best blend of solvents for the solubilization of a class of materials (*i.e.* natural resins).
6) Given a bottle of a chemical, which are the most important indications about the risks connected with its use, its correct handling and storage? Where it is possible to find them?

Answers

1) The f_d, f_p and f_h indicate the contribution of the different forces to the overall interaction between the molecules of a given substance:
   - f_d describes the contribution of dispersion forces to the intermolecular interactions;
   - f_p describes the contribution of polar forces to the intermolecular interactions. This contribution is higher for polar molecules, *e.g.* the f_p value for dimethylsulfoxide (36) is much higher than that of cyclohexane (1);
   - f_h describes the contribution of hydrogen bonding to the intermolecular forces. This contribution is maximum for all the substances characterized by the presence of a H atom bond to a much more electronegative atom (*e.g.* water or ammonia).
2) Hydrogen bonding is an attractive interaction typical of molecules containing hydrogen atoms covalently bound to more electronegative atoms; the electrons forming the covalent bond H–X (X is the more electronegative atom, such as oxygen, fluorine or nitrogen) are attracted by the X atom. This results in an accumulation of negative charge on the X atom and in a corresponding deficiency of electrons on the hydrogen atom. Thus, an attractive interaction between the positively charged hydrogen and the more electronegative atom of the nearby molecule can occur. Ammonia can produce hydrogen bonds because the H atoms (electronegativity = 2.1) are bound to an N atom that possesses a much higher electronegativity (3.0). In the case of hexane, the H atoms are bound to C atoms (electronegativity = 2.5), and the polarity of the bond is too low to allow the formation of H bonds.
3) Case 1: binary mixture of acetone–2-propanol 30 : 70. The solubility parameters for the two molecules are: acetone ($f_d = 47$, $f_p = 32$, $f_h = 21$), 2-propanol ($f_d = 38$, $f_p = 17$, $f_h = 45$).
   Contribution of acetone to f_d of the mixture $= 47 \times 30/100 = 14.1$
   Contribution of acetone to f_p of the mixture $= 32 \times 30/100 = 9.6$
   Contribution of acetone to f_h of the mixture $= 21 \times 30/100 = 6.3$
   Contribution of 2-propanol to f_d of the mixture $= 38 \times 70/100 = 26.6$
   Contribution of 2-propanol to f_p of the mixture $= 17 \times 70/100 = 11.9$

Contribution of 2-propanol to f_h of the mixture = 45×70/100 = 31.5
f_d of the mixture: 14.1 + 26.6 = 40.7
f_p of the mixture: 9.6 + 11.9 = 21.5
f_h of the mixture: 6.3 + 31.5 = 37.8

Case 2: ternary mixture composed of acetone–2-propanol–isooctane 20 : 50 : 30. The solubility parameters for the three molecules are: acetone (f_d = 47, f_p = 32, f_h = 21), 2-propanol (f_d = 38, f_p = 17, f_h = 45), isoctane (f_d = 100, f_p = 0, f_h = 0).
Contribution of acetone to f_d of the mixture = 47×20/100 = 9.4
Contribution of acetone to f_p of the mixture = 32×20/100 = 6.4
Contribution of acetone to f_h of the mixture = 21×20/100 = 4.2
Contribution of 2-propanol to f_d of the mixture = 38×50/100 = 19
Contribution of 2-propanol to f_p of the mixture = 17×50/100 = 8.5
Contribution of 2-propanol to f_h of the mixture = 45×50/100 = 22.5
Contribution of isooctane to f_d of the mixture = 100×30/100 = 30
Contribution of 2-propanol to f_p of the mixture = 0×30/100 = 0
Contribution of 2-propanol to f_h of the mixture = 0×30/100 = 0
f_d of the mixture: 9.4 + 19 + 30 = 58.4
f_p of the mixture: 6.4 + 8.5 + 0 = 14.9
f_h of the mixture: 4.2 + 22.5 + 0 = 26.7

4) The expression "*similia similibus solvuntur*" indicates that if a substance M has to be solubilized, the best solvent or solvent mixture is the one characterized by f_d, f_p, f_h values similar to those of M.
5) By using the free of charge application available online at the website http://iscr.beniculturali.it/flash/progetti/TriSolv/TriSolv.html, it is possible to determine the best solvent or blend of solvents for the solubilization of all classes of substances. Suppose we have a layer to be removed constituted by a natural resin. You move with the cursor into the centre of the area corresponding to this class of products. The program automatically calculates the set of f_d, f_p, f_h parameters (for natural resins f_d = 50, f_p = 25, f_h = 25). Then you can identify the best solvent or solvent blend to be used, taking into account their selectivity, their safety and their cost.
6) All the main information about the risks, the correct handling and the storage of a chemical are reported on the label stuck on the bottle or in the safety sheet. The Hazard statements (H phrases) indicate the hazards of the chemical substance and the Precautionary statements (P phrases) give recommendations about its correct handling; the hazard pictograms graphically describe the physical, health and environmental risks typical of the substance considered.

References

1. N. J. Israelachvili, *Intermolecular and Surface Forces*, Elsevier, Amsterdam, 3rd edn, 2011.

2. G. A. Jeffrey, *An Introduction to Hydrogen bonding*, Oxford University Press, Oxford, 1997.
3. V. A. Parsegian, *Van der Waals Forces: A Handbook for Biologists, Chemists, Engineers, and Physicists*, Cambridge University Press, Cambridge, 2005.
4. D. McQuarrie and J. Simon, *Physical Chemistry, A Molecular Approach*, University Science Books, USA, 1997.
5. R. Eisenschitz and F. London, *Z. Physik*, 1930, **60**, 491.
6. R. L. Feller, *Bull. Inst. R. Patrim. Artist.*, 1975, **15**, 135.
7. J. Hennig, *J. Kunstoffe Fortschriftsberidite*, 1978, **7**, 13.
8. K. Morimoto and S. Suzuki, *J. Appl. Polym. Sci.*, 1972, **16**, 2947.
9. R. L. Feller, in *Accelerated Aging: Photochemical and Thermal Aspects*, The Getty Conservation Institute, Los Angeles, 1994, p. 63.
10. C. Hansen and A. Beerbower, in *Kirk-Othmers Encyclopedia of Chemical Technology, supplementary volume*, ed. Kirk-Othmers, Interscience, Markham, Canada, 1971, p. 889.
11. A. F. M. Barton, *CRC Handbook of Solubility Parameters and other Cohesion Parameters*, CRC Press Inc., Boca Raton, USA, 4th edn, 1998.
12. J. Shibata, K. Fuji, N. Murayama and H. Yamamoto, *Kona*, 2002, **20**, 263.
13. C. V. Horie, *Materials for Conservation*, Butterworths, London, 1987.
14. J. P. Teas, *Journal of Paint Technology*, 1968, **40**, 19.
15. T. Suzuki, K. Ohtaguchi and K. Koide, *J. Chem. Engin. Japan*, 1991, **24**, 258.
16. J. A. Dean, *Lange's Handbook of Chemistry*; McGraw-Hill, New York, 1985.
17. http://iscr.beniculturali.it/flash/progetti/TriSolv/TriSolv.html (accessed 29 November 2012).

CHAPTER 6

Cleaning II: Surfactants and Micellar Solutions

P. LO NOSTRO

University of Florence, Italy
Email: pln@csgi.unifi.it

6.1 Surfactants

Surfactants are involved in everyday life and advanced chemistry, through a countless number of applications ranging from detergents and cleaning, food chemistry, oil recovery, pharmaceuticals, nanotechnology, *etc.* In this chapter we will discuss the structure and properties of surfactants, their ability to lower the interfacial tension, and the properties of the self-assemblies they produce upon aggregation in water. The chapter will end with a short review of the main techniques used to study the self-assemblies, and some aspects of their environmental impact. The information provided in this chapter grants the theoretical background for the applicative aspects related to the conservation of cultural heritage, which are dealt with in Chapter 7. Moreover, the chapter can be intended as a support for the formulation of systems for the cleaning of works of art (*e.g.* emulsions, microemulsions and micellar solutions), introducing some of the key parameters and criteria that underlie the stability of such systems.

6.1.1 Definitions and General Remarks

Surfactant molecules derive their name from their capability to modify the surface energy between two immiscible phases, such as air and water, or oil and

RSC Nanoscience & Nanotechnology No. 28
Nanoscience for the Conservation of Works of Art
Edited by Piero Baglioni and David Chelazzi

Published by the Royal Society of Chemistry, www.rsc.org

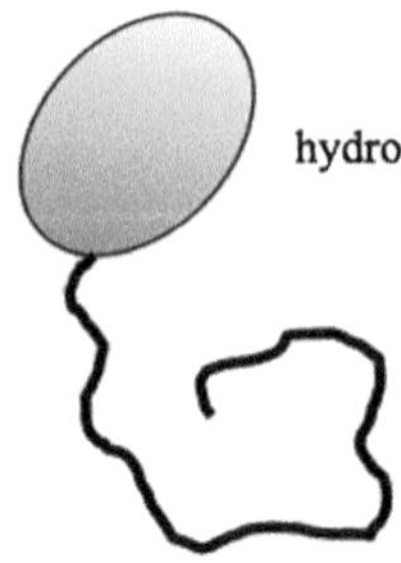

Figure 6.1 Schematic illustration of the molecular structure of an amphiphile, consisting of a hydrophilic head group covalently linked to a hydrophobic tail.

water. They belong to the more general class of *amphiphiles*, *i.e.* chemicals that combine in their structure both a hydrophilic and a hydrophobic moiety (see Figure 6.1) that have different interactions with a common selective solvent.

The hydrophilic (or solvophilic) part will interact favourably with water (or with the solvent), while the hydrophobic segment will avoid any contact with water. This is the main reason why surfactants produce in water a wide family of nanosized aggregates, in the attempt to escape from direct contact with the solvent. We will see later that this can be done in different ways, minimizing repulsions and maximizing attractive interactions.

6.1.2 Molecular Structure of Surfactants

6.1.2.1 Definition

In general, surfactants comprise a hydrophobic segment covalently bound to a hydrophilic head group. In this way the two incompatible units cannot separate, but are forced to remain bound in the same molecular entity, giving rise to peculiar behaviours and properties in aqueous dispersions. Phenomena such as adsorption at interfaces, formation of self-assembled particles, solubilization of hydrophobic materials, and much more derive from the particular architecture of surfactants or amphiphiles.

Water and an organic phase (such as a hydrocarbon or "oil") do not mix, and phase separate immediately. This is due to the different composition of their molecules, and thus to the different kinds of intermolecular interaction that are established by water molecules (dipolar interactions, hydrogen bonding) and by the oil molecules (mainly van der Waals or dispersion interactions), as described in Section 4.1.4. For similar reasons, water will dissolve polar compounds, while oils will generally mix with apolar chemicals, according to the "like dissolves like" rule of thumb (see Section 4.1.3).

The hydrophobic and the hydrophilic portions in a surfactant will then establish different interactions with a "selective" solvent. This means that, if the solvent is water, the polar head group will tend to maximize the contact (*i.e.* interactions with the solvent molecules), while the apolar segment will try to get away and hide from the solvent, in order to minimize any contact with it. Basically, this is the mechanism that drives the behaviour of amphiphiles in water solution, namely the formation of self-assembled aggregates such as micelles.

6.1.2.2 Classification and Composition

There are different ways to classify a surfactant, based on its structure, composition or applications. For example, amphiphiles can be classified as foaming or wetting agents, dispersants, and so forth. However the most commonly adopted criterion to classify amphiphiles considers the nature of the polar segment. For water-based systems we distinguish:

- anionic surfactants
- cationic surfactants
- non-ionic surfactants
- zwitterionic surfactants.

In *anionic* surfactants, the polar head group contains a carboxylate, sulfate, sulfonate or phosphate group. Some representative examples are depicted in Figure 6.2. Anionic surfactants such as sodium dodecyl sulfate (SDS), and sodium alkyl polyglycoside sulfosuccinates (AGESS, see Figure 6.2h), have been used to formulate nanostructured cleaning fluids for the removal of detrimental coatings from artistic surfaces, see Sections 7.5, 7.6 and 9.5.

Most ionic surfactants are 1 : 1 salts; the most common counterions are Li^+, Na^+, K^+, Cs^+, Mg^{2+}, Ca^{2+}, NH_4^+, and organic ammonium ions (such as tetramethyl- or tetraethylammonium).

A list of the most common cationic surfactants is shown in Figure 6.3. The head group, usually an ammonium residue, is positively charged, and the counterion is usually a halide (Cl^-, Br^- or I^-), SO_4^{2-}, NO_3^- or CH_3COO^-.

In gemini cationic surfactants the polar head group bears two positive charges, and the hydrophobic portion contains two or more hydrocarbon chains (see Figure 6.4).

Non-ionic surfactants bear a neutral head group, usually made up of hydroxyl groups, as depicted in Figure 6.5. For example Triton-X 100, a non-ionic surfactant, has been used for the formulation of nanostructured systems for the removal of black crusts (containing gypsum and insoluble hydrocarbons) from the surface of historical wall paintings (see Section 9.5). Polyethoxylated amphiphiles (C_iE_j) constitute a wide class of non-ionic surfactants that can be used for cleaning artefacts.

In zwitterionic surfactants the total charge of the molecule is zero, however there are two distinct opposite charges, usually due to a carboxylate or phosphate group and to an ammonium residue, as illustrated in Figure 6.6.

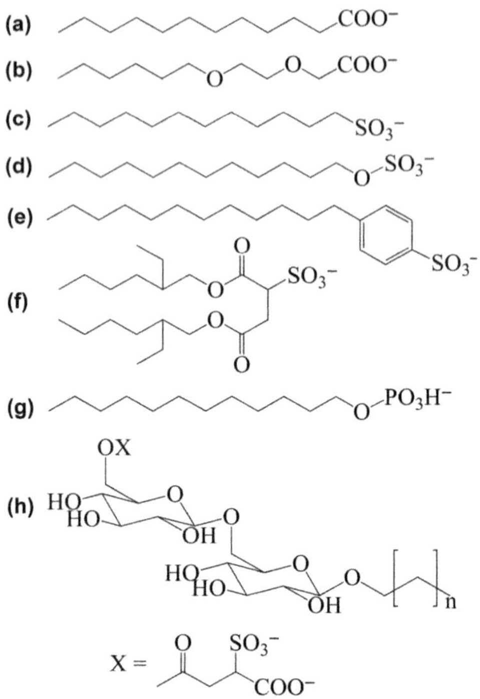

Figure 6.2 Anionic surfactants. Chemical structure of an alkyl carboxylate or fatty acid salt (a), alkyl ethercarboxylate (b), alkyl sulfonate (c), alkyl sulfate (d), alkylbenzene sulfonate (e), dialkyl sulfosuccinate (f), alkyl phosphate (g) and sodium alkyl polyglycoside sulfosuccinate (h).

Particular surfactants are those obtained from sugars or polyols such as those reported in Figure 6.7. The most interesting features of these compounds are the low aquatic toxicity and their tolerance to concentrated salt solutions.

The hydrophobic tail can be saturated or unsaturated, or can carry different functionalities such as in polydimethylsiloxane or in fluorinated surfactants. The latter, for instance, are used in the formulation of Bookkeeper®, a well-established product for the deacidification of paper, *e.g.* in the preservation of historical and documental heritage.

Besides this kind of classification, based on the nature of the polar head, other surfactants have been synthesized more recently:

1) *catanionic*, consisting in an equimolar mixture (ion pair) of cationic and anionic surfactants, *e.g.* in $H(CH_2)_{12}N(CH_3)_3{}^{+-}O_4S(CH_2)_{12}H$
2) *bolamphiphiles*, carrying two symmetrical or asymmetrical head groups, connected by a hydrophobic chain (see Figure 6.8)
3) *polymeric*, that is a polymer that possesses surface active properties, such as a copolymer of isobutylene and succinic anhydride (see Figure 6.9a)

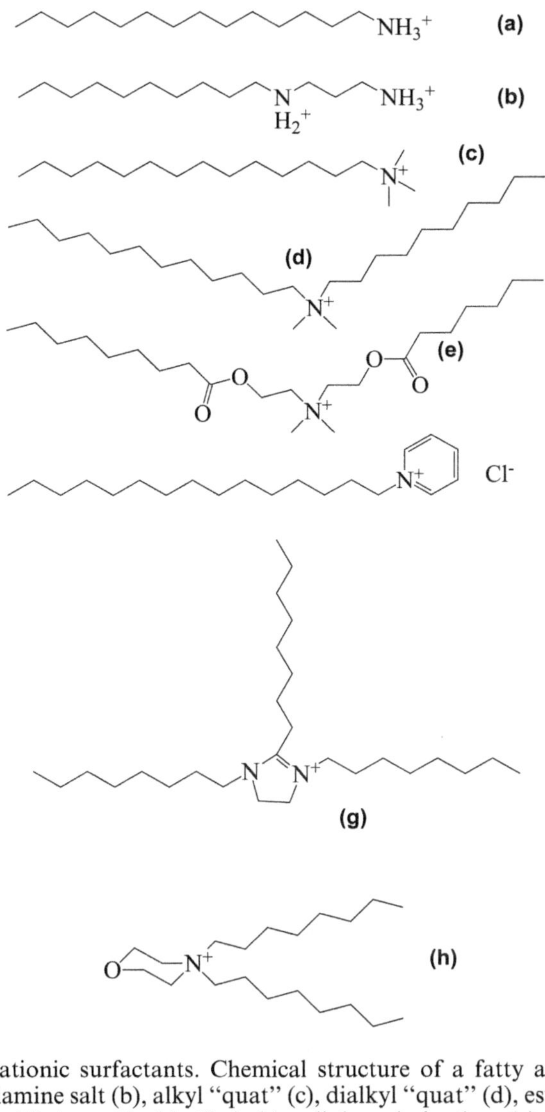

Figure 6.3 Cationic surfactants. Chemical structure of a fatty amine salt (a), fatty diamine salt (b), alkyl "quat" (c), dialkyl "quat" (d), ester "quat" (e), alkyl pyridinium chloride (f), imidazolinium derivatives (g), and morpholinium salts (h).

4) *polymerizable*, a surfactant that can homopolymerize or copolymerize with other components (see Figure 6.9b).

Some surfactants have a natural origin and possess important functions in the body (see Figure 6.10). For example they are present in the saliva, in the lungs, in tears, in the bile, in cell membranes and so forth. Interestingly, traditional cleaning methods included the use of physiological fluids, such as urine, bile fluids, blood or saliva.

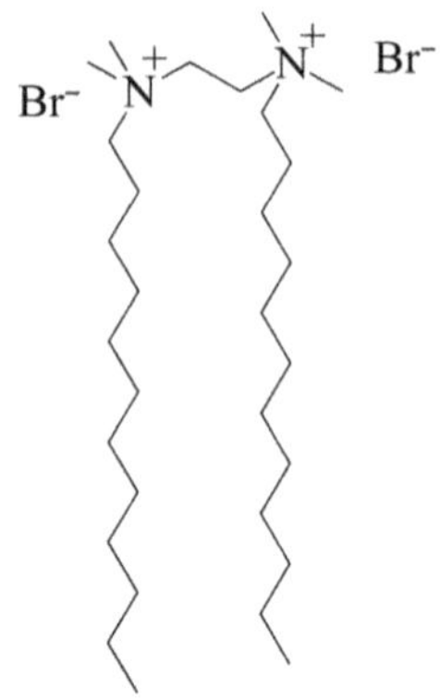

Figure 6.4 Structure of a gemini surfactant.

There are also non-hydrocarbon based surfactants, where for example the hydrophobic segment is made up by fluorocarbon or siloxanes (see Figure 6.11).

Fluorinated surfactants (see top image in Figure 6.11) can be either perfluorinated, that is, all the hydrogens are substituted by fluorine atoms, or partially fluorinated, with some CH_2 residues. They are strongly hydrophobic and show the lowest surface tension of any substance. Owing to the peculiar properties of fluorine, these surfactants adopt special structural helicoidal conformations, and possess great chemical and biological inertness and stability. The relatively high cost and environmental concerns are the major drawbacks in their use.

Siloxane oligomers (see bottom image in Figure 6.11) possess interfacial properties that are intermediate between those of the common hydrocarbon-based and fluorinated surfactants. Because of the presence of the Si–O bond, these surfactants do not behave as their hydrogenated analogues. They are very effective at the air–water interface, and are particularly used as antifoaming agents. For example, siloxanes can be used in defoaming formulations for waterborne paints and coatings. Moreover, both polysiloxanes and fluorinated compounds are used as additives in the preparation of inorganic nanosols for the preservation of wood (see Section 16.4).

6.1.3 Surface Tension

The origin of surface tension has already been discussed in Sections 1.1 and 1.2. We will thus recall here only some fundamental concepts, as necessary steps to introduce the following sections of this chapter.

All liquid surfaces spontaneously shrink, in order to minimize the area exposed to air. This leads, for example, to the formation of tiny spherical droplets. In a liquid all molecules are free to diffuse, but the cohesive forces that hold them together are quite strong, in fact a lot of energy is needed to compress a liquid or a solid.

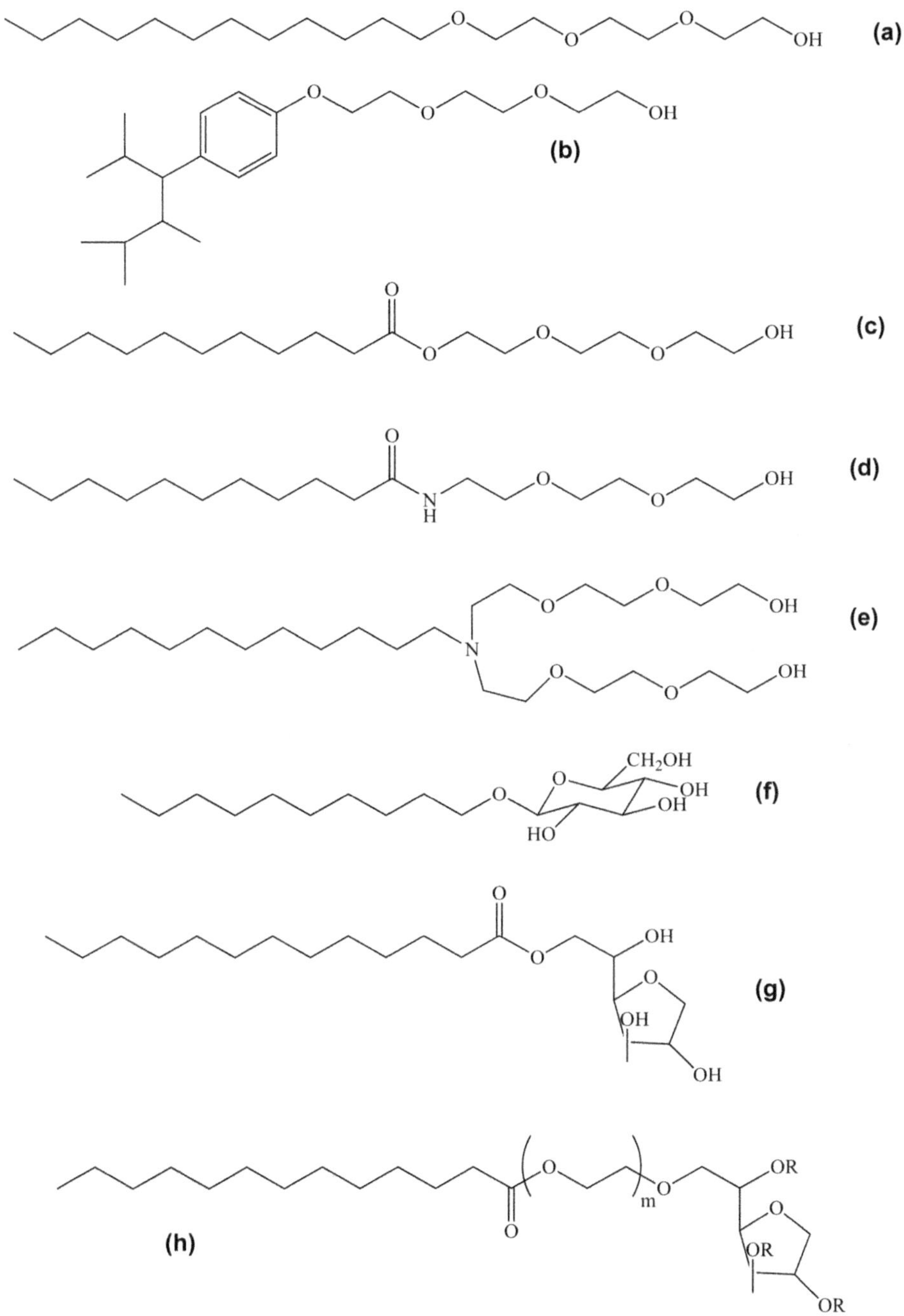

$R = O\text{-}CH_2\text{-}(CH_2\text{-}O\text{-}CH_2)_m\text{-}CH_2\text{-}OH$

Figure 6.5 Non-ionic surfactants. Chemical structure of a fatty alcohol ethoxylate (a), alkylphenol ethoxylate (b), fatty acid ethoxylate (c), fatty amide ethoxylate (d), fatty amine ethoxylate (e), alkyl glucoside (f), sorbitan alkanoate (g) and ethoxylated sorbitan alkanoate (h).

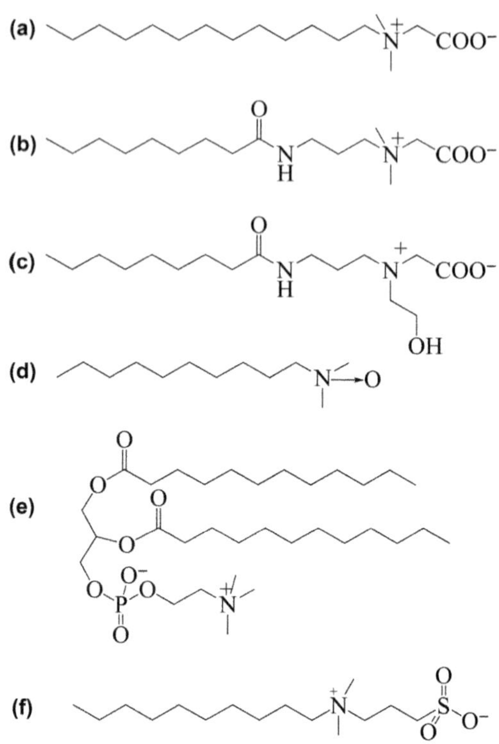

Figure 6.6 Zwitterionic surfactants. Chemical structure of a betaine (a), amidobetaine (b), imidazoline (c), amine oxide (d), dilauroyl phosphatidylcholine (e), and dodecyldimethyl(3-sulfonatopropyl)ammonium or lauryl sultaine (f).

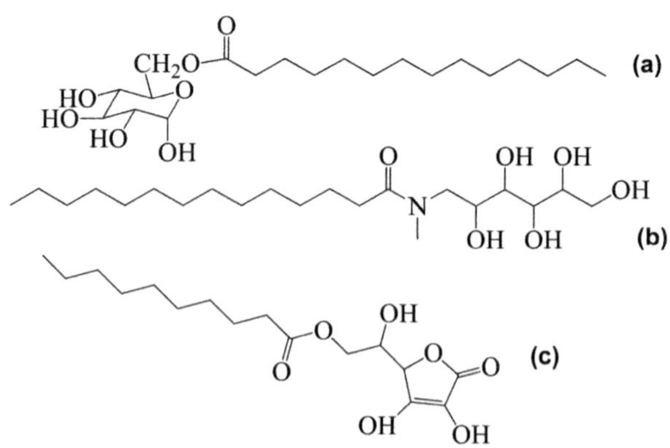

Figure 6.7 Sugar and polyol derived surfactants. Chemical structure of an alkyl glucose ester (a), glucamide (b), and of 6-*O*-ascorbyl decanoate (d).

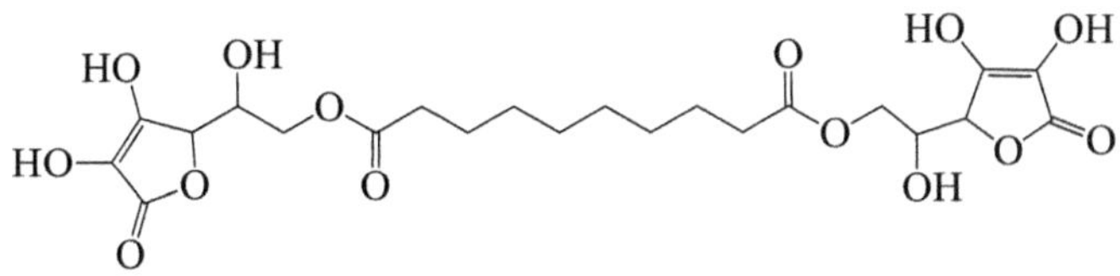

Figure 6.8 Structure of a bolaamphiphile, with two vitamin C groups as polar heads.

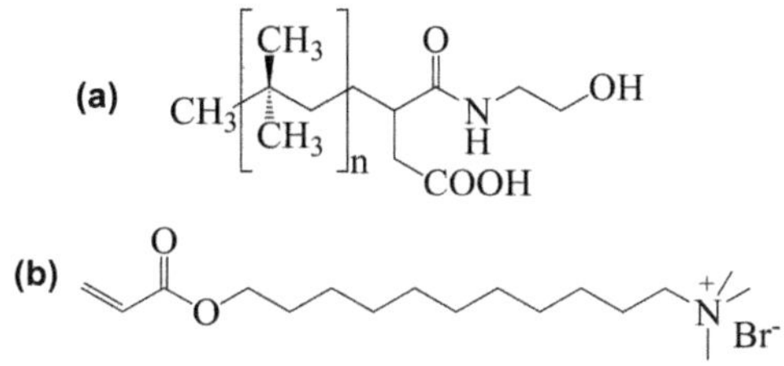

Figure 6.9 Polymeric (a) and polymerizable (b) surfactants.

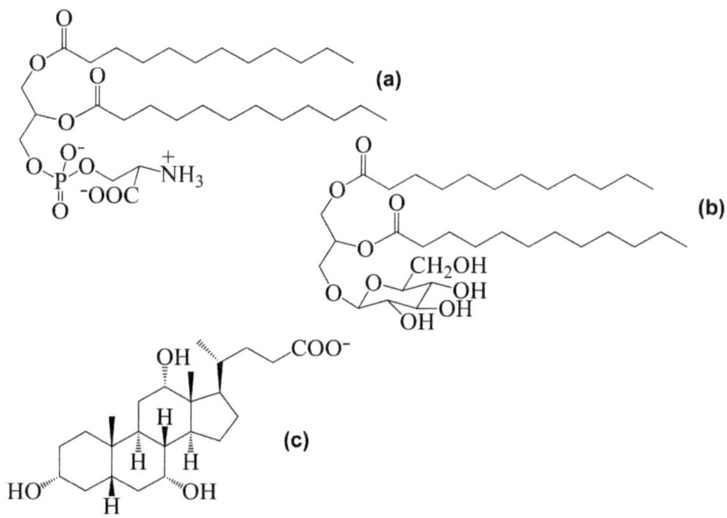

Figure 6.10 Naturally occurring surfactants. Chemical structures of a dialkanoyl phosphatidylserine (a), of glycolipid (b), and of cholate (c).

Recall the definition of surface tension (see Section 1.1.1):

$$\gamma = \left(\frac{\partial G}{\partial A}\right)_{T,P,N} \tag{6.1}$$

γ decreases with increasing temperature, because the molecules are freer to move at higher temperatures (owing to Brownian motion), as described by the Eötvös equation.

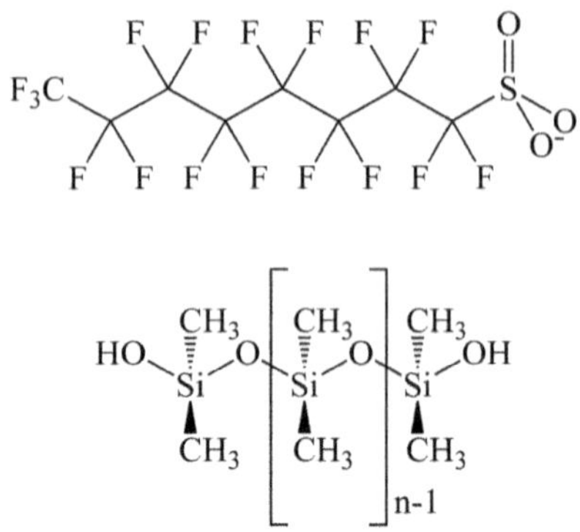

Figure 6.11 Structure of a fluorinated surfactant (a) and of a siloxane (bottom).

In the presence of surfactants, the surface tension at the interface between two immiscible phases (*e.g.* oil and water) is modified.

There are several methods to measure the surface tension of a liquid. The most common are:

- the Du Noüy ring
- the Wilhelmy plate
- the sessile drop and the stalagmometer
- the capillary rise.

The interested reader can find a detailed discussion in the main textbooks indicated at the end of this chapter (see Further Suggested Reading).

6.1.4 The Choice of the Right Surfactant for the Right Application

The uses and applications of amphiphiles in everyday life, science and industry are legion. They include agricultural applications, building materials and cement additives, coal fluidization, coating and levelling additives, dry-cleaning and detergents, electroplating, emulsion polymerization, leather processing, lubrication, mould-releasing agents, ore flotation, paper industry application, petroleum recovery, surface treatments and modifications, textiles, adhesives, food and beverages, laundering, pharmaceuticals, photographic processes, health care, waterproofing, and still others.

The problem of finding the proper surfactant for a given application is complex. It mainly depends on the balance between the lipophilic and hydrophilic blocks of the amphiphile, on its solubility, on its ability to lower the surface tension, on its critical micellar concentration, on its detergent power and wetting properties, on its foaming and thermal behaviour, on its environmental impact, and of course on the properties of the surfaces it is supposed to interact with. Moreover, the same surfactant could be applied in different formulations, for example as a simple aqueous dispersion or as an emulsion, or dispersed in a gel.

Besides trial-and-error procedures, there are some guidelines that help the formulator to find the appropriate surfactant for a specific application. For example, the packing parameter and the hydrophilic–lipophilic balance (HLB) number (see Section 6.2.5) are certainly important parameters for this purpose. However in the selection process we also need to take into consideration other "non-chemical" concerns, such as costs and environmental issues.

For these reasons a "universal" surfactant that can meet all the different requirements is still missing. For example a good detergent should possess a low critical micellar concentration (cmc) and strong stability against electrolytes and pH changes, and a good emulsifier should possess the correct HLB number, a reasonably safe environmental impact, low cost and good biocompatibility.

The different chapters of this book will discuss in greater detail the fundamental characteristics that a surfactant needs to possess in order to obtain a given result. Section 6.1.8 will briefly discuss the application of surfactants and soft matter to the conservation of cultural heritage, which will be dealt with in the following chapters (7 and 9).

6.1.5 Hydrophobic Effect and Thermodynamics

Water and "oil" do not mix. This trivial observation is the basis for the explanation of self-assembly. Water is a polar molecule, with an average dipole moment of about 1.85 D, while a hydrocarbon is almost completely apolar (see Section 4.1.3). Therefore water molecules can interact through a set of dipolar interactions including hydrogen bonding, while hydrocarbons experience only dispersion or London forces.

The introduction of an apolar molecule into a water phase will perturb the strong network established by the water molecules, and reduce the interaction energy. This leads to the formation of a layer of water molecules that face the apolar species directly, shielding the odd guest from the other solvent molecules. The energy necessary to create such a "cage" is provided by a loss in entropy, in fact the interfacial water molecules will suffer from a reduction in degrees of freedom, being frozen around the hydrocarbon. As a matter of fact cage-like structures such as clathrates are found in nature, at low temperatures and high pressures (at the bottom of the oceans), and are formed by an apolar species (methane or a noble gas) surrounded by a certain number of water molecules constrained in a cage-like well-ordered structure.

Association or self-assembly of apolar molecules reduces the interface with the water molecules, and partly relieves the perturbed situation, by allowing some of the interfacial water molecules to re-enter the bulk phase and reacquire their freedom of motion and hydrogen bonds, with a significant entropy increment.

The hydrophobic interaction between two apolar molecules can be of significant strength, and results from the balance between an enthalpic and an entropic term.

This mechanism, the *hydrophobic effect*, is shared by several systems: it occurs in the formation of micelles and other aggregates, in the folding of proteins and other supramolecular systems (for example in polyrotaxanes), in adhesion and wettability of hydrophobic surfaces (bio-fouling), in coagulation of lipophilic particles in water, and in froth flotation where hydrophobic particles interact with gas bubbles. In any case, the reason for such behaviours is the minimization of contact between water and apolar molecules, while the contacts between like molecules (water–water or oil–oil) are maximized.

6.1.6 Intermolecular Interactions

Depending on the structure of the surfactant, the intermolecular interactions that are established with the surrounding molecules can be: ion–ion, ion–dipole, dipole–dipole (orientation, induction, dispersion forces), or hydrogen bonding, as described in Section 4.1.4. In general the hydrophobic chains of the surfactant will experience attractive London interaction with the adjacent lipophilic regions, while the polar head groups interact with the water molecules and with the nearby hydrophilic heads. If the surfactant is non-ionic, repulsive steric interactions are established between the head groups, but if the amphiphile is ionic (for example SDS), strong coulombic repulsions will take place between the anionic heads. The structure of the self-assembled aggregate will thus result from the interplay of attractive (chain–chain and head–water) and repulsive (head–head) interactions, as depicted in Figure 6.12.

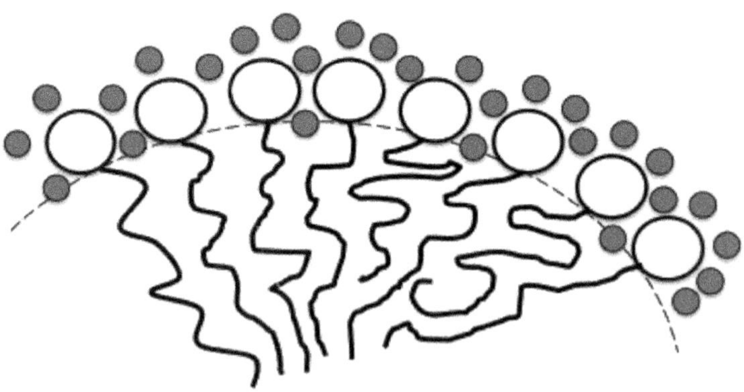

Figure 6.12 The surfactant molecules form an aggregate. The white circles represent the polar head groups, the full curved lines represent the hydrophobic chains, and the small grey circles are interfacial water molecules located near the interface (dotted line), which solvate the polar heads and may penetrate across the interfacial hydrophilic layer. The hydrophobic attractions are proportional to the interfacial area occupied by the amphiphile (a), while the hydrophilic repulsions are proportional to a^{-1}.

6.1.7 Phase Behaviour

Depending on the nature of the surfactant, and on the particular ambient conditions (concentration, temperature, pressure, *etc.*), amphiphiles can associate in different structures (spherical or rod-like or disc-like micelles, vesicles, planar bilayers, reversed micelles, *etc.*) and produce a plethora of different phases such as isotropic micellar, lamellar smectic, hexagonal and cubic phases. The addition of other components can modify the phase and induce a phase transition, as in the case of microemulsions upon addition of oil (see also Chapter 8).

Figure 6.13 illustrates a generic phase diagram of a non-ionic surfactant C_iE_j in water. This wide class of surfactants can also be used for the formulation of cleaning systems for cultural heritage (see also Section 9.5). Note that, on increasing the concentration of the surfactant at constant temperature, the system comprises a simple dispersion of monomers, a micellar solution, the formation of hexagonal aggregates and a lamellar phase. Different structures can affect the behaviour of these systems when they interact with materials, for instance if used for the cleaning and removal of detrimental coatings from artistic surfaces. The characterization of the system's phases is not straightforward, and often requires advanced analytical facilities (see also Section 6.2.6). The following is a quick overview of some typical structures.

Although these mesophases do not possess the rigid tridimensional lattice structure like that of crystalline solids, they retain a certain long-range orientation, and show a lesser order in the form of one, two or three degrees of translational freedom. This is why they possess physical properties that are

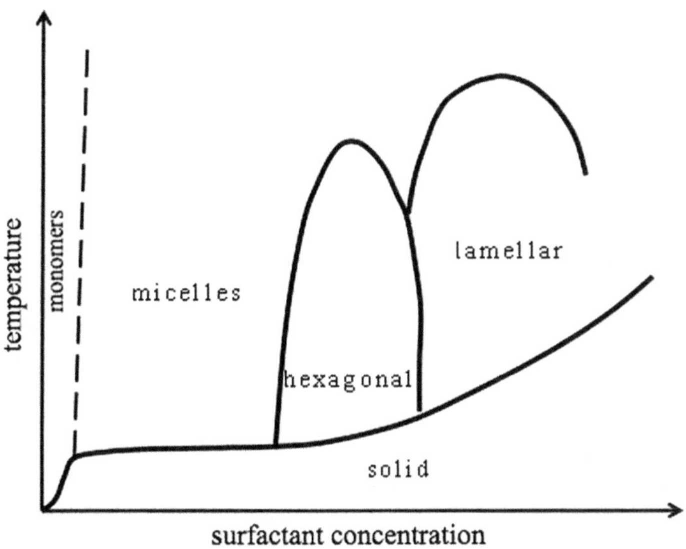

Figure 6.13 Generic phase diagram of a non-ionic surfactant C_iE_j in water. In the graph the temperature is plotted versus the surfactant concentration.

intermediate between those of the solids and those of the disordered liquids. For example liquid crystals display viscous flow, but respond in an anisotropic way to the presence of electrical or electromagnetic fields.

Liquid crystals can be either thermotropic or lyotropic. *Thermotropic* liquid crystals are single component systems that undergo phase transitions upon heating, before they melt and produce a normal liquid with its isotropic molecular distribution.

The phases formed by surfactant-based self-assemblies belong to the class of *lyotropic* liquid crystals. These systems comprise at least two components (*i.e.* water–surfactant, or water–surfactant–oil). The major difference from thermotropic liquid crystals is that in this case more phases can coexist at the same temperature. Here the lamellar and the hexagonal phases are the most important liquid crystalline phases. In the former, the amphiphilic molecules are arranged in bilayers, as depicted in Figure 6.14a, with the hydrocarbon tails in a liquid-like state.

In the hexagonal phase the rod-like micelles are arranged in a parallel hexagonal manner (see Figure 6.14b). If the solvent is oil, a hexagonal reversed phase will be formed, with the hydrophobic chains protruding into the organic solvent, and the polar heads confined in the hydrophilic core. Such a reversed system can hold up to 40% w/w in water.

The viscous *cubic* phases can form in systems such as monoglyceride aqueous dispersions, when the hydrocarbon chain contains at least 14 carbon atoms. Although the side-by-side intermolecular forces have a strong effect, the molecules in the mesophases still retain a certain degree of mobility in the direction of the main axis. Sponge phases are isotropic liquid phases containing disordered and interconnected bilayers.

6.1.8 Uses and Applications

A full discussion of the uses and applications of surfactants in daily life and industrial processes would require an entire book. The interested reader can refer to Section 6.1.4 and to the books listed at the end of this chapter for more detailed information (Further Suggested Reading).

We will briefly discuss here the application of amphiphiles for cleaning procedures and, more extensively, their use in cultural heritage conservation and restoration. In fact the removal of materials spread on the surface of an artefact during its creation or previous restorations, or contaminants accumulated over the years, is certainly one of the most important issues in cultural heritage conservation.[1]

For more specific concepts and applications of surface and interfacial chemistry to cultural heritage conservation, particularly for cleaning applications, the reader can refer to Chapters 2 (properties of artefacts materials), 3 (conservation treatments), 4 (properties of solvents and polymers), 5, 7 (cleaning of artefact), 8 (emulsions and microemulsions), and 9 (case studies).

For over 2000 years, cleaning has been the main application of surfactants, mainly soaps. In the last two centuries new synthetic products have been

(a)

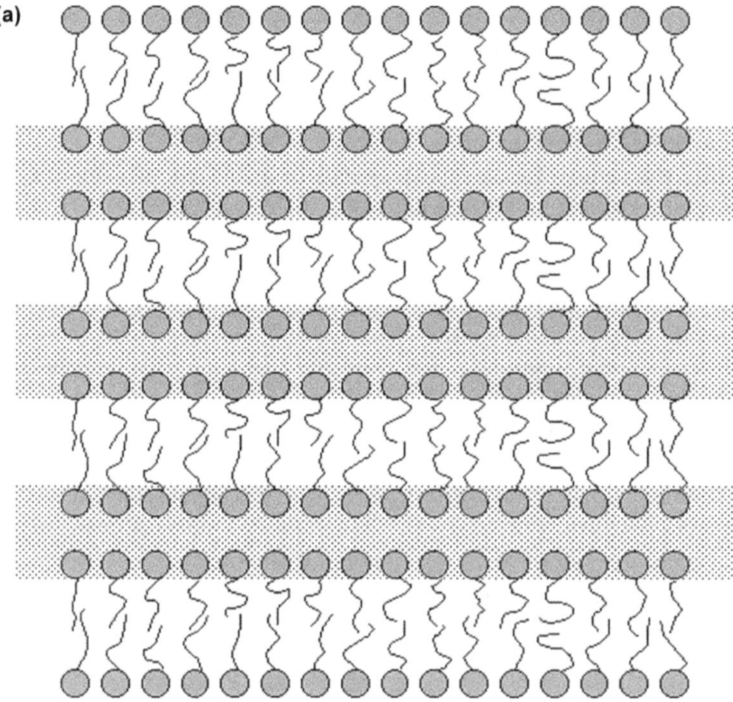

(b)

Figure 6.14 (a) Arrangement of surfactant molecules in a lamellar structure. (b) Arrangement of surfactant molecules in a hexagonal structure.

produced and used as detergents for clothes, dishes and houses. Detergents are generally aimed at removing dirt, oils and other undesired products, without producing irreversible damage to the substrate, possibly with a low-energy treatment (amount of water, temperature and agitation). More recently natural, bio-derived products with lower environmental impact, obtained from vegetable oils and animal fats, have been included in the list.

Very often accurate cleaning of the surfaces of works of art is needed, and this application involves the use of solvents (see Chapters 4 and 5), surfactants in different kinds of formulations (Chapters 8–11), *etc.*

Basically, the removal of a relatively hydrophobic undesirable material (for example wax, a protective layer, a consolidant, an adhesive, *etc.*) from a surface can be attained by using an aqueous dispersion of a surfactant (or of a surfactant and organic solvents). Other procedures may require in addition the use of gels and nanoparticles. The process is based on the solubilization of the hydrophobic chemical (*e.g.* the wax to be removed) in the lipophilic compartment of the aggregate, for example the core of a micelle or the oil droplet in a microemulsion.

In spite of the apparent simplicity, detergency involves a complex set of mechanisms (roll-up, emulsification and solubilization).[2]

In particular, during the roll-up process (see Figure 6.15), the surfactant solution wets the dirt, and wraps the hydrophobic material into tiny droplets that are eventually removed by mechanical means.

The solubilization of dirt in micelles, schematically depicted in Figure 6.16, is important for the removal of small amounts of oily dirt from surfaces and occurs when the surfactant concentration exceeds the cmc (critical micellar

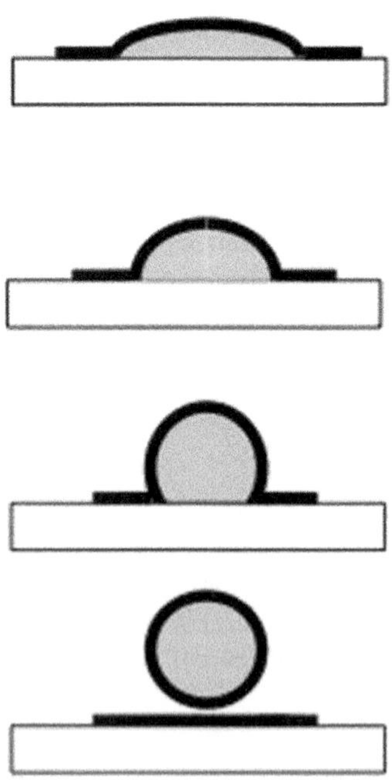

Figure 6.15 Schematic illustration of the roll-up process. The white area represents the substrate surface, the grey region is the dirt, and the thick black line is the layer of detergent.

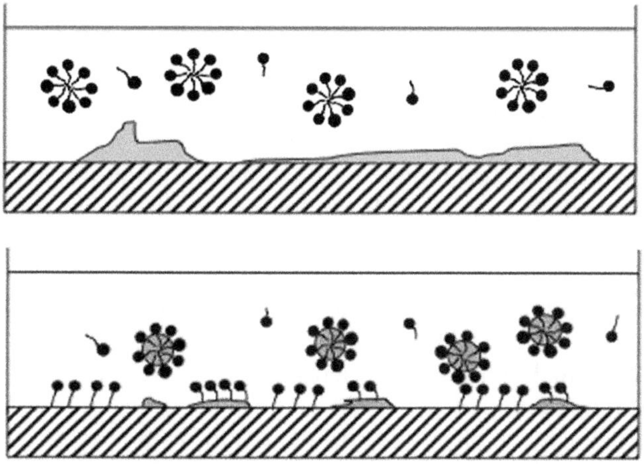

Figure 6.16 Schematic illustration of micellar solubilization. The dirt (grey region) is solubilized in the inner hydrophobic pool of the micelles. Surfactant monomers also adsorb at the substrate and at the soil surfaces.

concentration, see Section 6.2.1). The effectiveness of this process depends on the chemical nature of the amphiphile and on temperature (the solubilization increases sharply when the temperature approaches the cloud point of the surfactant).

Emulsification occurs when the interfacial tension between the oil droplets and the cleaning fluid is low, and requires mechanical work. However, the detached dirt droplets may remain stuck at the surface of the substrate, as opposed to micellar solubilization.

The best amphiphiles for cleaning applications are non-ionic and anionic surfactants. Given that substrates are usually negatively charged at neutral or alkaline pH, cationic detergents cannot work efficiently; the treatment often results in little removal of the dirt and significant re-deposition on the substrate surface.

6.1.9 Environmental Aspects

The concern for the environmental impact related to the global use of surfactants is a very serious issue that needs to be handled with care.

When a surfactant reaches surface water and land, it produces an ensemble of effects and phenomena such as sorption and degradation. The stability, extension and consequences of the sorption products depend on the natures of the surfactant and of the surfaces, and on the environmental parameters. In general, anionic surfactants produce less sorption than non-ionics and especially cationics.[3]

For applications in cultural heritage, the sensible general trend should be to minimize the amount of surfactants needed, whenever possible. In fact, besides the environmental aspects, in some cases surfactants may also remain on the treated artistic surfaces, producing undesired effects. For the same reasons, degradable (and biodegradable) surfactants are to be considered in basic research for the formulation of cleaning systems or other chemical tools. The use of low surfactant quantities and safety precautions by both producers (*e.g.* research laboratories) and end-users (conservators, restorers, *etc.*), are all factors that contribute to minimizing the environmental impact related to the production, handling and use of formulations for the preservation of cultural heritage. In the future, eco-toxicity aspects will be important when considering the production upscale of promising and effective formulations (such as those described in Chapters 7 and 9). As a matter of fact, the environmental issues related to surfactants are still mostly due to their historical usage and industrial production (also including the production of some art materials, *e.g.* paintings, varnishes and pigments).

More specifically, the effects of surfactants in water basins concern aquatic toxicity, biodegradability, and bioaccumulation.[4] Aqueous toxicity is assessed by measuring the lethal concentration (LC_{50}) or effective concentration (EC_{50}) in various species of fish, daphnia or algae. Environmentally friendly surfactants should possess values of about 10 mg L^{-1}.

In natural environments, bacteria take care of biodegradation. The so-called primary biodegradation breaks the structure of the surfactant, by separating the polar head from the side chain, leading to the loss of the surfactant properties. Mineralization involves the final conversion of the chemicals into CO_2, H_2O, SO_4^{2-}, NH_4^+ and other ions. If the surfactant does not undergo natural degradation, then it persists in the environment. The time required for total biodegradation usually spans between a few hours for soaps to some days for linear alkylbenzene sulfonates (LAS) and some months for branched alkylbenzene sulfonates. Biodegradation usually depends on the structure of the amphiphile (mainly on the presence of branched groups), on pH, temperature, and concentration. There are several procedures, such as the Sturm test, that assess the mineralization of a surfactant by measuring the formation of carbon dioxide as a function of time.[5] Some surfactants, such as LAS and cationics, can be persistent against degradation under anaerobic conditions.

While fatty alcohol sulfates (AS) and ethoxylates (AE) are easily degraded under aerobic and anaerobic conditions, the fate of quaternary ammonium cationic amphiphiles is complicated by the strong sorption of these chemicals onto suspended particulates and sludge (which are usually negatively charged). Moreover, cationics possess a significant biocidal activity, and can severely affect the capability of bacteria to carry out the biodegradation processes.

When a surfactant is discharged into the environment, it partitions into the different phases such as water, air, sediment and biota. Bioaccumulation involves the partitioning and accumulation of hydrophobic compounds in the fat of aquatic animals.

At equilibrium, the bioconcentration factor (BCF) is measured by the ratio between the surfactant concentration in the biota and in water, $BCF = C_{biota} : C_{water}$. This parameter and the octanol-to-water partition coefficient (K_{ow}) are necssary in order to quantify the uptake of a given chemical by an organism, and then to assess its potential environmental risk, without involving the use of living species.

For example, in the case of LAS, the value of K_{ow} increases – as expected – with the length of the hydrophobic chain, reflecting the greater lipophilicity. In general the threshold for $\log K_{ow}$ is 3, above which the surfactant is considered to bioaccumulate.

A serious environmental and health concern derives from the finding that some chemicals, such as nonylphenol and octylphenol, can disrupt the normal functioning of the endocrine system in wildlife and humans, and therefore they are called "endocrine disrupting chemicals" (EDCs).[6]

6.2 Micellar Solutions

6.2.1 Introduction to Self-assembly

We have already noted that amphiphilic molecules produce self-assembled aggregates, which is a successful way to minimize the contact between the solvent (usually water) and the solvophobic segment (usually an aliphatic chain). The peculiar behaviour of surfactants in water, which involves the formation of self-assembled particles, emerges when a particular physico-chemical property f(c) of its aqueous solution is recorded as a function of the surfactant concentration (c). Figure 6.17 illustrates the variation in the turbidity (τ), solubilization (S), osmotic pressure (Π), diffusion coefficient (D) and surface tension (γ) as a function of c. The relevant feature is that there exists an interval of concentration around which the property f(c) changes, more or less abruptly. At very low concentration, the surfactant behaves as a regular solute, with almost linear behaviours (an exception is surface tension, which shows a steep decrement as c increases). However, above a critical concentration the property f(c) changes: for example Π and γ reach almost constant values, while the solubilization and the turbidity of the dispersion increase significantly.

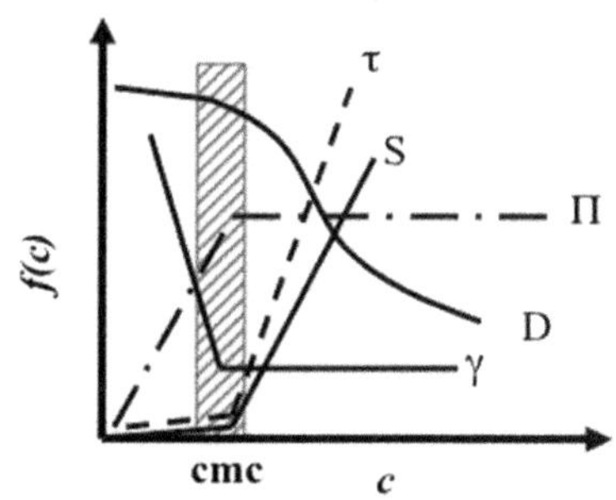

Figure 6.17 Variation in turbidity (τ), solubilization (S), osmotic pressure (Π), diffusion coefficient (D), and surface tension (γ) around the cmc of a surfactant at constant temperature.

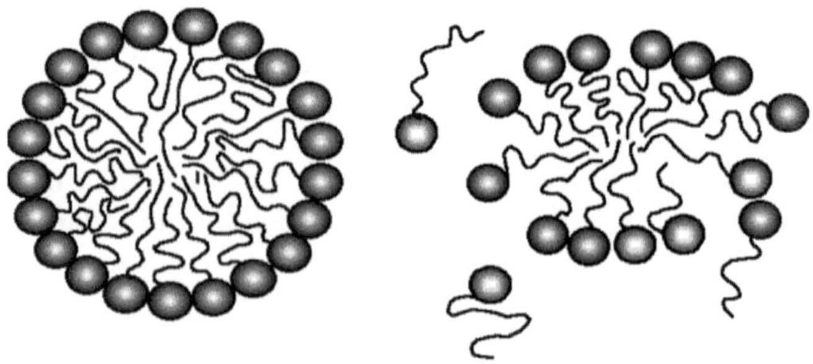

Figure 6.18 Idealized (1) and more realistic illustration (2) of a micelle.

This peculiar behaviour has been ascribed to the formation of self-associated objects in which a certain number (the "aggregation number") of monomers gather together. Such structures, called "micelles" are not static, in fact a surfactant monomer can easily escape the assembly and re-enter into the solution, leaving room for another amphiphilic molecule to temporarily join the aggregate. Usually the aggregates are depicted as golf balls, that is, as nearly spherical objects as depicted in Figure 6.18 (1). Instead, a more realistic picture represents the aggregate as a rather loose structure, as in Figure 6.18 (2).

The concentration at which the first aggregates form, *i.e.* the *critical micellar concentration* (cmc), depends on the ambient conditions (temperature, pH, ionic strength, presence of other cosolutes, *etc.*) and on the nature of the amphiphile.

The sequence of phenomena that occurs when we start adding an amphiphile to some water in a clean beaker is depicted schematically in Figure 6.19. Panel 1 shows a few surfactant molecules dissolved in the solvent in the monomeric state. The further addition of surfactant, above its temperature-dependent solubility limit, leads to the formation of the so-called "adsorption film" at the air–water interface (panel 2). This process ends with the formation of a saturated film at the interface (panel 3), before the first micellar aggregates are formed (panel 4).

A simple visual inspection of the dispersion of a surfactant in water may reveal the Tyndall effect, that is the onset of opacity due to the scattering of light produced by the aggregates, when the surfactant concentration exceeds the cmc. When the aggregate size is below a certain threshold, for example in very stable systems such as microemulsions (with typical size of about 5–50 nm, see Chapter 8 and Section 9.5), the dispersion can be optically transparent.

There are different ways to measure the cmc. For an ionic surfactant, conductivity is commonly employed. This property changes more or less abruptly when the concentration of the surfactant reaches the cmc, with a variation in the slope. Other common procedures to measure the cmc imvolve the measurement of surface tension, turbidity, refractive index, viscosity, density, fluorescence, nuclear magnetic resonance (NMR), and so forth.

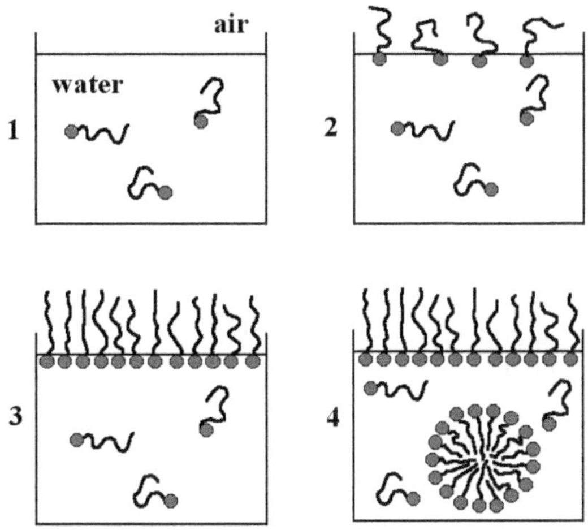

Figure 6.19 As surfactant molecules are added to water, they first dissolve up to the solubility limit (1), then they begin to form an adsorption monolayer (2), and keep accumulating at the interface until saturation is reached (3). Further addition of surfactant leads to the formation of the first micellar aggregates (4).

However, especially for short chain amphiphiles that form weaker aggregates, the evaluation of the cmc is not easy.

The plot of γ/c is one of the best methods to determine the cmc for a given single-chained surfactant. In fact the trend shows that, for small amounts of surfactant, the surface tension decreases steeply (left side) from the value for pure water (about 72 mN/m at 25 °C). At larger c values the decrement is less pronounced, and eventually γ reaches a more or less constant value (see Figure 6.17), corresponding to the formation of self-assembled structures in the bulk. Further addition of monomers increases only the number of aggregates, while the concentration of the monomers in solution remains constant. Moreover, this method also provides important information on the area per polar group, a_P. The experimental evaluation of a_P is very important, because it allows the calculation of the surfactant packing parameter (see Section 6.2.5).

We will now briefly discuss the main factors that affect the cmc of common surfactants. Such factors, together with their effectiveness in the formation of stable and performing systems, are important when selecting the appropriate surfactants for formulations such as micellar solutions and microemulsions, and underlie the choice of some of the surfactants found in the applicative chapters of this book. The following considerations are, therefore, more related to the preparation of cleaning systems than to applicative rules concerning their direct usage on works of art.

The first effect is that due to the aliphatic chain. The cmc decreases significantly with increasing length of the hydrophobic segment of the surfactant (see Table 6.1). In general, the cmc lowers by a factor of 2 for ionic species and by 3 for non-ionic, when a CH_2 residue is added to the side chain.

Table 6.1 Critical micellar concentration (cmc) of some surfactants at 25 °C. (Partially reproduced from Ref. 9 with permission. 2002 John Wiley & Sons, Ltd.)

Surfactant	*–log(cmc) (298 K)*
Dodecylammonium chloride	1.833
Dodecyltrimethylammonium chloride	2.693
Decyltrimethylammonium bromide	1.187
Dodecyltrimethylammonium bromide	1.807
Hexadecyltrimethylammonium bromide	3.036
Dodecylpyridinium chloride	1.833
Sodium tetradecyl sulfate	2.678
Sodium dodecyl sulfate	2.081
Sodium decyl sulfate	1.481
Sodium octyl sulfate	0.876
Sodium octanoate	0.398
Sodium nonanoate	0.678
Sodium decanoate	0.963
Sodium undecanoate	1.252
Sodium dodecanoate	1.556
Sodium p-octylbenzene sulfonate	1.833
Sodium p-dodecylbenzene sulfonate	2.921
Dimethyldodecylamineoxide	2.678
C_8E_4	2.071
C_8E_5	2.036
C_8E_6	2.004
$C_{10}E_5$	3.046
$C_{10}E_6$	3.022
$C_{10}E_8$	3.000
$C_{10}E_9$	2.886
$C_{12}E_5$	4.187
$C_{12}E_6$	4.167
$C_{12}E_7$	4.161
$C_{12}E_8$	4.149
C_8-C_6H_4-E_6	3.688
Potassium perfluorooctanoate	1.541
β-D-C_8 glucoside	1.602
β-D-C_{10} glucoside	2.658
β-D-C_{12} glucoside	3.721

For example the values of –log(cmc) at 25 °C for sodium dodecyl sulfate, sodium decyl sulfate and sodium octyl sulfate are 2.081, 1.481 and 0.876, respectively. The variation of the cmc with the number of carbons in the chain (n_C) is well approximated by the formula:

$$Log(cmc) = a - bn_C \tag{6.2}$$

For example, for sodium soaps at 20 °C, $a = 1.85$ and $b = 0.30$.

For non-ionic surfactants the cmc is much lower than for ionic surfactants with the same chain length. This means that a smaller amount of amphiphile is needed to prepare cleaning systems based on stable micellar phases suitable for conservation science applications.

In general, cationics have slightly greater cmc values than anionics, while for non-ionic surfactants the cmc increases moderately with the length of the hydrophilic block. This is another reason, besides environmental considerations, to prefer anionic over cationic surfactants.

The net charge of the counterion, for ionic surfactants, can affect the cmc considerably. In general, changing the counterion to another with the same valency (*e.g.* Na^+ and K^+) does not alter the cmc, but if a monovalent ion is substituted by a divalent ion (Na^+ with Ca^{2+}) the cmc can be lowered by a factor of 4. Organic counterions reduce the cmc with respect to inorganic cations, and the effect is larger if the hydrophobic moiety in the counterion is larger.

Regarding the hydrophobic chain, the presence of branched residues, double bonds and aromatic rings affects the cmc of the surfactant, but the effect is very significant when a hydrocarbon segment is replaced by a fluorinated chain, with a strong reduction in the cmc. Partially fluorinated chains instead lead to an increment in the cmc, owing to the incompatibility between fluorocarbons and hydrocarbons.

6.2.2 Effects of Salts on Micellization

This topic is of crucial importance for all kinds of applications of surfactants. The addition of salts and neutral solutes (for example urea) is commonly used to control the viscosity, stability, appearance and consistency of surfactant-based formulations.

As expected, salts affect the cmc of ionic surfactants significantly. For example, the addition of NaCl to alkyl sulfates results in a remarkable lowering in the cmc, even of an order of magnitude, and the effect is much larger for long-chain surfactants (carrying 10 carbon atoms or more). This effect also depends on the valency of the added ions, and is most sensitive to the valency of added counterions. For non-ionic surfactants the salt effect on cmc is rather small.

Neutral co-solutes such as alcohol or urea can modify the cmc. Small or moderate changes are recorded in the case of water-soluble polar compounds. Alcohols can act as co-surfactants and lower the cmc in a significant way, depending on their hydrophobicity. For example, 1-pentanol has been used as a co-surfactant in the formulation of cleaning systems (micellar solutions and microemulsions) that proved very effective in the removal of natural and synthetic polymers from wall-paintings (see Sections 7.5, 7.6 and 9.5).

6.2.3 Krafft Point and Cloud Point

The effect of temperature on micellization may be significant. In fact, the critical micellar concentration of a surfactant does depend on temperature. For example, in the case of SDS the cmc reaches a minimum of about 8.5 mM at 25 °C (see Figure 6.20), a behaviour that recalls the solubility of hydrocarbons in water.

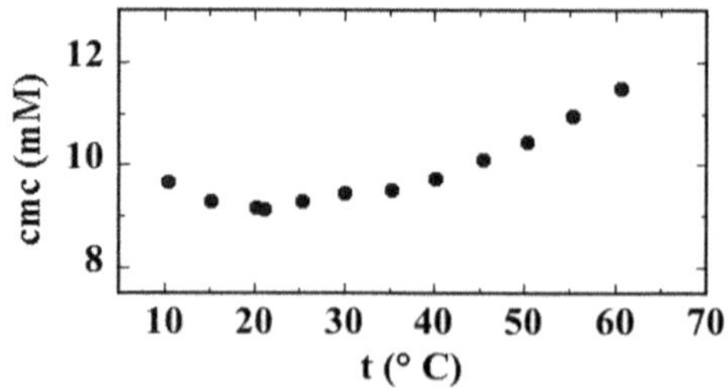

Figure 6.20 Variation in the cmc of SDS as a function of temperature. (Reprinted with permission from M. H. Ropers, G. Czichocki and G. Brezesinski, *J. Phys. Chem. B*, 2003, **107**, 5281. Copyright 2003 American Chemical Society.)

The solubility of surfactants depends on temperature in a way that has dramatic consequences for their applications. In the case of micellar solutions used for the cleaning of artefacts, the effectiveness of detergent action is related to the presence of micelles, and to their interaction with the layer that needs to be removed (*e.g.* dirt, aged coatings). If the system structure is altered as a result of, for example, a large change in temperature, then the effectiveness of cleaning can be negatively affected. This is what occurs with a cleaning formulation made up of water (88.5% w/w), SDS (5% w/w), and 1-pentanol (6.5% w/w). The same applies to aqueous dispersions of SDS below 34 °C. The parameters introduced in this section (Krafft point, cloud point) should thus be considered when selecting the appropriate surfactants for the design of cleaning systems that are be applied under particular environmental conditions (*e.g.* archaeological sites or outdoor restoration workshops).

For ionic amphiphiles the solubility is generally quite low at low temperatures, but it increases by orders of magnitude in a relatively narrow temperature interval. The temperature at which the surfactant solubility increases is denoted the Krafft temperature or *Krafft point* (see Figure 6.21).

At low temperatures the saturated solution is in equilibrium with the solid surfactant, the temperature is too low for micelles to form and the limiting solubility is small. When the temperature reaches the Krafft point, micelles form and the surfactant becomes much more soluble. While the cmc is mainly controlled by the hydrophobic chain length, the Krafft temperature also depends on the nature of the head group. Given that at higher temperatures the lipophilic chains are in a more fluid state, micellization becomes favoured. Since the freezing of the hydrocarbon chain is induced by the cohesion forces between hydrocarbon chains, which increase with their length, then the Krafft point increases with the chain length, as shown in Table 6.2.

However, when the cross-section of the polar groups is large enough, a close approach of the hydrophobic chains is hindered. The cross-section of a typical hydrocarbon chain is about 20 $Å^2$, therefore if the head group area is lower than about 50 $Å^2$ the polar groups will stabilize the solid hydrate and raise the Krafft temperature. The Krafft point of sodium hexadecylsulfonate is 45 °C and that of trimethyl hexadecylammonium bromide is about 26 °C.

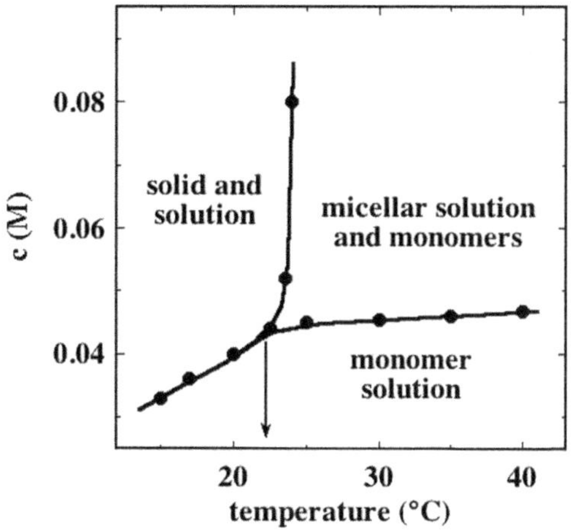

Figure 6.21 Determination of the Krafft point (indicated by the arrow) in the concentration–temperature plot.

Table 6.2 Krafft temperatures for some sodium alkyl sulfates and sulfonates. (Reprint from Ref. 7 with permission. Copyright 2006 John Wiley & Sons, Inc.)

Sodium dodecylsulfate	16 °C
Sodium tetradecylsulfate	30 °C
Sodium hexadecylsulfate	45 °C
Sodium dodecylsulfonate	38 °C
Sodium tetradecylsulfonate	48 °C
Sodium hexadecylsulfonate	57 °C

Non-ionic surfactants such as C_iE_j and also zwitterionic amphiphiles (*e.g.* lecithins) behave in a peculiar way. In fact they possess quite low cmc values if compared with ionic surfactants with the same chain length. For example, $C_{12}E_8$ has a cmc of about 10^{-4} M, and SDS of about 8×10^{-3} M, due to the fact that the repulsions between the neutral head groups are weaker than those between charged heads. Moreover, the addition of two methylene groups in the aliphatic chain lowers the cmc by a factor of 10.

The effect of temperature on the aggregation number depends on the nature of the polar groups, while salt effects are rather small. Interestingly, light scattering experiments performed on micellar aggregates produced by non-ionic or zwitterionic surfactants show the onset of attractive interactions between the micelles; if such attractions become strong enough, the system may undergo a macroscopic phase separation. This behaviour must have something to do with the interactions that the uncharged polar head groups establish with the adjacent heads and with the solvent molecules. In non-ionic surfactants the main interactions with water involve hydrogen bonding and dipolar interactions that are known to become weaker as temperature increases, *i.e.* water becomes a poorer solvent for the aggregates. As a result, the repulsions between the head groups decrease and aggregation is favoured. This is the mechanism that determines the onset of turbidity in an aqueous dispersion of non-ionic and zwitterionic surfactants upon temperature changes, a phenomenon that is commonly referred to as the *cloud point*, and that occurs even at low concentrations, such as 1% w/w. In the case of C_iE_j the dispersion becomes opaque and phase separation occurs when the sample is heated up, while dispersions of phospholipids (for example dioctanoyl phosphatidylcholine, diC_8PC) behave in a similar manner upon cooling. After phase separation, two phases in equilibrium are formed, one richer and one poorer in surfactant. Depending on the specific system under study, the cloud point temperature may depend on the surfactant concentration and on the composition of the solvent, more or less significantly. For C_iE_j surfactants the cloud point is essentially imposed by the length of the ethylene oxide (EO) segment, while the length of the alkyl chain (j) has a rather smaller effect. At constant concentration, $C_{12}E_8$ has a cloud point at 80 °C, $C_{12}E_6$ at 50 °C, and $C_{12}E_4$ at 10 °C.

Plotting the cloud point temperature as a function of the surfactant concentration or mole ratio results in the liquid–liquid coexistence curve, which presents a minimum in the case of C_iE_j or a maximum for phospholipid dispersions (Figure 6.22).

For non-ionic systems the region above the lower consolute temperature curve is biphasic, and the portion below the curve is monophasic. The opposite occurs for dispersions of zwitterionic amphiphiles. The addition of strong electrolytes lowers or increases the cloud point, depending on the nature of the specific salt (see Figures 6.23a and 6.23b).

The opposite effect of different salts has been ascribed to the different hydration properties of the ions, and to their different adsorption at the micellar–water interface. Salts that increase or decrease the cloud point of a

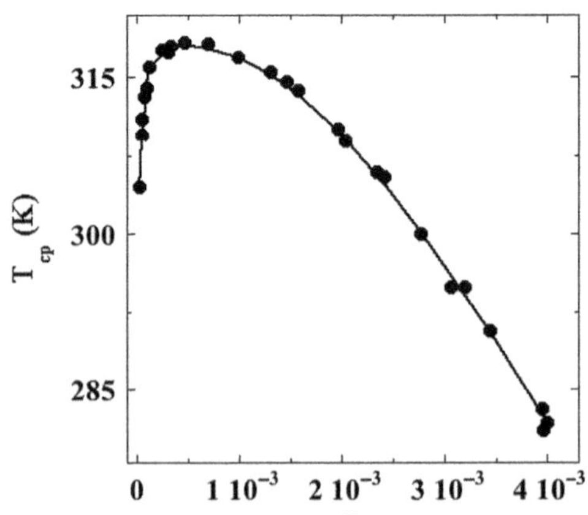

Figure 6.22 Cloud point temperature (T_{cp}) *vs.* the surfactant mole fraction (x) for dioctanoyl phosphatidylcholine.
(Reprinted with permission from P. Lo Nostro, N. Stubicarand S.-H. Chen, *Langmuir*, 1994, **10**, 1040. Copyright 1994 American Chemical Society.)

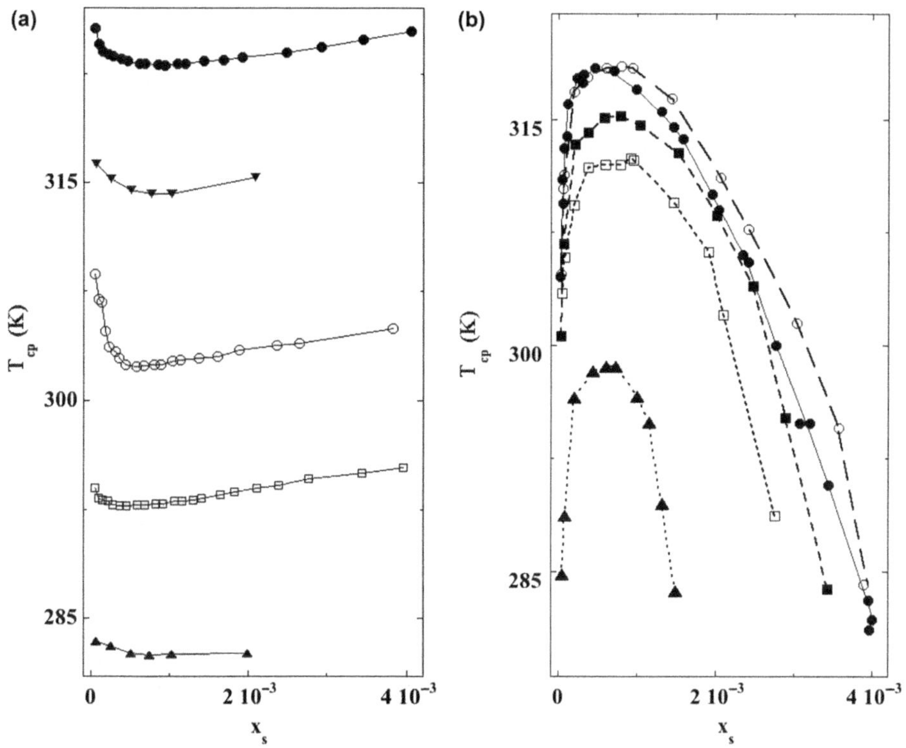

Figure 6.23 Variation of the cloud point temperature (T_{cp}) versus the surfactant mole fraction (x_s) for (a) $C_{12}E_5$ in pure water (○), and in the presence of 1 M NaCl (□), NaSCN (●), NaH_2PO_4 (▲), and NaI (▼); (b) dioctanoyl phosphatidylcholine in pure water (●), and in the presence of 0.05 M NaF (○), NaCl (■), NaBr (□), and NaI (▲).
(Reprinted with permission from M. Lagi, P. Lo Nostro, E. Fratini, B. W. Ninham and P. Baglioni, *J. Phys. Chem. B*, 2007, **111**, 589. Copyright 2007 American Chemical Society.)

C_iE_j dispersion are referred to as "salting-in or out". Similar effects are found when neutral co-solutes (sugars, alcohols or urea) are added to the solvent, or when H_2O is partially or totally replaced by D_2O. The nature and composition of the solvent are therefore useful tools for fine-tuning of the temperature behaviour of these dispersions. Finally, the addition of a small amount of ionic surfactants strongly increases the solubility and the cloud point, because of the formation of mixed micelles with the C_iE_j amphiphile.

Other compounds that contain EO residues, such as poly(ethylene oxide) (PEO) show similar behaviour to that of C_iE_j, depending on their molecular mass. When the molecular mass is very large, above 10^4 g mol^{-1}, a closed loop appears in the phase diagram.

Other examples are Pluronics, triblock copolymers that comprise EO and propylene oxide (PO) segments [$(EO)_n(PO)_m(EO)_n$, $(PO)_n(EO)_m(PO)_n$), $C_i(EO)_n(PO)_m$], and derivatives of glycerol and of ethylenediamine. In these cases the temperature effect on the phase behaviour of their dispersions is more complicated, owing to the high molecular mass of the copolymers, and to the fact that the PO residues have a higher degree of hydrophobicity than EO units.

6.2.4 Packing Parameter and HLB

The surfactant *packing parameter* is defined as:

$$p = \frac{v_H}{l_H \cdot a_P} \tag{6.3}$$

where a_P is the area per polar group, v_H is the volume of the hydrophobic chain and l_H is the length of the chain in its fully stretched conformation. These quantities are usually obtained from the Tanford formulae:[8]

$$v_H = 27.4 + 26.9 n_C \tag{6.4}$$

$$l_H = 1.5 + 1.265 n_C \tag{6.5}$$

where n_C is the number of carbon atoms embedded in the hydrophobic core.

The value of p can be used to predict the shape of the micellar aggregate that a particular surfactant is going to produce (see Figure 6.24).

Each surfactant in the micelle will have the polar head group occupying a portion of the outside surface (a_P), while the aliphatic chain fills the hydrocarbon core of the aggregate. If we assume that the hydrocarbon chain in the micelle has the same density as the pure corresponding hydrocarbon (for example assuming that the dodecyl chain in SDS has the same density as dodecane in its pure liquid state), then the length and volume of the surfactant lipophilic block can be obtained from Tanford's rules. Given that there is no vacuum inside the micelle, this implies that all hydrophobic chains must fully occupy the core of the aggregate, assuming a somehow curled conformation, whose length will be necessarily smaller than that in the fully stretched conformation.

Therefore, if g represents the aggregation number (the average number of monomers per micelle), the volume and surface area of the hydrophobic core will be given by:

$$\frac{4\pi R^3}{3} = g v_H \tag{6.6}$$

$$4\pi R^2 = g a_p \tag{6.7}$$

and therefore we obtain

$$R = \frac{3 v_H}{a_P} \tag{6.8}$$

If l_H is the length of the aliphatic tail in its fully stretched conformation, R will be smaller than l_H, and therefore:

$$p = \frac{v_H}{l_H \cdot a_P} < \frac{1}{3} \tag{6.9}$$

Packing parameters lower than 1/3 correspond to spherical micelles.

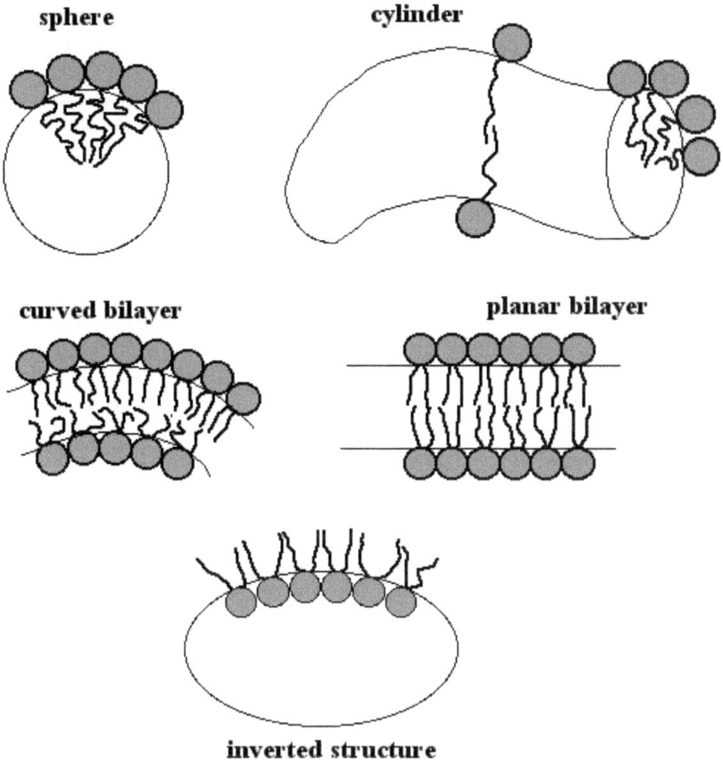

Figure 6.24 From left to right and top to bottom: schematic structure of a spherical and cylindrical micelle, of a curved and a planar bilayer, and of an inverted structure.

By repeating a similar calculation for cylinder, vesicles (curved bilayers), planar layers, inverted structures and cubic phases, we obtain different values, as shown in Table 6.3.

A common use of surfactants is in the formation of an emulsion between water and a hydrocarbon. This important topic will be discussed in Chapters 8 and 9. Here we recall the concept of the *hydrophilic–lipophilic balance* or HLB,[9,10] which will be further treated in Section 8.2.1. This parameter is helpful in predicting the structure of a surfactant aggregate (see Table 6.4). In fact, according to the Bancroft's rule, water-soluble surfactants (characterized by a high HLB) produce oil-in-water (o/w) emulsions, while oil-soluble surfactants (low HLB) produce water-in-oil (w/o) emulsions. Griffin introduced the HLB by considering the percentage weight (w%) of the polar head group and that of the hydrophobic segments. The calculation of the HLB depends on the general structure of the emulsifier. For example, for an alcohol ethoxylate or an alkylphenol ethoxylate the HLB is:

$$HLB = \frac{w\,\%\,(\text{ethylene oxide})}{5} \tag{6.10}$$

Table 6.3 Correlation between molecular structure, packing parameter and self-assembled structure preferentially formed by a generic surfactant.

Surfactant structure	*p*	*Self-assembled structure*
Single aliphatic chain, large polar head group	<1/3	Spherical or ellipsoidal micelles
Single aliphatic chain, relatively small head group – ionic at high ionic strengths	1/3–1/2	Cylindrical micelles
Two aliphatic chains, large polar head group	1/2–1	Vesicles and flexible bilayered structures
Two aliphatic chains, small polar head group	1	Planar rigid bilayer structures
Very large hydrophobic portion, small polar head group	>1	Reverse structures

Table 6.4 How the appearance of an aqueous solution and how the application depend on the hydrophilic–lipophilic balance (HLB) of a given surfactant.

HLB range	*Appearance*	*Application*
1–4	No dispersibility	
3–6	Poor dispersibility	w/o emulsifier
6–8	Milky dispersion (after stirring)	wetting agent
8–10	Stable milky dispersion	o/w emulsifier detergent
10–13	Translucent – clear dispersion	
13–20	Clear dispersion	dispersant

Further, Davies showed that the HLB number can be calculated by assigning a group number to chemical residues. For example sodium sulfate ($-SO_4Na$), sodium carboxylate (–COONa), a ternary amine, and a free –OH group have HLB numbers of 38.7, 19.1, 9.4, and 1.9 respectively. For the lipophilic segment, a $-CH_3$, $-CH_2-$ or –CH residue has an HLB number of –0.47, while a fluorinated $-CF_2-$ group has an HLB number of –0.87. The total HLB number of the entire surfactant will be given by:

$$HLB = 7 + \text{hydrophilic group numbers} + \text{lipohilic group numbers} \quad (6.11)$$

In summary, if a w/o emulsion is to be created, the selection of the surfactant should focus on hydrophobic chemicals with an HLB between 3 and 6 (for example Span80 or Brij72), while if an o/w emulsion is needed, the right surfactant should have an HLB number ranging between 8 and 18 (such as Tween85, Atlox4851B or sodium oleate).

For practical applications, it has been found that a combination of a more hydrophilic with a more hydrophobic surfactant produces a more stable emulsion that a single surfactant with an intermediate HLB number.

In conclusion, the most important issues for selecting a surfactant as a potentially good emulsifier are:

- strong migration of the surfactant at the oil–water interface;
- Bancroft's rule (oil-soluble surfactants form w/o emulsions and water-soluble surfactants produce o/w emulsions);
- a mixture of two surfactants with different HLB numbers works better than a single surfactant with an intermediate HLB number. For a mixture of emulsifiers, the total HLB number is calculated as the weight average of the single surfactants' HLB numbers;
- a more hydrophilic emulsifier works better if the oil phase is more polar;
- a good emulsifier should have the same HLB number as the phase to be emulsified.

The use of the HLB number has some drawbacks related to changes in temperature (especially for non-ionic surfactants), the presence of impurities in the oil phase and salts in the aqueous phase, and the presence of co-surfactants or other ingredients.

6.2.5 Experimental Techniques

We have already mentioned, in Section 6.1.3, the most common methods used to measure surface tension, from which the cmc of a given surfactant and the area per polar group can be determined. There are several other techniques that provide essential information on the structural and dynamic properties of self-assemblies. In this section we will briefly summarize the most important features related to these experimental methods. More information can be found in Chapter 8 and Section 10.5, and in the textbooks listed at the end of this chapter.

In this context, however, it is of fundamental importance to recall the role of techniques for the investigation of the properties of an assembly. For instance, the structure of self-assemblies helps to predict their behaviour as cleaning tools for cultural heritage (*e.g.* micelles can work as containers for solvents, modifying their effectiveness in the removal of dirt and coatings). Moreover, the structure of the self-assemblies is often altered in the very process of cleaning, and structural techniques may thus help us to understand the cleaning processes and mechanisms down to a very fine level of detail (nanoscale), allowing the refinement of useful restoration tools.

Given their complexity, diagnostic techniques must necessarily be carried out in specialized scientific facilities, by highly trained personnel. This aspect emphasizes the need for strong cooperation between scientists and end-users (conservators, restorers). The deep insight granted by the work of scientists, in fact, must be correlated with the practical information, needs and requirements that are developed by conservators through the restoration activity carried out in ateliers, restoration workshops, archaeological sites, *etc.* On the other hand, an awareness of the possibilities granted by science allows an end-user to develop useful collaborations. An interactive and iterative process should thus take place, resulting in the production of advanced, yet practical solutions for preservation issues.

6.2.5.1 Conductivity

The cmc of an ionic surfactant in solution and the degree of ionization (α) of the aggregates can be easily determined through conductivity measurements at constant temperature as a function of the surfactant concentration. As in the surface tension isotherm, the cmc is obtained from the intercept of the fitted lines in the regions where the conductivity changes linearly, as depicted in Figure 6.25.

The Krafft temperature of the surfactant is obtained by plotting the conductivity as a function of temperature, at constant surfactant concentration (see Figure 6.26). The Krafft point is determined by the middle point of the

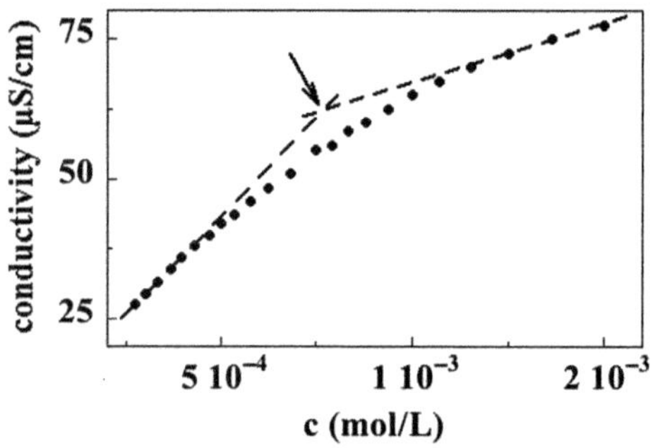

Figure 6.25 Conductivity measurements at constant temperature as a function of the surfactant concentration for 6-*O*-ascorbyl-decanoate. The arrow indicates the cmc.
(Reprinted from S. Palma, R. H. Manzo, D. Allemandi, L. Fratoni and P. Lo Nostro, *Colloids and Surfaces A: Physicochem. Eng. Aspects*, 2003, **212**, 163, Copyright 2003, with permission from Elsevier.)

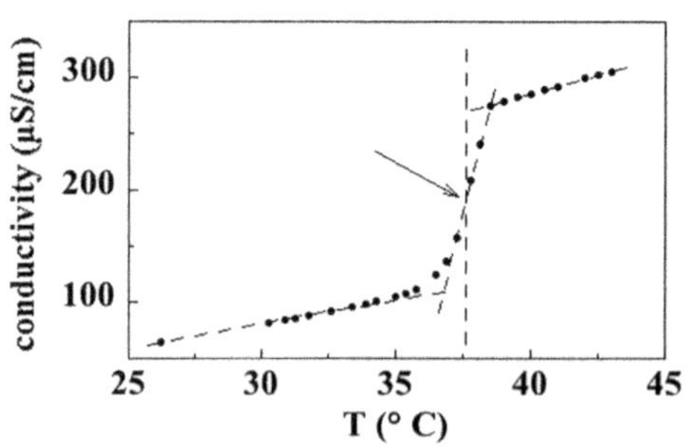

Figure 6.26 Conductivity measurements at constant concentration as a function of temperature for 6-*O*-ascorbyl-decanoate. The arrow indicates the Krafft point.
(Reprinted from S. Palma, R. H. Manzo, D. Allemandi, L. Fratoni and P. Lo Nostro, *Colloids and Surfaces A: Physicochem. Eng. Aspects*, 2003, **212**, 163, Copyright 2003, with permission from Elsevier.)

temperature interval (indicated by the arrow in the figure). The plot of ln(cmc) versus $1/T$, obtained by measuring the cmc at different temperatures, provides the value of ΔH°_{mic}, i.e. the standard enthalpy change of micellisation.

6.2.5.2 Fluorescence

This method, used for the determination of the cmc, exploits the different environment that a probe such as pyrene experiences when it is dissolved in water or in the hydrophobic micellar core.

Basically this method consists in the evaluation of the fluorescence intensity at different wavelengths for pyrene. In the presence of micelles – when the surfactant concentration reaches the cmc – the pyrene probe is dissolved in the lipophilic inner pool, and the fluorescence intensity reaches a constant value.

6.2.5.3 Differential Scanning Calorimetry

Differential scanning calorimetry (DSC) provides very important information on the thermal transitions that a system undergoes. In DSC measurements, the sample and a reference material are hosted in two separate cells. Upon heating or cooling, the differences in the thermal behaviour of the cells are recorded differentially. The heating–cooling process usually takes place at a rate varying between 0.1 and 10 K min^{-1}, while the kinetics of micelle formation are typically in the millisecond range and the phase separation occurs over some seconds. Therefore the system can reach equilibrium in the course of the DSC measurements. The micellization generally produces a characteristically pronounced endothermic peak in DSC thermograms, and the temperature of the peak gives the *critical micellar temperature* (cmt), which is the same as the Krafft temperature.

6.2.5.4 Scattering Techniques

In scattering techniques a beam of light, neutrons or X-rays impinges on the sample and the radiation is scattered in all directions.

A detailed description of the theoretical background of radiation scattering is beyond the scope of this work. It suffices to say that the intensity of the scattered radiation depends on the angle formed between the incident beam and the detector, on the temperature, and on the structural (size, shape) and dynamic properties of the aggregates and of their surface.

Scattering techniques are proving fundamental in understanding the structure of cleaning self-assembled systems (micellar solutions, microemulsions) at the nanoscale (see Section 7.6.1), and their interactions with detrimental layers of natural and synthetic polymers (*e.g.* acrylic–vinyl coatings) that have been applied on works of art such as wall paintings during past restorations (see, for instance, Section 9.5). This kind of advanced diagnostic procedure constitutes a typical example of the basic research carried out in scientific facilities devoted to conservation science, which allows a better comprehension of the mechanisms involved in the cleaning of artefacts, and in the refinement of nanostructured chemical tools for their preservation.

Further Suggested Reading

The following list provides support to the topics and concepts discussed in this chapter:

T. Cosgrove, *Colloid Science. Principles, Methods and Applications*, Blackwell Publishing Ltd, Oxford, 2005.
D. F. Evans and H. Wennerström, *The Colloidal Domain. Where Physics, Chemistry, Biology, and Technology Meet*, VCH, New York, 1994.
K. Holmberg, B. Jönsson, B. Kronberg and B. Lindman, *Surfactants and Polymers in Aqueous Solution*, John Wiley & Sons, Chichester, 2002.
H. Mollet and A. Grubenmann, *Formulation Technology*, Wiley-VCH, Weinheim, 2001.
D. Myers, *Surfactant Science and Technology*, John Wiley & Sons, Hoboken (NJ), 3rd edn, 2006.
B. Ninham and P. Lo Nostro, *Molecular Forces and Self Assembly. In Colloid, Nano Sciences and Biology*, Cambridge University Press, Cambridge, 2010.
R. M. Pashley and M. E. Karaman, *Applied Colloid and Surface Chemistry*, John Wiley & Sons, Chichester, 2004.

Review and Questions

The main topics discussed in the chapter may be summarized as follows:

- A surfactant contains a hydrophobic segment linked to a polar head group. Surfactants can be anionic, non-ionic, zwitterionic, or cationic, based on the charge on the polar hydrophilic head.
- Surfactants lower the surface tension of water and, above the critical micellar concentration (cmc) they form self-assembled nano-aggregates, whose shape depends on the geometric packing parameter (p) of the amphiphile.
- We introduced the hydrophobic effect as the driving force that leads to the formation of the aggregates, and the intermolecular interactions that are established between the hydrophobic chains and the polar heads.
- We have presented the uses and applications and the environmental concerns related to surfactant discharge.
- We have also presented the main effects of salts and temperature on the aqueous dispersions of amphiphiles, and defined the hydrophilic–hydrophobic balance (HLB) as a tool to predict the behaviour of a surfactant for a specific application.
- Finally, we summarized some of the most common techniques used to characterize the properties of surfactant aqueous dispersions.

Questions

1) Calculate the HLB number for a) sodium oleate, $C_{17}H_{33}COONa$; b) sorbitol monooleate; and c) sodium dodecyl sulfate.
2) Surface tension measurements on aqueous dispersions of a non-ionic surfactant have shown that γ decreases with log$c$ at the rate of 15 mN m^{-1} at 20 °C. Estimate the polar head group area a_P. Use the equation $a_P^{-1} = -(2.615/nRT) \cdot (\partial\gamma/\partial Logc)$.
3) The cmc of sodium dodecyl sulfate at 298 K is about 8.5×10^{-3} mol L^{-1}. Estimate the minimum amount of SDS to be dissolved in 100 mL of water in order to obtain a micellar dispersion.

Answers

1) $7 + 19.1 - (17 \times 0.47) = 18.1$; b) $7 + (5 \times 1.9) - (17 \times 0.47) = 8.5$; c) $7 + 38.7 - (12 \times 0.47) = 40.1$.
2) We make use of the reported equation to solve the calculation: $n = 1$, $T = 293$ K, and $(\partial\gamma/\partial Logc) = -15$ mN m^{-1}, therefore we obtain $a_P = 62$ Å^2.
3) SDS has molecular formula $C_{12}H_{25}NaO_4S$, and its molecular mass is 288.38 g mol^{-1}. Thus the cmc $= 8.5 \times 10^{-3}$ mol L^{-1} corresponds to $8.5 \times 10^{-3} \times 288.38 = 2.45$ g L^{-1}. Therefore in order to form micelles, we need to dissolve approximately at least 0.245 g in 100 mL of water.

References

1. R. Giorgi, M. Baglioni, D. Berti and P. Baglioni, *Acc. Chem. Res.*, 2010, **43**, 695.
2. M. Baglioni, R. Giorgi, D. Berti and P. Baglioni, *Nanoscale*, 2012, **4**, 42.
3. G. G. Ying, *Environ. Int.*, 2006, **32**, 417.
4. T. Cserháti, E. Forgács and G. Oros, *Environ. Int.*, 2002, **28**, 337.
5. P. T. Srinivasan and T. Viraraghavan, *Chemosphere*, 2000, **40**, 99.
6. http://www.endo-society.org/journals/ScientificStatements/upload/EDC_Scientific_Statement.pdf (accessed 29 Nov 2012).
7. D. Myers, *Surfactant Science and Technology*, John Wiley & Sons, Hoboken (NJ), 3rd edn, 2006.
8. C. Tanford, *J. Phys. Chem.*, 1972, **76**, 3020.
9. K. Holmberg, B. Jönsson, B. Kronberg and B. Lindman, *Surfactants and Polymers in Aqueous Solution*, John Wiley & Sons, Chichester, 2002.
10. H. Mollet and A. Grubenmann, *Formulation Technology*, Wiley-VCH, Weinheim, 2001.

CHAPTER 7

Cleaning II: Applications and Case Studies

E. CARRETTI* AND L. DEI

University of Florence, Italy
*Email: carretti@csgi.unifi.it; luigi.dei@unifi.it

7.1 Introduction: a Historical Perspective

Since the second half of the 20th century, surfactant solutions have played an interesting and progressively important role in many fields of industrial applications as a crucial component of research and development (R&D). In particular, surfactant solutions are involved in detergency. In the field of conservation and restoration of cultural heritage the use of micellar solutions as a possible tool for cleaning was quite belated, owing to the fact that for many works of art the use of water is forbidden, and also because it is easy to use organic solvents. For many years, when dealing with cleaning problems associated with the removal of hydrophobic substances, conservators referred to the triangular diagrams (Teas charts) and Feller tests (see Chapter 5) in order to prepare or choose the most suitable mixture of organic solvents to be used. With the advent of the environmental concerns and the strong request for both eco-compatible materials and "green chemistry" materials, many water-based systems assisted by surfactants were developed in several branches of industry (paints, varnishes, cars, *etc.*). Even in the field of conservation, several studies focused on the enormous potentialities of micellar solutions as possible detergents (cleaning agents) for the removal of hydrophobic substances ("foreign" compounds) that are present on the surface of cultural heritage objects. Indeed, one of the most critical cleaning problems to be solved was the

RSC Nanoscience & Nanotechnology No. 28
Nanoscience for the Conservation of Works of Art
Edited by Piero Baglioni and David Chelazzi

Published by the Royal Society of Chemistry, www.rsc.org

removal of degraded synthetic substances (polymers or copolymers) applied in previous conservation treatments, according to the supposed rule of reversibility (see Section 3.3).[1] As a matter of fact, the ageing of these polymeric materials revealed that the reversibility of the treatments was only a postulate, which unfortunately was incorrect. Thus, research pointed at the development of cleaning formulations able to remove these aged materials not only through traditional organic solvents (which are often unable to attack the strongly aged polymer coatings), but also by means of water-based systems. According to this new approach, micellar solutions, which may contain one or more co-solvents besides water, started to appear in the literature and became a possible research strategy in the growing field of "green chemistry". In the present chapter the use of micellar dispersions will be illustrated in strict association with the need to remove aged and degraded synthetic materials through swelling and solubilization.

7.2 The Degradation of Synthetic Materials

One of the most critical problems in the cleaning of the surfaces of works of art is the removal of degraded synthetic materials. Among these materials, acrylic polymers and copolymers have been widely used for the formulation of protective coatings, because they possess very good adhesion properties and succeed in suitable filmation processes. Often the use of these substances is strictly related to the need to change the hydrophilicity of the surfaces. As a matter of fact, the treated surfaces can eventually become quite hydrophobic and hydro-repellent.[2] For this reason synthetic polymeric coatings are applied on the surfaces of outdoor monuments to protect them from the deleterious effects associated with moisture. Moreover, an enhanced surface hydrophobicity can be important for the inhibition of the proliferation of moulds or other biological microorganisms (foulers); in fact, the presence of the acrylic polymer molecules introduces repulsive interactions between the surface of interest and the potential foulers.[3] One of the most important undesired side effects is the chemical and/or physical degradation upon ageing of these macromolecular compounds (see also Chapter 3).[4] In particular, owing to their photochemical reactivity and thermal degradation,[4,5] new materials based on fluorinated polymers[6,7] were developed, especially for stone monuments. Most of the research studies focused on the surface modification and on the strategy to produce polymeric coatings that are stable against thermal and photochemical decay.[2,3,8,9] Many of these coatings are still applied even when the same polymeric coatings applied in the past have clearly shown undesired effects.[10] Moreover, it has been clearly demonstrated[11] that there is a strong alteration of the physico-chemical properties at the interface between the cultural heritage masterpiece and the atmosphere. Owing to this evidence and to the degradation phenomena, conservation procedures on previously treated artefacts must very often deal with the removal of aged coatings, a step that makes the appropriate choice of cleaning agents critically important. Normally the removal is achieved through solubilization and/or swelling of the polymeric coatings by solvents, either pure or in mixtures. The next section will illustrate the swelling and solubilization phenomena.

7.3 Swelling and Solubilization

When a polymeric material comes in contact with a solvent or with a mixture of solvents, two different phenomena can take place: (i) complete solubilization occurs, leading to a macromolecular colloidal solution, or (ii) part of the solvent is adsorbed by the polymer strands through the swelling produced by the polymer chains, with a consistent volume increase of the solid polymer. A recent review[12] illustrates the basis of the theoretical treatment of hydrogel swelling. While in Chapter 5 all the aspects associated with the solubilization mechanisms have been extensively illustrated, no details have been given on the swelling phenomena. Swelling of polymers is one of the most important mechanisms through which "foreign" materials can be removed from the surfaces of works of art. Indeed, the well-known method based on ammonium carbonate solutions applied by means of wood cellulose poultices (the Ferroni method)[13] is a classical example of how many hydrophobic materials can be removed from wall paintings or stone surfaces as a result of their swelling in water, assisted and enhanced by the optimal pH conditions in the presence of ammonia and ammonium cations. When macromolecular compounds swell as a consequence of the adsorption of solvent into their structure, the increase in volume – which can be very large – generates a mechanical stress within the layer of the "foreign" material, leading to a relevant reduction of the adhesion forces between the hydrophobic polymeric chains and the substrate. This phenomenon leads to a softening of the substances to be removed, and a gentle mechanical action normally succeeds in the cleaning procedure without affecting the original layers underneath. Therefore, the choice of a suitable solvent to achieve a good swelling ratio is a crucial step in selecting the most appropriate cleaning technique, depending on the composition of the "foreign" materials. In the following sections we will show that swelling and solubilization are strictly related to the composition of the cleaning system, whose various components can act in a synergistic way.

7.4 Resin Soaps

In the second half of the 1980s, Richard Wolbers (conservator and researcher at the Wintherthur Museum of Wilmington in Delaware and teacher at the University of Delaware) and his staff set up the first formulation of a class of highly viscous fluids to be used as cleaning agents for the removal of oxidized natural varnishes from the surface of works of art: the resin soaps.[14–17]

These systems originated from a terpenoid-like molecule, abietic acid (ABA), which is characterized by a terpenic structure, and from a steroid derivative such as deoxycholic acid (DCA): both these acids can be neutralized by means of a base to obtain the ionized forms that behave as anionic surfactants. In fact, both molecules possess a non-polar part made up of 20 (ABA) and 24 (DCA) carbon atoms and a polar head formed by the carboxylic group. According to the original recipe indicated by Wolbers,[14,15] ABA and DCA are neutralized with triethanolamine (TEA), to form triethanolammonium abietate and triethanolammonium deoxycholate respectively (see Figure 7.1).

The most important property of these systems is related to the chemical nature of the surfactants, which makes them structurally similar to some

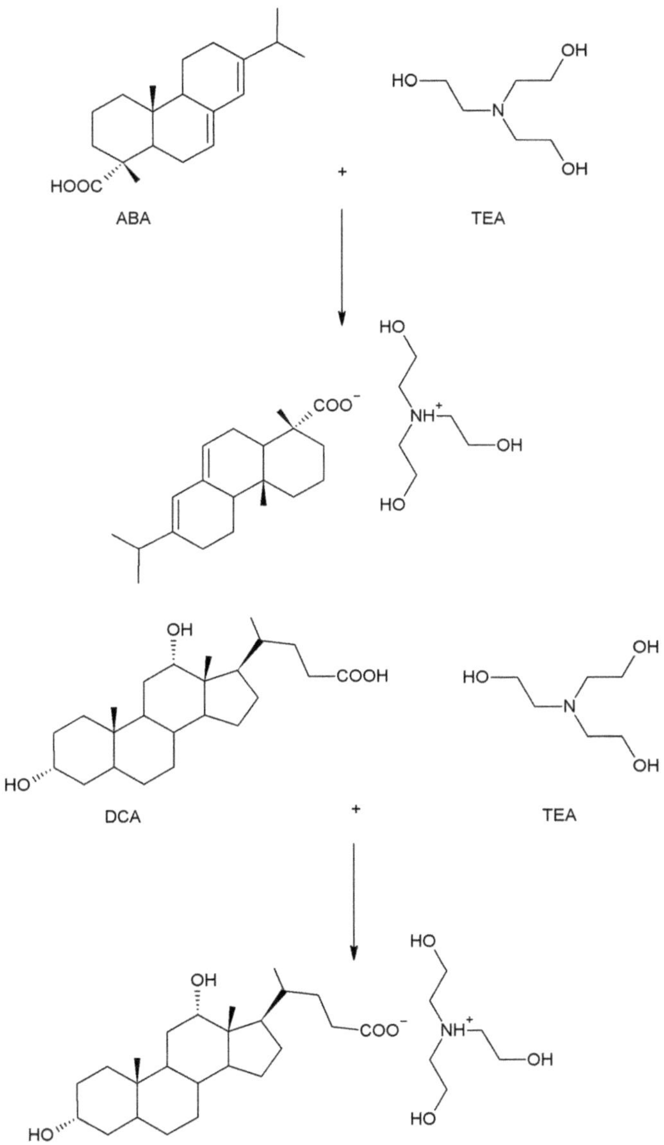

Figure 7.1 Chemical structures of abietic acid (ABA) and deoxycholic acid (DCA) and picture of their reactions with trietanolamine (TEA).

components of the terpenic natural varnishes largely used for the surface treatment of polychrome works of art (mostly easel and canvas paintings), as described in Figure 7.2.

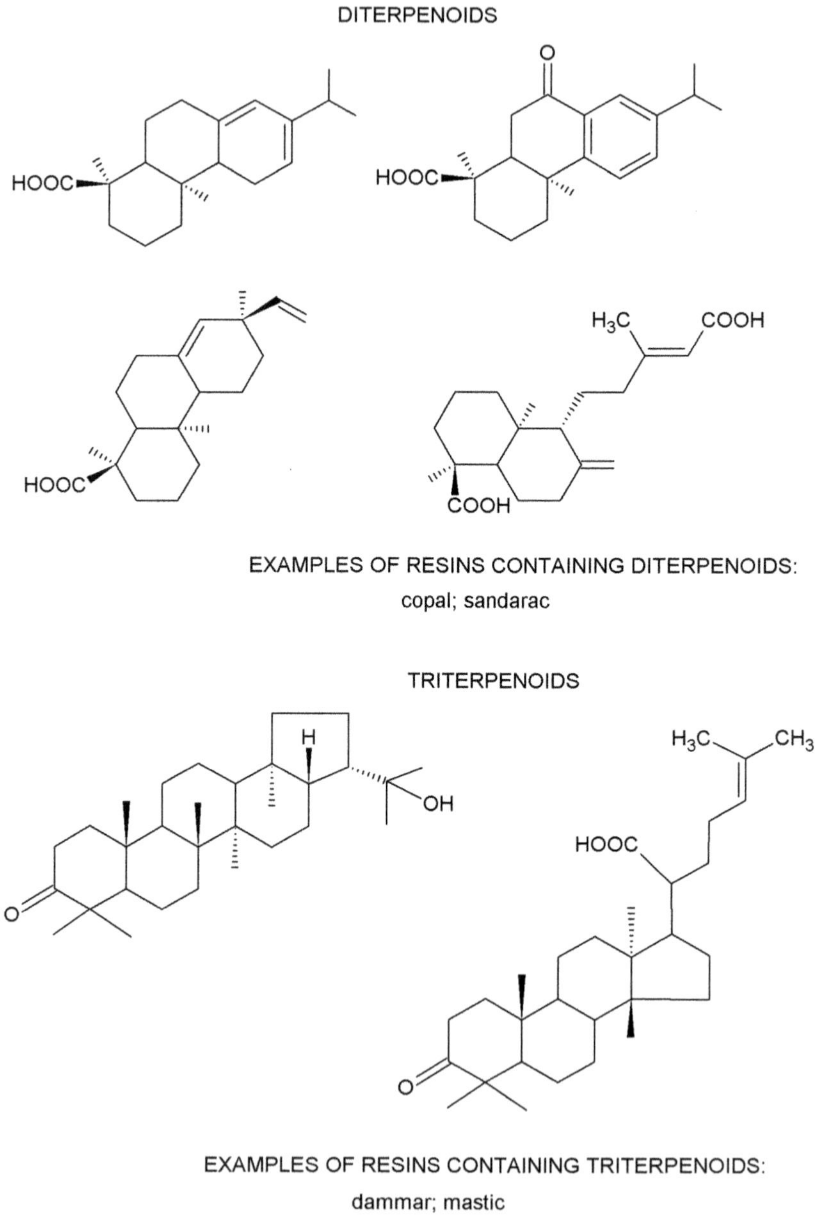

Figure 7.2 Molecular structure of some terpenoids contained into the natural resins commonly used as varnishes for paintings.

This similarity produces a specific interaction between the resin soaps and the varnish molecules, which – coupled to the detergent power typical of all surfactants (when the surfactant concentration is larger than the critical micellar concentration; see Chapter 6) – results in the selective removal of aged and non-aged natural varnishes from painted surfaces.[18,19] Sometimes the application of the resin soaps does not allow the complete solubilization of the varnish, but their action favours at least its partial break-up which, with an appropriate mechanical action, leads to the complete removal of the varnish layer.

Typically, for the preparation of the resin soaps, 2 g of ABA is dissolved in 100 g of demineralized water; once the surfactant is completely solubilized, 6 mL of TEA is added and the system is stirred until a transparent solution is formed. The pH is adjusted to about 8–8.5 with 1 M HCl solution. The only difference in the preparation procedure for DCA-based systems is the smaller amount of acid (0.5 g). The resin soaps are applied onto the surface of the painting with a cotton swab or a small brush, and the contact time is typically a few minutes. Once the system has carried out its function, the cleaned surface is washed several times with a solution of various surfactants such as bovine bile (2–3 w/w%), Tween 20 (2–4 w/w%), Brij 35 (2–4 w/w%) or artificial saliva. When the treated region is completely dried, the surface is washed with a mixture of low molecular weight hydrocarbons to completely eliminate all the lipophilic residues.

Another class of resin soap, based on the use of polycyclic aromatic hydrocarbons (PAHs), as indicated in Figure 7.3, was developed by White.[19] Unfortunately, while the planarity of the molecules increases their affinity with the oxidized terpenic resins, the main drawback is the extreme toxicity of the PAHs (some of the strongest carcinogenic substances belong to this class of compounds), and therefore their use was very limited.

The main limitations to the application of resin soaps are due to the residues that are left on the treated artistic substrates after the treatment. As a matter of fact, the use of solvents becomes necessary to remove any residue, with all the related issues that have been already discussed.

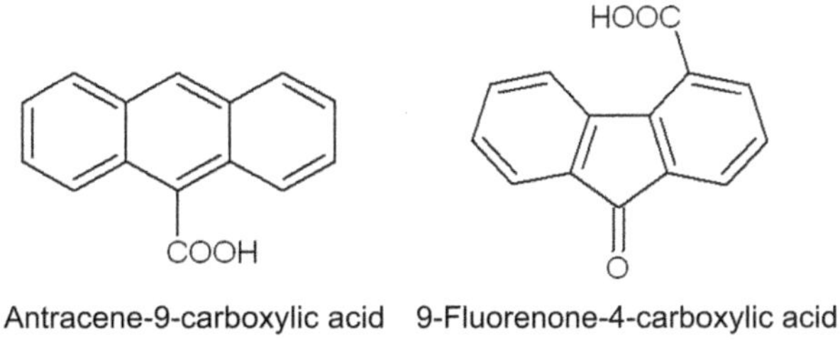

Figure 7.3 Structure of the compounds constituting the English resin soaps set up by Raymond White.[19]

7.5 Formulation and Application of Complex Micellar Solutions in Cultural Heritage Conservation

Before introducing the practical applications of micellar solutions, it is important to recall and highlight here the main advantages that this technology implies with respect to the use of neat solvents. The use of solvents for the removal of a polymer film from the surface of a porous matrix of historic, artistic or architectonic interest has many drawbacks.

1. The action of solvents is not completely selective and controllable: for instance, for wall paintings that are partially or totally realized with the "*a secco*" technique (see Section 2.1.1), partial solubilization or swelling of the organic components (*i.e.* binding media and organic additives) can occur, potentially causing severe damage to the work of art.
2. Most of the solvents rapidly dissolve the polymer layer giving a solution that, through capillarity, penetrates into the support, spreading the dissolved polymer within the matrix and depositing it in the pores.
3. Wall paintings are often placed in badly aerated environments, thus the level of safety associated with use of all the cleaning tools is a fundamental requirement for the safeguard of the operators' health. The toxicity of most of the common solvents used for the cleaning is a serious issue.

Aqueous nanostructured dispersed systems such as micellar dispersions and microemulsions overcome all the aforementioned limitations:

1. They are "green systems" with very low toxicity: they can contain up to 99 w/w% of water, with a drastic reduction in the impact on the health of the operators in comparison with the use of pure solvents.
2. The interaction between dispersed systems and polymers involves the interfaces between the nanodroplets, the aqueous continuous phase and the polymer surface. Given that micellar solutions and microemulsions possess an extended interface (up to hundreds of square metres per gram), the removal of surface polymer layers is highly effective. More details concerning the mechanism through which a cleaning process works are reported in Section 9.3.
3. Owing to the different mechanisms involved in the removal of coatings, nanostructured systems limit the penetration of the removed materials into the porous matrix of the artistic substrates. As a matter of fact, these systems provide rapid swelling of the coating layers at the coating–micelle interface, and promote the detachment of the coating owing to loss of adhesion. Since the cleaning fluids are aqueous (oil-in-water) systems, the re-deposition of swollen and detached polymers (hydrophobic material) within the substrate's pores is hampered by the presence of a "hydrophilic barrier".

From a practical point of view, the complete procedure that has to be followed for the formulation of a suitable complex micellar solution, to be used for the cleaning of a painted surface, is described schematically below and is similar to that commonly used for the set-up of microemulsions (whose possible use as cleaning tools in cultural heritage conservation is described in Chapter 9).

The four steps are:

1. Preliminary analytical investigation (carried out with spectroscopic and/or chromatographic techniques) is carried out to identify the materials to be removed.
2. Feller tests and Teas charts (see also Chapter 5) are used to determine the range of the f_d, f_p, f_h parameters corresponding to a suitable solvent for the selective solubilization of the foreign molecules in the external surface layer.
3. In order to set up the dispersed system, a surfactant [*e.g.* sodium dodecyl sulfate (SDS), Tween 20, Triton X100; see the examples reported in the case studies sections of Chapters 7 and 9] is dispersed in water. In some cases a co-surfactant is added. To minimize the amount of the residues left on the treated surface, the total amount of surfactant is usually kept below 5% (w/w).

   Once the solution has been obtained, the solvent (selected at point 2) is added drop-wise until the phase boundary is reached. The phase boundary can be visually identified when the system becomes opalescent upon the addition of an excess of solvent.

   The turbidity is a macroscopic sign of the instability of the system. In fact when the system is left undisturbed for a few hours or few days without any mechanical stimulation, the formation of two different phases appears. One layer is richer in the solvent and the other is mainly composed of water. From an applicative point of view, the phase separation represents an important limit because it does not allow the total control of the cleaning action, because the solvent-rich phase behaves in a way similar to the pure solvent. As a consequence, all the dispersed systems to be used for cleaning purposes for surfaces of artistic and/or historical interest should be thermodynamically stable, meaning that they should be perfectly transparent immediately after their preparation, for up to several months. When the maximum amount of solvent that can be added to the solution without the onset of phase separation is known, the final formulation suitable for applicative purposes can be prepared, using an amount of solvent slightly lower than the maximum limit (1–2 w/w% below), in order to avoid phase separation.

   The main difference between a microemulsion and a micellar solution is that in the first case the solvent (selected using Feller tests and Teas charts), which is not soluble in the continuous phase (*i.e.* water), is completely located in the hydrophobic core of the nanodroplets (Figure 7.4A). Instead, in a micellar solution (Figure 7.4B) the solvent is

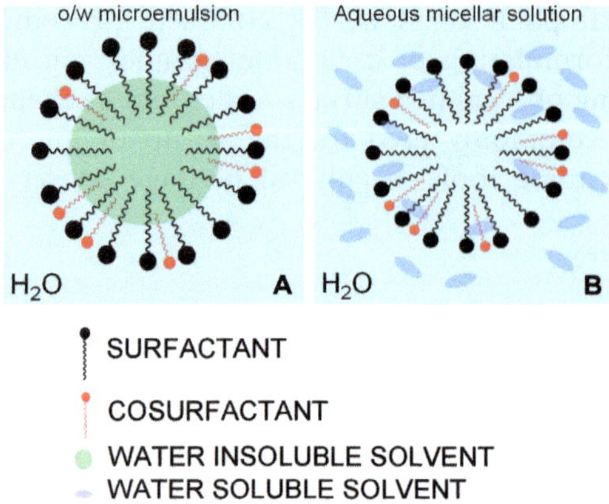

Figure 7.4 Schematic representation of the structure of a nanodroplet constituting an oil-in-water microemulsion (A) and a complex micellar solution (B).

partially soluble in water (*e.g.* ethyl acetate or propylene carbonate), therefore the solvent molecules will reside at the interface of the nanodroplets and in the continuous phase.

4. The application procedure is strictly related to the nature of the work of art to be treated and to its conservation status:
   - for wall paintings the wood poultice compress technique is commonly adopted;[20] this application methodology will be described in Section 7.6.2;
   - for easel paintings (canvas and wood paintings), besides the possibility of using solvent gels, the best solution is the confinement of the micellar solutions into a three-dimensional network such as a chemical gel, *e.g.* acrylamide–bisacrylamide.[21] The use of gels will be extensively discussed in Chapters 10 and 11.

As an example, the procedure adopted for the set-up of the system used for the removal of an aged layer of aged vinyl polymers from the surface of the wall paintings decorating the façade of the Conegliano Cathedral (Santa Maria dei Battuti) (see Section 7.6.1) is schematically reported in Figure 7.5.

Once the appropriate solvent (*e.g.* propylene carbonate, PC) has been selected, the system to be applied is set up as follows:

- 5 g of SDS solution (see Table 7.1) is weighed in a vial;
- propylene carbonate (PC) is added drop-wise;
- after the addition of each drop, the system is stirred for 30–40 s.

If, after such operation, the solution appears transparent, then an additional drop is added. This procedure is repeated until the phase boundary is reached (*i.e.* the solution becomes turbid upon stirring). In the particular case of the system prepared for the Conegliano Cathedral workshop (located in Northern

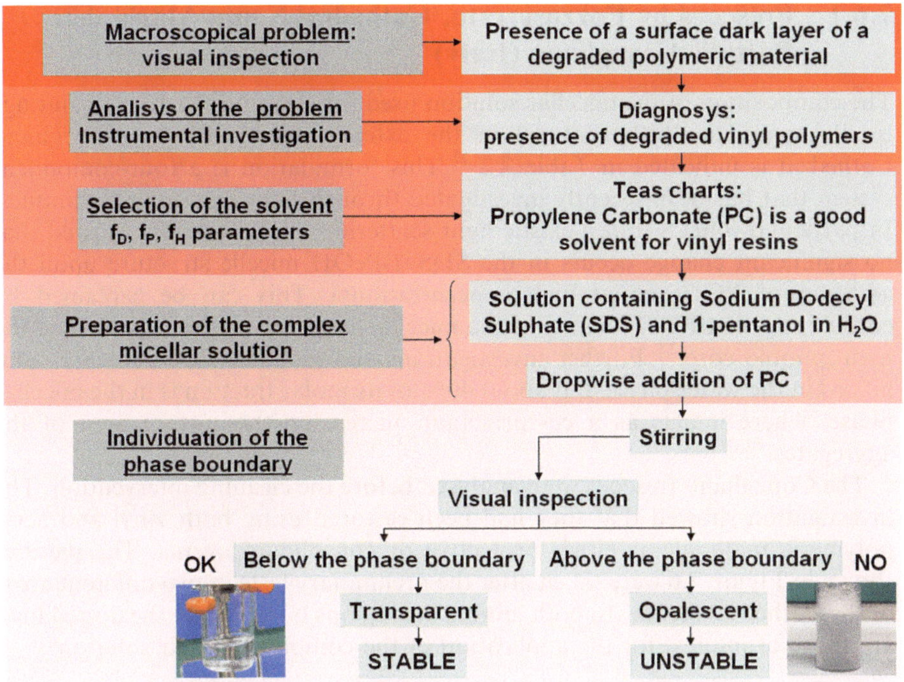

Figure 7.5 Schematic representation of the procedure for the set-up of a complex micellar solution to be used for the cleaning of a painted surface of artistic and/or historical interest.

Table 7.1 Composition of the aqueous solution used for the set up of the system used to remove vinyl aged polymer coatings from the surface of the external façade of the Conegliano Cathedral (Italy).

Composition	*Concentration wt%*
Water	88.5
Sodium dodecyl sulfate (SDS)	5.0
1-Pentanol (1-PeOH)	6.5

Italy, Veneto) the minimum quantity of PC necessary to reach the phase boundary was about 23 w/w%. After this, in order to be sure that the system applied onto the painting surface was stable, the final formulation contained a total amount of PC of about 22 w/w% (see Table 7.1).

7.6 Case Studies

In this section some important case studies where complex micellar solutions were successfully applied will be reported. The studies dealt with large surfaces, showing that the innovative "green" cleaning agents were and are ready for a true technological transfer.

7.6.1 Frescoes by Pozzoserrato, Cathedral Santa Maria dei Battuti, Conegliano (Italy)

The composition of the micellar solution used for cleaning the mural paintings by Pozzoserrato (16th century) in the external façade of the Conegliano Cathedral is indicated in Table 7.2.[22] This formulation is a four-component system that has been recently investigated through light scattering techniques. In particular, QELS (quasi elastic light scattering) measurements showed that no significant change occurs in the SDS–1-PeOH micelle structure upon the addition of PC (even at high concentrations). This can be explained by considering that PC is located at the micellar interface, rather than inside the hydrophobic core.[22] Further investigations showed that PC is mainly solubilized in the water phase, but about 40% of its molar fraction is in the micellar phase, where it acts as a co-surfactant, increasing the surface area of the aggregates.[23,24]

The Conegliano frescoes were analysed before the cleaning intervention. The investigation showed that they had been restored using both vinyl and acryl polymeric materials as either protective or consolidant agents. The pictures shown in Figure 7.6 were taken after the preliminary tests in two different areas of the Cathedral façade. In both images the regions bordered by the dotted lines are those of interest for the application of the complex micellar solution.

Table 7.2 Composition of the aqueous complex micellar solution used for the removal of vinyl aged polymer coatings from the surface of the external façade of the Conegliano Cathedral (Italy).

Composition	*Concentration wt%*
Water	69.0
Sodium dodecyl sulfate (SDS)	3.9
Propylene carbonate (PC)	22.0
1-Pentanol (1-PeOH)	5.1

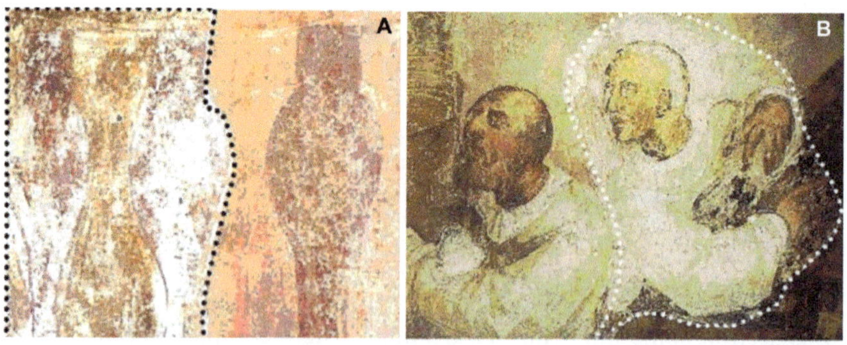

Figure 7.6 Detail of the two areas of the mural paintings of the external façade of the Conegliano Cathedral (Italy) where a complex micellar solution (see Table 7.2) has been tested (dotted line). (Courtesy of Conegliano's Diocese.)

The technique used was that of wood poultice compresses: a thin layer of wood poultice, previously wetted with the complex micellar solution, was evenly distributed onto the surface to be cleaned (using Japanese paper), and left in contact with the wall surface for at least three hours. The exact contact time was determined depending on the coating thickness. In order to eliminate the residual SDS (see Table 7.2) and to avoid its crystallization, after the removal of the wood poultice compresses, the fresco surface was rinsed several times with deionized water.

For the removal of oxalates and gypsum (which were present underneath the polymer coating), a modified polyacrylic acid based gel (1% w/w of gellant polymer)[25] containing an aqueous $(NH_4)_2CO_3$ saturated solution with ethylenediamine tetraacetic acid (EDTA) tetrasodic salt (0.1% by weight) was used.

The good results obtained in the tests (see Figure 7.6A and B) indicated this method to be "optimum" for the cleaning of the entire façade. It is worthwhile noting that the cleaning unveiled an opaque and "suffering" paint surface, due to the action of detrimental climate conditions for five centuries (the frescoes are on the external walls, exposed to high temperature excursions during the year and to photo-ageing effects). The coating by the vinyl and acrylic polymers of the previous conservation treatments made the colours more saturated and hid the true nature of the aged painting. Surprisingly the cleaning action returned the painting to its original natural aged state, but this could be the rule of a philological conservation procedure.

7.6.2 Wall Paintings by Vecchietta, Santa Maria della Scala Sacristy in Siena (Italy)

The Old Sacristy (also called *Cappella del Sacro Chiodo – Chapel of the Holy Nail*) of the Hospital of Santa Maria della Scala (Siena, Italy) was built in 1444 close to the Santissima Annunziata church to host some of the relics belonging to the Hospital. It was decorated by Lorenzo di Pietro, called "*il Vecchietta*", by means of an "*a secco*" technique (1446–1449). The painting was commissioned by the Rector Giovanni Buzzichelli, and represents a Christological cycle with histories of the Old and New Testaments. All the paintings were coated using a *scialbo* (a layer of slaked lime) in the 18th century and discovered only in 1930. Finally, at the end of the 1960s, the painted surface was treated with a thick layer of an acrylic copolymer [poly(EMA/MA) 70:30; ParaloidB72®] as a fixative and surface protective agent. The most relevant problems related to the use of this copolymer derive from its ageing, which results in three main drawbacks:

- the ageing polymer layer has induced a drastic alteration in the polychromy of the underlying painted surface and a decrease of its solubility in the solvents that are commonly used for the solubilization of the non-aged polymer;
- there has been alteration of the physico-chemical properties (particularly the water vapour permeability, the contact angle and the optical

Table 7.3 Composition of the aqueous complex micellar solution used for the removal of vinyl aged polymer coatings from the surface of the mural paintings by Vecchietta in the *Cappella del Sacro Chiodo* (Siena, Italy).

Composition	*Concentration wt%*
Water	77.5
Sodium dodecyl sulfate (SDS)	1.3
Propylene carbonate (PC)	18.0
1-Pentanol (1-PeOH)	3.2

properties) of the interface between the work of art and the environment. The image of a region of the paintings shown in Figure 7.7A (below) shows the glossy effect due to the presence of a thick layer of poly(EMA/MA);
- the penetration of the polymer into the residues of the *scialbo* made its removal very difficult.

The removal of both the aged polymer layer and the *scialbo* residues, while minimizing the spreading of the polymer into the porous matrix, was one of the most important targets of the restoration and conservation workshop. Thus, instead of using pure organic solvents or mixtures of them, a nanostructured system consisting of a complex aqueous micellar solution was chosen (see Table 7.3 for the formulation).

The micellar solution was applied *via* the wood poultice technique:[20] first it was added to a powder comprising fibres of pure cellulose, up to saturation. Then the pulp was used to make compresses that were applied onto the surface to be cleaned. Usually, a sheet of Japanese paper is interposed between the compress and the work of art surface, in order to avoid direct contact between the cellulose fibres and the surface. The application time was 2.5 hours, at a temperature between 20 and 25 °C. The cleaned surface was then washed with demineralized water in order to eliminate completely all the residues of surfactant (SDS; the foam due to the surfactant should disappear); this operation is usually done with the help of a cotton swab.

Figure 7.7 shows the image of a portion of the painting before (Figure 7.7A) and after (Figure 7.7B) the treatment with the micellar solution. The disappearance of the brown patina due to the removal of the surface layer of aged polymer is macroscopic evidence of the good performance of the applied system. The analytical investigations performed using Fourier transform infrared (FTIR) spectroscopy indicated the absence of a significant amount of polymer residue and of the surfactant used for the formulation of the cleaning system, but revealed the presence of gypsum, especially on the *scialbo* residues.[26]

In order to verify the effect of the cleaning on the physico-chemical properties of the interface between the paintings and the external environment, *in situ* capillary absorption measurements were carried out in two different areas of the paintings. The cleaning procedure was carried out in the first area, while the

Figure 7.7 Detail of the area of the mural paintings by Vecchietta in the *Cappella del Sacro Chiodo* (Siena, Italy) where micellar solution (see Table 7.3) has been tested. (A) Before cleaning. (B) After cleaning.
(Reprinted from S. Grassi, E. Carretti, P. Pecorelli, F. Iacopini, P. Baglioni, L. Dei, The conservation of the Vecchietta's wall paintings in the Old Sacristy of Santa Maria della Scala in Siena: The use of nanotechnological cleaning agents, *J. Cult. Herit.*, **8**(2), 119–125, Copyright (2007), with permission from Elsevier.) (Courtesy of the Municipality of Siena.)

second area was still affected by the presence of the surface layer of polymer. The data indicated that in the areas where the polymer was removed there was a drastic increase in both the absorption rate and in the total amount of adsorbed water (in the order of 80% after 10 minutes).

Further Suggested Reading

On swelling as an indicator of removability:

J. L. Down, *Reversibility – Does it exist?*, ed. A. Oddy and S. Carroll, British Museum Occasional Papers Nr. 135, The Trustees of British Museum, London, 1999, p. 111.

On the assessment of nanostructured complex fluids for the removal of hydrophobic coatings from painted surfaces:

E. Carretti, F. Rosi, C. Miliani and L. Dei, Monitoring of pictorial surfaces by mid-FTIR reflectance spectroscopy: efficiency of innovative colloidal cleaning agents, *Spectr. Lett.*, 2005, **38**, 459.
S. Grassi, M. Favaro, P. Tomasin and L. Dei, Nanocontainer aqueous systems for removing polymeric materials from marble surfaces: a new and promising tool in cultural heritage conservation, *J. Cult. Herit.*, 2009, **10**, 347.
B. Rosenzweig, E. Carretti, M. Picollo, P. Baglioni and L. Dei, Use of mid-infrared fiber-optic reflectance spectroscopy (FORS) to evaluate efficacy of nano-structured systems in wall paintings conservation, *Appl. Phys. A*, 2006, **83**, 669.

On the application of complex micellar solutions for conservation of wall paintings:

M. Baglioni, R. Giorgi, D. Berti and P. Baglioni, *Nanoscale*, 2012, **4**, 42.
M. Baglioni, D. Rengstl, D. Berti, M. Bonini, R. Giorgi and P. Baglioni, *Nanoscale*, 2010, **2**, 1723.
E. Carretti, B. Salvadori, P. Baglioni and L. Dei, Microemulsions and micellar solutions for cleaning wall painting surfaces, *Stud. Conserv.*, 2005, **50**, 128.
R. Giorgi, M. Baglioni, D. Berti and P. Baglioni, *Acc. Chem. Res.*, 2010, **43**, 695.

Review and Questions

The present chapter deals with the application of nanostructured complex fluids comprising an aqueous surfactant solution, a co-surfactant, and a suitable solvent that can interact with aged vinyl and acrylic polymers. The main aspects illustrated in this chapter were:

- The problems associated with the ageing of polymeric coatings used in the past as consolidant/protective agents, whose removal from the artistic or architectonic surfaces was necessary.
- The mechanisms according to which these aged materials can be removed when in contact with the selected nanostructured fluids, namely the solubilization and swelling phenomena.
- How to prepare these nanostructured fluids in order to have the best formulation in relationship to the polymer type.
- Two meaningful case studies where micellar solutions were applied to remove acrylic and/or vinyl polymers at different degrees of ageing, explaining the major results achieved and the problems associated with this particular nanotechnology.

Questions

1) What is the main problem, deriving from ageing, which can make the removal of a polymer coating problematic?

2) State in a concise way the main features of solubilization and swelling of a polymer by a solvent or by a mixture of solvents.
3) What are the various steps involved in the preparation of a complex micellar solution comprising water, a surfactant, a co-surfactant, and an organic solvent?
4) Why in the Pozzoserrato's case study was it impossible to remove gypsum and oxalates before the removal of the acrylic and vinyl coating?
5) How can one ascertain that, after the cleaning by these nanostructured fluids, some surfactant and polymer residues remain on the cleaned surface?

Answers

1) The main problem deriving from ageing, which can make the removal of a polymer coating problematic, is the high loss of solubility in most of the organic solvents. Indeed, the polymer ageing often leads to complex reticulation reactions that greatly change the solubilization properties of the applied polymer, hampering the removal of the polymer coating with the same solvents that were used for its application.
2) Solubilization is the complete dissolution of a polymer in a solvent, with the consequent formation of a colloidal macromolecular solution whose viscosity depends on both the polymer concentration and the molecular weight. On the contrary, swelling is a rather complex mechanism according to which the polymer adsorbs the solvent, and during the adsorption process the polymer chains stretch their structures and create interstices in which the solvent is adsorbed. Swelling leads to an increase of the polymer's volume, the extent of which depends on the amount of adsorbed solvent(s).
3) The various steps involved in the preparation of a complex micellar solution made up of water, surfactant, co-surfactant and organic solvent are the following: (i) preparation of a homogeneous and transparent solution of the selected surfactant, (ii) drop-wise addition of the co-surfactant, so that the system remains homogeneous without any phase separation, (iii) finally drop-wise addition of the selected solvent in the correct amount, with the same care as above.
4) In the Pozzoserrato case study it was impossible to remove gypsum and oxalates before removing the acrylic and vinyl coating, because the strongly hydrophobic coating inhibited the penetration of the aqueous ammonium carbonate and EDTA solutions, and the contact with the gypsum and the oxalate layers. The micellar solution succeeded in partly removing the impermeable coating, enabling the aqueous solutions to perform their action.
5) The best way to check whether some surfactant and polymer residues remain on the cleaned surface, after cleaning using nanostructured fluids, is to make FTIR measurements *in situ*. Another option is to sample some micro-fragments and analyse them with Pyr-GC-MS (pyrolyisis coupled

with gas-chromatography and mass spectrometry) for the polymer residues, ionic chromatography for ionic surfactants, or liquid chromatography for non-ionic surfactants.

Acknowledgements

The authors express their gratitude to Bruno Santi, Alessandro Bagnoli, Anna Maria Spiazzi and Vasco Fassina of the Local Offices of the Cultural Heritage Ministry, and to the conservators Paola De Santis, Saviano Bellè, Sabino Giovannoni and Fabrizio Iacopini for the cooperation during cleaning in the workshops.

References

1. L. Dei, P. Baglioni and M. Mauro, *Reversibility – Does it exist?*, ed. A. Oddy and S. Carroll, British Museum Occasional Papers Nr. 135, The Trustees of British Museum, London, 1999, p. 73.
2. W.-L. Chen and K. R. Shull, *Macromolecules*, 1999, **32**, 6298.
3. D. G. Walton, P. P. Soo, A. M. Mayers, S. J. S. Allgor, J. T. Fujii, L. G. Griffith, J. F. Ankner, H. Kaiser, J. Johansson, G. D. Smith, J. G. Baker and S. K. Satija, *Macromolecules*, 1997, **30**, 6947.
4. R. Feller, *Accelerated Aging*, The Getty Conservation Institute Publisher, Los Angeles, 1994.
5. A. Charlesby, *Atomic Radiation and Polymers*, Pergamon Press, Oxford, 1960.
6. F. Piacenti and M. Camaiti, *J. Fluor. Chem.*, 1994, **68**, 227.
7. A. Stempf, P. Muller, M. Pabon and J. M. Corpart, *Int. J. Restor. Build. Monum.*, 1999, **5**, 273.
8. M. Lazzari, M. Aglietto, V. Castelvetro and O. Chiantore, *Chem. Mater.*, 2001, **13**, 2843.
9. R. Winter, P. G. Nixon, G. L. Gard, D. G. Castner, N. R. Holcomb, Y.-H. Hu and D. W. Grainger, *Chem. Mater.*, 1999, **11**, 3044.
10. L. Tintori, *Stud. Conserv.*, 1963, **8**, 37.
11. E. Carretti and L. Dei, *Progr. Organ. Coat.*, 2004, **49**, 282.
12. F. Ganji, S. Vasheghani-Farahani and E. Vasheghani-Farahani, *Iran. Polym. J.*, 2010, **19**, 375.
13. P. Mora, L. Mora and P. Philippot, *The Conservation of Wall Paintings*, Butterworths, London, 1984.
14. R. C. Wolbers, *Cleaning Painted Surfaces. Aqueous Methods*, Archetype Publications, London, 2000.
15. R. C. Wolbers, *Notes for Workshop on New Methods in the Cleaning of Paintings*, The Getty Conservation Institute Publisher, Marina del Rey, CA 1990.
16. R. C. Wolbers, *Workshop on New Methods in the Cleaning of Paintings*, Getty Conservation Institute, Marina Rey, CA, United States, 1988.
17. D. Erhardt and J. J. Bischoff, *Stud. Conserv.*, 1994, **39**, 3.

18. D. Erhardt and J. J. Bischoff, *Restoration '92: Conservation, Training, Materials and Techniques: latest developments. Preprints to the Conference held at the RAI International Exhibition and Congress Centre, Amsterdam, 20–22 October 1992*, ed. V. Todd, J. Marsden, M. K. Talley, J. Lodewijks, V. Sluyterman van Loo and W. Koeno, IIC, London, 1992, p. 77.
19. A. Burnstock and R. White, *Cleaning, Retouching and Coatings: Technology and Practice for Easel Paintings and Polychrome Sculpture. Preprints of the Contributions to the Congress, 3–7 September 1990*, ed. J. S. Mills and P. Smith, IIC, London, 1990, p. 111.
20. E. Ferroni, *Ecological Physical Chemistry*, ed. C. Rossi and E. Tiezzi, Elsevier, Amsterdam, 1991, p. 345.
21. G. Pizzorusso, E. Fratini, J. Eiblmeier, R. Giorgi, D. Chelazzi, A. Chevalier and P. Baglioni, *Langmuir*, 2012, **28**, 3952.
22. E. Carretti, L. Dei and P. Baglioni, *Langmuir*, 2003, **19**, 7867.
23. G. Palazzo, D. Fiorentino, G. Colafemmina, A. Ceglie, E. Carretti, L. Dei and P. Baglioni, *Langmuir*, 2005, **21**, 6717.
24. G. Colafemmina, D. Fiorentino, A. Ceglie, E. Carretti, E. Fratini, L. Dei, P. Baglioni and G. Palazzo, *J. Phys. Chem. B*, 2007, **111**, 7184.
25. P. Cremonesi, A. Curti, L. Fallarini and S. Raio, *Progetto Restauro*, 2000, **7**, 25.
26. S. Grassi, E. Carretti, P. Pecorelli, F. Iacopini, P. Baglioni and L. Dei, *J. Cult. Herit.*, 2007, **8**, 119.

CHAPTER 8

Cleaning III: Emulsions and Microemulsions

D. BERTI

University of Florence, Italy
Email: berti@csgi.unifi.it

8.1 Introduction

This chapter will provide a simplified description of the main features of emulsions and microemulsions. Although many natural complex fluids are actually emulsions (*e.g.* milk), the fundamental research in the field started about 50 years ago, fostered by technical advances in the tools for characterization, such as nuclear magnetic resonance (NMR) and small-angle scattering techniques.

Despite its age, this field is still an active area of fundamental research, mainly as a result of its enormous and ubiquitous relevance in virtually all fields of technology,[1] spanning from medicine,[2] to detergents[3] and microetherogeneous synthesis.[4] Recently these nanostructured systems have gained the central stage as "smart" cleaning agents in the field of Cultural Heritage, as will be illustrated through several examples in Chapter 9.

Oil and water do not mix. Emulsions and microemulsions are macroscopically homogeneous mixtures of two immiscible liquids, *e.g.* water and oil, whose formation is necessarily mediated by at least one component, generally termed the *emulsifier*. From a microscopic point of view, these mixtures are, however, not as homogeneous as conventional solutions, because they consist of water and oil domains, while the emulsifier mainly covers the interface between the two immiscible liquids. One of the two liquid phases is dispersed in the so-called continuous liquid phase in the form of droplets of various sizes. An emulsifying

RSC Nanoscience & Nanotechnology No. 28
Nanoscience for the Conservation of Works of Art
Edited by Piero Baglioni and David Chelazzi

Published by the Royal Society of Chemistry, www.rsc.org

agent must have good surface activity and its diffusion rate to the interface must be comparable with the kinetics of dispersion.

From the point of view of microstructure, emulsions and microemulsions can be defined according to the nature of the droplets, *i.e.* w/o (water-in-oil, water droplets in oil), o/w (oil-in-water, oil droplets in water) or bicontinuous, where water and oil form two separate channels.

The most common emulsifying agents are surfactants, whose chemical and physical properties have been introduced in Chapter 6. However block copolymers and solid particles can stabilize liquid droplets in an immiscible dispersing phase, as shown in the scheme reported in Figure 8.1. Solid particles are present in the so-called Pickering emulsions, which will be briefly described in the following sections. Although an emulsion or a microemulsion can be prepared simply by following a recipe, such fluids are much more complex than the mere sum of their ingredients and a deep understanding of the underlying principles that rule their stability and phase behaviour are key to designing and using them correctly in every field of application.

As a first step, it is very useful to follow a straightforward exercise to evaluate the dispersion energetics in a two-phase system (*i.e.* two immiscible liquids without the presence of any emulsifier agent). The Gibbs free energy involved when two liquids that have negligible mutual miscibility are brought into contact can be written as:

$$\Delta G_{form} = \gamma \Delta A - T\Delta S_{form} \tag{8.1}$$

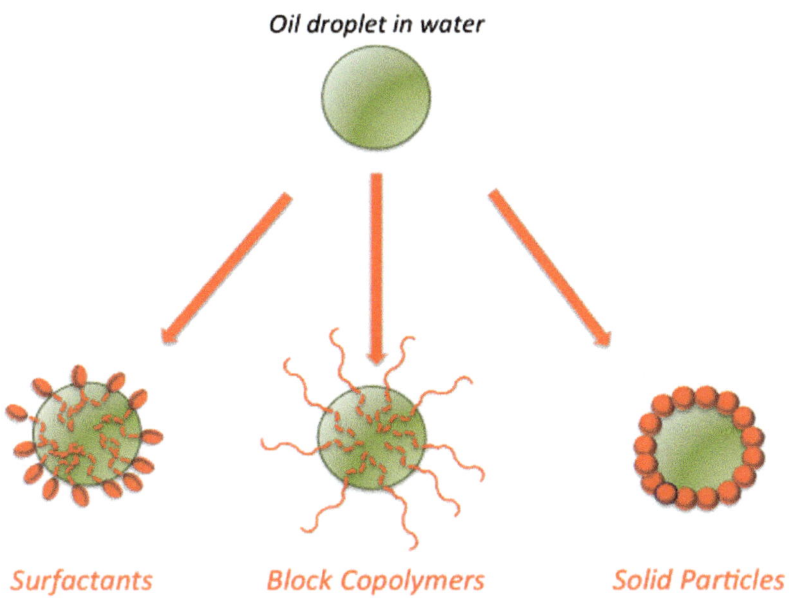

Figure 8.1 Emulsifying agents. The most commonly used for applicative purposes are undoubtedly surfactants and block copolymers.

where γ is the interfacial tension between the two liquids, ΔA is the change in interfacial area, ΔS_{form} is the change in configurational entropy and T is the absolute temperature. We recall that for any physico-chemical process to occur spontaneously, the associated ΔG must be negative.

The entropic term is expressed in terms of the volume fraction ϕ of one of the two liquids as:[5]

$$\Delta S_{form} = -nK_{\mathrm{B}}\left\{\ln\phi + \left[\frac{1-\phi}{\phi}\right]\ln(1-\phi)\right\} \tag{8.2}$$

here n is the number of droplets, and K_{B} the Boltzmann constant.

This term is always positive and therefore favours the dispersion of a volume fraction ϕ of liquid as droplets in a $(1-\phi)$ volume fraction of an immiscible liquid (recall that the volume fraction ϕ is always lower than 1). Moreover the entropic contribution scales linearly with the number of droplets, and the finer the dispersion, the greater the configurational gain.

On the other hand, a high number of droplets involves the creation of a large interfacial area, which conversely represents an impressive energetic cost, modulated by the interfacial tension, which is the first term of eqn (8.1).

If we consider that the interfacial tension between two immiscible prototypical liquids used in restoration workshops for standard cleaning procedures (*e.g.* water and hexane) is $50\,\mathrm{mJ\,m^{-2}}$, and evaluate both energetic contributions to the formation of an oil in water dispersion (5% of hexane with respect to the total volume), we obtain the results illustrated in Figure 8.2. The enthalpic term is never

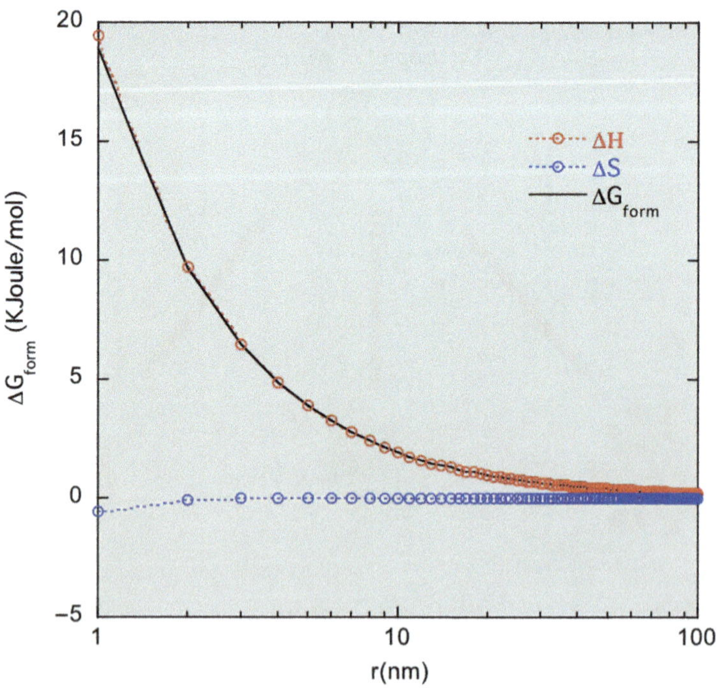

Figure 8.2 Energetics of the dispersion of hexane in water at a 5% volume fraction.

counterbalanced by the entropic gain, and the total energy is practically dominated by the energetic cost associated with the creation of a huge interfacial area between the two liquids. Moreover, as the size of the droplets (r) decreases, the enthalpic term drives the total balance to more positive values.

These considerations easily show why thermodynamic stability cannot be accomplished without the intervention of at least a third component, whose main effect is to decrease the interfacial tensions to ultra-low values and whose presence is a prerequisite, though not a guarantee, for spontaneous dispersion, as we will see in the following sections.

A very important concept is that the classification of dispersions performed according to the final size of the droplets, yielding emulsions, mini- (or nano-) emulsions and microemulsions, sketched in Figure 8.3, is also related to their thermodynamics of formation.

The visual appearance can be of considerable help in distinguishing the different cases of dispersion. Emulsions (such as milk) are very turbid, because of the multiple scattering of the visible light due to the droplets. For nanoemulsions, the aspect can be from opalescent to clear, depending on the size, while micro-emulsions cannot be distinguished from a classical solution by the naked eye. Also, the total specific interfacial area, *i.e.* the area per mass unit, varies by four orders of magnitude when passing from emulsions to microemulsions.

Figure 8.4 illustrates the energetic states of the separated phases in comparison to those of the dispersions. In the case of emulsions and nano-emulsions, the Gibbs free energy associated with dispersion is positive,

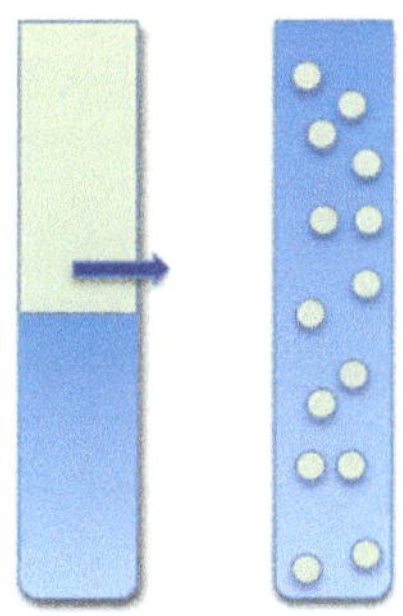

Name	Emulsions	Nanoemulsions	Microemulsions
Droplet Size	> 1 μm	1 μm–100 nm	< 100 nm
Appeareance	Turbid	Opalescent/Clear	Clear
Surface/mass [m^2/g]	0.07/70	70/330	130/1300
Stability	Metastable	Metastable	Stable

Figure 8.3 Schematic representation of the formation of a liquid–liquid dispersion and size-based classification.

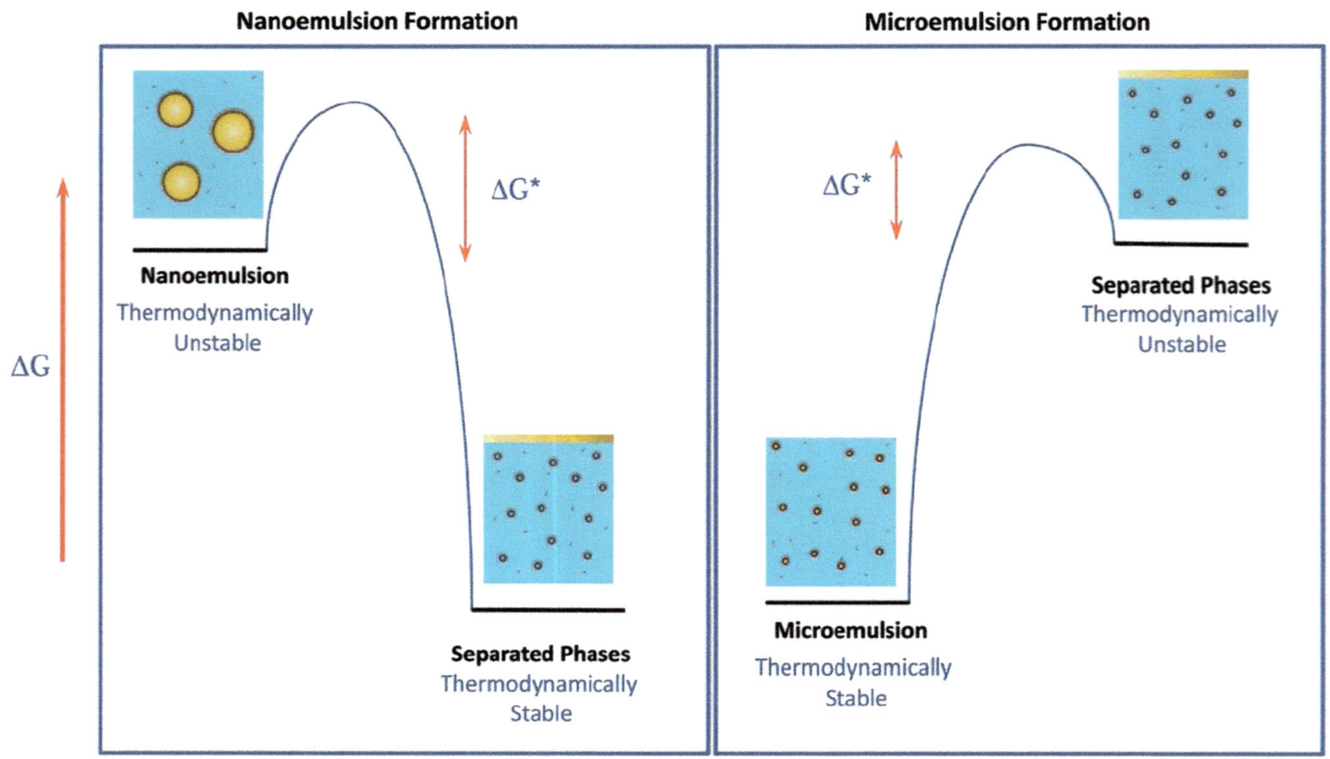

Figure 8.4 Gibbs free energy profiles for a nanoemulsion and a microemulsion. (Reproduced from D. J. McClements, *Soft Matter*, 2012, **8**, 1719–1729 with permission from The Royal Society of Chemistry.)

therefore such systems are metastable; on the other hand microemulsions, besides being characterized by the smallest size, are thermodynamically stable. It is important to underline that droplets in microemulsions are actually smaller than those in nanoemulsions, notwithstanding the "nano" prefix.

For emulsions, the thermodynamically stable state is the demixed one where the contact area between the two phases is minimized, a situation that is eventually reached through droplet coalescence. However, the system can persist in a metastable state for a considerable period of time if it is "kinetically stable".

This example clarifies why, in the case of emulsions and nanoemulsions, much of the effort has been devoted to defining and improving the kinetic stability, *i.e.* increasing ΔG^*, with respect to thermal energy RT (2.479 kJ mol^{-1} at room temperature) to avoid demixing during storage or use. The presence of the emulsifying agent has two different effects on these systems: for nanoemulsions, it raises the energetic barrier leading to separation (ΔG^*), while for microemulsions it allows thermodynamic stability.

In general the stabilization of liquid dispersions is achieved by lowering the interfacial tension; the major role is played by the amphiphiles, which produce stable dispersions either kinetically (emulsions) or thermodynamically (microemulsions). The surfactant behaviour at interfaces and aggregation has been reviewed in Chapter 6. The stabilization of the dispersed droplets basically involves the same affinity of surface-active agents for interfaces.

From an applicative standpoint, because emulsions and nanoemulsions cannot be formed spontaneously, their production requires an energy input, usually provided by mechanical devices. An example from daily life is the preparation of salad dressing, when the emulsification of oil in an aqueous medium (*i.e.* vinegar) is attained by vigorous mixing, but it eventually results in demixing when the dispersion is left at rest. For microemulsions, the thermodynamic spontaneity and the lowest size range occur at the same time. In this case the interfacial energy term, γ, must be extra-low in order to compensate for the huge increase of the surface of the interfacial layer.

Another kind of classification, which encompasses all the liquid–liquid dispersions, irrespective of their size, concerns the nature of the two immiscible liquids. Often one of the two liquid components is water, and the other is generically termed "oil", to stress its incompatibility with the aqueous medium. Based on the liquid phase that is finely dispersed into the other liquid phase, we may distinguish:

- a water-in-oil dispersion (**w/o**), when the dispersed phase is the aqueous pool;
- an oil-in-water dispersion (**o/w**) in the opposite case.

A process known as *phase inversion* turns one type of dispersion into the other, as we will see in the following sections.

We should mention here that some other kinds of liquid–liquid dispersions, especially microemulsions, have become popular in the last few years, *e.g.* microemulsions of supercritical CO_2 in water and various kinds of microemulsions based on ionic liquids. These systems will not be treated explicitly in this contribution, but the general guidelines outlined for w/o or o/w microemulsions still hold for such dispersions.

In order to summarize the nomenclature of liquid–liquid dispersion, besides the composition and the chemical nature of the droplets (oil or water), the prefix "micro" features thermodynamic spontaneity and the smallest size range.

One of the most thoroughly investigated areas in fundamental colloid science consists in the prediction of dispersion behaviour, once the components and the thermodynamic conditions are known. This aspect is of course of major importance for applicative purposes and will be extensively treated in the following sections.

The use of microemulsions in the field of Cultural Heritage, as smart and selective cleaning agents, is probably one of the most exciting applications. The scientific and technological challenges related to applications in this field, including environmental benignity and the complexity of interactions with precious multiphase solid substrates, just to name two, make these predictive tools even more important and central.

8.2 Emulsification

The formation of emulsions is a non-spontaneous process and as such it requires an energy input that drives the positive free energy variation. Moreover kinetic stability must be provided to the dispersion in order to prevent phase separation once the droplets are obtained.

The physical parameters characterizing an emulsion are, besides its composition, the average droplet size and the size dispersion. These structural factors may vary as a function of time, owing to insufficient kinetic stability. The emulsification method and the associated variable parameters determine the final structure of the emulsion.

The most popular preparation protocols involve mechanical energy, which can be provided in different ways. Normally the two liquid phases are stirred vigorously, or exposed to ultrasound, or forced to pass through tiny orifices in a homogenizer. In any case these operations result in a decrease in the mean diameter of the droplets of the dispersed phase from macroscopic to colloidal size range.

When the system is just composed of the two liquids, with no other additive, the higher the energy input, the smaller the final size of the droplets. Usually the volume fraction (ϕ) and the difference in viscosity (η) between the two phases determine which phase will act as the dispersing medium. In particular, the liquid with the lowest η and the lowest ϕ will be dispersed in droplets.

The presence of polymers, as solid particles or amphiphiles, can change the scenario radically. The first effect consists in the lowering of the interfacial tension and of the final size of the droplets, preventing their re-aggregation once they are formed. The kinetic stability against phase separation can be considerably high. This is the reason why it is so important to have some practical guidelines in order to prepare a stable emulsion.

8.2.1 Predicting Emulsion Type: Bancroft's rule and HLB Number

The main factors that determine the formation of an emulsion are the kind of emulsifier and the preparation temperature. The relative amounts and viscosities of the immiscible liquid phases also play a significant role.

In formulating an emulsion for a practical application, the choice of the surfactant is ruled by the nature of the disperse/continuous phases and by the working temperature. An intuitive and useful guiding principle is the *Bancroft Rule* (see also Section 6.2.4),[6] which states that the continuous phase is the one where the surfactant is more soluble. To verify that the correct kind of emulsion is formed, a very intuitive experiment consists in diluting aliquots of the emulsion with the two liquids alternately. While an emulsion can generally be diluted with the outer liquid without phase separation, on the other hand the two liquids will separate upon dilution with the inner phase.

In conclusion, a strongly water-soluble surfactant will stabilize oil-in-water droplets and *vice versa*. The surfactant hydrophilicity, and therefore its solubility in water, can be reckoned from some consideration of its chemical structure and charge. However, when two surfactants are both water-soluble, which one will work best in dispersing a given oil?

A more quantitative classification that solves this problem would be of more practical use. A very useful numerical index was introduced more than 60 years ago by Griffin for *non-ionic* surfactants, namely the hydrophilic–lipophilic balance (HLB).[7] This parameter has already been discussed in Section 6.2.4, and will be recalled here. The HLB number is a tool to predict the kind of emulsion and its stability; it is based on an empirical scale assigned to a given surfactant, on the basis of the balance between its hydrophilic and hydrophobic portions.

$$HLB = \frac{20 \times MW_{\mathrm{H}}}{MW} \tag{8.3}$$

where MW_{H} is the molar mass of the hydrophilic portion and MW the overall molar mass of the surfactant. This definition arbitrarily sets the lower and the upper limits of the scale (HLB = 0 for a completely apolar molecule, while HLB = 20 for a fully hydrophilic compound).

An extension of this definition that can be applied to *ionic* amphiphiles was suggested some years later by Davies and Rideal, who proposed assigning a specific contribution to each chemical group of the emulsifier [see Section 6.2.4, eqn (6.11)].

Generally, HLB is 0 for an emulsifier that cannot be dispersed in water at all, while for HLB = 18 a clear aqueous solution is expected to form; HLB numbers in between refer to milky or opalescent aqueous dispersions.

When, by a trial-and-error method, a given mixture is found to provide the best performance (*e.g.* minimizing the size of the droplets) in producing a w/o emulsion, then its HLB number is evaluated; it is thus assumed that comparable dispersions of the two liquid media should be obtained by surfactants or surfactant mixtures having the same HLB number. The HLB numbers have been tabulated for a large number of commercial surfactants, and can be found in several handbooks.[8]

We have seen that the overall shape of a surfactant molecule, expressed by the packing parameter introduced in Chapter 6 (Section 6.2.4), determines the geometry of the assemblies. We recall here that the packing parameter compares the cross-sections of the hydrophobic and hydrophilic portions, and its value determines the overall shape of the aggregates.

The HLB number is intimately connected to the shape of the amphiphile. Lower packing parameters (<1/3) imply a large cross-section of the hydrophilic group and therefore a significant water solubility and a large HLB number. As the hydrophobic cross-section increases, the packing parameter increases, while conversely the HLB number decreases, as shown in Table 8.1 for some very common amphiphiles.

The HLB classification is a useful rule of thumb, and it should be considered as a standard starting tool. For instance, to prepare a water-in-oil emulsion, one should select surfactants with an HLB spanning from 3 to 6, while an HLB number from 8 to 18 is suitable for oil-in-water emulsions (see also Section 6.2.4).

A more meaningful guide to surfactant-induced phase behaviour is the use of the Hildebrand solubility parameters, directly connected to molecular parameters and introduced in Chapter 4, which can be related to the van der Waals forces (see also Section 1.2) between the components; however, in spite of their empirical nature, the HLB numbers are widely used.

The most severe limitation of the HLB approach is that it is a "*static*" parameter and it does not take into account the effect of the surfactant concentration, the kind of oil, the presence of additives such as electrolytes and, most noticeably, temperature. This latter effect is indeed very marked for non-ionic surfactants, *e.g.* the C_iE_j series, where i denotes the length of the hydrocarbon chain, and j indicates the number of oxyethylenic groups of the hydrophilic portion.

As already introduced in Chapter 6, the water solubility of the hydrophilic portion displays a pronounced decrease as the temperature is raised (see Section 6.2.3). Aqueous solutions of this class of surfactant show phase separation above a given temperature (the cloud point). Basically the same phenomenon causes the "actual" HLB number to decrease as temperature is increased for the same surfactant. This means that the HLB parameter, evaluated from the chemical

Table 8.1 Hydrophilic–lipophilic balance (HLB) numbers of some common surfactants.

Surfactant	*HLB number*
Sodium dodecyl sulfate (SDS) $CH_3(CH_2)_{11}OSO_3Na$	40
Dodecyl hexaethylene glycol monoether ($C_{12}E_6$) $CH_3(CH_2)_{11}O(CH_2CH_2O)_5CH_2CH_2OH$	5.3
Glycerol monostearate $HOCH_2CH(OH)CH_2OOC(CH_2)_{16}CH_3$	3.7

composition as a fixed property of the emulsifier, is no longer adequate to predict the phase behaviour. At the *HLB temperature* (or phase inversion temperature, PIT) the amphiphile has balanced affinities towards oil and water, while above this value its preferential solubility is shifted towards the oil phase. On the original scale for non-ionic surfactants, this threshold corresponds to HLB = 10, while the packing parameter is instead 1. Therefore, w/o emulsions form above the PIT, while o/w dispersions are formed below this limit.

At the PIT the maximum possible solubilization of oil and water is attained with a given quantity of surfactant: two continuous liquid phases are separated by a locally planar surfactant film, as we will see later. Moreover, this temperature marks a minimum in the interfacial tension, which is around 10^{-3} mN m^{-1}. Therefore, relatively small droplets and a reasonably narrow size-distribution can be obtained with a low input of mechanical energy. This is the so-called "PIT method" for the preparation of an emulsion, and consists in bringing the ternary system (oil/water/surfactant) to the HLB temperature, providing a minimum of mechanical stirring, and then cooling or heating it very quickly to obtain o/w or w/o emulsions, respectively.

It is very important to underline that emulsions should be stored at a temperature as far away as possible from the HLB temperature. While the PIT method provides the minimum possible droplet size, the droplet size increases with time.

From a practical point of view, one should store o/w emulsions at temperatures 20–60 °C below the PIT and w/o emulsions 10–40 °C above the PIT. The issue of emulsion stability will be further addressed in the next section.

8.2.2 Factors Determining Emulsion Stability

As long as the energetic barrier ΔG^* for demixing is about $20RT$, emulsions can be considered kinetically stable during storage and use. The physical parameters that determine the stability of an emulsion are basically the same as those that govern the stability of solid dispersions. Remember that the typical size of the droplets is in the μm range. Therefore the action of *external* forces, such as gravity, can drive sedimentation (downward movement) or creaming (upward movement) of the droplets. In addition, *internal* inter-droplet forces, such as a net attractive interaction among them, can further drive droplet aggregation and ultimately result in macroscopic phase separation.

The "natural" competitor of buoyancy is Brownian motion, *i.e.* the ensemble of collisions with the outer phase molecules that ultimately leads to the erratic random walk of the droplets. The driving force of Brownian motion is thermal energy. Therefore sedimentation (or creaming) can be limited or absent if the following relation is fulfilled:

$$\frac{4}{3}\pi r^3 |(\rho_1 - \rho_2)| g H \ll k_B T. \tag{8.4}$$

here r is the droplet radius, ρ_1 and ρ_2 the densities of the two liquid phases, g the gravitational acceleration, and H the height of the tube where the emulsion is contained. As we have already recalled, tiny droplets guarantee higher stability.

Moreover, the lowering in $\rho_1 - \rho_2$ contributes to stabilizing the emulsion against sedimentation or creaming.

The sedimentation velocity for dilute non-interacting droplets is given by the Stokes law:

$$v_0 = \frac{2r^2 g|\rho_1 - \rho_2|}{9\eta} \tag{8.5}$$

here r is the droplet radius, ρ_1 and ρ_2 are the densities of the inner and the outer phases and η is the viscosity of the outer phase. The easiest way to decrease sedimentation or creaming is thus to increase the viscosity of the outer phases; thickeners (*e.g.* polysaccharides or acrylic polymers) are widely used for this purpose. We should mention that eqn (8.5) is strictly valid for non-interacting droplets, *i.e.* in the limit of infinite dilution. More complex equations account for the creaming behaviour of concentrated emulsions.

It is generally assumed that sedimentation and creaming keep the droplets as separate entities and therefore no change in droplet size occurs. For these reasons, it cannot be regarded as a phase separation, but rather as a concentration of the droplets of the dispersed phase. As such, in some cases this phenomenon is a desired event, for example in the separation of dairy cream from milk to produce butter.

Much more serious challenges to emulsion stability are droplet flocculation and coalescence, as illustrated in Figure 8.5. These phenomena lead to emulsion breakage in a time-range that can vary from minutes to years.

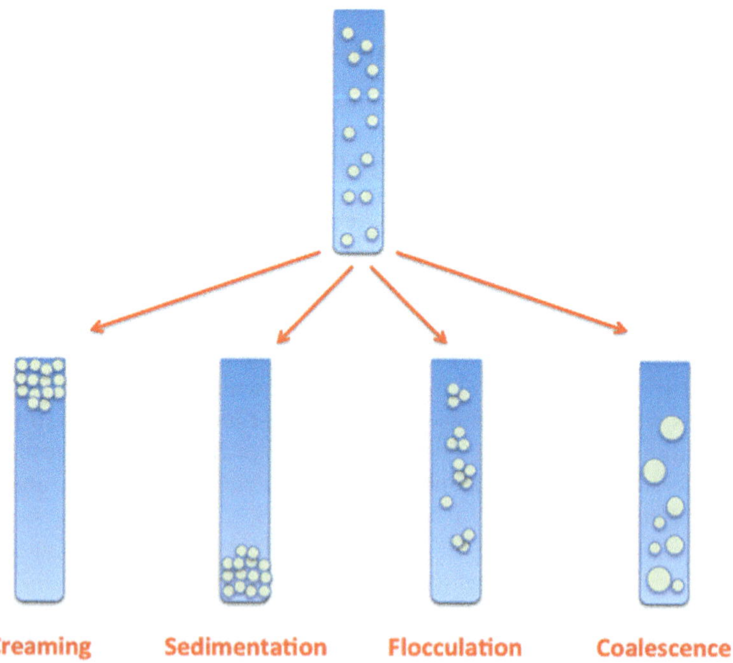

Figure 8.5 Principal mechanisms that lead to destabilization of emulsions.

These events are the result of internal forces, *i.e.* inter-droplet net attractive interactions. We should however mention that the close proximity of droplets due to creaming or sedimentation increases the likelihood of collision between particles, when attractive droplet–droplet van der Waals interactions dominate.

The aggregation of droplets (*flocculation*) can be prevented by adopting some counteractions for van der Waals forces, similar to the case of solid dispersions (*i.e.* in nanoparticle suspensions). We briefly recall that electrostatic stabilization, obtained with ionic surfactants, is the main kinetic factor preventing collisions between droplets and nanoparticles, while non-ionic stabilizers (either amphiphiles or polymers) cause steric interactions between the adsorbed layers, providing a thermodynamic contribution against van der Waals aggregation.

In the case of *coalescence*, *i.e.* the fusion of droplets following the collision event, a fundamental role is played by the physical nature of the interfacial film. The similarity here is to foams, rather than to solid dispersions. When two droplets are at the separation distance d, as a result of collision events or increases in local concentration (*e.g.* following sedimentation or creaming), the mechanical fluctuations of the liquid film between the droplets can eventually lead to rupture and coalescence. The presence of a stabilizer around the droplet will delay or inhibit this rupture, especially in the case of a rigid film. For this reason, it is usually the combination of two or more surfactants that produces emulsions that are most resistant to coalescence. High purity stabilizers are generally worse emulsifiers than lower-grade ones. For instance the addition of lauryl alcohol to sodium dodecyl sulphate (SDS) reinforces the mechanical strength of the film, by increasing the lateral interactions between the components.

Another effect, somewhat related to coalescence, but with a totally different origin, is *Ostwald ripening*. Strictly speaking, Ostwald ripening would not occur in emulsions characterized by monodispersed droplets. Conversely, an increase of the average size of the droplets is observed in a polydispersed emulsion, and this is due to the pressure gradient existing across a curved liquid–liquid interface. The difference in pressure is quantified by the Laplace Equation:

$$\Delta p = \frac{2\gamma}{r} \qquad (8.6)$$

where γ is the interfacial tension and r the droplet radius.

The pressure gradient is inversely proportional to the droplet radius, thus the liquid molecules of the inner phase have different chemical potentials if they belong to droplets with different sizes. More specifically, the liquid molecules residing in the inner phase of the smaller droplets possess a higher chemical potential, with respect to those located in the larger droplets. This imbalance drives a net migration towards the larger droplets and the consequent coarsening of the emulsion. Ostwald ripening can be prevented by reducing the width of the size distribution and by adding to the inner phase an inert component that is insoluble in the outer phase, *i.e.* a strongly hydrophobic substance in the case of o/w emulsions.

On the contrary, an excess of surfactant in the continuous phase might enhance the effect, in the case that micelles are formed: these aggregates can act as transfer intermediates from the smaller droplets towards the larger ones.

In all cases, kinetic stability of emulsions arises from an ensemble of factors, each properly weighted and balanced, starting from the choice of the stabilizer (HLB, purity grade), the preparation method (which affects the size distribution), the storage temperature, and leading to the inclusion of additives in the appropriate amount. Excess of surfactant and/or of thickener can eventually result in emulsion breakdown, instead of enhancing its stability.

8.2.3 Pickering Emulsions

In the final section of this brief introduction to emulsions, we address one of the most recent advances in the field, the study of Pickering emulsions,[3] in which the dispersed liquid phase is stabilized by the adsorption of colloidal particles at the oil–water interface, as shown in Figure 8.1. In this case the role of the HLB number, defined for amphiphiles on the molecular length scale, is replaced by the contact angle of the adsorbed particles (see Section 1.4.1).

If the contact angle is 90° (Figure 8.6, centre), the particles have balanced affinities for the two liquid phases and therefore the average curvature of the interface is zero. The same situation is obtained for an amphiphile with HLB = 10 or at the PIT temperature. For contact angles in the range 30–90° oil-in-water emulsions are formed, whereas larger contact angles lead to water-in-oil emulsions. Stabilization of emulsions using nanoparticles is not limited to the class of Pickering emulsions. In commercial products, particle attachment is controlled by using both surfactants and colloidal particles; this leads to a new class of emulsions.

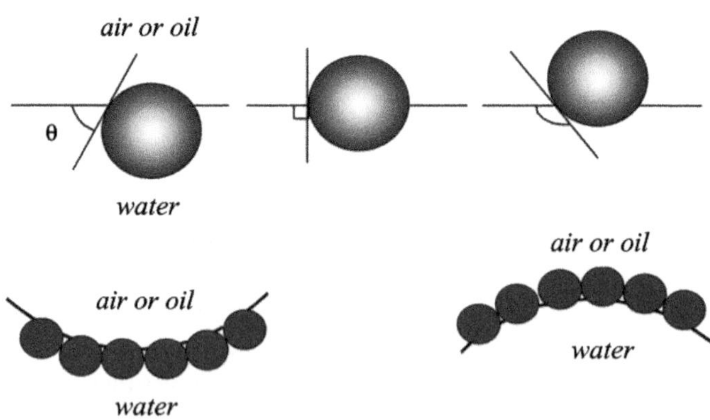

Figure 8.6 Affinity of solid particles towards the liquid phases and consequent formation of o/w (left), w/o (right) or balanced (centre) emulsions. (Reproduced from B. P. Binks, *Current Opinion in Colloid & Interface Science*, 2002, **7**, 21–41 with permission from Elsevier.)

8.3 Microemulsions

Microemulsions possess considerable advantages over emulsions and nano-emulsions in several applicative fields. According the scheme depicted in Figure 8.3, microemulsions could be viewed as the down-sized counterparts of emulsions, where the domains of the dispersed phase contain droplets with a radius spanning between 5 and 50 nm. However, as we have repeatedly mentioned throughout this chapter, this description does not capture all the salient features of the system, and in particular the fact that microemulsions form spontaneously in some experimental conditions, depending on composition, temperature and pressure. Their enormous interfacial area and thermodynamic stability, which implies ease of preparation, render microemulsions the ideal nanostructured systems for applications where the presence of oil and water is required simultaneously.

We anticipated that, based on eqn (8.1), the formation of nanometre-sized droplets can be spontaneous only if an ultra-low interfacial tension is reached. Given that formation and microstructure properties are ruled by thermodynamics, many efforts have been devoted to defining the experimental conditions under which the desired microemulsion is formed. This fundamental difference means that we can map the behaviour of ternary systems composed of water–oil–surfactant through phase diagrams, enabling a descriptive and predictive framework. Phase diagrams are the starting point and guideline in order to determine whether w/o or o/w nanometre-sized droplets will be formed, and at the same time they provide the basis for the design of novel formulations. Moreover, we can state that much of the physical chemistry of microemulsions stems from the pioneering work of phase mapping, integrated by the insights at the nanoscale made possible by the progress in structural characterization techniques, particularly electron microscopy,[9] scattering techniques such as dynamic light scattering (DLS), small-angle X-ray scattering (SAXS), and, most prominently, small-angle neutron scattering (SANS)[10] and nuclear magnetic resonance (NMR).[11]

The spontaneous formation of a microemulsion requires an ultra-low interfacial tension, of the order of $10^{-3}\,\mathrm{mN\,m^{-1}}$. This is occurs when the appropriate surfactant and the suitable environment are carefully selected.[5]

For charged single-chained hydrocarbon surfactants, such as SDS, the presence of a co-surfactant, in the form of a medium chain-length alcohol such as pentanol, is usually required to form o/w droplets. Generally, a co-surfactant is an additive that is incorporated into the surfactant monolayer and affects the spontaneous curvature of the interface, while a co-solvent remains in the bulk phase and improves the ability to solubilize further components. An illustrative example of such a system, formulated for the removal of acrylic films from pictorial surfaces, will be discussed in Chapter 9.

For ionic branched or linear double-chained surfactants, *e.g.* bis(2,2′)-ethylhexylsulfosuccinate (Aerosol-OT, AOT), or didodecyldimethylammonium bromide (DDAB), respectively, the presence of a cosurfactant is not necessary.

For several non-ionic surfactants, as for the C_iE_j series, spontaneous dispersion is achieved without the need of co-surfactants. In this latter case, which has been one of the most thoroughly studied in the past decades, the phase behaviour and the microstructure are controlled by temperature, as we will see in the following paragraphs. We recall that the reason for this behaviour relies on the temperature dependence of the HLB number, and of the packing parameter. All of these features are closely related and represent different approaches to looking at the same phenomenon.

Generally speaking, the smaller size of the droplets in a microemulsion amounts to a higher curvature that the amphiphilic film has to withstand, as opposed to the almost flat interface formed in emulsions. Therefore the curvature constraint and its deviation from the spontaneous value represent a further contribution to the energy balance of eqn (8.1), determined by the chemistry of the amphiphile and the experimental conditions.

The first experimental report of spontaneous emulsification dates back to the work of Hoar and Schulman in 1952.[12] This work describes the formation of an "oleophatic hydro-micelle" (w/o microemulsion), formed by oil, an alkali-metal soap, water and a long chain alcohol. In the same year, Winsor was observing the same phenomenon.[13] In the first example, upon titration with alcohol, the authors observed the formation of an optically transparent phase from a milky dispersion, while in the latter case, a change in salinity caused the nature of the dispersed phase to revert. The term microemulsion was used for the first time by Schulman in 1959,[14] referring to a system consisting of water, oil, surfactant and alcohol, which was later recognized to be *nano*structured.

The first thermodynamic and structural studies started in the 1970s, boosted by the application in petroleum extraction for tertiary recovery, during the oil crisis. The ultra-low interfacial tension between the oil excess phase and the water excess phase is the relevant feature that was exploited in this context. Later on, other applications were discovered, namely catalysis, inclusion in cosmetics and pharmaceuticals, preparation of submicron particles, solar energy conversion, and liquid–liquid extraction (minerals, proteins, *etc.*).

Nowadays, the fundamental research on microemulsions is still a lively scientific area, essentially driven by the search for and structural understanding of formulations tailored for specific applications. In the next chapter we will discuss some case studies related to the restoration of works of art.

8.3.1 Phase Diagrams

In Chapter 6, the phase behaviour of surfactant aqueous systems was introduced (see Section 6.1.7); similarly, in the case of microemulsions the structural and functional characteristics of the surfactant, at a molecular scale, affect the phase behaviour and define the phase regions of microemulsion stability.

Phase diagrams define the experimental conditions for stability and allow the mapping of the phase borders. Therefore they represent an essential tool for the correct formulation of already known microemulsions or else form the starting point for the design of novel formulations. The most obvious difference with respect to "simple" surfactant dispersions is that microemulsions are – at least – ternary systems, and therefore the thermodynamic description requires a phase triangle at fixed temperature, ionic strength and pressure, or a phase prism, if one of the above intensive parameters is also to be varied.

We remind the reader that, in a phase triangle at fixed pressure and temperature, each point defines a mole ratio composition with respect to each of the vertexes (pure components) that is directly proportional to the distance of the point from the side opposite to the vertex.

Figure 8.7 shows the phase diagram of AOT–water–iso-octane at room temperature and atmospheric pressure as an illustrative example of the use of phase diagrams. AOT is one of the few ionic surfactants that do not need the presence of co-surfactants to form microemulsions spontaneously. This system has been studied extensively in view of the extended region of the w/o micro-emulsion phase, which lends itself to numerous applications. Each monophasic region is denoted by a symbol. Below, we will guide the reader on how to extract the necessary information from a phase diagram.

The o/w region is called L_1, while the w/o phase is labelled L_2. F, D and I are liquid-crystalline mesophases and will not be considered here. Each point on the AOT–iso-octane side represents a reverse micellar solution of the amphiphile in the organic solvent. The shape of the AOT molecule resembles a truncated cone whose size is determined by the hydrophilic and hydrophobic parts. We recall that if ν is the volume of the hydrophobic part, a_0 the area per polar head, and l the length of the aliphatic region, the value $\nu/(a_0 l)$, called the "surfactant parameter" or "packing parameter", gives a good idea of the shape

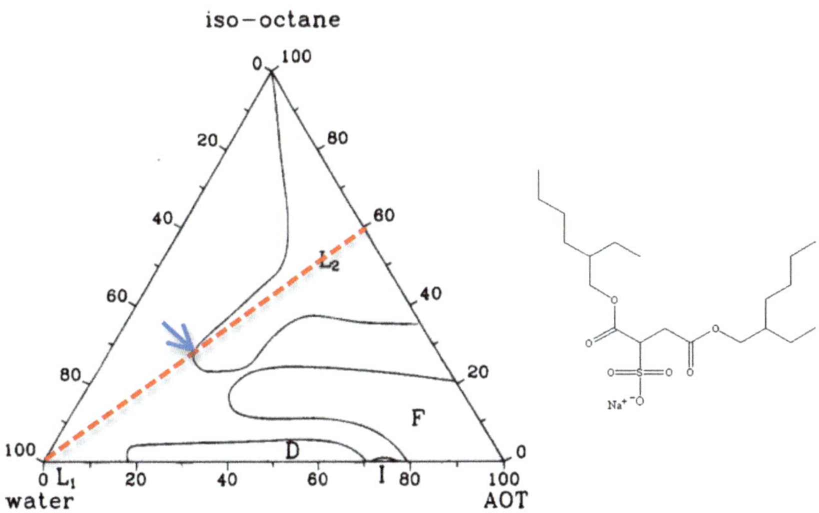

Figure 8.7 Phase diagram of the water–AOT–isooctane system at 25 °C. (Adapted with permission from G. J. M. Koper, W. F. C. Sager, J. Smeets and D. Bedeaux, *J. Phys. Chem.*, 1995, **99**, 13291–13300. Copyright 1995 American Chemical Society.)

of the aggregates that are going to be formed spontaneously (see Section 6.2.4). For AOT this is 1.1, which justifies the formation of reverse micelles.

If we depart from the AOT–iso-octane side and move towards the centre of the phase triangle, we enter the realm of w/o microemulsions, the L_2 phase. Conversely, the L_1 region is very limited (from about 1% AOT in water and 0.2% in iso-octane) and is too small to be visible on the scale of the picture.

Interestingly the water pools have a spherical droplet shape, whose diameter grows as the aqueous water volume is increased, in a considerably wide compositional range. Historically, different compositions within the L_2 region have been labelled with the parameter W_0, which is defined as:

$$W_0 = \frac{[H_2O]}{[AOT]} \tag{8.7}$$

This parameter was introduced because it has a straightforward connection with the size of the water pools. Assuming that the water-in-oil droplets are spherical, their radius can be expressed as $r = \frac{3V}{\Sigma}$, where V and Σ are the volume and the surface area of the sphere. If we assume that the volume of the droplets is entirely constituted by water and that all the surfactant is located at the interface, then it is very easy to demonstrate that

$$r_w = \frac{3V_{aq}[H_2O]}{a_0[S]} = \frac{3W_0V_{aq}}{a_0} \tag{8.8}$$

where V_{aq} is the volume of a water molecule (30 Å^3), a_0 is the area per surfactant polar head (Å^2), and [H_2O] and [S] are the water and surfactant concentrations in the organic solvent, respectively. For AOT, r_w(Å) = 1.5 W_0, therefore for spherical droplets the radius scales linearly along the dashed line.

These simple considerations highlight directly the structural control that can be exerted on the size of the dispersed pool. The dashed red line represents a so-called "dilution line", where, at a fixed surfactant : oil molar ratio, water is added stepwise at increasing volume fractions. For the sake of simplicity, only the monophasic regions are shown in the diagram, but it is clear that beyond the point indicated by the blue arrow, we leave the L_2 monophasic region and phase separation will occur. This composition marks the so-called emulsification failure, which depends on the experimental conditions and is strictly related to the structural properties of the surfactant film.

The dimensional control of the water pool has been one of the main reasons for the popularity of the AOT system and has found widespread use in important fields, such as template synthesis of inorganic nanoparticles, where the size control was achieved by confinement of the water-soluble reagent and control of nucleation and growth by the kinetics of droplet collisions[4] or nanosized reactors for enzymes acting on lipophilic substrates.[15]

8.3.1.1 Microemulsions from Ionic and Non-ionic Surfactants

The application of microemulsions for the conservation of works of art concerns mainly detergency-related aspects. In the vast majority of cases aqueous-based systems are required, for a series of reasons that will be widely discussed in the next chapter. Therefore the ionic systems described (DDAB and AOT) are very good for illustrative purposes, but they are not suitable for restoration purposes, as o/w microemulsions are generally required. These can be obtained with ionic surfactants (such as SDS) with low packing parameter through the addition of a medium-chained alcohol (such as pentanol) or electrolytes. If the presence of a co-surfactant or salts is to be avoided, the best option is a non-ionic surfactant. Some examples of the formulation of quaternary o/w microemulsions with ionic surfactants will be covered in detail in Chapter 9. In this section we will review the main physico-chemical properties of non-ionic microemulsions, where the main control parameter is the temperature. The variation in temperature affects the phase behaviour considerably and, to portray the system fully, one should consider a phase prism, where the third dimension is represented by temperature (for non-ionic surfactants).

Two important bidimensional phase diagrams that can be extracted from the ternary phase prism are the Shinoda[16] cut and the Fish-Eye cut (or Kahlweit-Strey cut),[17] sketched in Figure 8.8; these cuts undoubtedly have historical valence, but they are also important for applicative purposes. In the first case, the oil : water ratio is varied while the surfactant concentration is held fixed. Conversely in the Fish-Eye cut this ratio is constantly 1 and the surfactant concentration is varied. In both cases the phase behaviour is mapped as a function of the composition variable and temperature.

Depending on the particular end-use envisaged for the microemulsion formulation, we may need to know the phase behaviour in one of the two cases.

For applications in the field of cleaning of pictorial surfaces, the minimal amount of surfactant required to disperse a given oil efficiently is one of the most important criteria for the selection of a particular microemulsive system; this can be done by considering a Fish-Eye cut diagram, such as that illustrated in Figure 8.9, which illustrates the phase mapping of non-ionic microemulsions as a function of temperature and surfactant concentration for equal volumes of oil and water. The concentration γ_0 corresponds to the monomeric solubility of the surfactant in the solvents. If we increase the surfactant mass fraction beyond this limit, spontaneous emulsification will occur. Recall that water is generally denser than organic solvents and therefore it forms the bottom phase. Depending on the temperature, oil or water is partially dispersed into the other solvent. The $\tilde{X}$ point is defined by the surfactant concentration $\tilde{\gamma}$ for which at $\tilde{T}$ the one-phase region (1) is reached, *i.e.* the two liquid phases form two disconnected networks (bicontinuous microemulsion) separated by a surfactant monolayer. In the figure, $\underline{2}$ (also indicated as Winsor I), 3 (or Winsor III) and $\bar{2}$ (or Winsor I) denote water-, surfactant-, and oil-rich microemulsions in equilibrium with the corresponding excess phases, respectively.

From a practical standpoint, Figure 8.10 represents a simple procedure to map the phase diagram, when this is not already available in the literature.[18] A surfactant is added dropwise to a 1 : 1 oil : water biphasic system and one of three situations will occur.

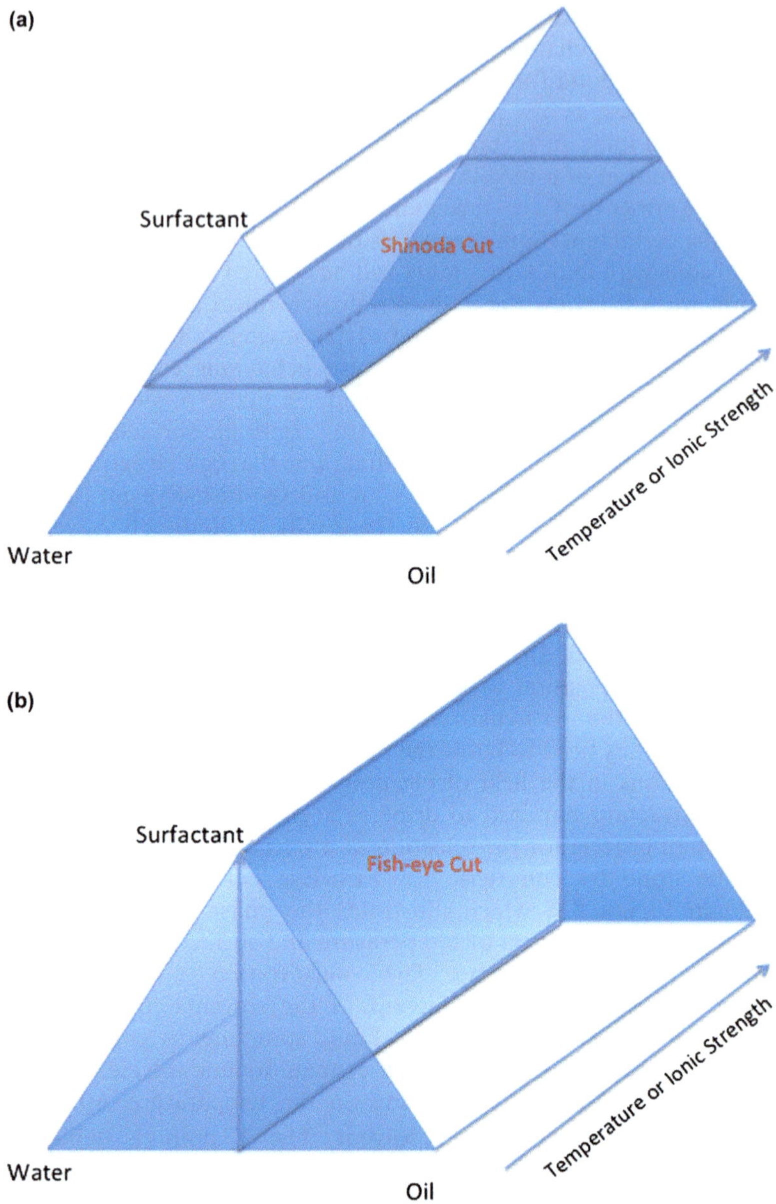

Figure 8.8 (a) Shinoda cut of a ternary microemulsion; surfactant concentration is held fixed, while the temperature T (or ionic strength) and the ratio of oil : water are the variables; conversely, in the Fish-Eye cut (b) this ratio is constant and equal to 1 and the phases are mapped varying the surfactant concentration and T.

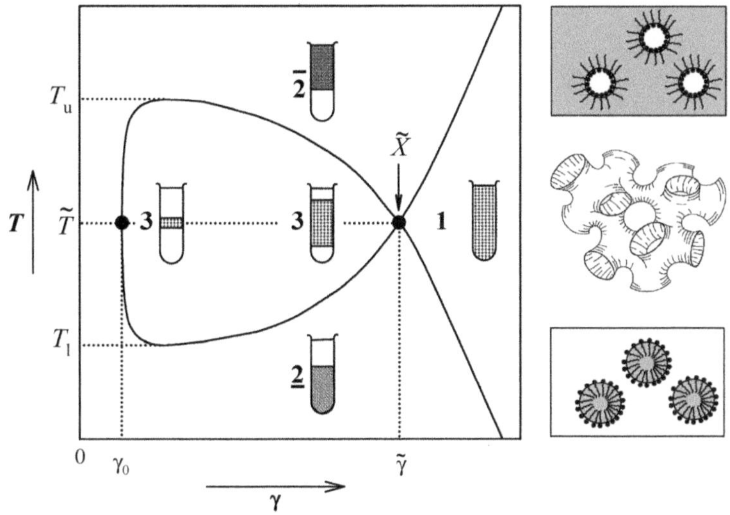

Figure 8.9 H_2O – alkane – C_iE_j, at a 1 : 1 water-to-oil ratio ($\alpha = 0.5$) as a function of the surfactant mass fraction γ and the temperature T. On the left-hand side the corresponding microstructures are sketched.
(Adapted from C. Stubenrauch, R. Tessendorf, R. Strey, I. Lynch and K. A. Dawson, *Langmuir*, 2007, **23**, 7730–7737. Copyright 2007 American Chemical Society.)

Of course the surfactant molecules possess a packing parameter, or equivalently an HLB number that will drive the phase behaviour towards one of the possible phase scenarios. This packing preference can be reverted (phase inversion) by a temperature or a salinity change. A temperature increment results in a partial dehydration of the oxyethylene chain of the C_iE_j molecules with two effects: (i) a decrease in the effective size of the polar head group of the amphiphile, *i.e.* an increase in the packing parameter; and (ii) an increase in the effective hydrophobicity of the surfactant, which enhances its oil solubility. This phase inversion is commonly related to the appearance of a bicontinuous microemulsion (Winsor III, see also Figure 8.9) in contact with excess oil and water phases (three-phase region). In the bicontinuous phase, ultra-low interfacial tension is achieved: this system is ideally suited for enhanced oil recovery. The thermal excursion that the microemulsion has to withstand is a critical parameter for non-ionic C_iE_j systems. Indeed the sensitivity to temperature can be a disadvantage in certain fields of application, because phase inversion can occur within the thermal range of applications.

Conversely, temperature-insensitive microemulsions can be obtained by using other non-ionic sugar-based surfactants such as alkyl glucosides, and alkyl polyglucosides (ApG or C_nG_m with $m \geq 1$, where C is the length of the alkyl chain, based on the number of carbon atoms and G is the number of sugar groups on the polar head). The insensitivity to temperature changes results presumably from the strength of the hydrogen bonds between the hydroxyl groups of the glucose moiety and water molecules, which prevents any

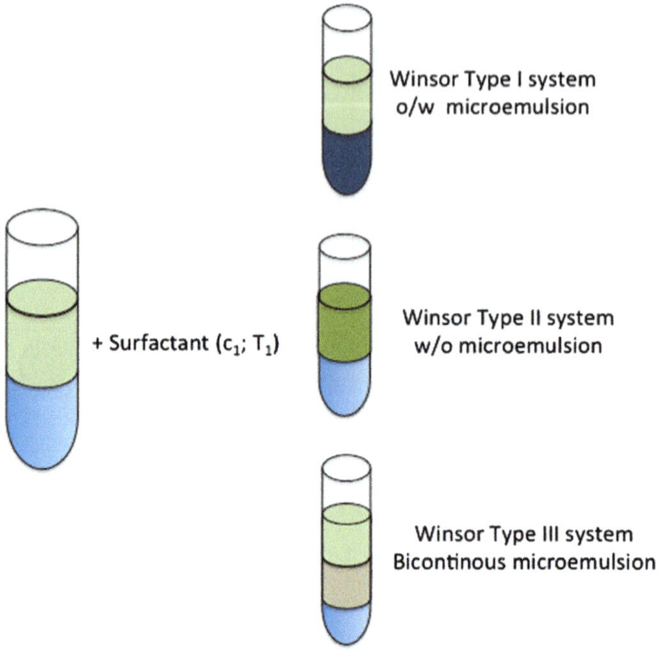

Figure 8.10 Possible phase scenarios obtained upon addition of a predetermined amount of nonionic surfactant of the kind C_iE_j to water and oil in equal volumes, at a temperature T_1. (Right top) An oil-rich phase (light green) is in equilibrium with an o/w microemulsion (dark blue). (Right centre) A w/o microemulsion (dark green) is in equilibrium with a water-rich phase (light blue). (Right bottom) A bicontinuous microemulsion (light brown) is in equilibrium with an oil-rich (light green) and a water-rich (light blue) phase.

significant dehydration of the head group in the relevant temperature range. For C_nG_m systems the phase inversion $\underline{2} - 3 - \bar{2}$ can be achieved by tuning the hydrophile – lipophile balance through the addition of a co-surfactant, such as an alkyl glycerol ether or an alcohol. In Chapter 9, a case study involving an ApG will be presented.

8.3.2 Properties of the Surfactant Film

At this stage we will briefly discuss an approach to the description of microemulsion phase behaviour, which allows a more rigorous and complete understanding of the thermodynamic properties of the surfactant and of the microstructural behaviour.

Regardless of the nature of the microemulsion, (o/w, w/o or bicontinuous), the oil and water phases are separated by a surfactant monolayer. Depending on the type of dispersion this monolayer is bent towards the oil phase (o/w), towards the water (w/o) or it shows no preferential bending (bicontinuous).

Surfactant molecules and their monolayers can be described in terms of three different concepts. The first and oldest criterion is based on the hydrophilic–lipophilic balance (HLB) of amphiphilic molecules; the second relies on the description of the surface monolayers in terms of the packing parameter (p), that has already been described in Sections 6.2.4 and 8.3.1. The third and last approach is related to the spontaneous curvature, H_0, introduced by Helfrich, that provides a more quantitative description of the phase diagrams. For a given composition, salinity and temperature, each amphiphilic film possesses a spontaneous curvature. Deviation from the spontaneous curvature, due for instance to compositional constraints, has an enthalpic cost, which depends on the rigidity of the film, just as the rigidity of a spring depends on the spring constant, *i.e.* on the material and its geometric features. In the simplest case, if we consider a spherical droplet of oil in water with a radius r and a curvature $H = 1/r$, once the radius of the microemulsion is given, the film curvature is known.

A full coverage of the Helfrich theory is outside the scopes of this chapter. However we would like the reader to become familiar with the language associated with this approach.

Generally, bending towards water has been conventionally assigned a positive sign. Therefore, spontaneous positive curvature means that the packing parameter is below 1, the HLB number is high and the amphiphilic will form o/w microemulsions, whose size deviation from the one dictated by the spontaneous curvature will depend on the energetic cost of bending the amphiphilic film. The reverse situation holds for w/o microemulsions, while for bicontinuous systems the average curvature is zero. We refer the interested reader to the specific textbooks.[19]

8.4 Characterization Techniques

Much effort has been devoted in the last decades to the phase and structural characterization of microemulsions.[20] The main issue related to liquid–liquid dispersions can be summarized as follows: which one is the internal phase? Oil or water?

Several simple methods, such as stepwise dilution with oil or water, provide an immediate answer. More sophisticated techniques, such as conductivity measurements, give a more quantitative answer. However, some consideration of the composition, the HLB number or the packing parameter of the amphiphile should provide a first hint of what should be expected.

- Is the dispersion stable from a thermodynamic point of view or is it only kinetically stable?

In the first case we are dealing with microemulsions, otherwise the dispersions are either nanoemulsions or emulsions. The visual appearance (transparent, translucent or milky) provides a valuable clue, which should be confirmed by the structural determination of the droplet size. Centrifugation can speed up the process of phase separation for emulsions, however these dispersions can have considerable stability, especially when the density of the two liquids is similar.

- What is the size of the dispersed droplets?

This parameter is connected to the previous question, but at the same time it gives information on the total interfacial area of the system. Nanoscale sized objects are suitably observed with small-angle scattering techniques, using neutrons or X-rays (see also Section 6.2.5.4). Moreover dynamic techniques, such as dynamic light scattering or pulsed field gradient NMR spectroscopy[9–11] can be used for this purpose. The instrumentation and the technical training required to perform the measurements and interpret the results are not routine ones.

- What is the thermodynamic stability range for a given microemulsion formulation?

This aspect can be investigated by inserting the system of interest in its phase diagram, provided it has already been investigated. Otherwise, one should concentrate on the parameters of interest (quantity of dispersed phase, temperature, salinity, *etc.*) to address the question, exploring the phase diagram in the neighbourhoods of the point in the phase space.

Many other instrumental techniques are available nowadays, but the essential tools are those that we have already described. For a more thorough coverage of characterization techniques, we refer readers to the more specialized textbooks.

Further Suggested Reading

J. Goodwin, *Colloids and Interfaces with Surfactants and Polymers*, John Wiley & Sons Ltd, Chichester, 2nd edn, 2009.

K. Holmberg, B. Jönsson, B. Kronberg and B. Lindman, *Surfactants and Polymers in Aqueous Solution*, John Wiley & Sons Ltd, Chichester, United Kingdom, 2nd edn, 2003.

C. Stubenrauch, *Microemulsions: Background, New Concepts, Applications, Perspectives*, Wiley-Blackwell, Chichester, 2009.

T. F. Tadros, *Applied Surfactants: Principles and Applications*, Wiley VCH, Weinheim, 2005.

Review and Questions

The main topics discussed in the chapter may be summarized as follows:

- The salient differences between emulsions, microemulsions and nanoemulsions.

- The size range and necessary ingredients of each of these systems.
- Main methods of emulsification.
- Factors determining emulsion stability.
- The main differences between ionic and non-ionic surfactants as microemulsion stabilizers.
- Phase prisms for microemulsions and salient cuts into two-dimensional diagrams.
- Correspondence between HLB numbers and packing parameters.
- The main differences between C_iE_j and C_mG_n surfactants.

Questions

1) Some alcoholic beverages, such as Ouzo or Pastis, are ethanol extracts of anis seeds, containing trans-anethole, a strongly hydrophobic oil with a low affinity for water. Traditionally, Ouzo is additioned with water just before drinking: upon water pouring, the ethanolic solution turns milky. Comment on this behaviour.
2) Explain in terms of water affinity the variation in the HLB numbers of the surfactants C_iE_j with temperature. What is the characteristic temperature where HLB = 10? What kind of dispersion (w/o or o/w) would you expect below this temperature? And above?
3) Three surfactants have the following packing parameters:

   a) 0.33; b) 0.5; c) 2

   Please assign to each amphiphile its corresponding HLB number, chosen from among 2, 20 and 40.

   Which amphiphile would you choose to form w/o dispersions?
4) How much water should be added to a w/o AOT (surfactant concentration 0.1 M) microemulsion to obtain an average droplet size of 3 nm?

Answers

1) Trans-anethole is kept in solution by the high ethanolic content, so that the Ouzo contained in the bottle is a "true" solution; as the alcohol content is lowered by the addition of water, an o/w emulsion forms. This effect has been studied also from a physico-chemical point of view, because emulsification occurs with minimal stirring and, more importantly, without the addition of a stabilizer. The presence of ethanol is however necessary.
2) The HLB number shows dramatic dependence on temperature for the C_iE_j series, and in particular it decreases when the temperature is raised. At the PIT, the surfactants have balanced affinity for water and oil. Below the PIT o/w dispersions preferentially form, and above the PIT inversion occurs.
3) 0.33, 40; 0.5, 20; 2, 2

   The best amphiphile would be that with $p = 2$ (HLB = 2). This is the typical value for AOT.

4) From eqn (8.8) it follows that $r_w(\text{Å}) = 1.5\ W_0$; therefore, in order to get a size of 3 nm (*i.e.* 30 Å) $W_0 = 20$; $W_0 = [H_2O]/[AOT] = 20$; therefore $[H_2O] = 2$ M. One litre of microemulsion contains 36 g of water, *i.e.* about 36 ml, 3.6% w/w of water.

References

1. M.-J. Schwuger, K. Stickdornt and R. Schömacker, *Chem. Rev.*, 1995, **95**, 849.
2. M. J. Lawrence and G. D. Rees, *Adv. Drug Del. Rev.*, 2000, **45**, 89.
3. N. Azemar, in *Industrial Applications of Microemulsions*, ed. C. Solans and H. Kunieda, Marcel Dekker Inc, New York, 1997, Chapter 18.
4. M. P. Pileni, *Langmuir*, 1997, **13**(13), 3266; J. Eastoe, M. J. Hollamby and L. Hudson, *Adv. Colloid Interface Sci.*, 2006, **128**, 5.
5. J. Eastoe, in *Colloid Science Principles, Methods and Applications*, ed. T. Cosgrove, John Wiley & Sons Ltd, Chichester, 2nd edn, 2010.
6. E. Ruckenstein, *Langmuir*, 1996, **12**(26), 6351.
7. W. C. Griffin, *J. Soc. Cosmet. Chem.*, 1949, **1**, 311–329; W. C. Griffin, *J. Soc. Cosmet. Chem.*, 1954, 5, 249; see also http://www.firp.ula.ve/archivos/historicos/76_Book_HLB_ICI.pdf (accessed 7 December 2012) for examples of calculation of HLB and a comprehensive list of HLB numbers for various surfactants.
8. K. R. Lange, *Surfactants: A Practical Handbook*, Hanser Gardner Publications, Cincinnati, OH, 1999.
9. W. Jahn and R. Strey, *J Phys Chem.*, 1988, **92**, 2294.
10. S. H. Chen, *Annu. Rev. Phys. Chem.*, 1986, **37**, 351.
11. O. Söderman and U. Olsson, *Curr. Opin. Colloid Interface Sci.*, 1997, **2**, 131.
12. T. P. Hoar and J. H. Schulman, *Nature*, 1943, **152**, 102.
13. P. A. Winsor, *Trans. Faraday Soc.*, 1948, **44**, 376.
14. J. H. Schulman, W. Stoeckenius and L. M. Prince, *J. Phys. Chem.*, 1959, **63**, 1677.
15. P. L. Luisi, M. Giomini and M. P. Pileni, *Biochim. Biophys. Acta*, 1988, **947**, 209.
16. K. Shinoda and S. E. Friberg, *Emulsions and Solubilization*, Wiley-Interscience, New York, 1986; B. Lindman, K. Shinoda, U. Olsson, D. Andersen, G. Karlström and H. Wennerström, *Colloid Surf.*, 1989, **38**, 205.
17. M. Kahlweit and R. Strey, *Angew. Chem. Int. Ed.*, 1985, **24**, 654.
18. N. M. van Os, J. R. Haak and L. A. M. Rupert, *Physico-Chemical Properties of Selected Anionic, Cationic and Nonionic Surfactants*, Elsevier, Amsterdam, 1993.
19. D. F. Evans and H. Wennerström, *The Colloidal Domain: Where Physics, Chemistry, Biology, and Technology Meet*, VCH publishers, New York, 1994.
20. M. Gradzielski, *Curr. Opin. Colloid Interface Sci.*, 2008, **4**, 263.

CHAPTER 9

Cleaning III: Applications and Case Studies

R. GIORGI* AND E. CARRETTI

University of Florence, Italy
*Email: giorgi@csgi.unifi.it; carretti@csgi.unifi.it

9.1 Introduction

In the field of restoration of works of art, the use of amphiphile-based colloidal systems has attracted a great deal of attention in the last decades because of the ability of these systems to solubilize otherwise insoluble materials, as in the case of synthetic materials (acrylic and vinyl polymers) applied in past restorations that have largely contributed in recent decades to the severe degradation of works of art due to ageing (see Chapter 3). This is the reason why amphiphile-based systems represent a significant step forward in conservation science.

Many conservation issues can still be faced by using simple solvent technology. However, increasingly difficult challenges have recently risen in the works of art conservation field, owing to the dramatic changes in the molecular properties of polymer coatings used in past restoration, which make their complete removal difficult. Therefore, it is mandatory to clean surfaces by using formulations with high efficacy, while preserving the integrity of the support. The cleaning or removal of soil, dirt/grime, and altered materials must be selective, through either swelling and/or solubilization process, without negative effects on the original layers of artefacts beneath the dirt and coatings. This can be achieved by using amphiphile-based colloidal systems such as microemulsions and/or micellar solutions.

RSC Nanoscience & Nanotechnology No. 28
Nanoscience for the Conservation of Works of Art
Edited by Piero Baglioni and David Chelazzi

Published by the Royal Society of Chemistry, www.rsc.org

The smaller droplet size, and therefore the larger surface area, of microemulsions (see Chapter 8) makes them more suitable than emulsions for these application purposes.

9.2 The Use of Emulsions: a Historical Perspective

The cleaning of mural and easel paintings presents several difficulties because of their porous structure (see also Chapter 7). In particular, there are some issues that make the classical solvent technology, *i.e.* the use of pure (or mixed) organic liquids, inadvisable.

We have already mentioned in Chapter 7 that the action of an organic solvent is often scarcely controllable. In fact, solvents with low surface tension quickly wet the substrate's matrix, dissolving the *patina* (this term traditionally includes dirt/grime, soil and altered coating materials) and spreading it through the pores. When solvents evaporate, the dissolved material remains in the pores and obstructs them deep below the surface, making any further attempt at removal practically impossible.

Another factor that has to be taken into account is the toxicity of most of the solvents that are used for these operations. Given that many wall paintings have wide surfaces and are conserved in poorly ventilated environments, the safety is of fundamental importance for the operators' health.

Selectiveness during cleaning is another fundamental feature: the principle "like dissolves like" suggests that the selective removal of the soiling layer is only possible if the material beneath presents totally different features. Otherwise, mostly depending on the painting techniques, the inner layers could be swollen and solubilized because of the solvents. In fact, it is quite common to face situations where the materials to be removed are soluble in solvents that also affect the supporting layers.

In principle, the combined use of solvents with different properties could overcome both health and selectiveness issues. The rationale for this approach is based on the possibility of joining the good properties of one solvent to those of a second that is not miscible at all.

For example, if a toxic solvent is dispersed in a non-toxic one, the impact on the end-users' health can be greatly reduced. Moreover, the dispersion of a solvent that solubilizes the outer layers, and also affects the inner paint layers, in a second solvent that is inert to the inner layers may grant good efficiency in the selective removal of soiling materials without detrimental effects on the paint.

Dispersion of two mutually immiscible liquids is defined as an emulsion (see Chapter 8). The stabilization of this dispersion is usually achieved through the presence of an emulsifier, *e.g.* a surfactant. The liquid in larger amount behaves as a dispersing medium (*i.e.* the continuous phase), while the other is the dispersed phase. Emulsions have been used for restoration purposes since the 1980s.[1]

A prominent example of the application of emulsion technology was reported by Richard Wolbers, regarding the removal of the darkened layer covering a late 19th-century bust of George Washington.[1] In fact, the patina included re-painting, soluble in aromatic solvents, that had been badly applied over a painted surface that still presented some soil materials and residual varnish. The use of polar solvents would have been the appropriate procedure to remove the soil, but it would have also induced the swelling of the sensitive underlying paint layers. On the other hand an apolar, aromatic solvent would have removed the re-painting without swelling the underlying paint layers, but would have been ineffective on the soil.

The cleaning issue was solved by dispersing a water solution of triethanolamine (3.3% w/w; with pH adjusted to 8.5 with 1 M HCl) in a xylene solution of the surfactant Triton X-100 (40% w/w) in the ratio 3 : 5 to obtain a thick, gel-like emulsion. This system allowed the selective removal of the patina, without any harmful effects on the painted layers. This case study provides a clear example of how emulsion technology can overcome limitations due to the use of miscible solvent blends, by mixing immiscible liquids.

It is worth noting that some confusion often surrounds the term "emulsion", especially within the conservator community. This is the case for the "waxy emulsion", or "stearate emulsion", a traditional tool used in cleaning interventions, which is also known as "*Pappina Fiorentina*" after the conservators of the Opificio delle Pietre Dure (OPD, a public institute of the Italian Ministry for Cultural Heritage) in Florence.[2] In fact, in this case the term "emulsion" should be replaced by "dispersion", because the internal phase is not liquid, but solid. It is worth recalling here the composition and mode of application of this system, as part of the historical background of amphiphile-based tools for artefacts cleaning.

The "waxy emulsion" is composed of a mixture of bleached beeswax, stearic acid, ammonia (30% w/w) and deionized water (weight proportion 500 : 1.2 : 12 : 750). Beeswax is made from a mixture of saturated fatty acid esters of long chain aliphatic alcohols (75%), fatty acids (13%) and minor amounts of hydrocarbons (1%), alcohols, and resins. Since it is insoluble in water, a surfactant is required to obtain an emulsion. A saturated fatty acid (stearic acid, $C_{18}H_{36}O_2$) is neutralized with ammonia, to obtain the "ammonium stearate" soap, or emulsifier.

The dispersion is applied directly to the artefact surface, and after its removal the residues are removed with a non-polar solvent such as white spirit. The peculiarity of dirt/grime removal by using non-polar solvents makes this system particularly suitable for cleaning surfaces that are very sensitive to water, such as gilding. In fact, the small amount of water present in the emulsion is sufficient to solubilize hydrophilic materials, but the water-sensitive support "sees" mostly a harmless non-polar system. However, the turbidity of the system does not allow the visual inspection of the restoration, and this is a serious disadvantage of the procedure.

It is also important to clarify the meaning of other traditional terms used by conservators: "fat" and "lean" emulsions. These terms are largely used, according to the teaching of the Italian school of restoration, to indicate a water-in-oil (w/o) and an oil-in-water (o/w) emulsion, respectively. The term "fat" is used to describe the viscous, creamy shape of some w/o systems; "lean" evokes the behaviour of water, fluid and optically clear, which is used as a dispersant.

It is also appropriate to recall the discussion in Chapter 8 about the classification of different liquid–liquid dispersions on the basis of the size of the droplets. Emulsions possess dimensions larger than 1 μm, while droplets spanning from 1 μm to 100 nm are typical of systems defined as mini- (or nano-) emulsions. Microemulsions are characterized by droplets with a size below 100 nm. In the scientific literature and in the following paragraphs, the term nanodroplet is usually used to indicate microemulsions (and not mini- or nano-emulsions).

9.3 Microemulsions in Cultural Heritage Conservation: Low Impact Cleaning Tools

To define a microemulsion is not an easy task because of the complexity of its nature. Nowadays the issue of whether a system can be defined as a complex micellar solution or a true microemulsion is still debated. The two definitions are:

1. A microemulsion is a liquid, stable and homogeneous, optically transparent, isotropic and "spontaneously" formed system, comprising two liquids mutually insoluble; one dispersed in the other in form of microspheres stabilized by at least a monolayer of amphiphilic molecules (surfactants).
2. A microemulsion is defined as a system of water, oil and amphiphile, which is a single optically isotropic and thermodynamically stable liquid solution.[3]

The latter is the most concise yet precise definition of such a system. From these definitions it is clear that microemulsions have some remarkable features: they are macroscopically homogeneous and optically transparent, although at the nanoscale they are dispersed discontinuous systems. Moreover, they are spontaneously formed and thermodynamically stable systems (see Chapter 8.1).

Nanodroplets of an o/w or w/o microemulsion develop a huge exchange surface area that is responsible for the enhanced interaction with detrimental coatings. This feature strongly enhances the performance of microemulsions with respect to simple emulsions.

The first application of microemulsions for the cleaning of works of art was performed in Florence by Enzo Ferroni and Piero Baglioni in 1986, during the restoration (1984–1990) of the Renaissance paintings by Masaccio, Masolino and Lippi in the Brancacci chapel in Florence.[4] Investigations of the paintings under ultraviolet (UV) light revealed the presence of a multitude of wax-spots due to the blowing out, over centuries, of votive candles that were kept close to the paintings. The removal of this hydrophobic material from the hydrophilic fresco painting required a smart system having two opposite physico-chemical characteristics: a polar solvent like water, but "capable" of dissolving apolar components.[5] The microemulsion used for Brancacci's paintings was very similar to the so-called "*French microemulsion*" and was formed of dodecane

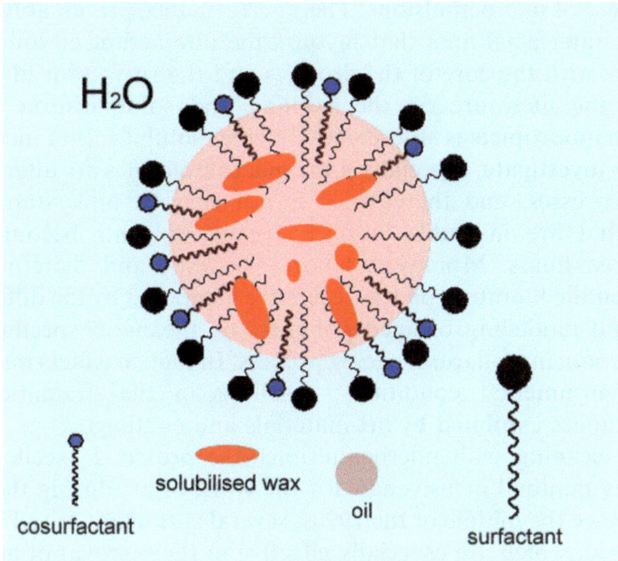

Figure 9.1 Scheme of the solubilization mechanism of the wax droplets by means of the nanodroplets constituting an o/w microemulsion.

nanodroplets, dispersed in an aqueous solution of sodium dodecylsulfate and 1-pentanol (more details are given in Section 9.5.1). This o/w microemulsion granted the removal of the hydrophobic material, leading to the effective cleaning of the wall paintings.

As a general feature, microemulsions provide several advantages.

- The spreading of the solubilized material into the porous matrixes is limited, because solubilization or swelling of hydrophobic material takes place in the dispersed phase, *i.e.* the core of the nanodroplets and/or at the droplet interface (as illustrated in Figure 9.1). The continuous aqueous phase acts thus as a hydrophilic barrier, preventing the re-deposition of the hydrophobic wax in the substrate porosity.
- Microemulsion systems are thermodynamically stable (see Section 8.1), and this ensures easy handling throughout a range of environmental operative conditions.
- The system's environmental impact is greatly reduced, owing to the minimal amounts of organic solvent necessary to formulate the microemulsion.
- By using nanostructured systems it is possible to achieve more controlled cleaning, because they often act more slowly than pure solvents.
- In most cases, microemulsions are more effective than other, conventional, cleaning systems.

Materials such as the altered polymers applied in past restorations, which became insoluble or hardly soluble in solvents or solvents blends, are often removed by use of microemulsions. Their performance presumably relies on the huge specific interfacial area that favours the interaction of soiling materials and polymers with the core of the droplets and the surfactant film.

However, the literature on the swelling and solubilization processes of coatings by nanodroplets is still scarce. Polymer solubilization mechanisms are not simple to investigate, especially if the macromolecules are altered as a result of ageing processes, and therefore it is not easy to understand why some substances that are insoluble in pure organic solvents become soluble in nanostructured fluids. Moreover, laboratory tests, and therefore the availability of scientific literature on this topic, are hampered by the difficulties in the formation and modelling of aged polymers by means of specific procedures aimed at reproducing a natural ageing process. In fact, artefacts may experience different environmental conditions, resulting in the dramatic variety of structural changes exhibited by art materials and coatings.

However, cleaning with microemulsions has provided excellent results to date, showing minimal invasiveness for the work of art during the restoration procedure. Since the middle of the 1990s, several surfactant-based systems have been developed, which are especially effective in the removal of acrylic and/or vinyl copolymers.

It is worth noting that a microemulsion may also form *in situ* during a cleaning operation performed using micellar solutions. Indeed studies of detergent mechanisms demonstrate that very good cleaning performance, *i.e.* the removal of a generic "oily" soil, is obtained using surfactant-based aqueous solutions which lead to the formation of microemulsions upon interaction with the oily soil. Research has been also carried out on microemulsions used as direct cleaning media for textile cleaning, hard surface cleaning, vehicle cleaning, and dry cleaning.

In conclusion, the removal of polymers using microemulsions is still a relatively unexplored field. In Section 9.5, several case studies will be reported that highlight the potential impact of microemulsions in this field, when used as direct cleaning agents.

9.4 Formulation and Application of Microemulsions in Cultural Heritage Conservation

The formulation of microemulsions obeys two guidelines: first, the choice of a dispersed phase that is effective for the removal of soil, dirt/grime or altered materials; second, the choice of a dispersant (continuous phase) that is inert to the supporting material.

For instance, o/w microemulsions can be used for the cleaning of stones and wall paintings, while in principle w/o microemulsions should be envisaged for the cleaning of canvas paintings, because both canvas and the preparation layer can swell in the presence of aqueous systems.

In order to prepare o/w microemulsions, surfactants with a hydrophilic–lipophilic balance (HLB; see Section 8.2.1) between 8 and 18 (water soluble) are selected. Non-ionic surfactants, characterized by a rather low critical micelle concentration (cmc, see Section 6.2.1), should be preferred because in this case the amount of surfactant is particularly reduced. However, lower quantities of the oil phase will be dispersed by these surfactants. Water-soluble surfactants make the preparation of o/w microemulsions quite simple. In fact, after their solubilization and the formation of further micelles, slow drop-wise addition of a co-surfactant (if necessary) and solvents (one or more, as a dispersed phase) is required according to the specific requirements of the system phase diagram (see Section 8.3.1). Obviously, the selection of one point in the phase diagram far away from the edges prevents the formation of phase separation due to temperature changes and/or to small changes in the chemical composition.

The application of microemulsions is quite simple because it combines the protocols and procedures commonly used by conservators when they handle, for different purposes, aqueous solutions or organic solvents. In the first case, either cellulose pulp (in the form of a compress) or hydrogels can be used to keep the aqueous solution in contact with the surfaces of stones and wall paintings. The same can be done with o/w microemulsions that behave like water. In the second case, thickeners and/or organogels are commonly used for the treatment of very delicate and water-sensitive surfaces. The same procedure can be used for the application of w/o microemulsions, provided that the features of the apolar phase (*i.e.* the dispersing medium) are compatible with the supporting materials.

On the occasion of the restoration of paintings in the Brancacci chapel (see Section 9.5.1), a specific apparatus was designed to ensure a continuous flux of the microemulsion onto the surface; this was realized under moderate pressure by using pipettes connected to a reservoir tank (see Figure 9.2).

Classical devices were used to protect the surface during contact with the cellulose pulp compress and to avoid the evaporation of the solvents (water): first, Japanese paper was applied over the painting; second, duplex paper (two sheets of paper or card-cover that are laminated together) covered with a polyethylene sheet was applied over the compress.

As already reported in Chapter 7, the use of detergents calls for much care in the evaluation of possible residues of surfactants left on the cleaned surface. However, most surfactants used in these formulations are soluble in water or other commonly used organic solvents (*e.g.* acetone, alcohol, and others). Therefore, a simple washing with water, or the use of thickened solvents for water-sensitive supports, is sufficient to avoid (below the experimental detectable limits) possible residues.

9.5 Case Studies

9.5.1 Frescoes by Masaccio and Masolino, Cappella Brancacci (Florence)

One of the first cleaning systems based on the use of microemulsions was set up in 1986 by Enzo Ferroni and Piero Baglioni during the restoration workshop

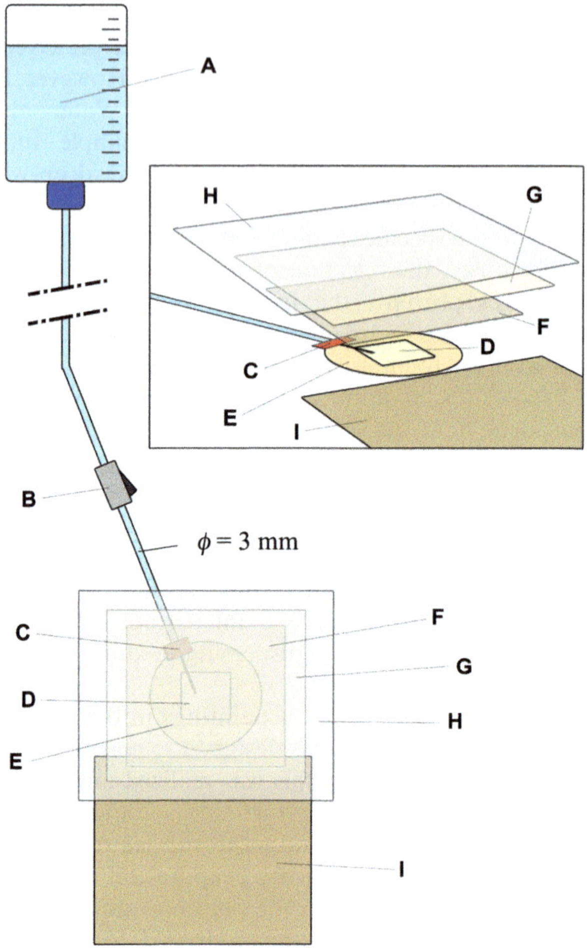

Figure 9.2 Scheme representing the application device set up for the removal of waxes from the fresco wall paintings in the Brancacci Chapel (Florence, Italy). The reservoir tank (A) contains the microemulsion, which passes through the flux regulator (B) and a pipette (C). The pipette releases the microemulsion over the wax-stained fresco area (D), passing through a sheet of Japanese paper with a cellulose pulp compress (F), placed over an area confined with cellulose pulp (E). The pulp compresses were soaked with water, and covered with a duplex paper made by a cellulose paper sheet (G) and a polyethylene sheet (H) to limit evaporation and protect the treated area. A double sheet (I), made from a polyethylene layer (in contact with the fresco) and a layer of adsorbing paper, is placed directly below the treated area to collect the excess microemulsion dripping down the wall from the cleaning spot. (Image by Michele Baglioni.)

for the Brancacci Chapel in Santa Maria del Carmine Church (Florence), performed by the restorers Ornella Casazza and Sabino Giovannoni from 1984 to 1990.[4,6] The Chapel was built at the end of the 14th century and frescoed by Masaccio and Masolino between 1423 and 1427; part of the Chapel was then completed in 1480 by Filippino Lippi. As clearly indicated by diagnostic investigations performed using spectroscopic techniques [in particular Fourier transform infrared (FTIR) spectroscopy], the surface of the frescoes was found to be locally affected by the presence of deposits containing drops of aged candle wax that were spread onto the surface by the monks who blew out the candles every day. This eventually had two main consequences:

- The wax adsorbed carbonaceous atmospheric particles that made the deposits brown, causing an aesthetic issue.
- The liquid wax penetrated into the porous matrix of the mortar and slowly crystallized, making its complete removal difficult.

The simple use of a pure solvent such as dodecane or hexadecane caused the formation of a solution that, being partially adsorbed by the porous support, favoured the spreading of part of the solubilized material into the wall matrix.

Ferroni and Baglioni provided a solution to this problem through the application of nanostructured systems (o/w microemulsions) whose composition is reported in Table 9.1. The dissolution of the hydrophobic deposits occurs into the nanodroplets of the microemulsions (see also Figure 9.1). The solubilization sites are mainly located close to the polar head groups near the first carbon atom of the alkyl chains of both the surfactant and the co-surfactant, and inside the hydrophobic core of the nanodroplet.

Preliminary tests performed on laboratory marble samples showed that it is possible to remove up to the 80 w/w% of the spread wax (see Table 9.1). However, in order to obtain good solubilization it was necessary continuously to renew the system that acted at the interface between the painting and the external environment. An application device was built in which the fresh microemulsion, stored in a tank, continuously reached the paint surface through a narrow drainage pipe (see Figure 9.2).[4]

The results of the application of the microemulsions to clean the surface of the Masaccio and Masolino frescoes are shown in Figure 9.3. Figure 9.3(A) is a UV light image collected before the cleaning. The yellow spots on the right side of the picture are due to the fluorescence of the organic matter present on the fresco surface (mainly crystallized wax). The image in Figure 9.3(B) was collected under UV light irradiation, after the treatment with

Table 9.1 Composition of dodecane- and hexadecane-in-water microemulsions for the solubilization of waxes.[6]

Composition	*A (wt%)*	*B (wt%)*
SDS	1.44	1.44
1-Pentanol	1.32	1.32
Dodecane	10.67	–
Hexadecane	–	10.35
Water	86.57	86.89
wt% of extracted paraffin	77	55

Figure 9.3 Image of a part of the fresco by Masaccio, *St. Peter Healing the Sick with His Shadow* (Brancacci Chapel, Firenze). Ultraviolet light images of the fresco before (A) and after (B) the removal of the crystallized wax spots by means of the o/w microemulsion. (C) Image of the restored painting under visible light. (Courtesy of the prefecture of Florence.)

the microemulsion A. The total absence of the spots is a visual indication of the complete removal of the wax.

For the preparation of the microemulsions, SDS is first dissolved in water. When the surfactant is completely solubilized, first 1-pentanol (the co-surfactant), and then the oil, are added drop-wise. During the addition of the liquid components (co-surfactant and oil) the system is stirred continuously until it looks completely transparent.

9.5.2 Wall Paintings in the Oratorio di San Nicola al Ceppo in Florence (Italy)

The "Oratorio [Oratory] di San Nicola al Ceppo" was built in the centre of Florence in the 16th century by the "Confraternita [Brotherhood] del Ceppo", which took care of the city children affected by congenital diseases. In the following centuries (mostly between the 17th and the 18th century) the interior of the building was decorated with wall paintings and many other works of art. These masterpieces were hit by the flood of the Arno river on the morning of November 4th 1966, when the river broke its banks, spreading a large amount of mud and debris all around the city. The most relevant problems caused for cultural heritage were due to the penetration of the floodwater (containing fuel oil and other substances) and of all the dispersed and solubilized substances into the porous matrixes of artefacts (wall paintings, sculptures, books, easel paintings, *etc.*). After the evaporation of the volatile components of the fuels, one of the most widespread phenomena was the precipitation of insoluble compounds as asphaltenes (mixtures of complex hydrocarbons) that gave rise to the formation of black crusts usually also containing gypsum, $CaSO_4 \cdot 2H_2O$.[7] The preservation of Oratorio's paintings took place almost 40 years after the Arno flood, and was particularly challenging owing to the cross-linking and polymerization (clearly detected by means of FTIR investigations) of the organic components present in the walls' matrix that occurred after that tragic event.

Although the solubilization of black crusts is usually achieved by the application of an aqueous saturated solution of ammonium carbonate $[(NH_4)_2CO_3]$,[8] in this case the treatment was not effective, and the application of a commercial mixture of organic solvents (nitro diluent®) only caused the solubilization of the organic fraction of the crusts (including asphaltenes). Thus, in order to obtain the complete removal of the deposits, simultaneous and combined action of two "active principles" (*i.e.* an organic solvent and ammonium carbonate) was necessary. An o/w microemulsion (whose formulation is reported in Table 9.2)[9] was the ideal system: $(NH_4)_2CO_3$ is solubilized into the continuous phase and *p*-xylene (a good solvent for asphaltenes) is dispersed into the hydrophobic core of the nanodroplets. The amount of *p*-xylene loaded into the nanodroplets is always kept below 1 w/w%, resulting in a considerable reduction in the environmental impact of the cleaning agents.

Table 9.2 Composition of the o/w microemulsion used for the solubilization of black crusts in the "Oratorio of Santa Maria al Ceppo" (Florence).[9]

Composition	*Concentration wt %*
Triton X-100	4.9
p-xylene	0.4
$(NH_4)_2CO_3$	2.1
Water	92.6

The microemulsion was applied by means of the wood poultice technique, as follows.[10]

1. A pulp composed of cellulose fibres (a 50 : 50 mixture of Arbocel BC1000® and Arbocel BC200®, with an average fibre length, respectively, of 700 and 300 μm) was saturated with the microemulsion reported in Table 9.2 and applied onto the paint surface as a compress (thickness of the compress ~0.5 cm); the application time was about two hours at a temperature between 20 and 25 °C.
2. To avoid the evaporation of the volatile components (mainly *p*-xylene) and the drying of the compresses, they were covered with a sheet of commercial polyethylene.
3. A sheet of Japanese paper (17 g m^{-2}) was placed between the compress and the paint surface in order to avoid the absorption of the cellulose fibres by the wall's porous matrix.
4. The macroscopic effect induced by the application of the o/w microemulsion was a softening of the crust so that it could be easily removed by means of a bistoury.
5. In order to remove the residues of the microemulsion (mainly the Triton X-100 surfactant) completely, the cleaned surface was treated for 30 minutes with a further cellulose poultice compress soaked with demineralized water, and then washed with a cotton swab soaked with demineralized water.

Figure 9.4(A) shows an image of the region of the wall paintings affected by the presence of the black crusts before the application of the microemulsion. After the treatment the deposits were removed [see Figure 9.4(B)]. The visual assessment of the system's performance was analytically confirmed *via* FTIR investigation: none of the collected spectra showed any absorption typical of gypsum or asphaltenes. The success of the cleaning was attributed to the simultaneous action of the *p*-xylene and of the $(NH_4)_2CO_3$ dissolved in the aqueous phase.

For the preparation of the microemulsions, the Triton X-100 is first solubilized in water and the system stirred continuously until a transparent solution

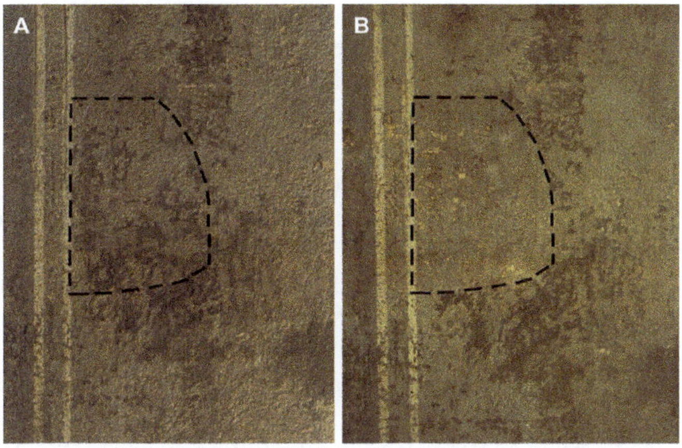

Figure 9.4 Oratorio of Santa Maria al Ceppo (Florence). Image of the area where the cleaning test for the removal of black crusts was carried out before (A) and after (B) the application of the o/w microemulsion indicated in Table 9.2. (Reprinted with permission from Carretti *et al.*, *Langmuir*, 2007, **23**, 6396. Copyright 2007, American Chemical Society.) Courtesy of the Compagnia di S. Niccolò detta del Ceppo.

is obtained; $(NH_4)_2CO_3$ is then added. After $(NH_4)_2CO_3$ has been completely solubilized, *p*-xylene is added drop-wise and the system is stirred until it becomes completely transparent.

9.5.3 Wall Paintings by Francesco Vecellio, Chiesa di San Salvador in Venice (Italy)

The San Salvador Church, dedicated to Jesus, was built in Venice in the 7th century. The original Romanesque building has been modified several times during the centuries and the present structure appeared only at the beginning of the 17th century, when the façade was finished. The sacristy, built in 1546, is decorated by mural paintings traditionally attributed to Francesco Vecellio, Titian's brother; on the roof there is a representation of the blessing Christ, while the walls are decorated with birds and floral patterns.

At the end of 2002, when the last restoration workshop began, the surface of the paintings exhibited a uniform thick layer of highly degraded acrylic copolymer that had been improperly applied (10% w/w solution in nitro diluent) as a surface protective at the end of the 1970s by the priest of the church. The coating had to be removed for two main reasons:

- Direct access to the paint surface was needed in order to eliminate all the salts (mainly sulfates and chlorides caused by infiltrations of the salty lagoon water) and retouches applied during a previous restoration workshop (at the beginning of the 20th century);

- The yellowing of the acrylic resin, due to its ageing, had greatly altered the polychromy of the paint underneath, partially compromising the readability of the work of art. As indicated in Figure 9.5 (right side), all the colours appeared darker and the coated surface was shiny.

Thus, in this case the removal of the copolymer layer was not only necessary in order to re-establish the original physico-chemical status of the materials, but also for aesthetic reasons.

In fact, the presence of degraded surface patinas made up of various acrylic copolymers is one of the most common problems with wall paintings, and the physico-chemical properties of the polymeric film on the Sacristy's paintings are similar to those commonly found on many other paintings. This case study can be considered as a reference for the use of o/w microemulsions in the removal of aged acrylic coatings from wall painting surfaces.[11,12] Nevertheless, it has to be stated that for every workshop the application procedure described in this section must be properly calibrated, especially in terms of the application time.

Figure 9.5 (left side) shows an area of the wall paintings on which a microemulsion (whose composition is reported in Table 9.3) containing *p*-xylene (a good solvent for degraded acrylic polymers) was applied by means of the wood poultice technique, following a protocol similar to that described in the Section 9.5.2.

Figure 9.5 Sacristy of the San Salvador Church (Venice). Wall paintings before (right side) and after (left side) the cleaning. (Courtesy of the Italian Cultural Heritage Department.)

Table 9.3 Composition of the o/w microemulsion used for the solubilization of the aged acrylic resin from the surface of the wall paintings by Francesco Vecellio in the Sacristy of the San Salvador Church in Venice.[11]

Composition	*Concentration wt %*
SDS	4.1
Water	85.4
1-PeOH	7.9
p-Xylene	2.6

1. A pulp composed of cellulose fibres (a 50 : 50 mixture of Arbocel BC1000® and Arbocel BC200®) was soaked with the microemulsion reported in Table 9.3, applied onto the paint surface as a compress (thickness of the compress ~0.5 cm) and coated with a sheet of commercial polyethylene.
2. The application time was between 1.5 and 2 hours, depending on the thickness of the copolymer layer.
3. A sheet of Japanese paper ($17\,g\,m^{-2}$) was placed between the compress and the paint surface.
4. Once the system had carried out its function, the surface was cleaned with demineralized water until the foam due to the residual surfactant had disappeared.

By comparing the left and the right sides of Figure 9.5 it is macroscopically evident that the acrylic coating was removed by the microemulsion. This evidence, along with the complete removal of the microemulsion components, was also confirmed by the profile of FTIR spectra collected on two micro-samples taken from the paint surface before and after the application of the microemulsion.

The presence of gypsum ($CaSO_4{\cdot}2H_2O$), detected in the FTIR spectrum of the sample collected after the cleaning test, suggested that the application of such a large amount of polymer was probably necessary to cover the gypsum efflorescence, in order to eliminate the whitening effect due to the scattering of the incident light by the $CaSO_4{\cdot}2H_2O$ crystals on the surface. The removal of the sulfates should be possible through either the application of anion exchange resins, or the Ferroni method.[13]

Before introducing new case studies, it is worthwhile to analyse here the mechanism of the interaction between the amphiphile-based system and the acrylic coating. This subject is not trivial, and investigation is still ongoing.

Many research efforts have been devoted in the past to understanding the interaction of surfactant solutions with partially soluble macromolecules, block copolymers,[14–18] or relatively low-molecular weight compounds, such as those contained in waxes and oily soils. For instance, in the case of the Brancacci

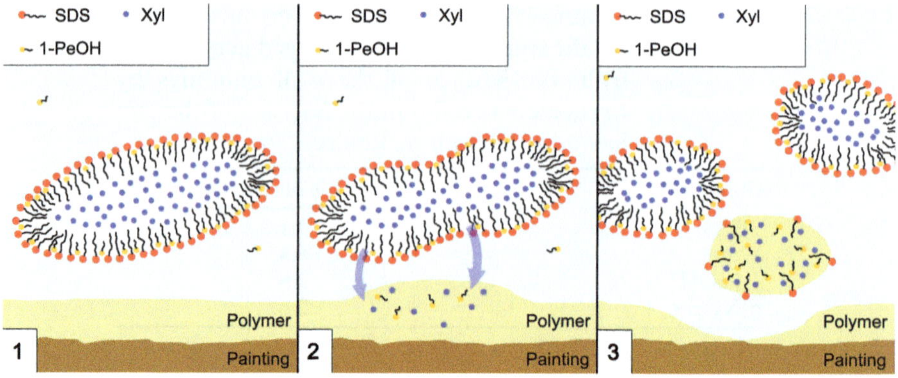

Figure 9.6 Schematic representation of the mechanism of the interaction between the detergent nanostructured system (xylene-based o/w microemulsion, see Table 9.3) and an acrylic polymer coating. (1) The o/w microemulsion in contact with the polymer coating. (2) Migration of the organic solvent from the droplets to the polymer. The migrated solvent swells the coating. (3) Reorganization of the nanodroplets and detachment of the swollen polymer. (Reproduced from Ref. 19.)

Chapel case study discussed in Section 9.5.1, the hydrophobic material (wax) was effectively dispersed into the surfactant micelles, and the process was enhanced by the presence of an apolar solvent.

However, the interaction of a high molecular weight acrylic coating with a w/o microemulsion, as considered in this section, follows a different model that was studied recently.[19–22] The results indicate that upon interaction with the polymer coating the xylene-based o/w microemulsion maintains its nanostructure,[20,22] but the nanodroplets become smaller and reorganize their structure owing to the migration of the organic solvent from the droplets to the polymer. The migrated solvent swells the coating, leading to its detachment from the surface, which is consequently cleaned. It was also concluded that 1-PeOH has a double role: besides being responsible for the formation of nanodroplets, it also promotes the removal of the acrylic coatings.[19] The polymer-interaction mechanism is depicted in Figure 9.6.

9.5.4 Wall Paintings, Villa del Barone in Prato (Italy)

The "Villa del Barone", located in Montemurlo (Prato, Italy), was built in the first half of the 16th century by Bartolomeo Valori in order to celebrate his appointment to the senate of the new princedom of Florence that was born after the capitulation of the Florentine Republic. During the following centuries the many owners of the Villa modified both the exterior and the interior of the building, which was decorated in neoclassical and baroque styles (18th and 19th centuries). After the Second World War, "Villa del Barone" was used as a hospice before it became neglected. During the following decades many acts of vandalism (mainly graffiti) were perpetrated especially on the 18th-century *a secco* paintings that decorate the internal walls of the building.

Traditionally, the removal of graffiti is carried out with nitro diluent, acetone, chloroderivatives or other organic solvents, either pure or in mixture. Unfortunately, as for the polymeric materials, the application of solvents presents two different drawbacks related to their toxicity and to the partial spread of the solubilized substances into the porous support.

Thus, the same amphiphile-based system as described in Section 9.5.3 (see Table 9.3) was used for the removal of part of the graffiti.[23]

Figure 9.7(A) shows the effects of the application of the o/w microemulsion for the removal of black oil paint [see Figure 9.7(A), left side; the image was obtained before the cleaning test]. The complete elimination of the paint was achieved simply by applying the microemulsion using the wood poultice technique already described in the previous sections, inducing a softening of the "graffito" and its partial solubilization. Then, the surface was treated with the microemulsion and washed several times with demineralized water until the disappearance of the foam, due to the presence of the surfactant, was achieved.

The microemulsion was also applied, following the same procedure, to remove some red "graffiti" from the same surface. Figure 9.7(B) (left side,

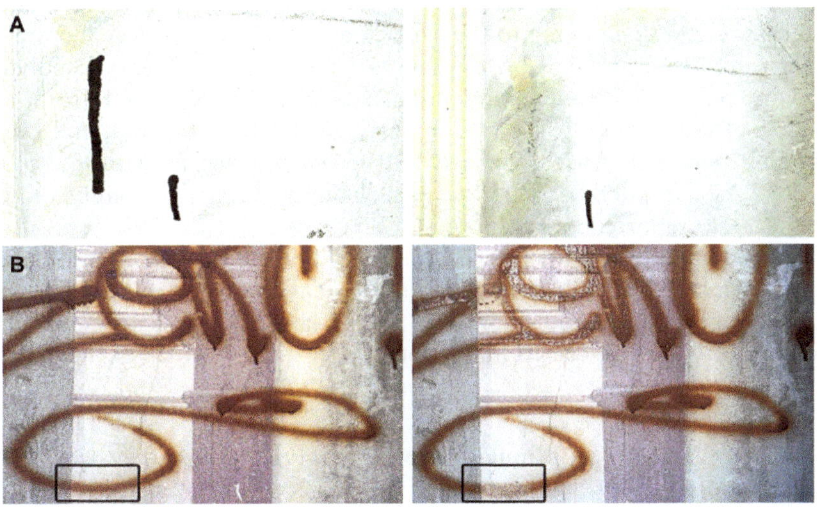

Figure 9.7 Detail of the wall paintings of the Villa del Barone (Prato, Italy) where the cleaning test using microemulsion has been performed and the surface has been tested for the removal of oil-paint. (A) Removal of oil paint (left side: before cleaning; right side: after cleaning); (B) removal of acrylic paint (left side, black box: before cleaning; right side, black box: after cleaning). (Reprinted with permission from Carretti *et al.*, *Studies in Conservation*, 2005, **50**, 128. Copyright 2005, International Institute for Conservation of Historic and Artistic Works.)

before the treatment with the microemulsion and right side, after a single application of the dispersed system) shows the region in which the test was performed. In this case the residual halo in the treated area is attributable to the fact that the paint had been applied by means of a spray, and small drops of the nebulized pigment solution penetrated into the porous matrix. Thus, their complete removal is possible only after repeated treatments.

The main drawback related to this applicative procedure is the mechanical action needed for the complete removal of the graffiti. The mechanical action is not feasible on a surface that is not perfectly cohered (*i.e.* owing to the presence of surface powdering and/or surface detachment), and the removal of the graffiti can be achieved only after a pre-consolidation procedure that must be performed through the application of $Ca(OH)_2$ nanoparticle dispersions (see Chapter 13).

9.5.5 Wall Paintings by Vecchietta, Santa Maria della Scala Sacristy in Siena (Italy)

Low impact o/w microemulsions were tested for the removal of the aged acrylic coating from the surface of the wall paintings in the Sacristy of the Santa Maria della Scala church (Siena, Italy; see Section 7.6.2).

The formulation of the microemulsion is reported in Table 9.4; AGE® and AGESS® (Lamberti, Milan, Italy) are respectively a non-ionic and an anionic surfactant belonging to the class of poly alkyl glucosides, and are used for the microemulsification of *p*-xylene, a good solvent for acrylic polymers. They are particularly interesting because they are synthesized from waste products of the food industry (mainly fatty alcohols and sugars), and because their environmental impact is greatly reduced in comparison with that of pure solvents. As a matter of fact, the microemulsion consists mainly of water (99 w/w%), as reported in Table 9.4.

For its preparation, after the complete solubilization of the surfactants in water, *p*-xylene is added drop-wise and the system is stirred continuously until a completely transparent and stable fluid is obtained. The application protocol is exactly the same as described in Section 9.5.3.

Table 9.4 Formulation of the polyalkyl glucoside-based o/w microemulsion used for the cleaning test on the surface of wall paintings by Vecchietta, Santa Maria della Scala Sacristy in Siena (Italy).[9]

Composition	*Concentration wt%*
AGE®	0.52
AGESS®	0.12
H_2O	99.00
p-Xylene	0.36

Figure 9.8 Santa Maria della Scala Sacristy, Siena, Italy, wall paintings by Vecchietta. (A) Area of the wall painting chosen for the cleaning test. (B) Grazing light image of the wall paintings showing the glossy effect due to the presence of the copolymer layer. (C) Grazing light image of the area of the wall paintings where the cleaning test was performed (black dashed line), after the application of the dispersed system whose formulation is reported in Table 9.4. (Courtesy of the municipality of Siena.)

Figure 9.8 shows the front view [Figure 9.8(A)] and the grazing light image [Figure 9.8(B)] of the treated area of the wall paintings. The application of the dispersed system was carried out with the wood poultice technique (contact time of two hours, followed by washing of the surface with demineralized water). Figure 9.8(C) shows the area of interest for the cleaning test (black dashed line) after the application of the microemulsion; the disappearance of the glossy effect macroscopically indicates that the surface layer of acrylic polymer has been successfully removed.

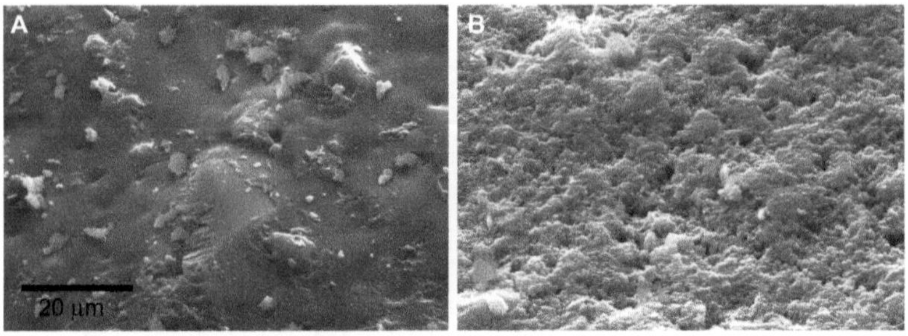

Figure 9.9 Santa Maria della Scala Sacristy, Siena, Italy. SEM micrograph of a microsample from the Vecchietta fresco before (A) and after (B) the application of the o/w microemulsion.

The FTIR investigation performed on the surface of two microsamples taken from the test area, before and after the application of the microemulsion, confirmed the effectiveness of the cleaning test.[24]

Scanning electron microscopy (SEM) investigation indicated that, after the application of the microemulsion, the treated surface is similar to that of a non-coated mortar. As indicated in Figure 9.9(B), most of the micropores are opened and the surface is very rough, while in the presence of the coating the surface appears smooth and homogeneous [see Figure 9.9(A)]. Thus, the analyses further confirmed the efficacy of the microemulsion-based method without any detectable negative side effect (*e.g.* formation of microfractures or local detachments).

9.5.6 The Maya Murals in Mayapan (Mexico)

The archaeological site of Mayapan was rediscovered in the 1940s by the explorer Stephens. The discovery brought to light one of the most important cities in the history of the Post-Classic Maya.[25,26]

Since then, the restoration of the site has been carried out slowly and only in 1996, with the "Proyecto Arqueologico Mayapan" work, were some wall paintings of great historical and artistic heritage recovered, ranging from 1300 to 1450 AD. After their uncovering, the conservation state of these paintings aroused great concern in the archaeologists. In Mayapan, in fact, most of the fragments of the wall paintings are located outdoors and, therefore, are exposed to driving rain, sunlight and wind. The climate conditions of the Yucatán Peninsula, with high temperatures, high humidity and abundant precipitation, have accelerated the degradation of the paintings.

Therefore, the restorers of the "Proyecto Arqueologico Mayapan" applied an organic polymer (Mowilith DM5®), as a protective, to the surface of several wall paintings.[27] Mowilith DM5 is a copolymer of vinyl acetate and *n*-butyl acrylate, in a 65 : 35 ratio, marketed in the form of an aqueous emulsion.[28] The application of an organic polymer on a mural painting drastically alters the physico-chemical properties at the painting–air interface, creating serious problems for conservation work (see Chapter 3). In fact a water-repellent film is formed, which is impermeable to water vapour and occludes the original porosity of the substrate.[29] As a matter of fact, the degradation of Mowilith DM5 has made the readability of several wall paintings difficult. In some areas the polymer layer has prevented the salts in the wall pores from crystallizing on the surface, resulting in the pulverization of the plaster just beneath the painted surface. In other areas, the polymer film is so thick as to provide the painting with a plastic-like visual appearance.

Nowadays, the removal of Mowilith represents one of the restoration priorities for Mesoamerican cultural heritage conservation.

Similarly to other case studies, the use of nano-structured systems such as micellar solutions and microemulsions was considered as an innovative alternative to the application of neat solvents. In fact, the murals in Mayapan were restored in 2009 by using the amphiphile-based "EAPC" system (see Table 9.5) to remove the polymer coatings that had altered the original painting surfaces (see Figures 9.10 and 9.11).[19]

Table 9.5 Formulation of the EAPC o/w system used for the cleaning test on the surface of wall paintings in Mayapan (Yucatan, Mexico).

Composition	*Concentration wt%*
SDS	3.7
Water	73.3
1-PeOH	7
Propylene carbonate	8
Ethyl-acetate	8

Figure 9.10 (A) View of the main building in Mayapan, Mexico. The dashed line indicates the location of the wall painting "de los discos solares". (B) A detail of the wall paintings after the removal of the polymer coating using the EAPC system. The dashed boxes highlight the small areas that were not cleaned and left untreated as a reference.
(Reproduced from Ref. 19.)

Figure 9.11 (A) The procedure for application of the EAPC system: a cellulose poultice impregnated with EAPC is applied over the surface of the painting, previously protected with a Japanese paper tissue. (B) The removal of the poultice after the cleaning procedure. Under glazing light the disappearance, after treatment, of the shining effect due to the coating can be appreciated. The presence of the polymer is evident in the narrow strip that was left untreated. (C) Application of the system for test experiments on the paintings. (D) Removal of the polymer coating. (Adapted from Ref. 19.)

All the tests carried out *in situ* confirmed the good performance of this system. From a practical point of view, the removal of the polymer coatings took place as a superficial peeling, upon slow and controlled swelling, which prevented the removed polymer from penetrating into the wall. Similar to the removal of acrylic coatings by xylene-based o/w microemulsions (see Section 9.5.3), the mechanism of the interaction between the amphiphile-based system (EAPC) and the polymer (acrylic–vinyl) was further investigated.

The first difference with respect to the previous case concerns the structure of the cleaning fluid itself. The xylene-based fluid is a "typical" o/w microemulsion, where the surfactant (SDS) and the cosurfactant (1-PeOH) stabilize the solvent nanodroplets. On the other hand the structure of the EAPC system is more complex. Small angle neutron scattering (SANS) measurements allowed a thorough analysis of this system, providing fundamental information on its nanostruscture.[30] The co-solvents (ethylene acetate, EA, and propylene carbonate, PC) were found to be partitioned between both the continuous aqueous phase and the dispersed droplets. The system thus cannot be defined as a simple microemulsion or micellar solution: the term "swollen micelles" would be in this case more appropriate to define this borderline structure.

These structural differences imply a difference in practical effectiveness, because it was found that the EAPC system is more versatile and effective than the xylene-based microemulsion. As a matter of fact, the fluid–polymer interaction mechanism is enhanced (see Figure 9.12); in this case, the solvents are present also in the continuous aqueous phase, and interact quickly with the polymer coating. Both EA and PC, in fact, migrate from the continuous phase to the coating, from the nanodroplets to the continuous phase, and from the nanodroplets to the polymer. As a result, swelling and detachment of the polymer coating occurs, paralleled by a reorganization of the swollen micelles.[19] It is worth noting that the role of the nanodroplets is crucial in granting the proper amount of solvents dispersed in water, even if they do not seem to be directly involved in the solubilization of the coating.

Finally, the presence of the continuous aqueous phase limits the penetration of the removed hydrophobic material into the pores of the substrate, as reported for the other amphiphile-based systems discussed so far.

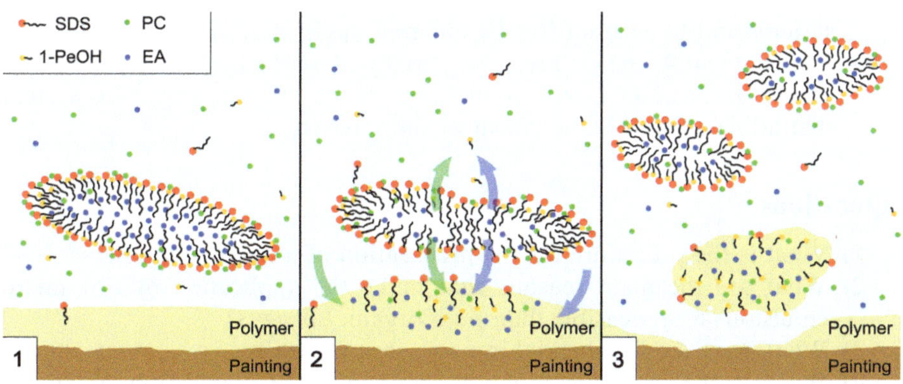

Figure 9.12 Schematic representation of the mechanism of the interaction between the detergent nanostructured system (EAPC, see Table 9.5) and an acrylic–vinyl polymer coating. (1) The amphiphile-based system in contact with the polymer coating. (2) EA and PC migrate from the continuous phase to the coating, from the nanodroplets to the continuous phase, and from the nanodroplets to the polymer. Swelling of the polymer coating occurs. (3) Detachment of the polymer, paralleled by a reorganization of the swollen micelles. (Reproduced from Ref. 19.)

Further Suggested Reading

C. Alba-Simionesco, J. Teixeira and C. A. Angell, *J. Chem. Phys.*, 1989, 91, 395.

Espinosa, Conservation and restoration of the Murals of the Temple of the Paintings in Bonampak. *Proceedings of in situ Archaeological Conservation, 6–13 April 1986, Mexico City*, ed. H. W. M. Hodges, INAH, Mexico City, 1987, p. 84.

H. Orea and V. A. Magar, *Preprints of the 13th Triennial Meeting ICOM Committee for Conservation, ICOM-CC*, Rio de Janeiro, 22–27 September 2002, p. 176.

J. Riederer, The Restoration of archaeological monuments in the tropical climate. *Proceedings of 7th Triennial ICOM Meeting, Copenhagen, 10–14 September 1984*, Paris 1, 1984, 21.

Review and Questions

The main topics discussed in this chapter are as follows.

- The conservation-specific glossary related to conservation studies is clarified in order to explain the concept of a nanostructured fluid, and the differences among emulsions, microemulsions, detergents and micellar solutions, terms that are sometimes misused.
- The detergency mechanism in the presence of nanostructured fluids is described with a special emphasis on the differences from the classical solubilization mechanism where blended solvents are used.
- Several examples are described in order to show the versatility of these systems and to account for the different application procedures.
- Environmental issues are also briefly described in order to make conservators more aware of the importance of using dispersed systems that allow a stable confinement of the solvents.

Questions

1) What is the procedure for the preparation of a microemulsion?
2) What are the main reasons that make the application of o/w micro-emulsions more feasible than pure organic solvents?
3) What does "environmental impact" mean?
4) What is the most common procedure for the application of an o/w microemulsion for the cleaning of the surface of a wall painting?

Answers

1) A microemulsion can be prepared by mixing the appropriate components according to the following procedure. First, the surfactant has to be dissolved in the continuous phase (*i.e.* water for o/w microemulsions and

oil for w/o microemulsions). A co-surfactant, if necessary, has to be added drop-wise to the surfactant solution at room temperature, stirring until a transparent solution is formed. The dispersed phase (*i.e.* water for w/o microemulsions and oil, *e.g.* *p*-xylene, for o/w microemulsions) is finally added drop-wise at room temperature; the system is kept under continuous stirring until it becomes transparent and homogeneous.

2) The main reasons that make the application of o/w microemulsions more feasible than pure organic solvents for the solubilization of hydrophobic organic polymers and copolymers are: (i) the spread of the solubilized hydrophobic material into the porous matrix is avoided because solubilization or swelling takes place at the interface between the nano-droplets and the continuous aqueous phase in the microemulsion. Thus, the continuous aqueous phase acts as a hydrophilic barrier which prevents the hydrophobic polymer from penetrating further into the porous structure; (ii) the dispersed systems are thermodynamically stable throughout a wide range of environmental operating conditions; (iii) the impact of the treatment on the environment and on the health of the conservators is drastically reduced because o/w microemulsions are mainly composed of water (up to 99%) and the amount of the oil fraction (an organic apolar solvent) is typically less than 15% by weight.
3) The traditional cleaning procedures that involve solvents often require the use of toxic neat solvents or blends. Conservators, usually aware of the health risks in handling some solvents, still rely on the use of effective solvents blends. The use of small amounts of hazardous organic solvents (*e.g.* aromatic or halogenated) confined in a stable microemulsion strongly reduces both the health risks for the operator and the amount of solvent that is actually released into the environment.
4) For the cleaning of the surface of a wall painting, microemulsions are usually applied by means of the wood poultice technique. First, the cleaning fluid is added to a powder made up of fibres of pure cellulose (a 50:50 mixture of Arbocel BC1000® and Arbocel BC200®), until saturation. The pulp is converted into compresses that are applied onto the surface to be cleaned (typical application time of about 2.5 hours, at a temperature between 20 and 25 °C). Usually, a sheet of Japanese paper is located between the compress and the work of art surface to avoid direct contact between the cellulose fibres and the surface. In order to avoid the evaporation of the volatile components (mainly the oil phase) and the drying of the compresses, the latter are covered with a sheet of commercial polyethylene. Once the compresses are removed, the cleaned surface is washed with demineralized water in order to eliminate all the residues of surfactant completely.

References

1. R. C. Wolbers, *Cleaning Painted Surfaces: Aqueous Methods*, Archetype Publications, London, 2000.

2. R. Bellucci, E. Buzzegoli, M. Matteini and A. Moles, in *Metodi e scienza – Opertività e ricerca nel restauro*, Sansoni editore, Firenze, 1982, p. 261.
3. B. Lindman and I. Danielsson, *Colloids Surf. A*, 1981, **3**, 391.
4. E. Ferroni, G. Gabrielli and G. Caminati, in *La Cappella Brancacci. La scienza per Masaccio, Masolino e Filippino Lippi*, Olivetti, Milano, 1992, p. 162.
5. P. G. De Gennes and C. Taupin, *J. Phys. Chem.*, 1982, **86**, 2294.
6. L. Borgioli, G. Caminati, G. Gabrielli and E. Ferroni, *Sci. Technol. Cultural Heritage*, 1995, **4**, 67.
7. L. Yan Zhang, S. Lawrence, Z. Xu and J. H. Masliyah, *J. Colloid. Interface Sci.*, 2003, **264**, 128.
8. E. Ferroni and P. Baglioni, in *Scientific Methodologies Applied to Works of Art: Proceedings of the Symposium*, Florence, Italy, 2–5 May 1984, Montedison progetto cultura, Milano, 1986, p. 108.
9. E. Carretti, R. Giorgi, D. Berti and P. Baglioni, *Langmuir*, 2007, **23**, 6396.
10. E. Ferroni, in *Ecological Physical Chemistry*, Elsevier, Amsterdam, 1991, p. 345.
11. E. Carretti, C. Miliani, L. Dei and P. Baglioni, *Progress in Colloid and Polymer Science*, 2011, **118**, 63.
12. E. Carretti, L. Dei and P. Baglioni, *Langmuir*, 2003, **19**, 7867.
13. E. Ferroni and D. Dini, in *Scritti di storia dell'arte in onore di Ugo Procacci*, Electa Editrice, Firenze, 1977, p. 17–22.
14. M. Muller, B. Stuhn, K. Busse and J. Kressler, *J. Colloid Interface Sci.*, 2009, **335**, 228.
15. A. Holmberg, P. Hansson, L. Piculell and P. Linse, *J. Phys. Chem. B*, 1999, **103**, 10807.
16. A. Holmberg, L. Piculell, P. Schurtenberger and U. Olsson, *Colloids Surf. A*, 2004, **250**, 325.
17. K. J. Mutch, J. S. van Duijneveldt, J. Estoe, I. Grillo and R. K. Heenan, *Langmuir*, 2008, **24**, 3053.
18. K. J. Mutch, J. S. Van Duijneveldt and J. Estoe, *Soft Matter*, 2007, **3**, 155.
19. M. Baglioni, R. Giorgi, D. Berti and P. Baglioni, *Nanoscale*, 2012, **4**, 42.
20. M. Baglioni, D. Rengstl, D. Berti, M. Bonini, R. Giorgi and P. Baglioni, *Nanoscale*, 2010, **2**, 1723.
21. R. Giorgi, M. Baglioni, D. Berti and P. Baglioni, *Accounts Chem. Res.*, 2010, **43**, 695.
22. E. Carretti, E. Fratini, D. Berti, L. Dei and P. Baglioni, *Angewandte Chemie*, 2009, **48**, 8966.
23. E. Carretti, B. Salvadori, P. Baglioni and L. Dei, *Stud. Conserv.*, 2005, **50**, 128.
24. E. Carretti, L. Dei, C. Miliani and F. Rosi, *Spectrosc. Let.*, 2005, **38**, 459.
25. A. Barrera Rubio and C. Peraza Lope, in *La pintura mural prehispánica en México II, área Maya - tomo IV*, UNAM-IIE Press, Mexico City, 2001, p. 419.

26. D. Magaloni Kerpel, in *La pintura mural prehispánica en México II, área Maya- tomo III*, UNAM-IIE Press, Mexico City, 2001, p. 155.
27. C. A. García Solís, C. A. García Solís, V. G. Vierna and A. V. Veiga, *Intervención*, 2010, **2**, 49.
28. L. Borgioli and P. Cremonesi, in *Le resine sintetiche usate nel trattamento di opere policrome - Collana I Talenti n.17*, Il Prato, Padova, 2005, p. 63.
29. E. Carretti and L. Dei, *Prog. Organic Coat.*, 2004, **49**, 282.
30. M. Baglioni, D. Berti, J. Teixeira, R. Giorgi and P. Baglioni, *Langmuir*, 2012, **28**, 15193.

CHAPTER 10

Cleaning IV: Gels and Polymeric Dispersions

E. FRATINI* AND E. CARRETTI

University of Florence, Italy
*Email: fratini@csgi.unifi.it; carretti@csgi.unifi.it

10.1 Introduction

Gels are soft materials consisting of interconnected long polymer chains, organic molecules or colloidal particles that entrap a fluid. As a result, they exhibit physical properties that are intermediate between those of a solid and of a liquid. In this regard, a colloidal particle is intended as a continuous entity in the dimensional range from 1 to 1000 nanometres (*i.e.* the size typical for colloidal systems).

Usually, a gel is made up of a solid (called a gelator or gellant) that, dispersed into a fluid, forms a three-dimensional (3D) skeleton or, according to the terminology of the International Union of Pure and Applied Chemistry (IUPAC), a "non-fluid colloidal network or polymer network that is expanded through its whole volume by a fluid".[1] In general a gel can be considered as a bicontinuous fluid phase, where continuity means that it is possible to go across the whole solid phase from one side of the system to the other, without entering into the liquid. Similarly, one can make the same path into the liquid without touching the solid phase.

From a historical point of view gels were already known in prehistoric times. For instance, as described by McCreesh *et al.*,[2] the ancient Egyptians used various fat-based systems to ensure that their hair was kept in place for long time. Gels and gel-like systems are still omnipresent in our daily life, with many

RSC Nanoscience & Nanotechnology No. 28
Nanoscience for the Conservation of Works of Art
Edited by Piero Baglioni and David Chelazzi

Published by the Royal Society of Chemistry, www.rsc.org

different applications from cosmetic products (hair gels, body care creams, toothpastes, *etc.*) to medical liniments used after trauma, foods (jams, chocolate paste, ketchup, mayonnaise, puddings, whipped cream, *etc.*), fillers for prostheses (breast implants), soft contact lenses, just to cite some examples.

In the field of Cultural Heritage gels are widely used to "structure" water-based detergent systems (oil-in-water microemulsions or micellar solutions) or organic neat solvents or mixtures, to obtain innovative cleaning tools. For instance, a gel-based cleaning tool can control and tailor the diffusion phenomena of the liquid phase, reducing the aggressiveness of organic solvents. As a result, the swelling or leaching of the polymeric materials constituting the paint (*e.g.* binding media in the painted layers), and the overall cleaning stress, are all mitigated.

The mechanical properties of these systems, which depend strictly on their chemical composition and nature, are some of the key factors that control their performance. Although gels have been used since ancient times, the description and definition of a gel are still open to discussion. J. Lloyd, about 70 years ago, wrote that "the colloid condition, the gel, is easier to recognize than to define".[3] Empirical descriptions of the gel state, including the first by T. Graham back in 1861,[4] were related to the available qualitative macroscopic observations. The first attempt to link the macroscopic and microscopic properties of gels was made, several years later, by P. H. Hermans, who described a gel as "a coherent system of at least two components, which exhibits mechanical properties characteristic of a solid, where both the dispersed component and the dispersion medium extend themselves continuously through the whole system".[5]

In order to provide the reader with a comprehensive overview of gels, this chapter will begin with the basic concepts and definitions. Then a systematic classification of the types of gel will be presented. Their properties, uses and applications will be described and some techniques for characterization will be reviewed. All the applicative aspects of the use of gels in the Cultural Heritage field will be disclosed in Chapter 11, where some real case studies will be illustrated.

10.2 Definition of Gels and Gelation Process

Before going into a detailed description of the properties, uses and major applications of these intriguing systems, a precise definition of what we mean by "gel" needs to be stated. This is not an easy task owing to the intrinsic complexity of the system, as testified by all the different approaches used in the literature. Some phenomenological definitions of gel were given by H. K. Henish in 1970. He referred to a gel "as a two component system of a semisolid nature, rich in liquid, and no one is likely to entertain illusions about the rigor of such a definition"[6] and by Almdal *et al.* in 1993:[7] "A gel is a soft, solid or liquid-like material of two or more components, one of which is a liquid, present in substantial quantity".

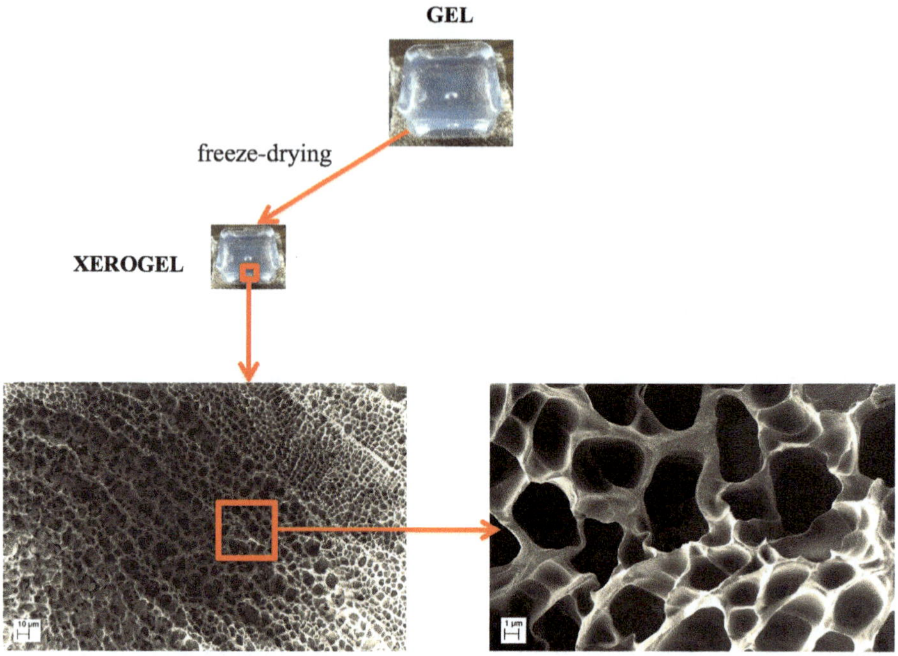

Figure 10.1 (Top) Images of a gel and the corresponding xerogel after freeze-drying. (Bottom) Electron microscopy images at two different magnifications of the xerogel sample evidencing the macroporosity of the polymer architecture. Bar is 10 µm (bottom left) and 1 µm (bottom right).

In the case of polymer-based gels (see Figure 10.1 as an example), the major class of gel, the formation of the 3D solid-like network requires the polymerization of monomers/comonomers with a cross-linker, usually activated by a radical initiator. Other gels can be prepared by dissolving the polymer chains in a solvent at high temperature, and the final network is achieved by lowering the temperature (down to about 25 °C).

Several statistical approaches have been developed for the prediction of the gel point and the modelling of the gel formation process, depending on the experimental conditions. One of the most important approaches from a practical and historical point of view is the "classical statistical theory" postulated by Flory and Stockmayer.[8] Suppose you have a z-functional molecule (*i.e.* in the sense that each monomer of "gelator" can form at a maximum z "interactions" with the other monomers through hydrogen (H)-bonds, van der Waals, London forces or covalent bonds; the chemical reactivity of each group is the same), if $z > 2$ the molecule is referred to as polyfunctional and hence is able to form a 3D network. The formation of intramolecular bonds between monomers appertaining to the same polymer molecule is considered negligible, and only interactions between different monomer/polymer molecules are taken into account. In this way, a 3D network (called a Cayley tree or Bethe lattice), such as the one presented in Figure 10.2 for a

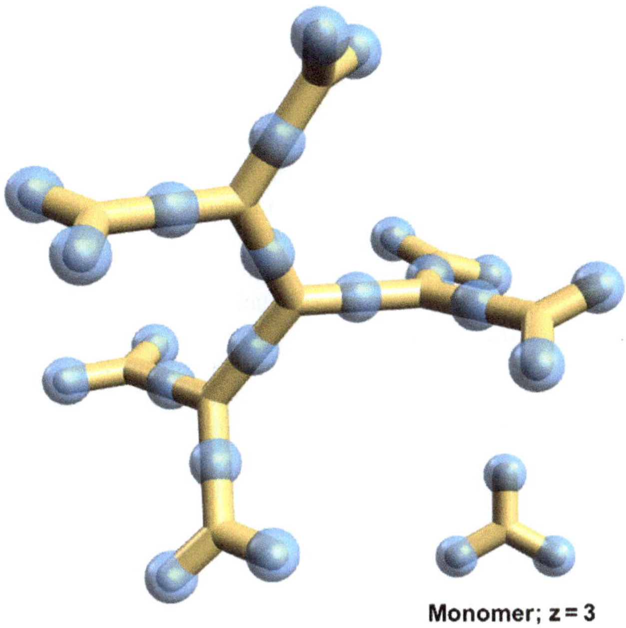

Figure 10.2 Three-dimensional representation of the Cayley tree or Bethe lattice for a system obtained starting from a monomer characterized by a functionality, $z = 3$. The spheres highlight the interaction points in the monomer. Image realized by the Avogadro software v.1.0.3.[10]

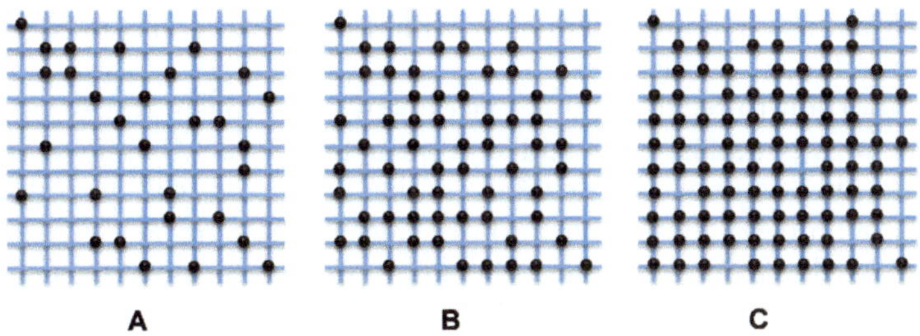

Figure 10.3 Site percolation model. The images represent: A) 25, B) 50 and C) 75% of filling of the lattice.[11]

monomer with $z = 3$, is obtained. There are no limits to the dimension of the macromolecular structure obtained. If the growth of one of these macromolecules occurs in a solvent, extending throughout the solution, the system obtained is called a gel.[9]

Another theory that describes the formation process of a gel is the percolation model described by Zallen[11] and Stauffer.[12] This second statistical approach is depicted in Figure 10.3 using, for simplicity, a two-dimensional (2D) representation. Here the formation of intramolecular bonds is not excluded *a priori*. The intersection between vertical and horizontal lines in the

2D grid represents a potentially reactive point. Each black spot indicates a point filled by a reactive molecule. The formation of a bond occurs when two neighbouring sites are filled. This model is called site percolation.

When the fraction of filled sites overcomes a critical value referred to as the "percolation threshold", the formation of a cluster with an average size approaching infinity is observed.[11] The increase of the average size to infinity corresponds to a system where all the monomers are linked together and participate in the formation of a unique 3D network. When the growth of this macromolecular structure occurs into a liquid, the system is defined as a gel.

Owing to the structural complexity of a gel and to the different typologies of the "gelator" and of the continuous phase, several different features such as the hydrophobicity of the gelator can be used to classify gels.

Hydrophobic gels are 3D networks in equilibrium with a non-polar solvent (these systems are also referred to as organogels), while hydrophilic gels have a hydrophilic solvent as the continuum phase. If the hydrophilic solvent is water, the final gel is commonly called hydrogel. If the fluid is air (or any other gas medium) the system is known as aerogel.

The intimate knowledge of the affinity of the gelator for different solvents can be very helpful from an applicative point of view. When a hydrophilic gel is immersed in a hydrophobic solvent it usually tends to shrink in order to minimize the interactions with the organic solvent, rendering the system completely unusable. Both hydrogels and organogels can be applied to works of art cleaning: highly retentive hydrogels are useful to control the cleaning action on water-sensitive substrates (see Section 11.6), while organogels could be considered to support waterless fluids for the cleaning of substrates that would be damaged upon interaction with water (*e.g.* collagen-based objects such as parchment and leather).

It is important to recall the differences between aerogels and xerogels, which are both porous materials obtained from a liquid gel. Owing to their characteristics, they cannot be included in the family of gels according to the Almdal's definition,[7] but other authors suggest the contrary.[5,8] Aerogels are prepared starting from a gel where the liquid phase is replaced by a gas (usually air) by means of the supercritical drying technique. This step ensures no modification of the structure of the starting gel. On the other hand, xerogels are solids formed by drying a gel without controlling its shrinkage; therefore, during dehydration, the structure of the gel partially or totally collapses, with a change in porosity. The freeze-dried gels are xerogels, where usually the macro and mesoporous structure are retained, while the micro (*i.e.* nano) structures have collapsed (see Figure 10.1).

Gels can also be classified by their response to various stimuli. Temperature, pH, and magnetic responsive gels have been prepared and used in Cultural Heritage conservation, and several examples will be described in Chapter 11.

Another way to classify gels is based on the nature of the gellant and on the preparation methods (see Figure 10.4). For a gel of type **A** the gellant is a polymer (polystyrenes, polyacrylates, proteins, gelatins, celluloses, *etc.*)[13] whose molecules can cross-link one with each other through different types of

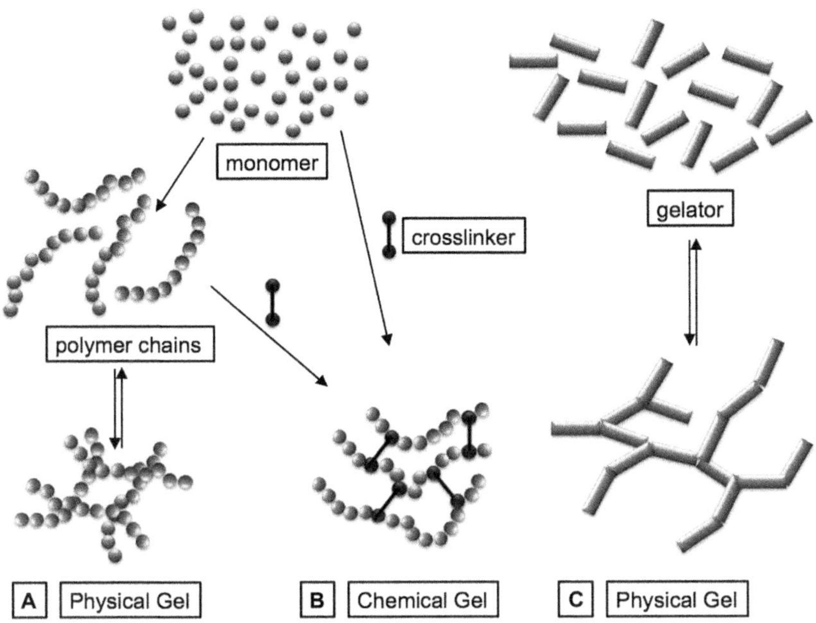

Figure 10.4 Three types of gel. A: Gel made by directly cross-linking preformed polymer chains in a solution. B: Gel made by polymerization and cross-linking of a mixture of monomer (open circles) and cross-linker (solid circles). C: Physical gels made by self-aggregation of low molecular weight compounds.

non-covalent interaction (physical gels). Gels **B** are obtained through monomer cross-linking: a reactive monomer is polymerized in the presence of a 3D structure builder and/or a cross-linker.[14] In gels **C**, the gellant is a low mass organo gelator (LMOG: steroid derivatives, organometallic compounds, surfactants of different nature, *etc.*), usually an organic compound completely soluble in the continuous phase only at high temperature.[15] One characteristic feature of this kind of gel is that the network does not consist in long-polymer chains but simply in the self-association of gelator molecules in a monodimensional fibrillar structure.

The fundamental classification, however, is usually based on the nature of the cross-linkers connecting the network primary units (see Figure 10.5). In this regard, chemical and physical gels are the two main categories that profoundly differ if we consider both their mechanical stability and their ability to be residue-free in view of a restoration application. For instance, chemical gels are characterized by strong cohesive forces (due to the covalent bonds), thus they do not leave residues on a treated substrate. On the other hand, physical gels can be easily shaped, granting a perfect and "homogeneous" interaction with the treated substrate, but they often leave residues that must be cleaned using solvents, which might redisperse the solubilized material within the porous matrix of the substrate.

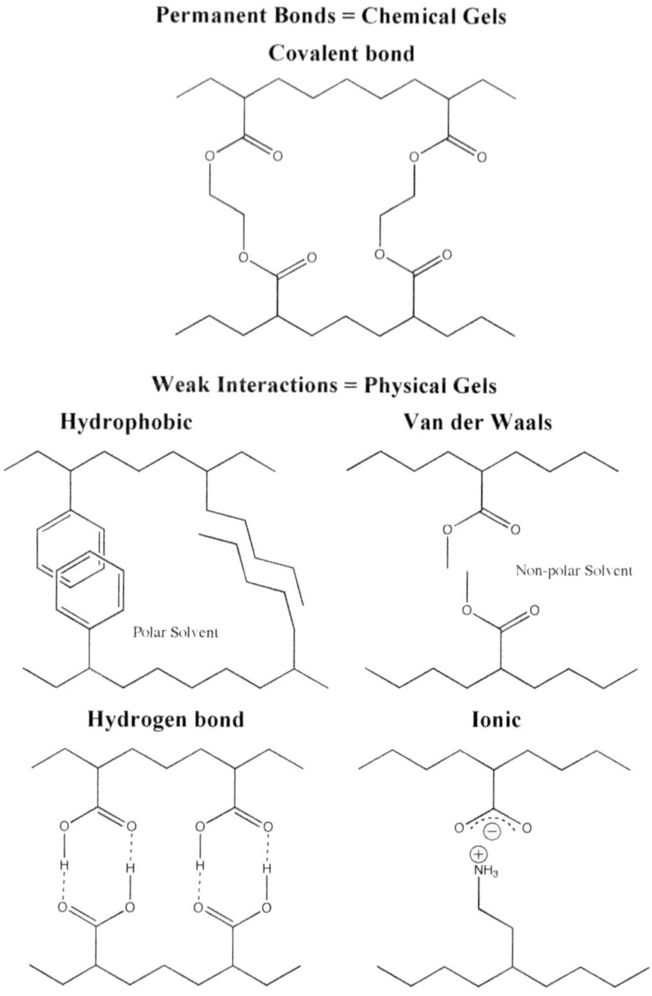

Figure 10.5 Schematic representation of the possible interactions stabilizing chemical and physical gels.

10.2.1 Physical Gels

In physical gels the bonds among the gellant molecules usually involve hydrophobic, electrostatic (ionic), van der Waals interactions or hydrogen bonds. These kinds of interaction have an energy spanning roughly between 1 and 120 kJ mol^{-1}, which is much lower than that of a typical covalent bond, therefore these systems are thermoreversible. This is one of the main properties typical of physical gels: by heating them up to a critical temperature Tg (gelation temperature, whose value depends on the chemical nature of both the gellant and the gelled liquid) the supramolecular structure melts and the single molecules, or the clusters composed of aggregates of molecules, are re-dispersed in the solvent bulk phase. The final liquid system is called a "sol". On cooling to room temperature, physical gels turn to the original condensed system; this process is called the sol–gel transition.

One of the most diffused physical gels is the creamy Italian sugar pudding (the "panna cotta", a famous dessert from Piemonte). In its preparation the sol–gel process plays a fundamental role: first the sugar is added to the whipped cream in the presence of a thickener. Usually the thickener is a natural polymer insoluble in water at room temperature, for example agar-agar (a cellulose derivative) or fish-glue (a protein). As for many other gels the preparation is completed by the solubilization of the gellant (*i.e.* the thickener) by heating the mixture up to 60–70 °C. Then the mixture is slowly cooled to room temperature until a jelly-like system is obtained. In the final stage the liquid whipped cream is supported by the 3D network formed by the polymer molecules that interact through non-covalent interactions (*i.e.* hydrogen bonds or van der Waals interactions that include Keesom, Debye and London forces, see Chapter 4).

In the field of works of art cleaning, an example of a physical gel is the "solvent gel" developed by Wolbers (see Sections 11.1 and 11.2).

10.2.2 Chemical Gels

In a chemical gel the links among the building blocks occur through the formation of covalent bonds that usually involve high energy, ranging between *ca.* 200 and 650 kJ mol^{-1} (roughly 10 times stronger than a hydrogen bond).

In this case, all the polymer chains cross-link through covalent bonds, and the chemical gel can be considered as one single giant molecule. Chemical gels are thermally irreversible systems: a temperature increment above a critical value results in the breaking of the bonds and the system does not produce the original structure by cooling.

One example of chemical gels is the acrylamide–bisacrylamide (ACRY/BIS) hydrogel, used as substrate for the gel-electrophoresis technique for protein separation and analysis. Reformulated hydrogels based on the same monomer and cross-linkers have been used recently as detergent containers in cultural heritage treatments.[16] In particular, their mechanical properties and water release capacity can be tuned by changing the porosity of the hydrogel for applications on water-sensitive artefacts.

Another example of a chemical gel is polyhydroxyethylmethacrylate cross-linked with ethylenglycol dimethacrylate (HEMA/EGDMA), which was the first hydrogel used in the production of soft contact lenses.[17]

A summary of the main differences between physical and chemical gels is reported in Table 10.1, however it is worth noting that these differences become less substantial as the energy involved in the interaction forces for physical gels increases (hydrophobic < van der Waals < hydrogen bond < electrostatic forces).

Table 10.1 Summary of the principal differences between physical and chemical gels.

	Physical gels	*Chemical gels*
Cross-linking	Weak interaction	Covalent bond
Thermal behaviour	Reversible	Irreversible
Solvent dilution	Eventually solubilize	Swell but not solubilize
Shape	Mechanically adjustable	Fixed
Elastic modulus	Intermediate values	High values

10.3 Structural and Mechanical Properties of Gels

The structural and mechanical properties of a gel, as well as its chemical nature, are of fundamental importance in determining its response to stimuli and hence in defining the fields of applicability. In this context, for a particular restoration event, the gel typologies are selected keeping in mind that they must interact with artistic substrates in a certain way to give an expected result, but also considering that the substrates are extremely complex (in composition) and sensitive to different factors.

10.3.1 Structural Properties

In order to characterize a gel from a structural point of view it is important to consider at least three parameters:[18]

1. The polymer volume fraction in the swollen state ($v_{2,s}$);
2. The molecular weight of the polymer chain between two neighbouring cross-linking points (M_c);
3. The characteristic mesh size (ζ) (see Figure 10.6).

These three parameters can be tuned by changing the chemical nature of the monomers/polymer and the solvent, their concentration and the concentration of the linking points.

The polymer volume fraction in the swollen state ($v_{2,s}$) is a measure of the amount of fluid imbibed and retained by the gel. It can be determined from the volume of the dry polymer, V_P, and the gel volume in the swollen state, V_G:

$$v_{2,s} = V_P/V_G = 1/Q \tag{10.1}$$

The reciprocal is known as the volumetric swollen ratio, Q. From a practical point of view, as $v_{2,s}$ decreases the affinity of the polymer for the solvent increases, leading to gels that are more prone to swell. As a result more solvent or detergent can be included in the gel formulation if the binding is weaker. As an example we mention the case of the ACRY/BIS hydrogel with $v_{2,s}$ values of about 0.05 (*i.e.* $Q = 20$), while the standard HEMA/EGDMA (2-hydroxyethyl

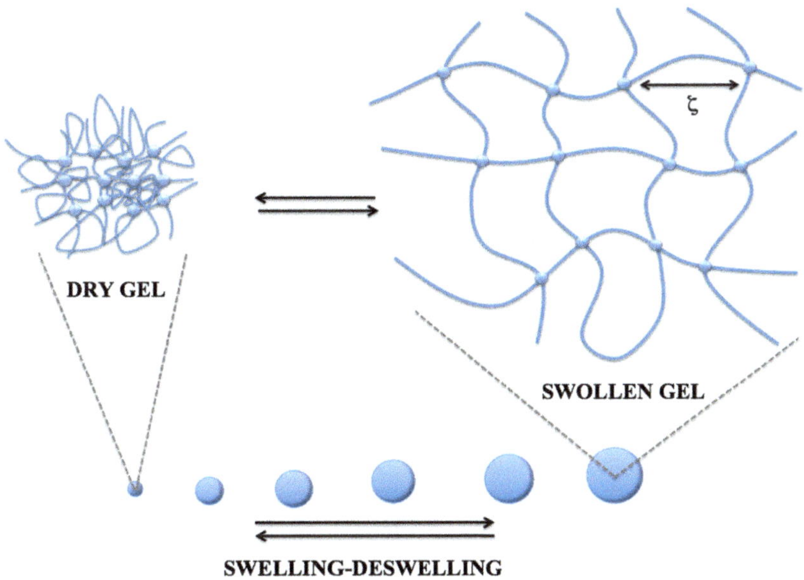

Figure 10.6 Microscopic and macroscopic representation of the swelling–deswelling equilibrium that can be induced by several stimuli: temperature, pH, ionic strength, solvent composition, light, electric field, magnetic field, *etc.* The mesh size, ζ, is shown as defined by the cross-link points, which can be generated by labile interactions or real chemical bonds.

methacrylate–ethylene glycol dimethylacrylate) hydrogels possess $v_{2,s}$ values of about 0.6 (*i.e.* $Q = 1.6$). Given that this property is strictly connected to the molecular affinity between the monomer and the solvent, an easy way to tune it is to introduce a fraction of a different monomer (co-monomer) in the polymerization batch.

The $v_{2,s}$ can be determined by using equilibrium swelling experiments. A parameter strictly linked to $v_{2,s}$ usually found in the gel literature is the equilibrium solvent content (ESC), which in the case of hydrogels is the equilibrium water content (EWC). The ESC can be calculated by knowing the weight of the dry gel, W_P, and the weight of the swollen gel, W_G:

$$ESC = \frac{W_G - W_P}{W_G} \cdot 100 \tag{10.2}$$

This parameter gives important indications of the gel state, and is useful when checking the ageing and performance during swelling–dehydration–swelling (s-d-s) cycles which mimic the application procedure. Furthermore, the ESC can be evaluated more easily than the $v_{2,s}$. The ESC (or $v_{2,s}^{-1}$) is proportional to the gel porosity, so when differences are observed during an s-d-s cycle these changes (which are usually negative because the ESC decreases after dehydration), they can be directly connected to the volume of pores that collapsed irreversibly during the dehydration step.

Porosity is one of the most important characteristics of gels because it can strongly affect some other important features of the system such as the solvent retention power, and the optical properties of the system (transparency, translucency or opacity). The terms porous or macroporous are defined in different ways and sometimes are used as synonyms of other terms such as microporous and molecular porosity. Here, according to the IUPAC recommendations, pores will be classified according to their dimension in three different classes, even if this old definition is sometimes misleading:[19]

1. Micropores: pore dimension up to 2 nm;
2. Mesopores: the dimension lies between 2 and 50 nm;
3. Macropores: pore dimension greater than 50 nm.

Thus micropores are indeed pores of nanometre size, mesopores account for tenths of nanometres, and macropores cover all the rest. These limits are to some extent arbitrary and have been imported mainly from sorption isotherm terminology where the pore filling mechanism evolves gradually, starting from the primary physisorption process (or micropore filling), passing through physisorption in mesopores (monolayer–multilayer adsorption) and ending with macropore filling (or capillary condensation). The sorption isotherm technique is the classical method used for porosity quantification, and it has been fully described elsewhere.[19]

As we will see in Section 10.5, small angle scattering techniques show that when a gel turns into a xerogel a loss of the micro- and mesoporosity occurs. In contrast, macroporosity is observed with scanning electron microscopy (SEM).

The average molecular weight between two consecutive cross-links, M_c, defines the degree of cross-linking of the gel network, X:

$$X = \frac{M_0}{2M_c} \tag{10.3}$$

where M_o is the molecular weight of the repeating units in the polymer chains.

Knowing the values of M_c (also referred to as the cross-linking density), the network pore size can be determined by calculating the end-to-end distance of the swollen polymer chains between two consecutive cross-linking points. In the case of a chemical gel the cross-linking point is fixed, and M_c cannot vary with the swelling degree as occurs for the mesh size value (*i.e.* the microporosity) because it is strictly connected to the chemical topology.

The network mesh size (ζ) is the actual distance between two consecutive cross-linking points, as illustrated in Figure 10.6. Moreover, it intrinsically provides a measure of the average microporosity of the network, which can be directly determined by a small angle scattering technique. In chemical gels this quantity depends on ESC if M_c does not change. At this point it is important to stress that, due to the random nature of the polymerization process itself, M_c and ζ have a certain statistical distribution around average values of the 3D gel network. From now on we will refer to these average values by using simply M_c and ζ.

The $v_{2,s}$, M_c and ζ are inter-related and can be either estimated theoretically or evaluated through a variety of experimental techniques. The theoretical assessment derives from complicated statistical calculations as the Equilibrium Swelling Theory and the Rubber Elasticity Theory.

10.3.2 Mechanical Properties

The mechanical properties of gels play a primary role in their use and handling. From a mechanical point of view, gels are viscoelastic materials, that is they possess a mechanical behaviour typical of the liquid state (purely viscous materials, see below) intimately coexisting with the pure elastic behaviour typical of solids (see below). As we will see in detail, the elastic character of a gel always prevails over the viscous behaviour. Therefore in the case of gels two characteristic parameters should be considered to define their performance in various technological applications: the elastic modulus and the viscosity.

The elastic (or Young's) modulus E is defined as the proportionality coefficient between the applied tensile stress T and the relative elongation L_R (or tensile strain):

$$E = \frac{T}{L_R} = \frac{F/A}{\Delta L/L_0} \tag{10.4}$$

where F is the applied force perpendicular to the surface A, ΔL is the change in the sample length and L_0 is the length at rest.

The region where this relation holds is called the elastic domain and a straight line is obtained if tensile stress is plotted against the deformation (see Figure 10.7). At some relative elongation, which is characteristic of the gel considered, this relation fails (*i.e.* at the elastic limit), and the material enters into the plastic region. Before reaching the elastic threshold, the process is reversible and the original dimensions are restored as soon as the load or stress is removed. On the contrary, beyond the elastic limit the shape of the sample is permanently modified. Eventually the material will break under a certain stress.

The values of the elastic modulus for various materials are reported in Table 10.2. In general, the elastic modulus increases by increasing the cross-linking density (*i.e.* reducing M_c) and decreases as the gel swelling increases (*i.e.* enlarging the mesh size). It has been recently shown that in order to impart peelable properties to a polyvinyl alcohol (PVA)–borax formulation the threshold value of 200 Pa must be met in the elastic modulus of these systems (see also Section 10.5).[20] Peelable systems are particularly desirable in Cultural Heritage conservation because they can be easily removed (peeled) from artistic substrates, after their cleaning action, without leaving residues.

It is important to note that, for anisotropic materials, the value of E may depend on the direction of the external force. In the case of gels, which are usually isotropic materials, the value of E is constant in all directions.

Viscosity is a parameter that indicates the resistance of fluids to external mechanical perturbations.[24] Once a force (stress) is applied to a fluid, it starts

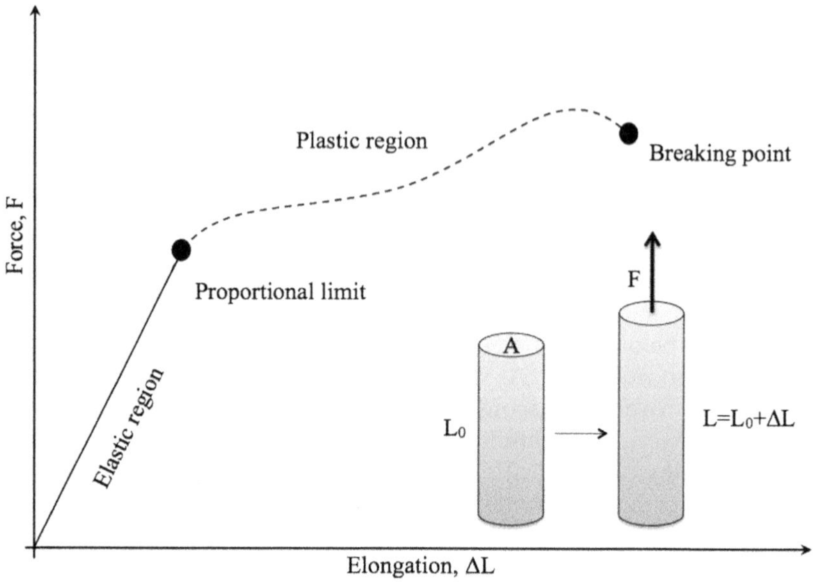

Figure 10.7 General response of a material to tensile stress. The inset on the right shows the experimental set up applied for the *E* measurements.

Table 10.2 Elastic modulus for various materials and gels.[21–23]

Material	*Elastic Modulus* (10^9 *N* m^{-2} *or GPa*)
Diamond	1220
Nylon	2–4
Polyethylene terephthalate (PET)	2–2.7
Glass	50–90
Aluminium	69
Concrete	30
PMMA	3
HEMA/EGDMA	0.3–0.5 MPa[22]
ACRY/BIS	0.2–40 kPa (2–10% BIS)[23]

moving. The entity of this displacement, in addition to pH and temperature, mainly depends both on the magnitude of the applied stress, on the nature of the fluid, and particularly on the molecular interactions occurring among the components of the system: the stronger and more numerous these interactions, the higher the viscosity.

In principle, by tuning the nature of the monomers and the amount of cross-linking it is possible to tailor the viscoelastic properties of the final gel formulation to achieve peelable gels, high stress resistant gels (gels that can tolerate several application cycles), highly deformable gels (to guarantee the prefect interaction even with very rough substrates), *etc.* All these aspects will be described in Chapter 11.

10.3.3 Viscoelastic Materials

Gels present marked viscoelastic properties that determine and characterize the response of the material to external mechanical stimuli.[25] In the case of conservation science treatments, the mechanical stimuli are mainly due to the gel handling before, during and after the application. Hence, it is extremely important to have a clear definition of the meaning of the term viscosity.

If we have two surfaces separated by a fluid (*e.g.* a gel as shown in Figure 10.8 by the green area) and we apply a force F parallel to the free surface, we define the shear stress σ and the shear strain γ as:

$$\sigma = \frac{F}{A_0} \tag{10.5}$$

$$\gamma = \frac{\Delta x}{z_0} \tag{10.6}$$

where A_0 is the area of the mobile surface, Δx is the fluid displacement in the direction parallel to the force (*e.g.* the x direction) and z_0 is the distance between the mobile and the fixed parallel surfaces.

A constant shear stress must be applied to the mobile surface in order to move it at a constant rate u, whose value depends on the position on the z axis: $u = u(z)$. A decrease in the modulus of the fluid rate is observed moving toward the fixed surface, as shown in Figure 10.8, and we define the shear rate as:

$$\dot{\gamma} = \frac{du}{dz} \tag{10.7}$$

According to Newton's law, the proportionality constant η between the shear stress and the shear rate is defined as the dynamic viscosity, whose dimension is pressure × time (usually Pa·s):

$$\eta = \frac{\sigma}{\dot{\gamma}} \tag{10.8}$$

If we plot the viscosity value as a function of the shear stress, the curve is usually referred to as the flow curve. The trend of the flow curve depends on the fluid under investigation. The previous equation means that, for the same

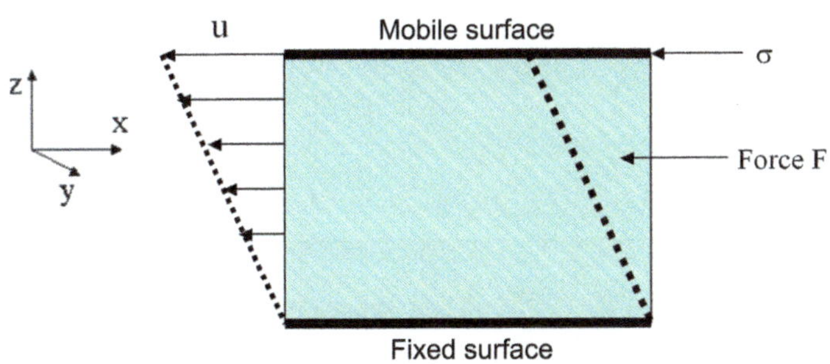

Figure 10.8 A rate u gradient produced when a fluid (green area) is sheared.

applied shear stress in high viscosity fluids, the internal layers must be slower than in a low viscosity fluid.

From the η value the kinematic viscosity v is calculated as:

$$v = \frac{\eta}{d} \tag{10.9}$$

where d is the density of the fluid at the given temperature; v is commonly expressed in Stokes (1 Stoke $= 1\,m^2\,s^{-1}$).

If the dynamic viscosity value is constant over the entire shear stress range explored (see Figure 10.9A), the material behaves as a Newtonian fluid, and follows Newton's law:

$$\sigma = \eta\dot{\gamma} \tag{10.10}$$

Examples of this class of liquids include water and all the solvents commonly used for the cleaning of painted surfaces. In Table 10.3 some values of η for a series of Newtonian fluids are reported.

For non-Newtonian fluids, the viscosity is strictly dependent on the applied shear stress σ. Gels usually belong to this category of material.

Non-Newtonian fluids can be divided into two different classes: shear thinning and shear thickening materials. In systems that exhibit shear thinning behaviour, the viscosity decreases with an increase of σ (see Figure 10.9B). All the materials characterized by shear thinning are also called pseudoplastics.

This behaviour is typical of polymer solutions and systems such as paints, blood, ketchup and shifting sands. In particular, the reason why shifting

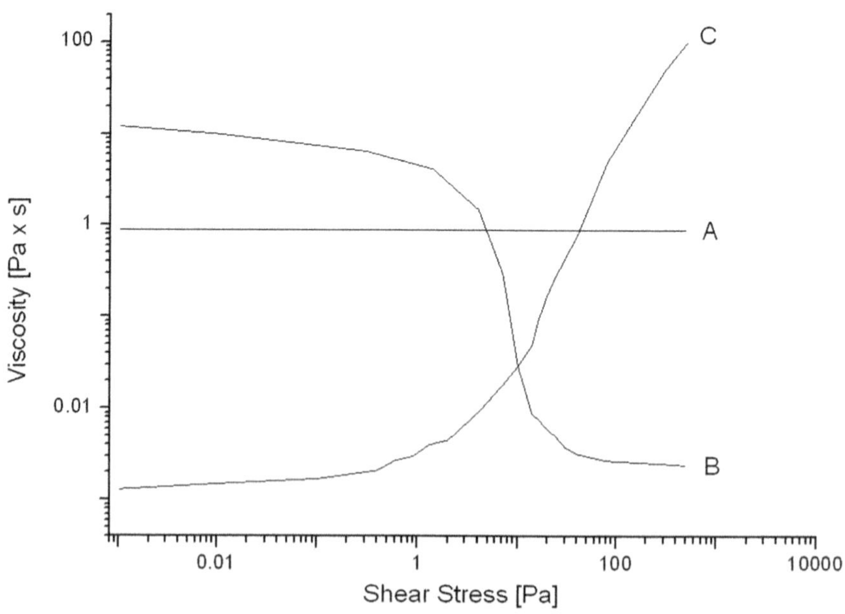

Figure 10.9 Flow curves of generic Newtonian (A), shear thinning (B) and shear thickening (C) fluids.

Table 10.3 Viscosity values for some Newtonian fluids.[26]

Fluid	*Dynamic viscosity [Pa × s] × 10³*
Water	1.0
Methyl ethyl ketone	0.43
Acetone	0.36
Isopropyl alcohol	2.4
Toluene	0.59
Microemulsion used for the cleaning of painted surface[27]	0.84

sands are so dangerous is because they are a shear thinning system. Once a person falls into shifting sands and begins to move rapidly, this perturbation produces a decrease of the viscosity and the person sinks into the sand. Similarly, when a bottle of ketchup is inverted, the fluid does not pour. But once it is shaken, the ketchup sauce rapidly flows out of the bottle (in fact the applied mechanical stress produces a decrease in the η value), although maintaining its shape once poured onto the food surface.

Instead, in shear thickening materials the viscosity increases upon an increment in the applied shear stress (see Figure 10.9C). For example, this behaviour is typical of corn-starch aqueous solutions and Silly Putty, a toy made from a synthetic rubber obtained from silicone oil and boric acid.

Materials can be also classified according to their capacity to recover their initial shape once the applied stress that induces the deformation is removed. There are three different classes:

- Elastic materials completely recover their shape when the perturbation is removed (mechanical memory effect). A purely elastic substance is modelled by an ideal spring (a purely elastic device), whose behaviour is quantitatively described by Hooke's law, according to which the elongation is proportional to the applied force through the spring constant.

  When any form of strain γ is applied on the surface of a purely elastic solid or fluid, the shear stress responds proportionally to the strain, but in this case the response of the spring can be described as a function of the shear modulus, G_0:

  $$\sigma = G_0\gamma \tag{10.11}$$

  where σ and γ are the shear stress and the strain respectively. The shear modulus G_0 indicates how elastic is the solid or fluid, and it corresponds to the strength of the spring with which the material can be ideally represented.
- Viscous materials can be represented as a cup filled with a liquid of a given viscosity, η. The top of the material is in contact with a piston. When a stress is applied to these materials the response of the sample is well described by the Newton's law [see eqn (10.10)]. Once the perturbation is

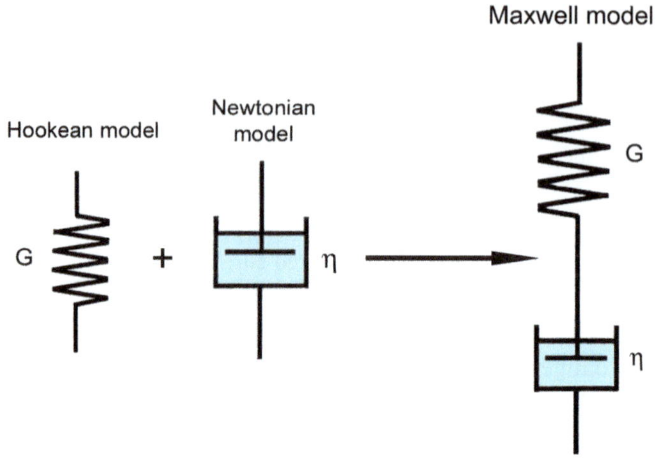

Figure 10.10 Mechanical equivalents for the definition of various kinds of materials: elastic spring for pure elastic materials and Newtonian dashcup for pure viscous materials. From their combination the mechanical equivalent of a Maxwell fluid is obtained.

removed, the system does not recover its initial shape, and all information about the original state is lost (*i.e.* there is no mechanical memory effect).

- Viscoelastic materials have an intermediate behaviour. To better understand what viscoelasticity means, the viscoelastic behaviour of fluids will be discussed by referring to combinations of Newtonian dashpots and Hookean springs as indicated in Figure 10.10 for these materials, also known as Maxwell fluids.[24]

Quantitatively, the viscoelastic behaviour is described by two different parameters: the storage modulus G' and the loss modulus G''. The storage modulus takes into account the energy stored by the system during the perturbation and reflects the elastic nature of the material. The loss modulus, also called the viscous modulus, indicates the energy dissipated by the system, and reflects the viscous nature of the material. A purely elastic material exhibits a non-zero elastic modulus and a viscous modulus $G'' = 0$. In contrast, a purely viscous material has a zero elastic modulus and a non-zero viscous modulus.

10.4 Uses and Applications

Uses and applications of gel formulations move on a parallel track with the high variety of their tunable properties. Apart from the use in Cultural Heritage conservation, which is the subject of the present book, gels in general and hydrogels in particular play a significant role in many other fields ranging from medicine and pharmaceuticals to the food industry. Because of their high water content and their softness they mimic natural living tissues better than any other class of synthetic biomaterials, and provide high biocompatibility. For these reasons hydrogels are used as materials for contact lenses and artificial skin, scaffolds for tissue engineering, membranes for biosensors, linings for artificial hearts, pharmaceutical wound dressing and drug delivery devices, and even in less relevant applications including diet aids, moisture traps (soil moisture maintenance), disposable nappies (super sorbent polymers) and sanitary towels, just to cite some examples.

In conservation of Cultural Heritage, gels are mainly used as vehicles for carrying the active cleaning components to the surface to be restored.[28] They can replace cotton swabs or poultices of cellulose pulp if better retention is required. Moreover, organic solvents tend to evaporate and wet porous matrices very quickly, coming into contact with materials that are not supposed to be dissolved in the conservation treatment. For these reasons cleaning operations that involve solvents are very difficult to control, and are dangerous for both the conservator and the artefact. By loading the solvent into a gel, the rate of solvent evaporation is lowered and its penetration into the artistic substrate is controlled. This imparts selectivity in removing undesired covering layers, thus providing a versatile and selective cleaning device, which can also respond to external stimuli.

Recently developed gel systems possess pH-responsive rheological behaviour, resulting from simply adding weak acid[29–31] or an intrinsic magnetic response, that can help in the removal step.[32,33] The use of chemical gels instead of standard physical gels can also avoid the contamination of the artistic substrate during the application or the removal steps, owing to the fraction of gellant that can remain on the treated surface.[17]

10.5 Characterization Techniques

As previously described, the macroscopic recognition of a gel state is somewhat ambiguous, and various classifications have been proposed based on the morphology, consistency, or upon chemical considerations.[5,34]

Rheology studies the mechanical properties of a system by measuring the deformation of the material after the application of a force.[24] This investigation is experimentally conducted by applying a controlled force to the sample, and measuring the resulting deformation. The detailed knowledge of the response to the applied forces is of great value in predicting how the gel formulation will behave during its application, including restoration interventions.

The quantitative study of the viscoelastic behaviour of a material can be done through a frequency sweep measurement. In these tests an oscillating perturbation is applied to the sample, and the storage modulus $G'(\omega)$ and the loss modulus $G''(\omega)$ values (see Section 10.2.3) are plotted as a function of the perturbation frequency ω (the common frequency range is between 10^{-3} and 100 Hz, while the amplitude of the perturbation is constant). Figure 10.11 shows the frequency sweep for an aqueous polymeric dispersion made up of PVA and borax (cross-linker), and containing 7 w/w% of cyclohexanone. For high frequency values, G' tends to an asymptotic value that can be

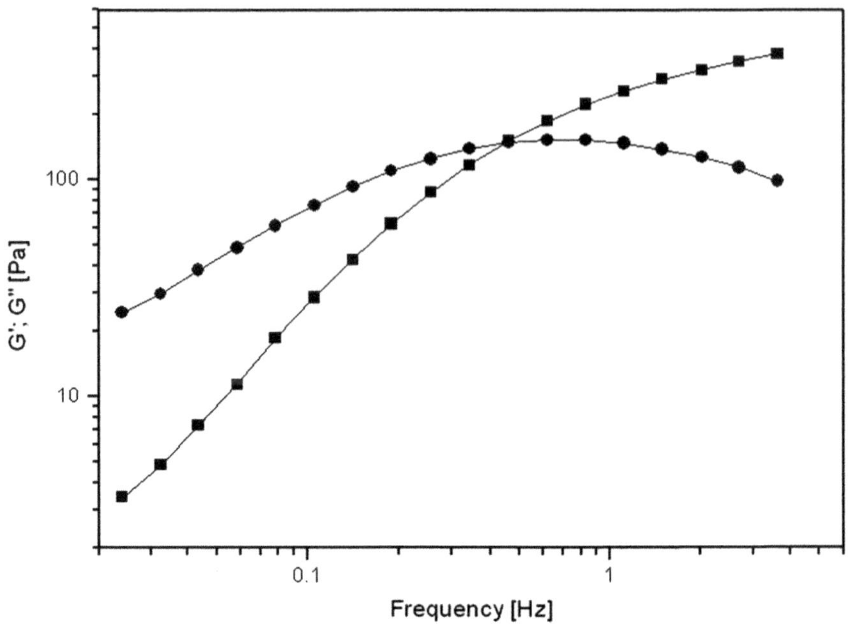

Figure 10.11 Frequency sweep of a polymeric dispersion constituted by PVA (2 w/w%)/borax (0.4 w/w%) and water containing 7% w/w of *cyclohexanone*. Storage modulus G' (■), loss modulus G'' (●).

approximated to the system shear modulus G_0, which is correlated to the density of the cross-linking between the PVA molecules. The value of G_0 has a crucial applicative importance; as a matter of fact, the correlation between the analytical rheological data and the macroscopic behaviour of the system, exhibited during cleaning tests on artistic surfaces, showed that a minimum value of 200 Pa is needed in order to grant easy peeling of the system from treated surfaces.[20] The importance of this formulation will be further highlighted in Chapter 11 when its application to Cultural Heritage conservation treatments will be discussed.

Although all gels are viscoelastic materials, from a strictly mechanical point of view a system can be rigorously considered a "gel" only if $G' \gg G''$ over a wide range of frequencies.[35] For this reason, in the case of PVA–borax systems the term "highly viscous polymeric dispersions (HVPDs)" is preferred, rather than "gels".

An example of the frequency sweep of a polyacrylic acid-based hydrogel is reported in Figure 10.12. This system, composed of polyacrylic acid (1 w/w%), 1-propanol (20 w/w%) and water, belongs to the class of solvent gels, commonly used for the cleaning of painted surfaces.[36]

The presence of a gel point is another way to characterize a gel. The gel point indicates the formation of a semisolid (gelled) system from the liquid state. From the thermal point of view the gel point is the temperature Tg at which the gel-to-sol transition occurs. Other variables can be used to define the gel point,

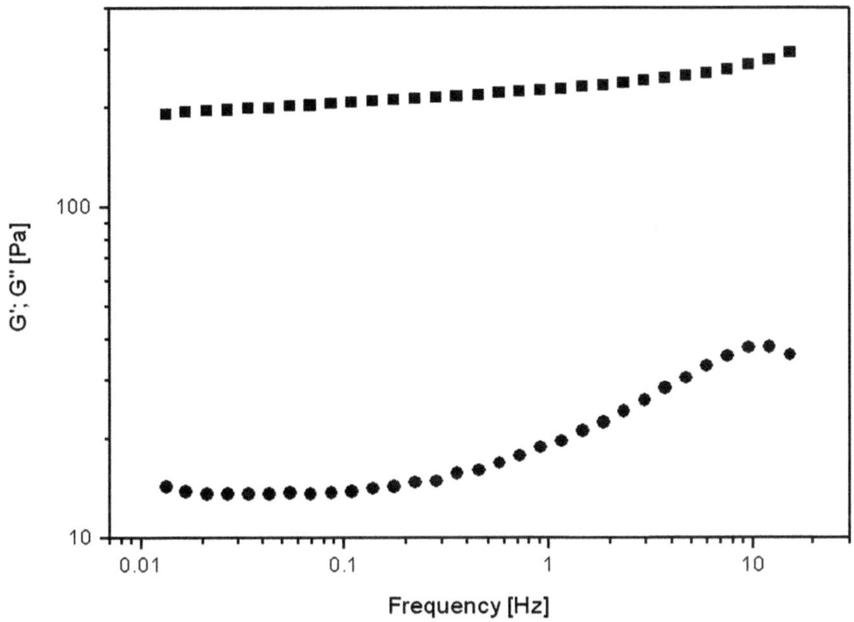

Figure 10.12 Frequency sweep of a gel constituted by polyacrylic acid (1 w/w%), 1-propanol (20 w/w%) and water. Storage modulus G' (■), loss modulus G'' (●).

e.g. the concentration of the gelator, the polymerization time, *etc.* The gel point can be determined in different ways, and some examples are reported below.

The simplest methods are the "tilted test tube method" and the "falling ball method" (see Figure 10.13). In the tilted test tube method, the solution or the gel is observed while the tube is rotated during a temperature change, or during the addition of gelator at constant temperature. In the falling ball method a ball is located inside the fluid. The gel point of the solution or the melting temperature of the gel is defined as the point at which the system respectively stops or starts to flow, or when the ball either begins to fall across the material or is stopped within the system.

The temperature range of the gel formation (or melting) can be alternatively monitored by differential scanning calorimetry experiments (thermal method) and through mechanical methods. The latter are based on the determination of the viscosity or of the storage modulus at a given temperature and frequency, as a function of time. The instant or the temperature at which the storage modulus increases abruptly from low values indicates the gel point, as shown in Figure 10.14. Another way is to determine the point at which the storage and the loss moduli cross each other as a function of the temperature or the time.

The size of the network pores (named also mesh size or correlation length, ζ) is a very important parameter for gels, especially when they are used as smart containers for liquids. The mesh size, which is related to the porosity at the nanometre scale, can be determined theoretically from the cross-linking density

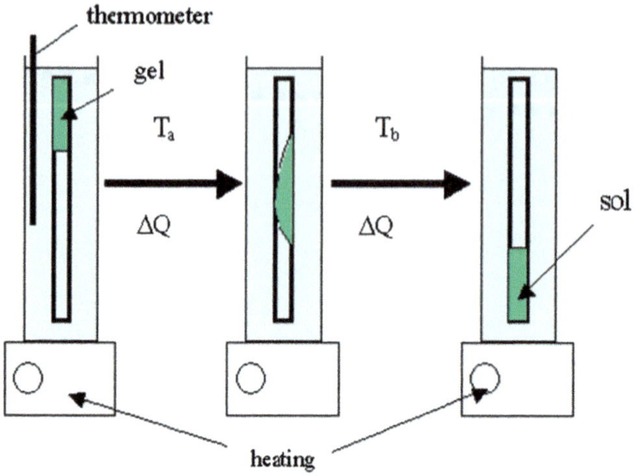

Figure 10.13 Graphical illustration of the tilted test tube method.

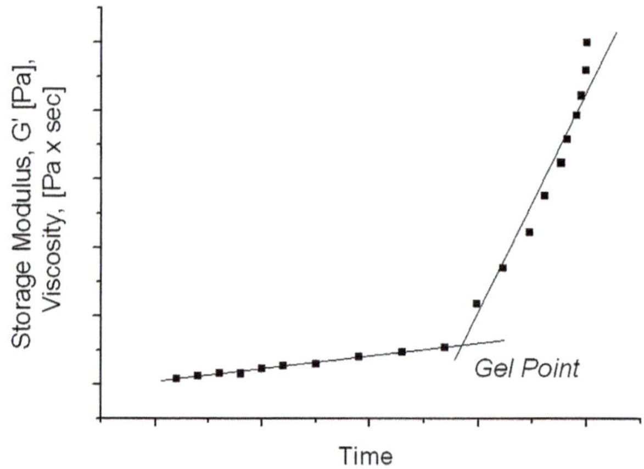

Figure 10.14 Gel point determination for a generic gel following the G' value or the viscosity γ as a function of time.

and the molecular weight of the chain segments that separate two adjacent cross-links. It can be measured experimentally through small angle X-ray scattering (SAXS) and small angle neutron scattering (SANS).

It is important to recall that, because electron microscopy requires dried samples (xerogels), this technique can only reveal meso- and macropores, as the micropores usually collapse during standard drying procedures.

However, in a small angle scattering experiment a probe (usually visible light, X-rays or neutrons) interacts with the sample and produces an angular distribution of the scattering intensity, which depends on the inhomogeneities at the nanometre scale (*i.e.* 1–100 nm) related to a change in the electronic density of the material (for X-rays and visible light) or to the isotopic composition (in the case of neutrons). Such changes are directly related to the system porosity in the specific dimensional range investigated by that probe. The information on the porosity and surface area is thus contained in the angular distribution of the scattered intensity.

Usually in a scattering experiment the scattering angle is replaced by the scattering vector, q:

$$q = \frac{4\pi}{\lambda} \sin\theta \tag{10.12}$$

where λ is the probe wavelength and θ is half the scattering angle (2θ) defined by the incident and the scattered beams.

The details of a small angle scattering experiment are reported in Figure 10.15, where the incident and scattered wavevectors (k_i, k_s) and the resultant scattering vector (q) are shown.

The apparatus for both SAXS and SANS requires a source of radiation, a monochromator, a collimation system, and a sample position and detection system.[37] Scattering techniques are non-destructive and have particular advantages in the characterization of the surface and porous properties of materials containing either closed or molecule-sized pores. In fact, the outgassing pre-treatment may result in irreversible changes in the microstructure as in the case of swollen gels, where the removal of the solvent to perform standard electron microscopy is known to irreversibly alter the sample.[16] In the case of scattering techniques the sample preparation is minimal, and usually limited to the transfer of the sample in a holder, with no need to treat the sample under vacuum, or metallize it as in the case of electron microscopy.

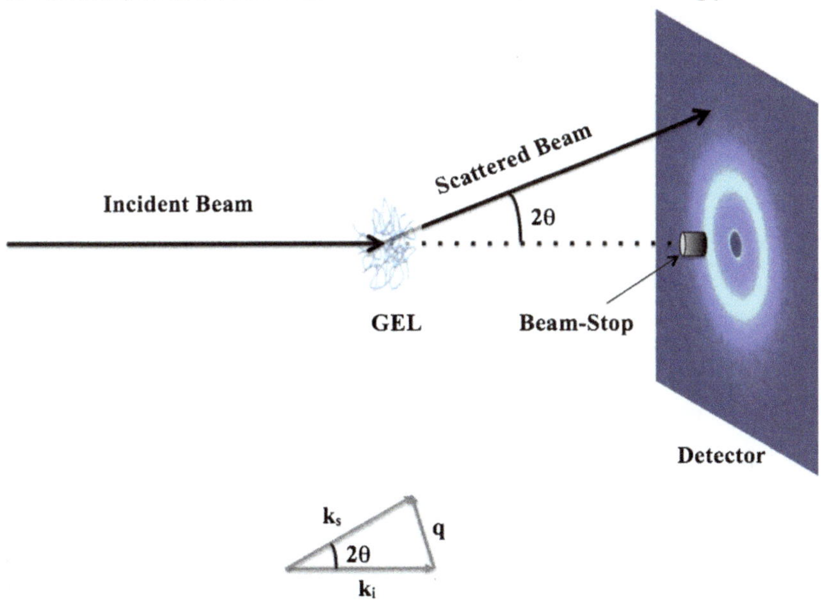

Figure 10.15 Schematic representation of a small angle scattering experiment. The scattering vector is also reported as the difference between the scattered and incident wave vectors associated with the probe.

Figure 10.16A shows the comparison of SAXS curves obtained for the ACRY/BIS hydrogel and the corresponding xerogel, highlighting the differences in the microstructure. In particular, the xerogel curve shows a defined q^{-4} power-law trend, which indicates the complete collapse of the microporosity proper of the hydrogel. In this last case, the scattering curve needs to be analysed according to specific fitting models to retrieve the structural parameters defining the hydrogel that are the average mesh and inhomogeneity sizes (see below).

In general for a given polymer gel the scattering intensity distribution, $I(q)$, can be written as:

$$I(q) = I_{sol}(q) + I_{ex}(q) + I_{bkg} \quad (10.13)$$

where I_{bkg} is the instrumental background, $I_{sol}(q)$ is the contribution of the solution term and $I_{ex}(q)$ is the excess term counting for the deviations between a simple polymer solution and the chemical cross-linked equivalent. For a swollen gel, the polymer concentration is typically in the semi-dilute regime and, in this case, $I_{sol}(q)$ has the functional form given by the Ornstein-Zernike equation, that is a lorentzian function:[16]

$$I_{sol}(q) = \frac{I_{lor}(0)}{1 + \zeta^2 q^2} \quad (10.14)$$

where $I_{lor}(0)$ is the intensity for the lorentzian part at $q = 0$ and ζ is the average mesh size of the polymer network as already defined. Several functions have

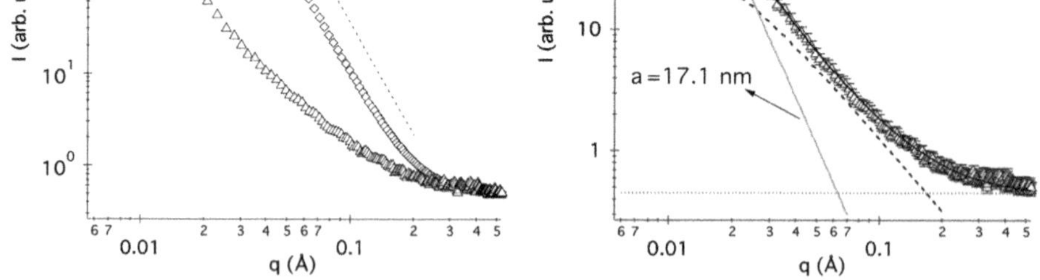

Figure 10.16 A) Comparison of SAXS curve obtained from a hydrogel (open triangles) and the corresponding xerogels. A q^{-4} power-law (dashed line) is reported for comparison. B) Fitting of the hydrogel curve according to the Debye-Bueche approach (bold continuous line). The contribution of the lorentzian (dashed line), excess (light grey continuous line) and background (grey dotted line) terms according to equations 10.13–10.15 are also shown.[16]

been proposed in literature for $I_{ex}(q)$, including another lorentzian function, a stretched exponential function and a Debye-Bueche function (see Further Suggested Reading). In recent papers the Debye-Bueche function has been found to be a good approximation for the excess scattering in the case of ACRY/BIS hydrogel[16] and the magnetically doped equivalent,[33] well describing the additional fluctuation and solid-like inhomogeneities:

$$I_{ex}(q) = \frac{I_{ex}(0)}{(1 + a^2q^2)^2} \tag{10.15}$$

where $I_{ex}(0)$ is the intensity for the excess term at $q=0$ and a is the average dimension characterizing the inhomogeneities in the gel. This extra term is introduced to describe additional fluctuations and/or solid-like inhomogeneities imposed by the presence of the cross-linker.

For example, Figure 10.16B shows the three contributions behind the scattering intensity distribution in the case of the ACRY/BIS hydrogel.[16] In this particular case, the fitting results disclose that the 3D network has an average mesh size of 7.5 nm and an average dimension of the inhomogeneities which is more than double (17.1 nm).

A thorough discussion of the application of scattering techniques to characterization of the various gel systems is beyond the scope of this chapter. The interested reader can refer to the vast literature on this subject (see Further Suggested Reading).

Further Suggested Reading

On Equilibrium Swelling Theory and Rubber Elasticity Theory:

A. M. Lowman and N. A. Peppas, Hydrogels, in *Encyclopedia of Controlled Drug Delivery*, John Wiley & Sons, Inc., New York, 2nd edn, 1999, vol. 1, pp. 397–418 and references therein.

On polymerization techniques:

J. W. Nicholson, in *The Chemistry of Polymers*, RSC Publishing, Cambridge, 4th edn, ch. 2, 2012.

On gels and gelation:

C. J. Brinker and G. W. Scherer, in *Sol–Gel Science*, Academic Press, inc., San Diego, 1990, ch. 5, pp. 303–356.

On Dynamic Mechanical Analysis:

K. P. Menard, *Dynamic Mechanical Analysis: A Practical Introduction.* CRC Press, Boca Raton, (1999).

Background on SANS with specific examples on polymers:

B. Hammouda, *Probing nanoscale structures – The SANS toolbox*, 2009, available online at http:www.ncnr.nist.gov/staff/hammouda/the sans toolbox.pdf, accessed May 2012.
J. S. Higgins and H. C. Benoit, *Polymers and Neutron Scattering*, Clarendon Press, Oxford, 1994.

On applications of scattering techniques:

S. Candau, J. Bastide and M. Delsanti, Structural, elastic, and dynamic properties of swollen polymer networks, *Adv. Polym. Sci.*, 1982, **44**, 27.
J. S. Higgins and H. C. Benoit, *Polymers and Neutron Scattering*, Clarendon Press, Oxford, 1994.
F. Horkay, A. M. Hecht, S. Mallam, E. Geissler and A. R. Rennie, Macroscopic and microscopic thermodynamic observations in swollen poly(vinyl acetate) networks, *Macromolecules*, 1991, **24**, 2896.
S. Mallam, F. Horkay, A. M. Hecht and E. Geissler, Scattering and swelling properties of inhomogeneous polyacrylamide gels, *Macromolecules*, 1989, **22**, 3356.
M. Shibayama, in *Soft Matter Characterization*, ed. R. Borsali and R. Pecora, Springer, Netherlands, 2008, pp. 783–832.
W. Wu, M. Shibayama, S. Roy, H. Kurokawa, L. D. Coyen, S. Nomura and R. S. Stein, Physical gels of aqueous poly(vinyl alcohol) solutions: a small-angle neutron-scattering study, *Macromolecules*, 1990, **23**, 2245.

Review and Questions

The main points discussed in the chapter can be summarized as follows:

- The definition of the term "gel" is provided and discussed.
- The difference between chemical and physical gels is detailed; the use of these two classes of materials has important consequences for the cleaning of artistic substrates.
- The main structural and mechanical properties of gels are presented, and examples of their applications are provided.
- Viscoelastic materials are introduced, in particular highly viscous polymeric dispersions (HVPDs), which have important applications for the cleaning of artistic substrates.
- Both standard and advanced characterization techniques are briefly introduced.

Questions

1) Define a gel.
2) Define the difference between a physical and a chemical gel.

3) Relate the volumetric swollen ratio Q to the equilibrium solvent content, assuming that the densities of the solvent (ρ_G) and of the polymer (ρ_P) are known.
4) Describe a Newtonian fluid. How many classes of non-Newtonian fluids do you remember?
5) Describe qualitatively the most important features of a viscoelastic substance.

Answers

1) A gel is a material comprising a 3D solid network immersed in a fluid. The network must be percolative, *i.e.* only one single aggregate can be recognized. The fluid is usually a liquid: water for a hydrogel, and an organic solvent for an organogel.
2) The main difference between physical and chemical gels is the nature of the crosslinks that stabilize the 3D network. Physical gels are stabilized by weak interaction forces, while in a chemical gel the polymer chains are connected by covalent bonds. For this reason, chemical gels have higher elastic and storage moduli, they swell in the solvent but do not solubilize in it, they are not thermally reversible, and possess a defined shape.
3) Given that Q is defined as $\boldsymbol{V_G}/\boldsymbol{V_P}$ and ESC is defined as $(\boldsymbol{W_G} - \boldsymbol{W_P})/W_P$ 100, it is straightforward to show that:

$$ESC = \frac{W_G - W_P}{W_G}100 = 100 - \frac{100 W_P}{W_G}$$

$$= 100 - 100\frac{\rho_P}{\rho_G}\frac{V_P}{V_G}$$

$$= 100 - 100\frac{\rho_P}{\rho_G}\frac{1}{Q}$$

or

$$\frac{ESC}{100} = 1 - \frac{\rho_P}{\rho_G}\frac{1}{Q}$$

$$1 - \frac{ESC}{100} = \frac{\rho_P}{\rho_G}\frac{1}{Q}$$

$$Q = \frac{\rho_P}{\rho_G}\frac{1}{1 - \frac{ESC}{100}}$$

4) In a Newtonian fluid the viscosity is constant in a broad range of applied shear stress. Two classes of non-Newtonian fluids were treated in this chapter: shear thickening substances show an increase in the viscosity value as the applied shear stress increases. In contrast, shear thinning

materials exhibit a decrease in the viscosity value when the applied shear stress increases.

5) From a mechanical point of view, a viscoelastic body can be considered as the combination of a purely viscous (*i.e.* a liquid, such as water) and a purely elastic material (*i.e.* a solid, such as a crystal). Upon external stress, a viscoelastic body undergoes a deformation whose entity mainly depends on its physico-chemical properties. Once the perturbation is removed, the recovery of the original 3D conformation, driven by its elasticity, is not completely achieved owing to its viscous character.

References

1. J. Alemán, A. V. Chadwick, J. He, M. Hess, K. Horie, R. G. Jones, P. Kratochvíl, I. Meisel, I. Mita, G. Moad, S. Penczek and R. F. T. Stepto, *Pure Appl. Chem.*, 2007, **79**, 1801.
2. N. C. McCreesh, A. P. Gize and A. R. David, *J. Archaeol. Sci.*, 2011, **38**, 3432.
3. D. J. Lloyd, in *Colloid Chemistry*, ed. J. Alexander, The Chemical Catalogue Company, New York, 1926, vol. 1, p. 767.
4. T. Graham, *Phil. Trans. Roy. Soc. London*, 1861, **151**, 183.
5. P. H. Hermans, in *Colloid Science*, ed. H. R. Kruyt, Elsevier, Amsterdam, 1949, vol. 2, pp. 483–520.
6. H. K. Henish, in *Crystal Growth in Gels*, Pennsylvania State University Press, University Park, 1970, pp. 41–60.
7. K. Almdal, J. Dyre, S. Hvidt and O. Kramer, *Polymer Gels Netw.*, 1993, **1**, 5.
8. P. J. Flory, in *Principles of Polymer Chemistry*, Cornell University Press, New York, 1953, ch. 9.
9. C. J. Brinker and G. W. Scherer, in *Sol–Gel Science*, Academic Press, San Diego, 1990, pp. 2–11.
10. http://avogadro.openmolecules.net/wiki/ accessed on 31 May 2012.
11. R. Zallen, *The Physics of Amorphous Solids*, Wiley, New York, 1983, ch. 4.
12. D. Stauffer, A. Coniglio and M. Adam, *Adv. Polymer Sci.*, 1982, **44**, 105.
13. J. M. Guenet, *Thermoreversible Gelation of Polymers and Biopolymers*, Academic Press, San Diego, 1992.
14. M. Y. Kizilay and O. Okay, *Polymer*, 2004, **45**, 2567.
15. P. Terech and R. G. Weiss, *Chem. Rev.*, 1997, **97**, 3133.
16. G. Pizzorusso, E. Fratini, J. Eiblmeier, R. Giorgi, D. Chelazzi, A. Chevalier and P. Baglioni, *Langmuir*, 2012, **28**, 3952.
17. O. Wichterle and D. Lim, *Nature*, 1960, **185**, 117.
18. A. M. Lowman and N. A. Peppas, Hydrogels, in *Encyclopedia of Controlled Drug Delivery*, John Wiley & Sons, Inc., New York, 2nd edn, 1999, vol. 1, pp. 397–418.
19. K. S. W. Sing, D. H. Everett, R. A. W. Haul, L. Moscou, R. A. Pierotti, J. Rouquerol and T. Siemieniewska, *Pure Appl. Chem.*, 1985, **57**, 603.

20. I. Natali, E. Carretti, L. Angelova, P. Baglioni, R. G. Weiss and L. Dei, *Langmuir*, 2011, **27**, 13226.
21. The Engineering ToolBox. http://www.engineeringtoolbox.com/young-modulus-d_417.html accessed May 2012.
22. M. Monleon Pradas, J. L. Gomez Ribelles, A. Serrano Aroca, G. Gallego Ferrer, J. Suay Anton and P. Pissis, *Polymer*, 2001, **42**, 4667.
23. O. Okay, in *Hydrogel Sensors and Actuators*, Springer Series on Chemical Sensors and Biosensors 6, ed. G. Gerlach and K.-F. Arndt, Springer-Verlag Berlin, Heidelberg, 2009, pp. 1–14.
24. H. A. Barnes, J. F. Hutton and K. Walters, *An Introduction to Rheology*, Elsevier, Amsterdam, 1989.
25. J. D. Ferry, *Viscoelastic Properties of Polymers*, Wiley, New York, 3rd edn, 1980.
26. *CRC Handbook of Chemistry and Physics*, 73rd edn, 1992–1993.
27. E. Carretti, L. Dei and P. Baglioni, *Langmuir*, 2003, **19**, 7867.
28. E. Carretti, M. Bonini, L. Dei, B. H. Berrie, L. V. Angelova, P. Baglioni and R. G. Weiss, *Acc. Chem. Res.*, 2010, **43**, 751.
29. E. Carretti, L. Dei, P. Baglioni and R. G. Weiss, *J. Am. Chem. Soc.*, 2003, **125**, 5121.
30. E. Carretti, A. Macherelli, L. Dei and R. G. Weiss, *Langmuir*, 2004, **20**, 8114.
31. E. Carretti, L. Dei and R. G. Weiss, *Soft Matter*, 2005, **1**, 17.
32. M. Bonini, S. Lenz, R. Giorgi and P. Baglioni, *Langmuir*, 2007, **23**, 8681.
33. M. Bonini, S. Lenz, E. Falletta, F. Ridi, E. Carretti, E. Fratini, A. Wiedenmann and P. Baglioni, *Langmuir*, 2008, **24**, 12644.
34. P. S. Russo, in *Reversible Polymeric Gels and Related Systems* (ACS Symposium Series 350), ed. P. S. Russo, American Chemical Society, Washington, DC, 1987, ch. 9.
35. K. Almdal, J. Dyre, S. Hvidt and O. Kramer, *Polymer Gels and Networks*, 1993, **1**, 5–17.
36. R. Wolbers, *Workshop on new Methods in the Cleaning of Paintings*, Getty Trust Publications, Marina del Rey, 1988.
37. J. Rouquerol, D. Avnir, C. W. Fairbridge, D. H. Everett, J. M. Haynes, N. Pernicone, J. D. F. Ramsay, K. S. W. Sing and K. K. Unger, *Pure Appl. Chem.*, 1994, **66**, 1739.

CHAPTER 11

Cleaning IV: Applications and Case Studies

E. CARRETTI* AND R. GIORGI

University of Florence, Italy
*Email: carretti@csgi.unifi.it; giorgi@csgi.unifi.it

11.1 Introduction

The application of liquid organic solvents for the cleaning of easel paintings produces several problems related to their penetration into the paint structure. As described in Figure 11.1, once the solvent is applied onto the paint surface (Figure 11.1, step ①), wettability and capillary forces (see Sections 1.4.1, 1.4.2 and 2.4.1) drive its penetration into the paint layer (the average rate is of the order of some $\mu m\ s^{-1}$). This induces the swelling (and, as a consequence the softening) of most of the common polymeric materials in the paint (mainly varnish layer and binding media; see Figure 11.1, step ②) (see also Chapters 1, 7 and 9). For instance, toluene induces a volume increase in linseed oil films of about 28%; the corresponding values for acetone and xylene are 36% and 29% respectively.[1] The removal of undesired materials from the original work of art is thus carried out by means of a cotton swab roll (Figure 11.1, step ②). Unfortunately, this operation can also cause the extraction of the part of the softened components constituting the paint layer, because it is very difficult to control the mechanical action completely, especially for very fragile and degraded supports. Once the solvent evaporates, the molecules of the polymeric components of the paint can assume a three-dimensional (3D) conformation different from the original one, potentially inducing a thinning effect (Figure 11.1, step ③) and lowering the mechanical strength of the paint layer.

RSC Nanoscience & Nanotechnology No. 28
Nanoscience for the Conservation of Works of Art
Edited by Piero Baglioni and David Chelazzi

Published by the Royal Society of Chemistry, www.rsc.org

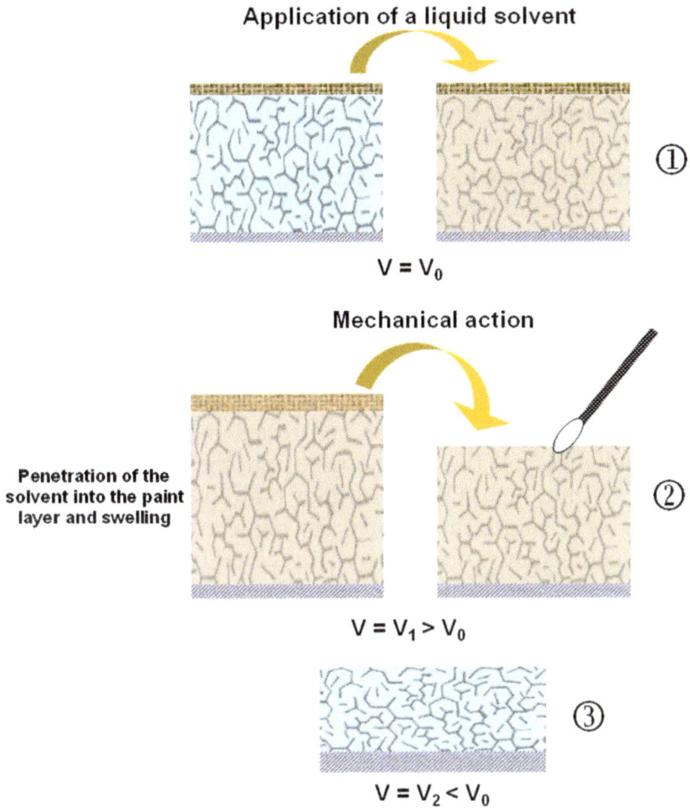

Figure 11.1 Scheme of the effects of the application of a liquid solvent for the cleaning of a paint surface.

In order to minimize these drawbacks, Wolbers proposed at the end of 1980s the application of the same solvents confined in a 3D very viscous matrix (gelled solvent or solvent-gels).[2] Using this approach, owing to the high viscosity of gels, many problems related to the traditional cleaning procedures are avoided. In particular:

1. As indicated by the Washburn equation (11.1), the increase in the viscosity of a fluid drastically lowers the extent and the rate of its penetration into the porous matrix.

$$L^2 = \frac{\gamma D t}{4\eta} \tag{11.1}$$

Here L is the penetration depth, t is the time, γ is the surface tension of the fluid, D is the diameter of the generic pore into which the fluid is penetrating and η is the viscosity of the fluid. For example for water at 25 °C ($\eta = 0.894$ mPa·s; $\gamma = 72.8$ mN m^{-1}) in contact with a porous matrix (with an average D of approximately 100 μm), after 10 s the L value is about 14 cm. If water molecules are confined into a highly viscous polymeric dispersion (HVPD) network made up of polyvinyl acetate 70% hydrolysed (70PVAc, 3 w/w%) and crosslinked with borax (0.6 w w%), the η value increases up to 405 Pa·s, and the surface tension decreases (the addition of a polymeric substance such as 70PVAc usually causes a slight decrease in the surface tension). Assuming that the resulting γ value is about 30 mN m^{-1} (*i.e.* the typical value of a 0.02 M aqueous solution of sodium dodecyl sulfate, SDS), then after 10 s of contact time the L value will be about 1.1×10^{-2} cm, indicating that the increase in the viscosity value results in a decrease of the penetration depth of almost three orders of magnitude, leading to a significant reduction of the previously mentioned negative effects. A primary consequence is that the cleaning action is selectively limited at the interface between the paint surface and the environment.

2. The cleaning procedure can be controlled by changing the application time (usually for gels the contact time with the surface to be cleaned is of the order of minutes). Owing to their high viscosity, the mobility of the molecules of a solute inside of a gel phase (quantitatively expressed as diffusion coefficient, D) is lowered by several orders of magnitude with respect to a liquid. Thus the rate of solubilization of any undesired material into the bulk of the gel is much lower in comparison with a liquid (see Figure 11.2).

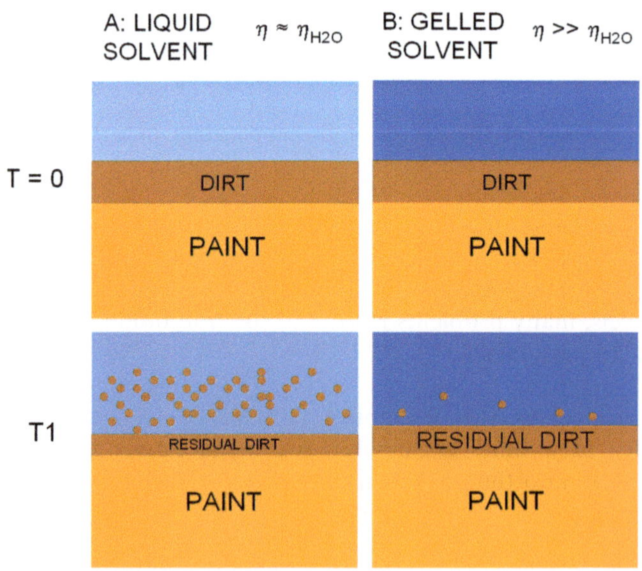

Figure 11.2 Effects of the application of a liquid (A) and of a gelled solvent (B) for the cleaning of a painted surface. (A) The low viscosity of the liquid accounts for a fast solubilization of the dirt layer, hindering the possibility of completely controlling the cleaning action. (B) The gelled solvent, owing to the high viscosity of the system, allows slow and much more controllable cleaning.

Once the solvent turns into a gel, its viscosity can increase by 4–5 orders of magnitude (see Figure 11.2, light and dark blue). When the liquid and the gel state of the same solvent are applied onto the paint surface, after the contact time T1, part of the surface dirt (brown layer) is solubilized into the liquid or into the gelled state (small brown spots). As shown in Figure 11.2, because of the low diffusion coefficient of the solubilized particles into the gelled phase, at T1 the total amount of solubilized material is much higher for the liquid phase. Thus, the much lower rate of the solubilization kinetics typical of highly viscous systems allows perfect control of the surface cleaning action.

3. A further advantage of the application of gels is the drastic decrease in the evaporation rate of the confined solvents. This effect will result in a strong reduction in the impact on the health of restorers. Figure 11.3 reports the evaporation curves of a mixture of water and acetone (15% w/w) both liquid and inside an HVPD of polyvinyl alcohol (PVA) molecules cross-linked by borate ions (see Section 11.5).
4. The variegated nature of undesired materials, especially varnishes applied as protective agents in previous restorations, requires the use of gels with solvents that possess different polarities, such as acetone, benzene, water, *etc.* (*vide infra*).
5. The versatility of this class of system also allows the gelation of different highly selective cleaning agents such as enzymes or chelates. Lipases are

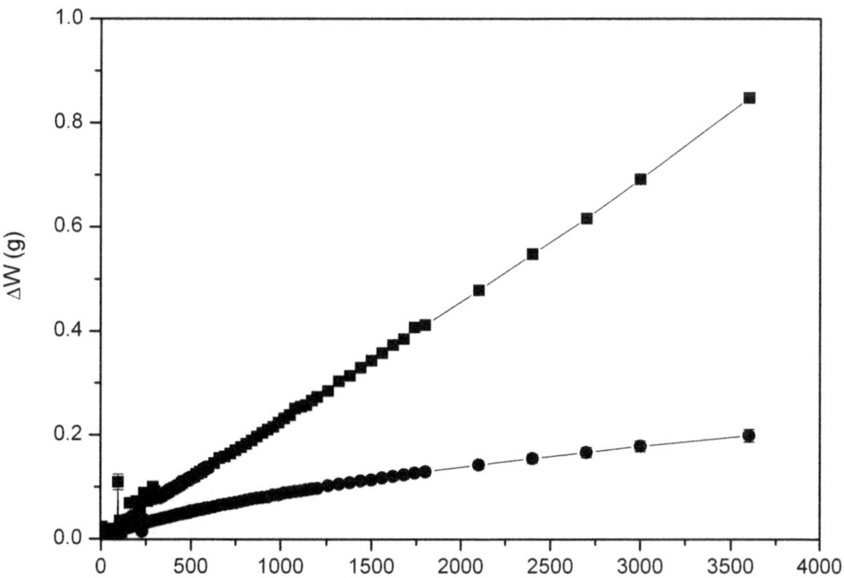

Figure 11.3 Evaporation curves of free acetone (black squares) and of acetone (15 wt% in water) embedded in a PVA–borax 3D matrix as a function of time. Error bars are the standard deviations.

employed in gelled water solutions for the selective removal of darkened coatings containing degraded natural triglycerides, with excellent results.[3] Similarly, proteases and carbohydrases are used for the degradation of protein-based coatings and carbohydrate-based polymers, respectively. In the case of enzymes, the environmental conditions (namely pH and temperature) must be controlled during the application.[2,3] Chelating agents such as ethylenediamine tetraacetic acid (EDTA) or citric acid are useful for the selective removal of salts (their action is usually specific for a narrow range of ions, especially metal cations).

6. Gels can be applied to a wide range of supports. Although gels were introduced in the works of art conservation field as specific cleaning agents for easel paintings, they have successfully also found application in the restoration of works of art made from different painted materials such as glass,[4] metals[5] and feathers.[6]

The great potential of this approach has induced many researchers to address their work across this new border of applied science.[7]

11.2 Classical Formulations of Gels in Cultural Heritage Conservation

The most widely used gelators for the preparation of gels to be applied for the cleaning of paintings, with the exception of the wax emulsion (see Section 9.2), belong to the family of water-soluble polymers. They belong to two different categories: the polymers obtained by modifying natural products, mainly cellulose ethers, and synthetic polymers such as polyacrylic acid (Carbopol®; see Figure 11.4).

For the selection of the appropriate gellant, the main requirements that should be considered are:

- A relative chemical inertness: none of the components in the cleaning system should interact with the work of art;
- Physical and chemical stability, so that no gellant residue may represent a potential cause of degradation.

The most diffused cleaning agents in the gel form are a family of physical gels (see Section 10.2.1) called *solvent gels*, developed by Wolbers between the end of the 1980s and the beginning of the 1990s.[2] The gelator (polyacrylic acid; see Figure 11.4) has two different important features: (i) a high molecular weight

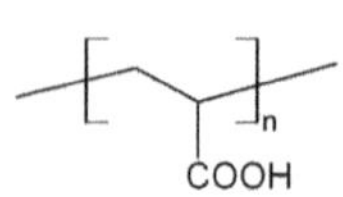

Figure 11.4 Chemical formula of polyacrylic acid.

(up to 4×10^6 Dalton) that allows the formation of an extended and stable 3D network; (ii) the presence of many carboxylic groups that, in the protonated form, interact with each other through hydrogen bonds favouring the folding of the polymer chains (see Figure 11.5A).

When the molecules possess this conformation, the solubility of polyacrylic acid in water is very low. As indicated in Figure 11.5, the acidic functions can be neutralized by adding a base, and this produces negatively charged carboxylate groups that repel each other through electrostatic interactions. This effect results in a significant extension of the chains and in the formation of the gelling network (see Figure 11.5B).

The deprotonation is usually achieved by means of non-ionic surfactants (Ethomeen C12 or C25, also known as cocoamines) with weak basic properties. Their structure is reported in Figure 11.6. Both surfactants have a long alkyl chain and form a salt with the carboxylic groups of the polyacrylic acid chains through the amine group (Figure 11.7). The gelation properties are related to the alkyl chain length and to the *x* and *y* values. For C12 [hydrophilic–lipophilic balance (HLB) = 10], *R* usually contains a number of carbon atoms between 8 and 20 (molecular mass 217–385 Da). This compound is used in the gelation of solvents with low polarity. In contrast, the higher HLB value for C25 (HLB = 19, molecular mass 789–957 Da) makes this surfactant useful in the gelation of highly polar solvents.

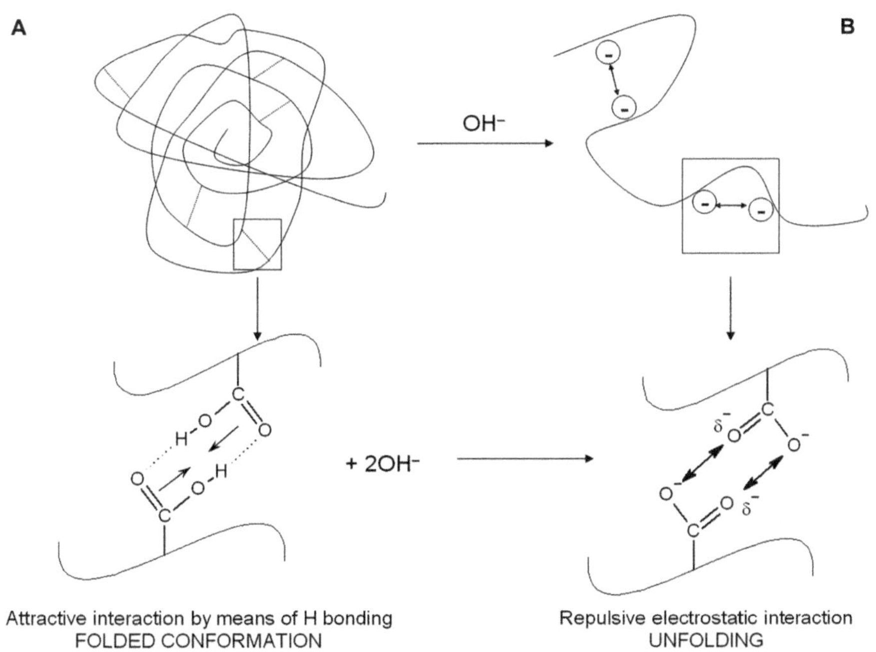

Figure 11.5 Mechanism of unfolding of polyacrylic acid molecules induced by the addition of a base.

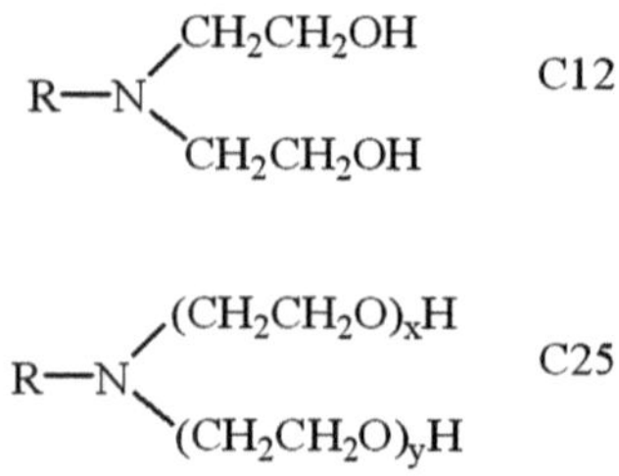

Figure 11.6 Structure of Ethomeen C12 (top) and C25 (bottom). *R* is usually between 8 and 20 carbon atoms.

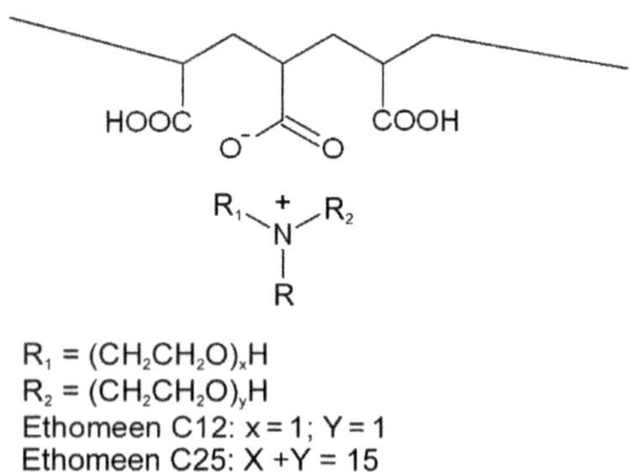

Figure 11.7 Formation of the salt between the surfactant and the carboxylic groups of the polyacrylic acid molecules.

Solvent gels are usually prepared by dispersing polyacrylic acid (1 w/w%) into the solvent to be gelled by means of a magnetic stirrer until a homogeneous system is obtained. In this process the formation of lumps should be strictly avoided. Then the surfactant is added (10–15 w/w%) while the system is maintained under stirring. The following addition of a small amount of water (~5 w/w%) induces the formation of the gel (macroscopically an instantaneous drastic increase of viscosity is observed). Thus, the liquid cleaning agent is entrapped in the 3D polymeric matrix, producing an increased viscosity and a great retention power. The solvent gels must be applied a few hours after their preparation, in order to allow the system to reach the equilibrium conformation.

Other gel formulations are known especially for polar solvents and water solutions. In this case polymeric gellants such as cellulose derivatives are employed, for example methyl cellulose, ethyl cellulose and hydroxyethyl cellulose, however they show a reduced ability in the gelation of low polarity media. Their preparation is achieved by adding the polymer (1–2 w/w%) to the solvent or the solution to be gelled, under stirring at 45–50 °C until a homogeneous system is obtained. The gelation occurs when the mixture is slowly cooled down to room temperature.

For all the gels that we have mentioned, the application can be carried out in two major ways. The first procedure involves the application of the gel without any further operation (contact time with the paint surface of the order of few minutes); the second method requires the use of a swab roll and some mechanical action on the surface of the work of art.

Recently, two natural gums based on polysaccharide polymer chains have found some practical application in the field of surface cleaning. The first is gellan gum, also branded as Phytagel, used to retain water for the gentle cleaning of paper and parchment documents;[8] as a food additive, it is used as a thickener, emulsifier and stabilizer. It is a very clear gel, and is considered to be inert. The second is Agar, used to support solutions for the hydrolysis and solubilization of the painting surface.[9]

Gellan gum is a water-soluble polysaccharide produced by the bacterium *Pseudomonas elodea*. The repeating unit of the polymer is a tetrasaccharide, which consists in two residues of D-glucose (Glc), one of L-rhamnose (Rha) and one of D-glucuronic acid (GlcA). The tetrasaccharide repeating unit has the following structure: [D-Glc (β1→4) D-GlcA (β1→4) D-Glc (β1→4) L-Rha (α1→3)]. The tetrasaccharide units are connected by (α1→3) glycosidic bonds.

Agar or agar-agar is a gelatinous substance derived from an unbranched polysaccharide (gelling agent) that accumulates in the cell walls of agarophyte red algae, primarily from the genera *Gelidium* and *Gracilaria*, or seaweed (*Sphaerococcus euchema*). It consists of a mixture of agarose and agaropectin. Agarose is the main component, and is a linear polymer made up of agarobiose (a disaccharide produced by *D*-galactose and 3,6-anhydro-*L*-galactopyranose). Agaropectin is a heterogeneous mixture of smaller molecules. Although both polysaccharide classes share the same galactose-based backbone, agaropectin is heavily modified with acidic side-groups, such as sulfate and pyruvate.

Typically, agar is sold commercially as a powder that can be mixed with water and prepared similarly to gelatin. As a gel, an agarose medium is porous and the gel porosity is directly related to the concentration of agarose in the medium. Thus, various levels of effective viscosity can be obtained.

11.3 Removal of Gels from Painted Surfaces

The main drawbacks involved in using physical gels for the cleaning of painted surfaces are related to their complete removal from the treated artistic substrates once the cleaning action is over. In particular, the removal of the gellant and of all the non-volatile components is problematic.[10] In fact even if the recommended washing procedures are followed, the presence of non-volatile gel residues (even in traces) can potentially produce unpredictable degradation phenomena. As a matter of fact, while the power of a solvent to dissolve and swell a given material decreases as it evaporates, the concentration of non-volatile cleaning agents increases. Thus, one of the most pressing concerns is the presence of gel residues on the treated surfaces and whether the residue may pose a risk to the surface of the work of art. Risks would include an increased solubility of the paint or an alteration of the paint's chemistry, with a potential acceleration of its deterioration. A number of institutions have undertaken studies on various aspects of possible long-term effects due to the use of gel systems. However, the issue concerning the presence and the effects of the residues has not been fully answered and the further widespread use of solvent-based gels is impeded.

Furthermore, in order to achieve the complete removal of gel residues, washing with a solvent or a mixture of solvents is usually carried out with a cotton swab roll. The most effective method is based on the application of the same solvent mixture as that confined in the gel phase. However, this would imply that some of the problems related to the use of liquid solvents (described in Section 11.1), would still persist. Furthermore, the traditional procedure usually involves a mechanical action that, depending on the conservator's experience, can potentially cause a loss of selectivity and control in the cleaning action, leading to the removal of some artefact material.

Gas chromatography–mass spectrometry (GC-MS), Fourier transform infrared (FTIR) spectroscopy and scanning electron microscopy (SEM) analysis have indicated that, after the treatment of a painted surface with solvent gels, some residues may be detected, mainly in little holes, small cracks and in all the rough portions of the treated work of art.[10,11] Furthermore, investigations performed on the surface of naturally aged paintings indicated that the residual Ethomeen (both C12 and C25) degrades to low molecular weight compounds that, upon accelerated ageing, evaporate only partially from the paint layer (this behaviour is particularly important for Ethomeen C25). So far the effects of the residual substances have not been investigated.[11]

Given that solvent gels are very versatile tools for the cleaning of paintings, the previously described drawbacks attracted the interest of many scientists to the development of new gel systems that combine the good cleaning performance typical of solvent gels with a complete, easy, and non-invasive removal procedure, avoiding the use of solvents or aggressive mechanical action.

11.4 Responsive Gels

In this section, a conceptually different approach to solvent gel cleaning is presented. The word "responsive" here refers to gels that can be easily and rapidly removed owing to their responsiveness to either a "chemical switch" (rheoreversible gels), to a "physical switch" such as an external magnet (a nanosponge containing magnetic nanoparticles), or to a "mechanical switch" such as peeling (highly elastic gels).

The properties of these systems thus make their removal from artistic substrates easier and potentially risk-free, and minimize (or completely eliminate) the presence of residues left on the surface. Because of the potentially great impact on restoration practice, many efforts have been recently carried out in this research area.

"Latent" gellants are reported to be effective substances used to obtain systems that can be switched from a solution-type into a gel-type through a simple chemical action. In particular, polyamine (*i.e.* polyallilamine, PAA, or polyethyleneimine, PEI; see Figure 11.8A) is a good example. In fact bubbling CO_2 through a PAA solution converts it into a gellant (see Figure 11.8B). The polyallylammonium carbamate ($PAACO_2$) formed exhibits strong inter-chain

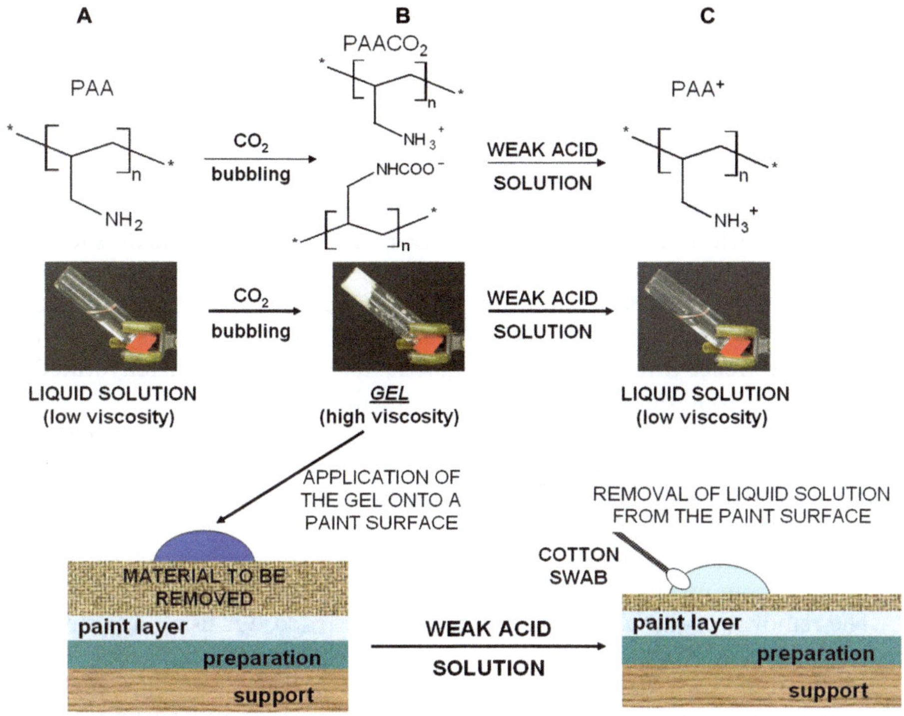

Figure 11.8 Synthesis (A), application (B) and removal (C) of polyallylamine rheoreversible gels on the surface of a painting.
(Adapted with permission from Carretti *et al. Langmuir*, 2004, **20**, 8414–8418. Copyright (2004) American Chemical Society.)

electrostatic attractive interactions,[12] that make the gel directly applicable onto a painted surface.[13] The removal of the system after the cleaning action can be performed by adding few drops of an aqueous solution of a weak acid that promotes the chemical decarboxylation of $PAACO_2$ and the formation of polyallylammonium ions (PAA^+). Correspondingly, the viscoelastic system is switched to a free-flowing liquid that can be wiped away with a dry cotton swab (Figure 11.8C). Unfortunately, owing to problems related to the complete removal of the liquid solution after the acidification of the gel, this technology has not had, at the present date, any further developments.

Further, it is possible to remove gels from a painted surface without leaving residues by incorporating functionalized nanoparticles into a polyacrylamide gel network in order to obtain a chemical sponge.[14,15] These systems were prepared by embedding ferrite nanoparticles [coated with a dicarboxylic derivative obtained through the esterification of poly(ethylene glycol) (PEG) with maleic anhydride] in a polyacrylamide matrix. The nanoparticles are chemically linked to the PEG molecules through the carboxylate functional groups, while the two double bonds per molecule resulting from the esterification anchor the nanoparticles within the gel matrix (see also Section 12.5).

In fact, radical copolymerization of these functionalized nanoparticles with acrylamide and *N*,*N*′-methylene bisacrylamide produces a nanomagnetic gel in which both the physico-chemical properties of acrylamide-based gels and the magnetic responses of ferrite nanoparticles are retained.

The viscoelastic nature and structure of the magnetic gels are very similar to those of conventional acrylamide gels, with nanosized mesh, inhomogeneous domain sizes of a few tens of nanometres, and micrometric pores. The chemically anchored magnetic nanoparticles act as entanglement sites and increase the value of the elastic modulus, G' (see Sections 10.3.3 and 10.5).

These gels behave as "containers" for aqueous systems, including microemulsions and micellar solutions (see Chapters 7 and 9); as a matter of fact, they can be freeze-dried to obtain magnetic xerogels and then rehydrated to "sponges" up to about 10 times their dry weight.

Moreover, these gels can be shaped as desired and, even in their hydrated state, they can be cut with a knife to a desired shape and moved with an external magnet (see Figure 11.9). Overall, these features ensure a fine spatial control of the cleaning area during the application onto a painted surface.

The microemulsion-loaded nanomagnetic gels have been used to remove Paraloid® from marble, frescoes and painted surfaces.

The removal of the nanomagnetic sponge is carried out using a permanent magnet: in the experiments that have been performed, no residue of the cleaning tool was detected on the treated surface through analytical measurements [microreflectance FTIR analysis and SEM/EDS (energy dispersive spectrometry) investigations].

The above-mentioned nanosponge and rheo-reversible gels share one feature that makes them highly suitable for the cleaning of works of art. Depending on different forces (chemical and physical), they are both responsive. In the next section, an innovative gel with interesting mechanical properties is presented. Although it is not considered as a responsive tool, its extraordinary viscoelasticity makes it very versatile for the treatment of irregular and chemically sensitive surfaces.

11.5 Polyvinyl Alcohol Based Systems: Peelable Gels

Poly(vinyl alcohol) (PVA) is a well-known hydrophilic polymer that is able to form highly viscous systems in the presence of a cross-linking agent such as borax.[16–19] Borate anions induce the formation of covalent intermolecular bridges through a di-diol condensation reaction, as shown in the scheme reported in Figure 11.10 [according to the International Union of Pure and Applied Chemistry (IUPAC) gold book,[20] a condensation reaction is defined as "a reaction in which two or more reactants yield a single main product with accompanying formation of H_2O or some other small molecules"].

A 3D PVA network is thus obtained, whose mechanical properties, as extensively reported in the literature, are strictly related to the concentrations of borax and PVA, and to the temperature, pH and composition of the continuous embedded aqueous phase. Although the macroscopic appearance

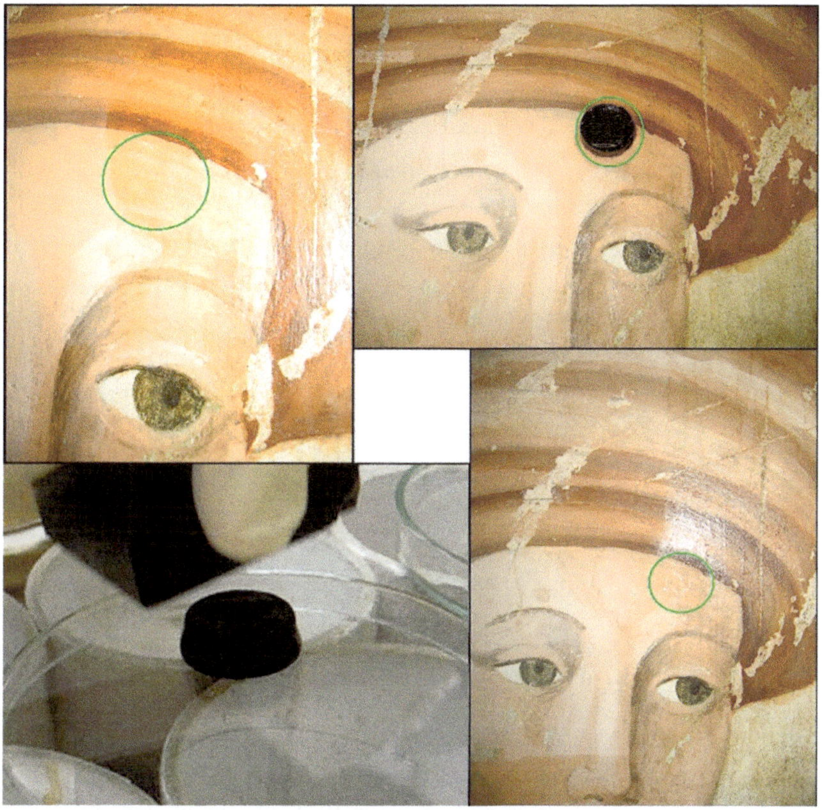

Figure 11.9 Application of a microemulsion-loaded nanomagnetic sponge. The gel is made of an acrylamide–bis-acrylamide copolymer, functionalized with cobalt–ferrite nanoparticles. The picture shows the removal of an acrylic coating (top left corner of the picture) after the application of the nanosponge over a wall painting model sample. The removal of acrylic coating (bottom right corner) is highlighted by using glazing light that enhances the surface reflectivity of the acrylic-coated area. A 1.4 Tesla permanent magnet (bottom left corner of the picture) is used to remove the gel. (Images by Michele Baglioni and Giacomo Pizzorusso.)

of these materials is similar to that of gels, they cannot be rigorously considered as such from the rheological point of view. As a matter of fact, as indicated in Section 10.5, when an oscillating shear stress is applied to the surface of a gel, the profile of the frequency sweep curve should be similar to that shown in Figure 10.12 ($G' \gg G''$ over a wide range of frequencies). On the other hand, for all PVA–borax based systems a crossover between the G' and G'' curves is almost always observed (see Figure 10.11; this behaviour is typical of water and water–co-solvent based systems). Therefore, these systems can be more properly named highly viscous polymeric solutions (HVPDs). An important property of the PVA–borax HVPD is the formation of thermodynamically stable systems also in the presence of significant amounts of organic co-solvents

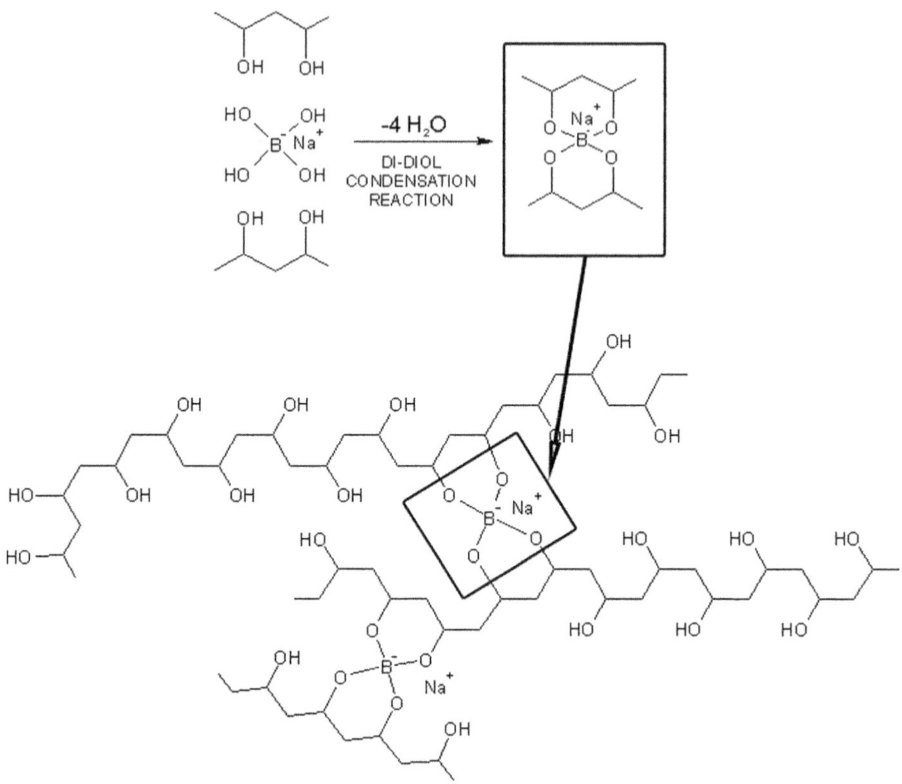

Figure 11.10 Condensation reaction between PVA and borax, and structure of the PVA–borax network formed.

(up to 50% in the case of 1-propanol, 2-propanol and *N*-methyl-2-pyrrolidone).[21] Therefore, the polarity of the liquid confined in the 3D PVA–borax network (this factor determines the selectivity and the efficacy of the cleaning action that depends on the f_d, f_p and f_h values of the loaded solvent, see Chapter 5) and the mechanical properties (especially the shear modulus G_0, see Section 10.5) can be carefully controlled. G_0 is strictly related to the nature and the amount of the chosen solvent as indicated in Figure 11.11.

The large value of the shear modulus G_0 (Figure 11.11) typical of the PVA–borax HVPDs allows easy removal of the HVPDs simply by peeling them off with pincers (see Figure 11.12). In practical applications, the removal is carried out without leaving detectable (within the limits of an FTIR analysis) amounts of residue on the treated surface. This feature makes these systems particularly effective for the cleaning of painted surfaces.

In order to optimize the quality of the cleaning action for any surface to be cleaned, it is mandatory to conduct a preliminary test in order to identify the correct balance between the following parameters:

- The efficacy, strictly related to the choice of the organic solvent embedded in the HVPD (f_d, f_p and f_h values);

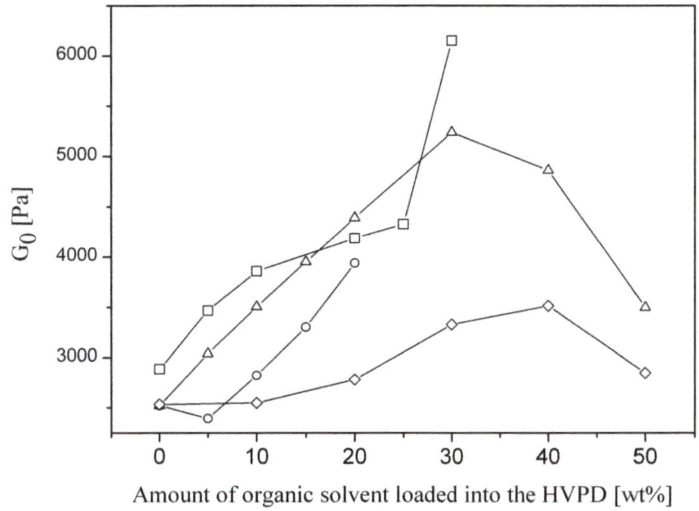

Figure 11.11 Trend of the shear modulus G_0 as a function of the amount of different solvents added to dispersions containing 4 wt% PVA (hydrolysis degree of the polymer ≈80%) and 1 wt% borax: (◇) 1-propanol, (□) NMP, (△) acetone, (○) MEK.
(Reprinted with permission from Natali *et al.*, *Langmuir*, 2011, **27**, 13226–13235. Copyright (2011) American Chemical Society.)

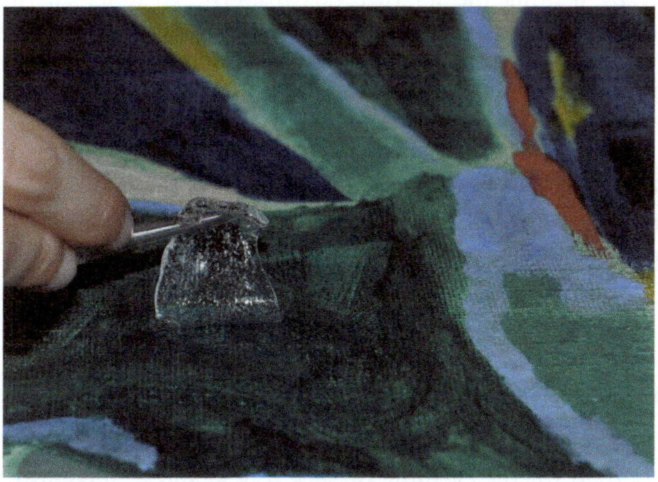

Figure 11.12 Removal of a PVA–Borax HVPD by peeling off using a pincer.

- Ease of application and removal by peeling, which is determined by the nature and the amount of PVA and borax (and their ratio) and by the nature of the co-solvent loaded in the aqueous phase.

The preparation procedure for the HVPD is strictly related to the chemical nature of the PVA and particularly to its degree of hydrolysis. PVA is usually

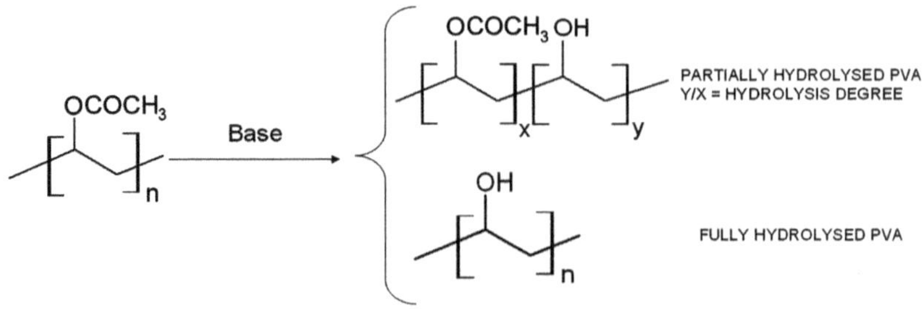

Figure 11.13 Synthesis of partially (top) and fully (bottom) hydrolysed PVA from PVAc.

synthesized *via* the hydrolysis of polyvinylacetate (PVAc) in the presence of a base (see Figure 11.13). The synthesis allows the control of the number of acetate groups that are left on the reactant molecule, by creating a random copolymer from vinyl–acetate and vinyl–alcohol.[22] The proportion of the alcohol groups (*i.e.* the hydrolysis degree, $HD = y/x$) is used for the identification of the polymer type: by changing its composition, its properties will change as well. For instance, the solubility in water depends on the degree of hydrolysis and on molecular weight.

The HVPDs are usually prepared by solubilizing the polymer in water (PVA is for the most part insoluble in water at room temperature; the solubilization occurs only by heating the mixture up to 85–90 °C under stirring for three hours), adding the required amount of solvent (this operation is usually done once the PVA–water solution is cooled to room temperature) and then the borax. Only 80% hydrolysed PVA is water-soluble at room temperature; in this case the PVA is solubilized in water without heating, then the solvent and the borax are added in sequence.

In both cases the HVPDs are obtained after vigorous shaking of the mixture, and their application is usually performed at least 24 hours after preparation in order to ensure complete equilibration.

11.6 Chemical Hydrogels for Water-Sensitive Materials

Chemical hydrogels are some of the most attractive tools that have been recently proposed for the cleaning of very fragile and water-sensitive surfaces. The mechanism through which the gel is formed is also responsible for its compactness and very high water retention capability.

The gel chain network is held by covalent bonds that provide the system with stronger cohesion forces with respect to those typical of physical gels or viscous cross-linked polymer solutions such as the solvent gels, in which the junctions between the polymeric chains are due to weaker interactions. This feature makes chemical gels particularly suitable for cleaning procedures, because they do not leave residues upon application.

Given that the viscosity and density of the system can be tuned to range from "softer" to "harder" materials, a variety of substrates can be treated according to several applicative methodologies.

Moreover, the hydrophilicity and good water retention shown by these gels allows their loading with water-based cleaning fluids, such as micellar solutions and microemulsions, whose main advantages with respect to pure solvents were widely discussed in Chapters 7 and 9. The supported fluids may range in polarity for the removal of different undesired coatings, such as hydrocarbon-based layers (waxes) or synthetic polymers (*e.g.* acrylic and vinyl coatings). As a matter of fact, the controlled release of aqueous cleaning fluids by chemical gels is an appealing perspective for the treatment of water-sensitive substrates, because it prevents swelling and the consequent mechanical damage.

It has already been shown how the co-polymerization reaction of acrylamide–bis-acrylamide monomers, supported by functionalized magnetic nanoparticles, leads to the formation of a nanosponge that is a clear example of chemical gel (see Section 11.4). More recently, excellent mechanical properties, in terms of adhesion capability and water retention, were achieved through the co-polymerization reaction of 2-hydroxyethyl-methacrylate (HEMA) and *N*-vinyl-1-pyrrolidone (VP).[23]

The synthesis procedure and the mechanism of gel formation are very similar to those reported for acrylamide gels. Obviously, handling acrylamide monomers during the gel preparation requires much care because of their toxicity, which is nonetheless strongly reduced when the reaction ends, and no possible release of unreacted monomers is detectable. HEMA and VP are also unstable monomers, so they need to be handled with care using personal protective equipment, but they are much less toxic then acrylamide (oral rat 50% lethal dose (LD_{50}) 124 mg kg^{-1} for acrylamide *vs.* 1470 mg kg^{-1} for VP and 5564 mg kg^{-1} for HEMA)[24]. However, the polymers formed from these monomers are not toxic but bio-compatible. In fact, chemically cross-linked hydrogels of poly(2-hydroxyethyl methacrylate), p(HEMA), are some of the most studied hydrophilic polymers for several biomedical and pharmaceutical applications, such as the production of materials for soft contact lenses[25] or drug delivery systems.[26] The potential use of p(HEMA) for contact lens material was first published by Wichterle and Lim in a 1960 article in *Nature*.[27] The oxygen-containing acrylic group and the hydroxyl pendant group make the polymer relatively hydrophilic; the maximum water content can vary in the range of 40–50%, values that are much lower than those typical of acrylamide based hydrogels. This is the reason why the VP co-monomer was added for the preparation of p(HEMA) based hydrogels. Vinylpyrrolidone is a highly hydrophilic monomer and its polymer, PVP, is largely used in pharmaceutical preparations and in many technical applications (as an adhesive, emulsifier, thickening agent, *etc.*). The copolymerization of HEMA and VP increases the EWC (equilibrium water content; see Section 10.3.1) of the gels. Other characteristics (such as mechanical stability, porosity and optical transparency) can also be gradually changed, by tuning the weight ratio of the two monomers. Because of these important features, HEMA–VP hydrogels have been recently studied as new nanostructured containers for microemulsions used for the removal of polymers from canvas paintings.

The different solvent affinity of the pHEMA and PVP fractions makes the hydrophobic–hydrophilic character tuneable, depending on the HEMA : VP ratio. Moreover, these gels are transparent. This feature allows conservators to check in real time the swelling and the initial removal of soling materials.

The original preparation method, based on a co-polymerization reaction, has been recently modified in order to reinforce the mechanical resistance of these gels, which tend to break when exposed to high mechanical stress. This result was achieved by using an already polymerized PVP (with a molecular weight of about 10^6 Da), and by improving the yield of the HEMA polymerization reaction. A perfectly clear gel was thus obtained, where all the covalent bonds connected HEMA monomers to the entrapped PVP (which functions as a sort of skeleton). It has been demonstrated that the polymerization reaction is quantitative and that no PVP is lost in the process. Presumably this effect occurs owing to the compactness of the 3D structure formed after the reaction.

According to these findings, these formulations [semi-interpenetrated networks of p(HEMA) with PVP] are highly interesting because the main physico-chemical and mechanical properties of the gel can be easily tuned (see Figures 11.14 and 11.15). In fact, by simply modifying its chemical composition it is possible to change the hydrophobic : hydrophilic ratio, the compactness and, therefore, the water retention capability. This makes these gels very versatile containers for detergent solutions or solvents with different polarity. Moreover, these gels are also completely inert to pH and/or to chelating agents, with a vast variety of possible applications in the field of works of art cleaning.

Figure 11.14 Chemical gel based on a semi-interpenetrated network of p(HEMA) with pVP after completion of the polymerization reaction. The shape of the gel is determined by the container used during the reaction. The water content is about 85% w/w.

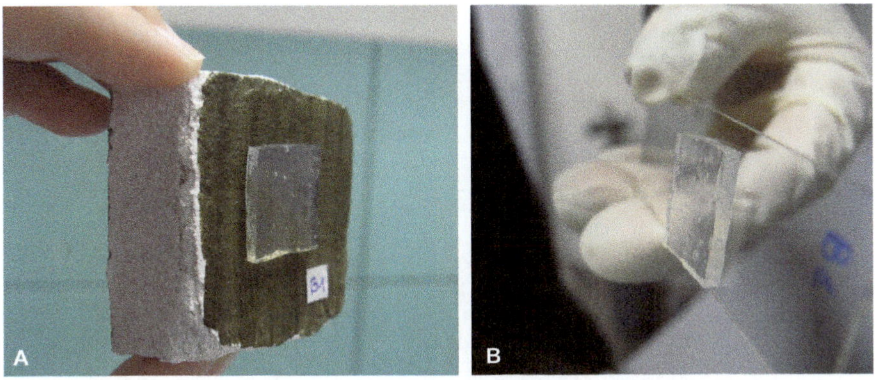

Figure 11.15 Application of the semi-interpenetrated network of p(HEMA) with pVP over different surfaces. The pictures show the high adhesion of the gels over vertical surfaces (glass and plaster).

11.7 Case Studies

11.7.1 Coronation of the Virgin with Saints by Neri di Bicci, Florence (Italy)

The "Incoronazione della Vergine con Santi" (Coronation of the Virgin with Saints, see Figure 11.16A) is an egg tempera wood panel (15th century, 150 × 150 cm) by Neri di Bicci conserved at the Galleria degli Uffizi in Florence, and presently on loan to the Civic Museum of Pescia (Pistoia), Italy. The surface of the panel was affected by a highly degraded and darkened layer of natural resins (Figure 11.16B). FTIR analytical investigations revealed that the layer contains mainly shellac and other natural components (presumably oils). The anti-aesthetic effects of this degraded coating are particularly evident, especially in the red mantle of Saint Catherine and Saint Jerome (lower part of the painting, in red vermilion). The texture of this surface layer at a microscopic level is shown in Figure 11.16B (the image was collected in the area indicated by the black box in Figure 11.16A). The shrinkage of the varnish results in the formation of an inhomogeneous patina comprising two oxidized overlapped layers: a dark texture, directly in contact with the paint surface (Figure 11.16B right side), and a second layer with brown clots that cannot be easily removed.

According to a preliminary Feller test (see Chapter 5) the best solvent for the removal of the patina was a liquid with an f_d value ranging between 47 and 48 (see Sections 5.2 and 5.5), and in particular a mixture of 90% w/w acetone ($f_d = 47$; $f_p = 32$; $f_h = 21$) and 10% w/w benzyl alcohol ($f_d = 48$; $f_p = 16$; $f_h = 36$). However, the application of this blend produced a severe softening of the red pigment layer, due to significant swelling. Thus, the selection focused on highly viscous systems, and particularly on PVA-based HVPDs (see Section 11.4). Given that it is not possible to produce stable transparent HVPDs by

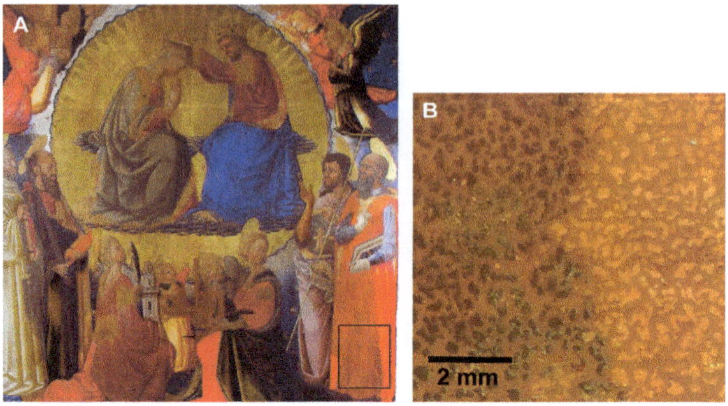

Figure 11.16 (A) "Coronation of the Virgin with Saints" (Neri di Bicci, 15th-century egg tempera wood panel), Galleria degli Uffizi, Florence, Italy. The black box indicates the region involved in the cleaning test performed by means of PVA–borax–acetone HVPD. (B) optical micrograph of the test area before cleaning. The degraded varnish is composed of two superimposed layers: brown clots (left) and a brown texture underneath (right). (Reprinted from *Journal of Cultural Heritage*, **11**(4), E. Carretti, I. Natali, C. Matarrese, P. Bracco, R. G. Weiss, A. Salvini, P. Baglioni, L. Dei, A new family of viscoelastic gels for cleaning easel paintings, 373–380, Copyright (2010), with permission from Elsevier. Courtesy of the Italian Cultural Heritage Department.)

Table 11.1 Composition of the HVPD used for the cleaning of the surface of the "Coronation of the Virgin with Saints".

Composition	*Concentration wt%*
Fully hydrolysed PVA	2
Acetone	15
Demineralized water	82.6
Sodium tetraborate decahydrate	0.4

embedding a 90 : 10 mixture of acetone and benzyl alcohol in the PVA 3D network, a system containing only acetone was prepared and tested.[28] This choice was made because hydrophobic materials can be dissolved or swollen in aqueous solutions of hydrotropes (hydrotropes are chemicals that comprise a hydrophilic and a hydrophobic block, but the hydrophobic segment is relatively too small to drive self-aggregation, as in the case of surfactants).[29] The composition of the HVPD obtained is indicated in Table 11.1.

After the application of the HVPD (see below for the preparation and the application procedures), the clots on the paint surface were softened, and easily removed by a gentle mechanical action performed with a scalpel (see Figure 11.17B). A second application allowed the complete removal of the dark texture underneath, as shown in Figure 11.17C.

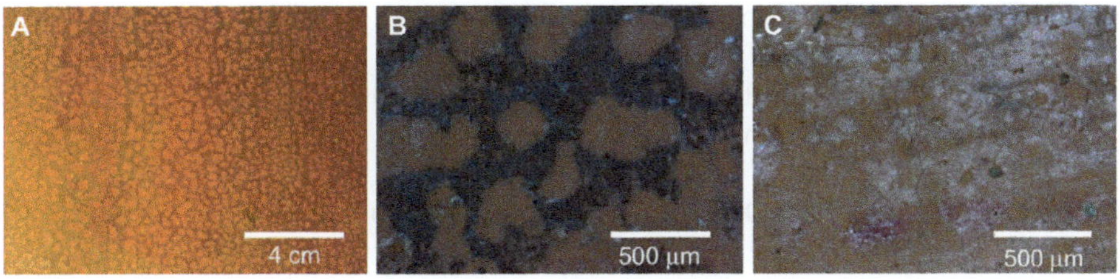

Figure 11.17 (A) Image in true colours and (B) false colour micrographs of the residue after the removal of the clots constituting the degraded surface patina affecting the "Coronation of the Virgin with Saints" by Neri di Bicci. (C) USB false colour micrograph obtained after a second application of the PVA–borax–acetone HVPD.
(Reprinted from *Journal of Cultural Heritage*, **11**(4), E. Carretti, I. Natali, C. Matarrese, P. Bracco, R. G. Weiss, A. Salvini, P. Baglioni, L. Dei, A new family of viscoelastic gels for cleaning easel paintings, 373–380, Copyright (2010), with permission from Elsevier.)

Table 11.2 Maximum amount of solvents loadable into the HVPD composed of fully hydrolysed PVA (2 w/w%) and borax (0.4 w/w%).

Solvent	*Maximum loadable amount (w/w%)*
1-Propanol	25
2-Butanol	5
1-Pentanol	10
Acetone	10
Propylene carbonate	10
Cyclohexanone	5

The procedure for the preparation of the HVPD is strictly related to the nature and to the amount of the solvent, and particularly to its volatility. In general it can be schematically summarized as follows:[30]

1. First the borax is dissolved in water, then the desired amount of solvent is added. For scarcely volatile liquids such as propylene carbonate, cyclohexanone, 1-pentanol and 2-butanol, the solvent should be added immediately after point 1. The solvents that can be loaded into the HVPD and the corresponding maximum amounts are reported in Table 11.2.
2. The fully hydrolysed PVA is added.
3. The mixture is stirred at 80–90 °C for almost three hours, until a perfectly homogeneous and transparent system is obtained. This operation should be done in a sealed container in order to minimize the evaporation of water.
4. The sample is cooled to room temperature.
5. The solvent is added. For volatile substances such as acetone or 1-propanol, the solvent should be added after point 4, in order to minimize its evaporation during the heating of the specimen at 80–90 °C (point 3).

6. The system is mechanically stirred (*e.g.* with a vortex mixer), until the solvent is completely absorbed.
7. The HVPD should be used at least 48 hours after its preparation to make sure that the system has reached equilibrium.

Once ready, the HVPD is applied and modelled onto the paint surface in the following manner:

1. A small amount of HVPD (1 g is effective for the cleaning of about 2–3 cm^2 of surface) is applied onto the paint surface and its shape modelled with a spatula.
2. The contact time is usually kept between 3 and 6 minutes, depending on the conservation status of the area to be treated.
3. The HVPD is easily removed with a pincer as indicated in Figure 11.12.
4. If necessary, the removal of softened residues of the undesired layer can be carried out by gentle mechanical action, using either a cotton swab or a scalpel.

11.7.2 Gilded Stucco by Pietro da Cortona, Palazzo Pitti (Florence)

At the end of 1640, Ferdinando II de' Medici commissioned Pietro da Cortona to decorate five rooms (the rooms of Venus, Apollo, Mars, Jupiter and Saturn, named after the corresponding decorative themes) of the Grand Duke's winter apartment in Palazzo Pitti in Florence. The rooms are currently part of the Palatina Gallery. The work was intended to exalt the virtues, and celebrate the glory, of the Medici dynasty in the figure of Grand Duke Ferdinando II, depicted as the ideal prince and associated with Hercules in the decorations.

The rooms are decorated with mural paintings and stucco work. The gilded stuccoes are affected by the presence of a superficial degraded organic patina, which alters both the decoration colours and the brightness (see Figure 11.18). Preliminary cleaning tests were performed using solvents on restricted areas (4 × 4 cm, see inset in Figure 11.18). The complete removal of the resin was obtained using 1-propanol (f_d=40, f_p=16, f_h=44). However the use of such solvent produced weakening of the surface gilded layer owing to its partial penetration inside the porous matrix.

In order to avoid this drawback an 80% hydrolysed PVA–borax–water–1-propanol HVPD was used (see Table 11.3).

When the cleaning has to be carried out on a vertical surface the following three requirements must be met: (i) the HVPD must remain on the surface without falling down for at least 3–6 minutes (the minimum time needed for a good cleaning action); (ii) perfect adhesion between the HVPD and the treated surface must be granted; (iii) the shape of the HVPD should not be altered during the application so that the cleaning action can be monitored.

Figure 11.18 Detail of the stuccos by Pietro da Cortona (17th century, Room of Jupiter, Palazzo Pitti, Florence, Italy). Visible light image of the treated region after the cleaning with a PVA–Borax–1-propanol HVPD. The black circle indicates the area where the cleaning was carried out. The inset shows a small area (4 × 4 cm) where cleaning tests were performed. (Courtesy of the Italian Cultural Heritage Department.)

Table 11.3 Composition of the 80% hydrolysed PVA–borax–water–1-propanol HVPD used for the cleaning of gilded stuccoes of the Galleria Palatina (Florence).

Composition	*Concentration wt%*
80% hydrolysed PVA	3
Borax	0.75
Demineralized H_2O	66.25
1-PrOH	30

Table 11.4 Maximum amount of solvents loadable into the HVPD composed of 80% hydrolysed PVA (4 w/w%) and borax (1 w/w%). (Adapted with permission from Angelova *et al.*, *Langmuir*, 2011, **27**, 11671–11682. Copyright (2011) American Chemical Society.)

Solvent	*Maximum loadable amount (w/w %)*
Ethanol	50
1-Propanol	50
2-Propanol	50
Acetone	40
Methyl ethyl ketone (MEK)	20
N-Methylpyrollidone (NMP)	50

The system reported in Table 11.3 gave excellent results in terms of cleaning efficacy (see Figure 11.18 for the complete recovery of the original brightness of the surface), HVPD peelability and amount of residues left on the treated surface. Furthermore, the HVPD could be applied to the vertical surface for the required time (10 minutes) and no meaningful alteration of its shape was observed.

In general, the formulation of an HVPD based on 80% hydrolysed PVA should follow this procedure:

1. The best organic solvent is selected according to the Feller test, from among the liquids reported in Table 11.4.
2. The best performances (cleaning efficacy, peelability and presence of residues left on the treated surface) are usually obtained with systems containing 3 w/w% of PVA (80% hydrolysed) and 0.75 w/w % of borax. The mechanical properties of the final system can be tuned by controlling the ratio or the amount of ingredients, depending on the specific case study. For instance, an increase in the amount of PVA results in an increase of the system stiffness and in a decrease of both the adhesion to the support and the cleaning power. This expedient can be useful when highly de-cohesive paint layers are present.
3. The total amount of the organic solvent loaded into the HVPD can be modulated in order to optimize the control of the cleaning action (the maximum amounts of solvents loadable into the 80% hydrolysed PVA-based HVPD are reported in Table 11.4).

The HVPD was prepared according to the following steps:

1. Preparation of the water/solvent mixture.
2. Solubilization of the PVA (80% hydrolysed) in a mixture of water and the selected solvent. This step can be done at room temperature, but the mixture can be heated up to 45–50 °C for 10–15 minutes with stirring in order to improve the solubilization of the polymer,
3. Once the system is homogeneous, sodium tetraborate decahydrate is added and the mixture is kept under vigorous stirring for a few minutes until the HVPD is formed.

4. The HVPD is applied at least 48 hours after its preparation, in order to achieve the maximum performance.

The performance of an 80% hydrolysed PVA-based HVPD in the cleaning tests was comparable or better with respect to that achieved with an HVPD based on 99% hydrolysed PVA (see Section 11.7.1), owing to the larger amount of organic solvents that can be loaded into the systems.[30] Moreover, the preparation procedure is much easier (99% hydrolysed PVA-based HVPD requires vigorous stirring for 2–3 hours at 80–85 °C, while 80% hydrolysed PVA-based HVPD can be prepared at room temperature and the preparation is quicker), with no appreciable change in the physico-chemical properties of the two formulations.

Thus, the peelable systems used for the cleaning of the gilded stuccoes of the Galleria Palatina can be regarded as the HVPD reference system.

11.7.3 Wall Paintings, Sala dell'Eneide, Palazzo Buonaccorsi (Macerata)

Palazzo Buonaccorsi was built in Macerata (Italy) between 1700 and 1720 by Count Raimondo Buonaccorsi and his son, Cardinal Simone Buonaccorsi, in a project by Giovanni Battista Contini. The rooms and the corridors of the main floor of the palace were decorated by two Italian artists (Carlo Antonio Rambaldi and Antonio Dardani) with a series of Rococo paintings representing scenes of the Aeneid. After the devastating earthquake of 1997 that affected the centre of Italy, many emergency repairs were necessary in order to ensure the stability of the damaged cultural heritage and to avoid the loss of the detached fragments of the wall paintings. All the paintings of the Sala dell'Eneide (Aeneid room) were fastened with pieces of roller bandage soaked with a solution of an acrylic copolymer (Paraloid B72®, 10–15 w/w% in nitro diluent) in order to guarantee the cohesion of the detached part of the wall paintings. However, during the restoration workshop it was necessary to remove the drapery completely in order to recover the adhesion of the detached fragments to the structure of the wall by means of an aesthetically non-invasive method.

Preliminary cleaning tests indicated that an oil-in-water (o/w) microemulsion, composed of water–SDS–PeOH–*p*-xylene, was effective in producing swelling of the copolymer for the removal of the gauzes. This system, indicated as M1, has a composition similar to that used for the cleaning of the frescoes in the sacristy of the San Salvador Church in Venice (see Section 9.5.3). The main drawback was the high penetration of the liquid into the porous matrix (mainly due to the high porosity of the damaged areas of the wall), which induced the conservators to examine a new class of highly viscous gel-like systems obtained from a cellulose derivative (hydrophobically modified hydroxy ethyl cellulose, hmHEC, ≈2% w/w) and the liquid M1 microemulsion (see Table 11.5).[31] For the preparation of this highly viscous microemulsion based system (microE/hmHEC), M1 was first prepared following the procedure described in Section

Table 11.5 Composition (w/w %) of the microemulsion based highly viscous system.

Composition	*Concentration wt%*	
SDS	3.3	M1 system (liquid)
Demineralized H_2O	87.3	
1-PeOH	6.0	
p-Xylene	1.5	
Hydrophobically modified hydroxyethylcellulose (hmHEC) (NEXTON®)	1.9	

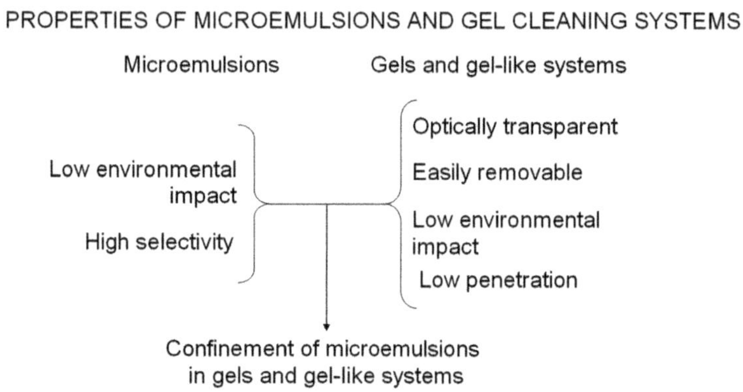

Figure 11.19 The cleaning properties of gels and gel-like systems and microemulsions are synergistically enhanced by embedding a microemulsion in a gel or gel-like phase. This confinement maximizes the cleaning effectiveness and the control of the cleaning action in a dramatic way.

9.5.3. Briefly, SDS, 1-pentanol and *p*-xylene were added in sequence to deionized water. The mixture was then stirred until a transparent solution was obtained. Finally the hmHEC was added and the sample stirred overnight until a transparent and homogeneous system was obtained. The preparation had to be carried out in a sealed container in order to avoid the evaporation of the volatile components (*i.e.* *p*-xylene and 1-pentanol).

As Figure 11.19 suggests, the most important applicative technological advantages of microemulsions (see Section 9.3) can be combined with those typical of highly viscous gels and gel-like systems (see Section 11.1) by using a microE/hmHEC system. As a matter of fact, the application of microE/hmHEC (contact time 10–15 minutes) results in the selective and complete removal of the gauzes owing to the progressive and perfectly controllable softening of the acrylic adhesive used to fix the drapery to the damaged wall (see Figure 11.20).

It is important to recall that, although the microE/hmHEC system exhibits macroscopic properties typical of gels, it cannot be rigorously considered as a gel. In fact, as discussed in Section 10.4, there is a crossover between the elastic modulus G' and the viscous modulus G'' in the system frequency sweep curve.

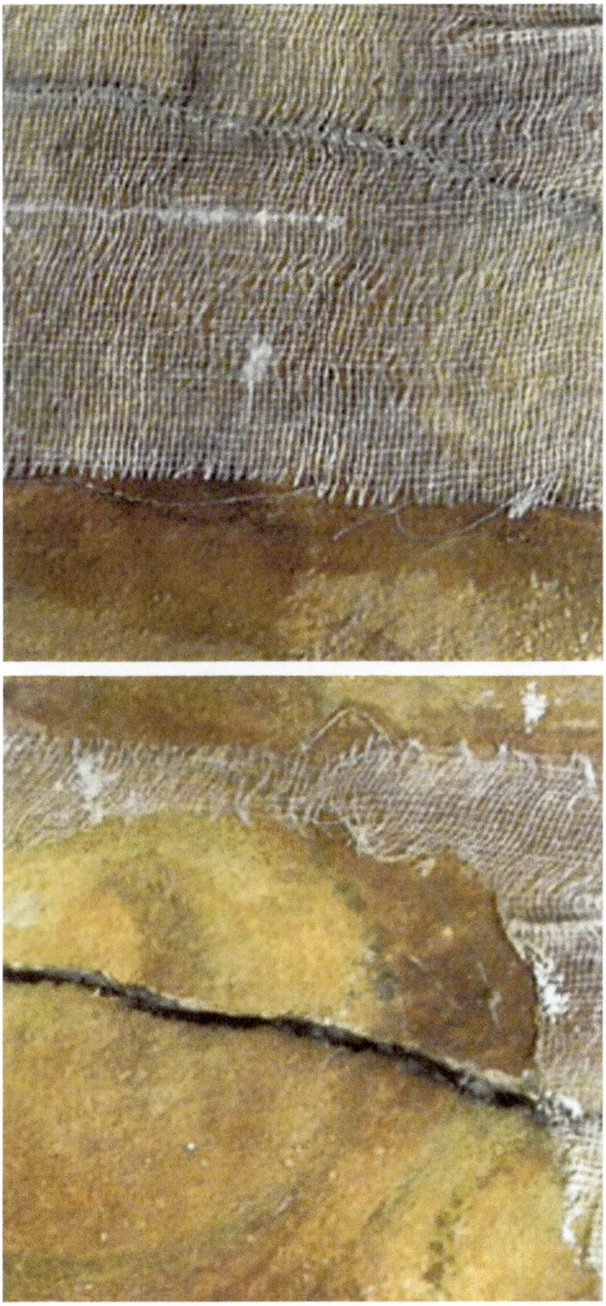

Figure 11.20 Wall painting in the "Sala dell'Eneide" (Aeneid room), Palazzo Bonaccorsi, Macerata (Italy). Test area for the application of the microE/hmHEC system, before (top) and after (bottom) cleaning.

We recall that for a gel G' should be higher than G'' over a wide range of frequencies.[32]

It is well known that synthetic and natural polymeric coatings strongly alter the wetting properties of solid surfaces.[33] Thus, in order to evaluate further the polymer extraction capability of the microE/hmHEC, a layer of poly(EMA/MA) (Paraloid-B72®) was spread onto glass surfaces, and the contact angle formed by a drop of water on these surfaces was measured before and after the removal of the coating with the emulsion-based system.

Figure 11.21 shows the behaviour of the contact angle on the surface of a glass treated according to two different procedures. Coating the surface with a poly(EMA/MA) film (Paraloid-B72®) resulted in a strong increase of the contact angle, indicating an enhanced hydrophobicity of the surface. After the application of the microE/hmHEC a strong increase in wettability is observed, and the contact angle value is equal to that of the original substrate before the application of poly(EMA/MA). This indicates good swelling of the acrylic copolymer with the microE/hmHEC system, followed by the detachment and removal of the hydrophobic film from the surface of the glass substrate. Gravimetric measurements further confirmed the quantitative extraction of the polymer (about 95–98%), and the effectiveness of the cleaning procedure was assessed using FTIR spectroscopy data.

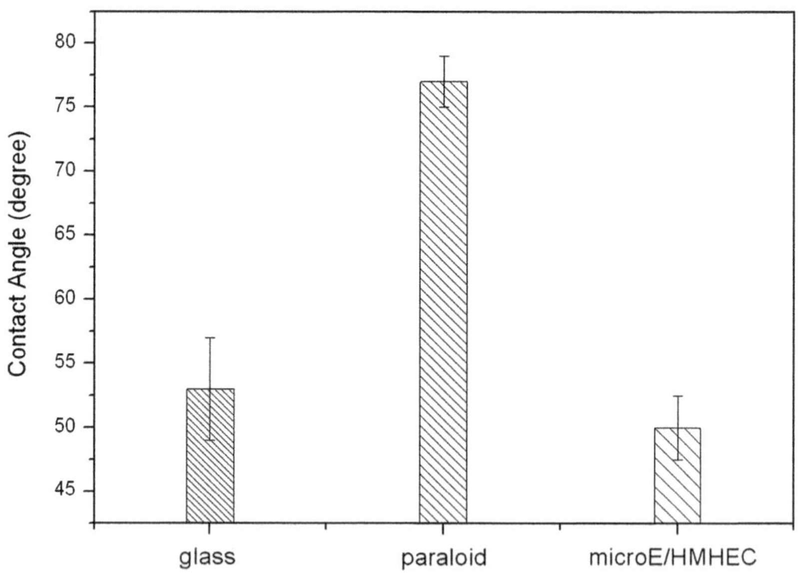

Figure 11.21 Contact angle behaviour for water droplets on the surface of a microscope glass. From left to right: "glass" is the free glass surface, "paraloid" is the surface coated by a poly(EMA/MA), and "microE/HMHEC" is the surface after the application of the microemulsion based highly viscous system.

11.7.4 Removal of Relining Adhesives from Canvas Supports

Under natural ageing conditions the photo- and thermal oxidation of easel paintings produces a yellowing in the varnish layer and, depending on the environmental conditions, drastic changes in the mechanical properties of the painted layer. The appearance and readability of the painting is strongly altered by oxidation, which also favours the cracking of the protective varnish film.[33–36] Moreover, the ageing of varnish and drying oil produces acids that, in the case of canvas paintings, favour the acid-catalysed de-polymerization of the canvas cellulose fibres.[37,38]

Thus, aesthetic alterations of the varnishes are often paralleled by a severe degradation in the support, leading eventually to laceration and rupture.

The traditional technique of "relining" (or lining) has been widely used to counteract the poor mechanical properties of aged canvas. This procedure is based on the application of a new linen (or cotton) canvas on the back of the paint; natural materials or synthetic adhesives are used to glue the two sides. However, both synthetic and natural adhesives undergo ageing processes that favour the formation of by-products (*e.g.* acids) that may promote further degradation.[39] For this reason, several attempts were made to formulate efficient systems for the selective removal of aged adhesives from lined paintings, without alterations of the original painting materials. Traditionally, this could be performed by using detergent solutions or, in some specific cases, mixtures of solvents. More recently, the use of enzymes (*e.g.* proteases) has been proposed to remove natural glues, such as animal glue, from textiles.[40]

Nanostructured fluids such as oil-in-water (o/w) microemulsions or micellar solutions are greatly effective in the removal of different synthetic polymers[41–45] from wall paintings. The ability of these systems to swell, solubilize, and remove the synthetic adhesives largely used in works of art conservation (*e.g.* acrylate–methacrylate and acrylate–vinyl acetate copolymers), makes them potentially effective for the removal of such materials from the back of lined canvas.

An o/w microemulsion loaded in an acrylamide–bis-acrylamide chemical hydrogel has been recently used in the removal of adhesive polymers from the back of relined canvas paintings.[46] Figure 11.22 shows an EAPC[43] cleaning system loaded gel used on a canvas model sample treated with an acrylate–vinyl adhesive and artificially aged (see also Section 9.5.6). The application, which usually can take up to a couple of hours, leads to a consistent softening and swelling of the adhesive coating, so that the mechanical removal of the adhesive film can be easily carried out with a scalpel, without strong mechanical stress applied to the cellulose fibres (see Figure 11.22B). The compactness and density of the gel are tuneable and depend on the monomer : cross-linker ratio. This feature can be exploited to modify the properties of the cleaning systems in order to meet the specific properties of the substrate.

As shown in Figure 11.22, these gels provide an extremely limited spread of the loaded fluids along the plane of the canvas surface; moreover, the very high water retention capability minimizes the wetting of water-sensitive canvas. For this reason these systems are valid alternatives to the use of waterless cleaning systems.

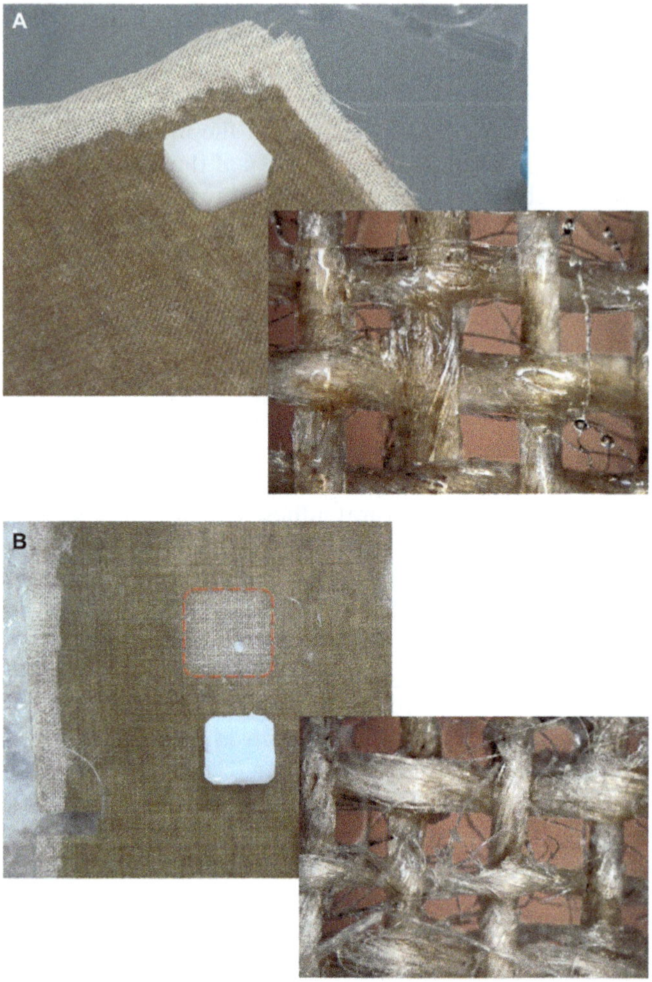

Figure 11.22 (A) EAPC microemulsion loaded-acrylamide gel put on top of a model sample of canvas, treated with an acrylate–vinyl adhesive. (B) Cleaned area after gel removal. Details of canvas under optical microscope show the fibres before and after cleaning. (P. Baglioni, D. Berti, M. Bonini, E. Carretti, M. Carmen Casas Perez, D. Chelazzi, L. Dei, E. Fratini, R. Giorgi, I. Natali, M.C. Arroyo, *Gels for the Conservation of Cultural Heritage*, MRS Online Proceedings Library, 2012, 1418, reproduced with permission.)

Figure 11.23 shows the FTIR spectra of an aged canvas sample before and after the removal of Plexisol® P550, a widely used adhesive based on *n*-butyl methacrylate and applied as a hydrocarbon solution.[47] The absorption band at 1724 cm^{-1} corresponds to the stretching of the C=O group of the acrylate chains, and its strong decrease accounts for the quantitative removal of the adhesive from the substrate, according to the detection limit of the FTIR

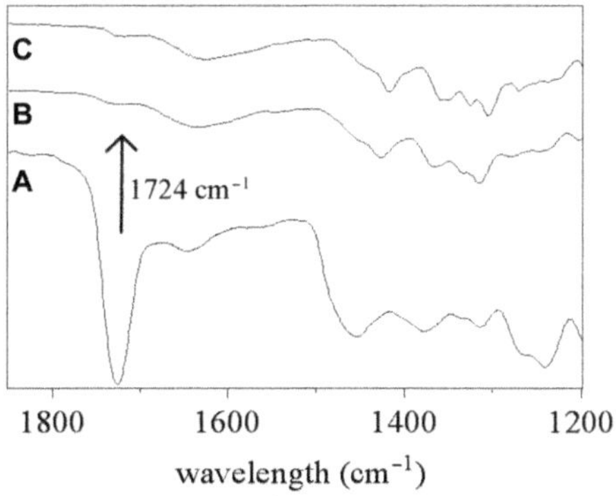

Figure 11.23 Aged canvas sample, treated with Plexisol® P550. FTIR spectra before (A) and after (B) cleaning. Plexisol® P550 was removed by application of an EAPC microemulsion (2 hours contact time) loaded in the acrylamide gel, and further gentle mechanical removal. (C) Spectrum of reference canvas. (P. Baglioni, D. Berti, M. Bonini, E. Carretti, M. Carmen Casas Perez, D. Chelazzi, L. Dei, E. Fratini, R. Giorgi, I. Natali, M.C. Arroyo, *Gels for the Conservation of Cultural Heritage*, MRS Online Proceedings Library, 2012, 1418, reproduced with permission.)

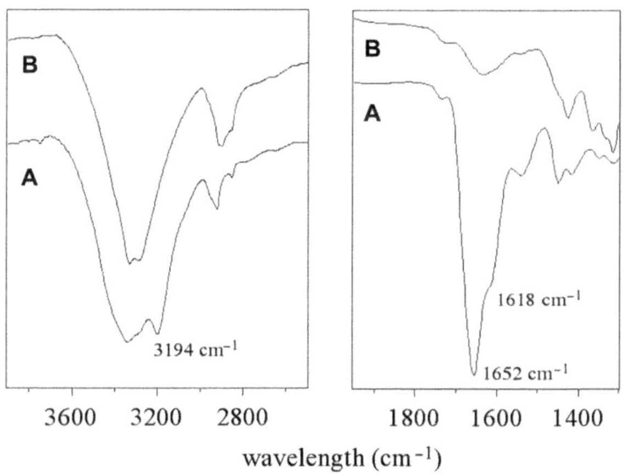

Figure 11.24 Details of the FTIR spectra of (A) dried (in oven under vacuum) acrylamide gel, and (B) the cleaned area of an aged canvas sample. (P. Baglioni, D. Berti, M. Bonini, E. Carretti, M. Carmen Casas Perez, D. Chelazzi, L. Dei, E. Fratini, R. Giorgi, I. Natali, M.C. Arroyo, *Gels for the Conservation of Cultural Heritage*, MRS Online Proceedings Library, 2012, 1418, reproduced with permission.)

technique. The spectra of the cleaned sample and the reference untreated canvas provided clear evidence for the cleaning efficiency.

An accurate analysis of the FTIR spectra, collected on canvas after the cleaning procedure, provided evidence of the lack of gel residue on the cellulose substrate. Figure 11.24 compares the FTIR spectrum of the dried acrylamide gel and the spectrum of the canvas after the removal of the adhesive with the gel. The spectrum of the acrylamide gel shows two peaks, at 1652 and 1618 cm^{-1}, due to the C=O stretching (amide I) and to the NH_2 bending (amide II) modes respectively, and a peak at 3194 cm^{-1}, due to the symmetrical NH_2 stretching.[48] In the presence of gel residues, some shoulders should appear in the canvas spectrum at the corresponding wave numbers. In fact, the partial overlapping of amide I and amide II bands with that at 1635 cm^{-1} (due to water adsorbed on linen fibres) should be expected.[49] Similarly, a partial overlapping of the NH_2 stretching peak with the broad –OH stretching band of cellulose should be detected. Instead, as shown in Figure 11.24, no peak (or shoulder) was observed in the 3600 – 2900 and 1690 – 1550 cm^{-1} range of the cleaned canvas spectrum, excluding the presence of FTIR-detectable acrylamide residues on the cellulose substrate.

Further Suggested Reading

P. Baglioni, L. Dei, E. Carretti and R. Giorgi, Gels for the conservation of cultural heritage, *Langmuir*, 2009, **25**, 8373.

E. Carretti, M. Bonini, L. Dei, B. H. Berrie, L. V. Angelova, P. Baglioni and R. G. Weiss, New frontiers in materials science for art conservation: responsive gels and beyond, *Accounts of Chemical Research*, 2010, **43**, 751.

E. Carretti and L. Dei, in *Molecular Gels: Materials with Self-Assembled Fibrillar Networks*, ed. P. Terech and R. G. Weiss, Springer, New York, 2005, p. 929.

E. Carretti, L. Dei and R. G. Weiss, Soft matter and art conservation. Rheoreversible gels and beyond, *Soft Matter*, 2005, **1**, 17–22.

M. G. Zepeda Martinez, *Uso de geles en la limpieza gradual de pinturas sobre lienzo*, Instituto Nacional de Antropología e Historia, Mexico City, 1997.

Review and Questions

The main points discussed in the chapter may be summarized as follows:

- Gels and highly viscous systems in general are the most technologically advanced tools for the cleaning of painted surfaces, especially wood and canvas paintings. Owing to their high retention power and viscosity, the penetration of liquid solvents into porous matrices is minimized, and therefore the drawbacks related to the swelling of the paint layer and to the scarce control of the cleaning action are avoided.
- The most relevant limitation to the use of traditional gels is the complete and selective removal of the gel components.
- In order to minimize both the presence of gel residues and the invasiveness of the gel removal, different solutions are offered by nanotechnologies, with a reasonable level of performance. In particular:
  a. Responsive systems, *i.e.* families of gels that can be easily and completely removed from the paint surface by exploiting some of their

peculiar physico-chemical properties (*e.g.* with nanomagnetic sponges and rheoreversible gels);

b. Highly viscous polymeric dispersions, particularly effective for the gradual, controlled and selective removal of natural varnishes, that can be simply removed with a pincer without leaving detectable amounts of residue on the paint surface;
c. Chemical gels.

Questions

1) What is the procedure for the preparation of a HVPD, using either fully hydrolysed PVA or 80% hydrolysed PVA?
2) What are the main drawbacks related to the use of liquid solvents for the cleaning of easel paintings? How can these drawbacks be completely or partly eliminated with the application of gels (or highly viscous systems)?
3) What are the main advantages related to the use of HVPDs?
4) Illustrate the main macroscopic differences between HVPDs and traditional solvent gels.
5) Indicate the main advantages of including an o/w microemulsion in a highly viscous matrix.
6) Indicate the main approaches based on nanotechnologies for the minimization of gel residues on the paint surface.

Answers

1) HVPDs with fully hydrolysed PVA are prepared by dissolving PVA in an aqueous solution of sodium borate, followed by the addition of the solvent. The system is kept at 90 °C for 3 hours under stirring. Then, in order to obtain the gel, the system is slowly cooled to room temperature. For systems containing 80% hydrolysed PVA, the polymer is dissolved in water (or in a water–solvent mixture) at room temperature. The appropriate amount of a 3 w/w% borax solution is added drop-wise to the solution while the system is stirred continuously. The polymeric network becomes rigid after a few minutes.
2) There are two main problems related to the direct application of pure organic solvents for the removal of foreign materials from the surface of paintings:
   - The poor control of the liquid media, and particularly of their spread onto the paint surface, and their penetration and diffusion into the porous structure, which can cause the swelling of the binding media and of the original varnish layers;
   - The incomplete selectivity of the cleaning action.

   Gels, owing to their high power of retention and high viscosity, minimize the penetration of liquid solvents into the paint layer. Furthermore, because most of these gels are completely transparent, the cleaning action can be easily controlled during the contact time between the gel and the paint surface.

3) The main advantages in using HVPDs are related to the elasticity of these systems, which can be easily removed from the paint surface with a pincer without leaving any detectable residue on the treated surface.
4) From the physical point of view, HVPDs are much more elastic than solvent gels, which have the consistency of a hair gel. As a consequence, while the complete removal of HVPDs is easily carried out, the removal of solvent gels requires the application of a liquid solvent and a potentially invasive mechanical action.
5) As in the case of pure organic solvents, the confinement of an o/w microemulsion in a highly viscous matrix (*i.e.* a gel or an HVPD) minimizes the penetration of the cleaning system into the porous matrices, as well as the impact of the cleaning action on the support.
6) The most important approaches that minimize the amount of residue left on the paint surface are three in number:
   - The use of highly elastic systems such as HVPDs that are very viscous and can be easily removed simply by peeling them off from the paint surface with a pincer;
   - The use of chemical gels. For example, sponges functionalized with magnetic nanoparticles allow simple removal of the cleaning agent with a permanent magnet, avoiding any contact between the operator and the work of art;
   - Alternatively, HEMA-VP hydrogels can be used as new nanostructured containers for microemulsions used in the removal of polymers from canvas paintings.

References

1. S. Michalski, *Cleaning, Retouching and Coatings: Technology and Practice for Easel Paintings and Polychrome Sculpture: Preprints of the contributions to the Congress, 3–7 September 1990*, ed. J. S. Mills and P. Smith, International Institute for Conservation of Historic and Artistic Works, London, United Kingdom, 1990, p. 85.
2. R. C. Wolbers, in *Workshop on New Methods in the Cleaning of Paintings and other Decorative Surfaces*, 1988, Marina del Rey, CA, Canadian Conservation Institute. Ottawa, Canada, 1990, p. 188.
3. N. Butazzoni, A. Casoli, P. Cremonesi and P. Rossi, *Progetto restauro*, 2000, **7**, 11.
4. N. Valentin, A. Sánchez and I. Herraez, in *Preprints of the ICOM committee for conservation, 11th triennial meeting in Edinburgh, Scotland, 1–6 September 1996*, ed. J. Bridgland, James & James Ltd., London, United Kingdom, p. 851.
5. M. Tomozei and Z. Balta, in *Metal 98: Proceedings of the International Conference on Metals Conservation. Draguignan-Figanières, France, 27–29 May 1998*, James & James, London, United Kingdom, p. 188.

6. L. da Silveira, *Stud. Conserv.*, 1997, **42**, 11.
7. N. Khandekar, *Rev. Conserv.*, 2000, **1**, 10.
8. S. Iannuccelli and S. Sotgiu, in *Preprints of GDWG Interim Meeting ICOM-CC "Choices in Conservation practice versus Research", 6–8 October 2010, Copenhagen*, ed. L. Watteeuw, M. Vest, J. Palm and D. van der Reyden, ICOM-CC, 2010, p. 47.
9. E. Campani, A. Casoli, P. Cremonesi, I. Saccani and E. Signorini, *L'uso di Agarosio e Agar per la preparazione di "Gel Rigidi" – Use of Agarose and Agar for preparing "Rigid Gels", Quaderni del Cesmar7*, number 4, Il Prato, Padua, 2007.
10. D. Stulik, D. Miller, H. Khanjian, N. Khandekar, R. Wolbers, J. Carlson and W. C. Petersen, in *Solvent Gels for the Cleaning of Works of Art, The Residue Question*, ed. V. Dorge, Getty Publications, Los Angeles, 2004, pp. 18–83.
11. A. Burnstock and T. Kieslich, in *Preprints of the ICOM committee for conservation, 11th triennial meeting in Edinburgh, Scotland, 1–6 September 1996*, ed. J. Bridgland, James & James Ltd., London, United Kingdom, p. 253.
12. E. Carretti, L. Dei, P. Baglioni and R. G. Weiss, *J. Am. Chem. Soc.*, 2003, **125**, 5121.
13. E. Carretti, A. Macherelli, L. Dei and R. G. Weiss, *Langmuir*, 2004, **20**, 8414.
14. M. Bonini, S. Lenz, R. Giorgi and P. Baglioni, *Langmuir*, 2007, **23**, 8681.
15. M. Bonini, S. Lenz, E. Falletta, F. Ridi, E. Carretti, E. Fratini, A. Wiedenmann and P. Baglioni, *Langmuir*, 2008, **24**, 12644.
16. H. Ochiai, Y. Kurita and I. Muratami, *Macromol. Chem.*, 1984, **185**, 167.
17. M. Tsujmoto and M. Shibayama, *Macromolecules*, 2002, **35**, 1342.
18. I. Sakurada, *Polyvinyl Alcohol Fibers*, Marcel Dekker, New York, 1985.
19. M. Shibayama, M. Adachi, F. Ikkai, H. Kurokawa, S. Sakurai and S. Nomura, *Macromolecules*, 1993, **26**, 623.
20. http://goldbook.iupac.org/C01238.html (accessed November 29 2012).
21. I. Natali, E. Carretti, L. Angelova, P. Baglioni, R. G. Weiss and L. Dei, *Langmuir*, 2011, **27**, 13226.
22. F. M. Herman, *Encyclopedia of Polymer Science and Technology*, Wiley, San Francisco, 2004.
23. G. Pizzorusso, Synthesis and characterization of systems for the micro-confinement of detergents for the cleaning of canvas paintings, PhD thesis, University of Florence, 2010; J. A. L. Domingues, N. Bonelli, R. Giorgi, E. Fratini, F. Gorel and P. Baglioni, *Langmuir*, 2013, In press.
24. University of Oxford, *Chemical and Other Safety Information*, 2010, http://msds.chem.ox.ac.uk/.
25. B. Tighe, Hydrogels as contact lens materials, in *Hydrogels in Medicine and Pharmacy. Properties and applications*, CRC Press, Boca Raton, Florida, 1987, vol. III.
26. E. J. Mack and S. W.Kim, Biomedical applications of poly(2-hydroxyethyl methacrylate) and its copolymers, in *Hydrogels in Medicine and Pharmacy*, CRC Press, Boca Raton, Florida, 1987, vol. II, pp. 65–93.

27. O. Wichterle and D. Lim, *Nature*, 1960, **185**, 117.
28. E. Carretti, I. Natali, C. Matarrese, P. Bracco, R. G. Weiss, P. Baglioni, A. Salvini and L. Dei, *J. Cult. Herit.*, 2010, **11**, 373.
29. P. Bauduin, A. Renoncourt, A. Kopf, D. Touraud and W. Kunz, *Langmuir*, 2005, **21**, 6769.
30. E. Carretti, S. Grassi, M. Cossalter, I. Natali, G. Caminati, R. G. Weiss, P. Baglioni and L. Dei, *Langmuir*, 2009, **25**, 8656.
31. E. Carretti, E. Fratini, D. Berti, L. Dei and P. Baglioni, *Angewandte Chem.*, 2009, **48**, 8966.
32. K. Almdal, J. Dyre, S. Hvidt and O. Kramer, *Polymer Gels Netw.*, 1993, **1**, 5.
33. I. Pastorova, K. J. van der Berg, J. J. Boon and J. W. J. Verhoeven, *Anal. Appl. Pyrolysis*, 1997, **43**, 41.
34. E. Marengo, M. C. Liparota, E. Robotti and M. Bobba, *Vibr. Spectrosc.*, 2006, **40**, 225.
35. G. A. Van der Doelen and J. J. Boon, *J. Photochem. and Photobiol. A*, 2000, **134**, 45.
36. E. R. De la Rie, *Stud. Conserv.*, 1988, **33**, 53.
37. A. M. Seves, S. Sora, G. Scicolone, G. Testa, A. M. Bonfatti, E. Rossi and A. Seves, *J. Cult. Herit.*, 2000, **1**, 315.
38. G. Testa, A. Sardella, E. Rossi, C. Bozzi and A. Seves, *Acta Polym.*, 1994, **45**, 47.
39. A. Chevalier, D. Chelazzi, P. Baglioni, R. Giorgi, E. Carretti, M. Stuke, M. Menu and R. Duchamp, in *Proceedings of the ICOM-CC 15th Triennial Conference, 22–26 September 2008*, New Delhi (India), 2008, Vol. 2, pp. 581–589.
40. H. E. Ahmed and F. N. Kolisis, *J. Appl. Polym. Sci.*, 2012, **124**, 3565.
41. E. Carretti, L. Dei and P. Baglioni, *Langmuir*, 2003, **19**, 7867.
42. E. Carretti, R. Giorgi, D. Berti and P. Baglioni, *Langmuir*, 2007, **23**, 6396.
43. R. Giorgi, M. Baglioni, D. Berti and P. Baglioni, *Acc. Chem. Res.*, 2010, **43**, 695.
44. M. Baglioni, D. Rengstl, D. Berti, M. Bonini, R. Giorgi and P. Baglioni, *Nanoscale*, 2010, **2**, 1723.
45. M. Baglioni, R. Giorgi, D. Berti and P. Baglioni, *Nanoscale*, 2012, **4**, 42.
46. G. Pizzorusso, E. Fratini, J. Eiblmeier, R. Giorgi, D. Chelazzi, A. Chevalier and P. Baglioni, *Langmuir*, 2012, **28**, 3952.
47. L. Osete-Cortina and M. T. Doménech-Carbó, *J. Chromatogr. A*, 2006, **228**, 1127.
48. R. Murugan, S. Mohan and A. Bigotto, *J. Kor. Phys. Soc.*, 1998, **32**, 505.
49. P. Garside and P. Wyeth, *Stud. Conserv.*, 2003, **48**, 269.

CHAPTER 12

Inorganic Nanomaterials: Synthesis and Properties

M. BONINI,* P. BAGLIONI AND D. CHELAZZI

University of Florence, Italy
*Email: bonini@csgi.unifi.it; baglioni@csgi.unifi.it; chelazzi@csgi.unifi.it

12.1 Nanostructured Materials: Introduction to Size-dependent Properties

More than 50 years after the seminal talk given by the Nobel Laureate Richard Feynman, we can say that the expected nanotechnology revolution has finally taken place. Nanostructured materials (or simply nanomaterials) have at least one dimension in the nanometre range, *i.e.* 10^{-9} m. Such a definition includes a wide range of structures that can be classified according to their dimensionality. If a bulk tridimensional (3D) material is made thinner and thinner, until the thickness falls into the nanometre range, the sheet of material obtained is a bidimensional (2D) nanomaterial. Examples of this class are quantum wells (sandwiched layers of semiconductors), superlattices, supramolecular networks, Langmuir–Blodgett films, membranes, graphene, *etc.* If this layer is further made thinner along a direction perpendicular to its surface, until the nanoscale is reached again, a monodimensional (1D) nanomaterial is obtained. Carbon nanotubes, metal nanowires, nanorods and cylindrical micelles can all be assigned to this class. When the only dimension not yet in the nanometre range is reduced to the nanoscale, zerodimesional (0D) nanostructures are finally obtained, such as nanoparticles, colloids, nanodots from the gas phase, *etc.* In Figure 12.1 a schematic representation for each class is given.

RSC Nanoscience & Nanotechnology No. 28
Nanoscience for the Conservation of Works of Art
Edited by Piero Baglioni and David Chelazzi

Published by the Royal Society of Chemistry, www.rsc.org

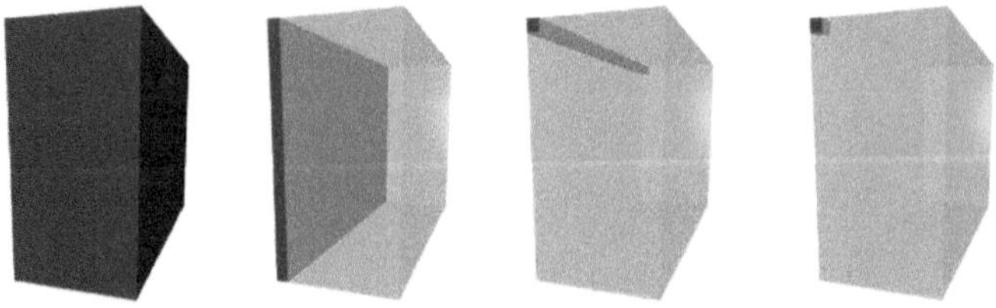

Figure 12.1 Sketch highlighting the sequential path from 3D to 2D, 1D and 0D structures.

The size reduction brings about the peculiar properties of nanomaterials, which lie in between those of atoms and bulk materials. In the case of 0D nanostructures, as the size is reduced the particles consist of a very limited number of atoms. Figure 12.2 (top) shows the variation in the number of atoms as a function of the number of atomic shells, *i.e.* of the number of layers assembled on top of a single atom.

In the case of gold clusters, the radius of a nanocluster consisting of seven layers is slightly less than 2 nm. Such a nanoparticle is made of 1415 atoms.[1] Of these, about 500 lie on its surface. Surface atoms differ from bulk atoms because of their coordination numbers, *i.e.* the number of nearest neighbour atoms surrounding each atom. This leads to significant changes in many properties related to the differences in reactivity and in the energetic scenario between bulk and surface atoms. Some examples are briefly described in the following sections.

12.1.1 Melting Point

The lower coordination number of surface atoms corresponds to a decrease in their binding energy compared with bulk atoms. This is clearly reflected in the lowering of the melting point as a function of the size. In the case of gold nanoparticles the melting point of nanoparticles smaller than 3–4 nm is nearly halved with respect to that of the bulk material (see Figure 12.2, centre). In fact, surface atoms can be separated more easily from the particle and rearranged in a less-ordered structure, corresponding to the onset of the melting process at lower temperatures.

12.1.2 Surface Area

A reduction in particle size, together with changes in the number of surface atoms, results in a greater surface area. A size reduction from a microparticle (for instance, 5 μm in radius) to a nanoparticle (for instance, 50 nm in radius) leads to an increase in the surface area per unit volume of a factor of 10^2. The relationship between the surface area per unit volume and the radius of the particles is schematically shown in Figure 12.2 (bottom), where the surface area

Full-shell Clusters		Total Number of Atoms	Surface Atoms (%)
1 Shell		13	92
2 Shells		55	76
3 Shells	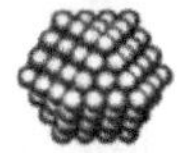	147	63
4 Shells		309	52
5 Shells	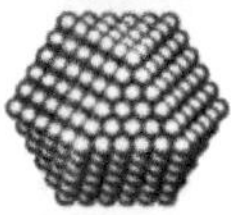	561	45
7 Shells	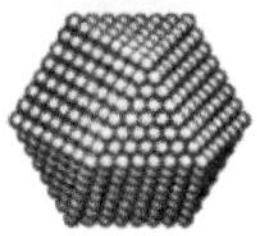	1415	35

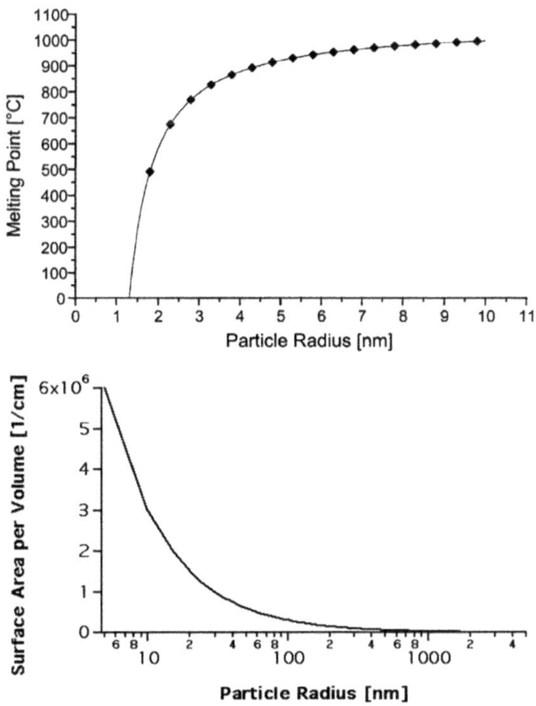

Figure 12.2 (Top) Total number of atoms and % of surface atoms in clusters of nanoparticles consisting of one to seven full shells. (Centre) Decrease of melting point in gold nanoparticles as the size is decreased. (Bottom) Increase in the surface area per unit volume in nanoparticles as the size is decreased. (Top and middle figures are reproduced from Ref. 1 with permission of Wiley VCH.)

per unit volume is plotted against the radius of a corresponding spherical particle. An example of size reduction, in the framework of the conservation of historically valuable paper documents, is related to the use of particle dispersions for the deacidification of cellulose (see also Chapter 15). Recently, the efficiency of nanoparticles of $Mg(OH)_2$ (about 150 nm) dispersed in alcohol has been compared to that of microparticles of MgO (slightly less than 1 μm) dispersed in fluorinated solvents (Bookkeeper® Deacidification Process[2]). The two methods show comparably good deacidifaction properties, but the reduced size of magnesium hydroxide nanoparticles allows the treatment of low-porosity paper granting a deeper penetration, without leaving undesired particles as deposits on the treated surface. Moreover, nanoparticles are stable in alcohol (*e.g.* propanol), without the need for stabilizers that could leave residues on the treated artefacts. On the other hand, the Bookkeeper method has possible drawbacks due to the use of large amounts of fluorinated surfactants, while granting the advantage of a more inert solvent with respect to alcohols.[3] This is a clear example where the reduction of the particle dimensions allows for an innovative formulation where drawbacks are minimized, but still preserving optimal performance (see also Section 15.6.5).

12.1.3 Catalytic Activity

Such a high proportion of atoms on the surface modifies the chemical behaviour of the materials for several reasons. As the size gets smaller, the number of defect sites on the surface is increased (such as anion or cation vacancies; see also Section 1.4), as well as the number of atoms lying on edges and corners; in fact, crystalline nanoparticles are not spherical, but polyhedral (see Figure 12.2, top). As a result, the chemical activity is strongly enhanced, which is especially important for catalytic applications. These effects are of interest in the field of restoration and consolidation. Dispersions of $Ca(OH)_2$ particles are widely used for the deacidification and consolidation of many types of works of art,[4–6] owing to their carbonation reaction with CO_2 (see Chapters 13 and 16). This process is of fundamental importance for both deacidification and consolidation, and it depends on the surface area of the particles, *i.e.* their size: smaller particles exhibit an enhanced reactivity.[7]

12.1.4 Solubility

In a similar manner, the presence of atoms that can be separated more easily from the particles leads to significantly increased dissolution rates, making the nanostructuration of drug crystals an effective approach to pharmaceutical formulations showing enhanced bio-availability. Furthermore, the different solubility of atoms belonging to different crystalline planes is often exploited to prepare nanoparticles of different shapes. This is the case in the preparation of silver nanoprisms, which are of particular interest in the field of protection of artefacts whose surface is exposed to the growth of bacteria (see Section 12.7).

12.1.5 Sintering

A direct consequence of the decrease in the melting point as a function of the particle size is the preparation of sintered materials from nanostructured powders at much lower temperatures than those with conventional micropowders as precursors. The advantage of sintering nanopowders is twofold, because a significant amount of energy can be saved and materials with peculiar properties are obtained.

12.1.6 Mechanical Properties

The mechanical properties of materials are related to their atomic structure. It is not surprising therefore that the mechanical properties of nanomaterials are often greatly different from those of the corresponding bulk materials, and bring significant improvements to many applications. This is the case for many nanostructured ceramics, which show higher strength and hardness, as well as carbon nanostructures, as demonstrated by the high strength plus extraordinary flexibility and resilience of carbon nanotubes. In the field of Cultural Heritage conservation, very often it is necessary to recover the mechanical properties of the artefact to its original state. For this reason the cohesive properties of dispersions of calcium hydroxide nanoparticles are of great importance in the consolidation of deteriorated works of art (see Chapter 13).

12.1.7 Electronic Properties

Metal electrons can be excited by electromagnetic radiation. In particular, when the size of the metal particle is smaller than or similar to the mean free path of the electrons, visible light causes collective oscillations, referred to as plasmons. These electronic oscillations lead to fluctuations in surface charge, resulting in dipole, quadrupole and higher multipole moments.[8] In silver and gold nanoparticles, plasmons result in strong absorbance in the visible region. The intensity, wavelength and width of the absorption peaks depend on the composition, size and shape of the nanoparticles, as well as on the surrounding medium. Figure 12.3 shows a *rainbow* of colours from dispersions of silver nanoparticles of different sizes and shapes.

12.1.8 Semiconductors: Quantum Dots

Nanoparticles made of a semiconductor are often referred to as quantum dots. Rigorously, a quantum dot is a particle whose excitons are confined in all three dimensions. Consequently, such materials have electronic properties that are neither those of bulk semiconductors nor those of discrete molecules. In a bulk semiconductor, the valence and conduction bands correspond to a large number of discrete electronic quantum states, separated by an energy bandgap of 4 eV at most. When the size of the crystal is reduced, the difference in energy between the highest valence and the lowest conduction band becomes larger,

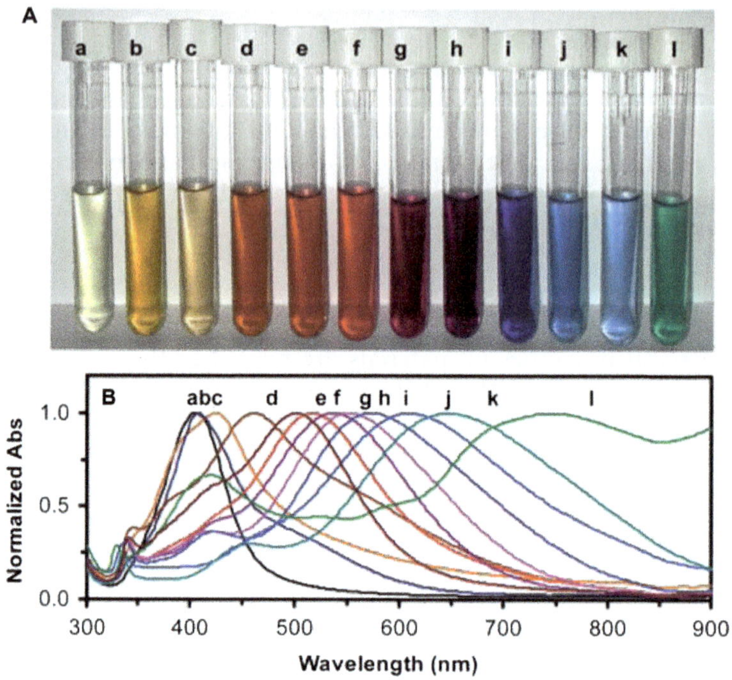

Figure 12.3 Plasmonic optical properties of Ag nanoparticles. (A) Photos of colloids showing: (a) light yellow; (b) yellow; (c) light orange; (d) orange–red; (e) red; (f) dark red; (g) purple; (h) purple–violet; (i) violet; (j) blue; (k) light blue, and (l) green colours. (B) Normalized absorbance of UV/Vis absorption spectra of the Ag nanoparticles pictured in (A). (Reproduced from Ref. 9 with permission from The Royal Society of Chemistry.)

owing to an increase in the energy needed to excite the dot (see Figure 12.4). As a consequence, the absorption and emission wavelengths of light shift towards smaller wavelengths; thus semiconductor nanoparticles typically absorb in the visible–ultraviolet (Vis/UV) region. In analogy to the definition of quantum dots, quantum wires are those structures confining electrons or holes in two spatial dimensions and allowing free propagation in the third, while structures able to confine electrons or holes in one dimension and allow free propagation in two dimensions are defined as quantum wells. Owing to these unique properties, semiconductor nanostructures have been already studied in LEDs (light emitting diodes), transistors, solar cells, and as markers for medical imaging.

12.2 Methods of Synthesis of Inorganic Nanoparticles

Inorganic nanoparticles can be prepared through several different methods. In this section we provide a short review of the most common strategies pursued

Figure 12.4 Pictures of three colloidal suspensions of quantum dots. On the right, the corresponding energetic scenario is sketched, highlighting the decrease in the band gap between conduction and valence bands as the nanoparticle dimensions are increased.

to fabricate these materials, with a special emphasis on advantages and drawbacks. Each method will be also discussed in terms of its suitability for different production scales (small or industrial scale) or if it is easily scalable to different amounts. In the specific case of the synthesis of oxide and hydroxide nanoparticles from alkaline earth metals, more details will be provided in Section 12.4. In the last two decades these materials have found fundamental applications for the compatible preservation of both immovable (*e.g.* consolidation of wall-paintings) and movable works of art (*e.g.* deacidification of paper and wood artefacts). A comprehensive discussion of the most recent and promising applications of these classes of nanomaterials in different restoration fields is given in Chapters 13–16.

A first basic difference between *top-down* and *bottom-up* approaches resides in the starting material. In top-down methods a material whose features are not in the nanoscale is worked to produce a nanostructured material. On the other hand, in bottom-up methods the nanostructures are built atom by atom (or molecule by molecule) from scratch. Currently, the advantages of different strategies and their better suitability for laboratory or industrial scales depend strongly on the material to be produced. Different methods are very often combined to produce materials with unprecedented properties, highlighting the convergence of top-down and bottom-up methods in structures with sizes between a few nanometres and a few tenths of micrometres (see Figure 12.5).

12.2.1 Top-down Methods

These methods start from a bulk material and break it into small pieces using mechanical, chemical or other forms of energy. Top-down fabrication resembles the process of sculpting from a block of stone: the excess material is gradually removed from the starting block until the desired shape is achieved.

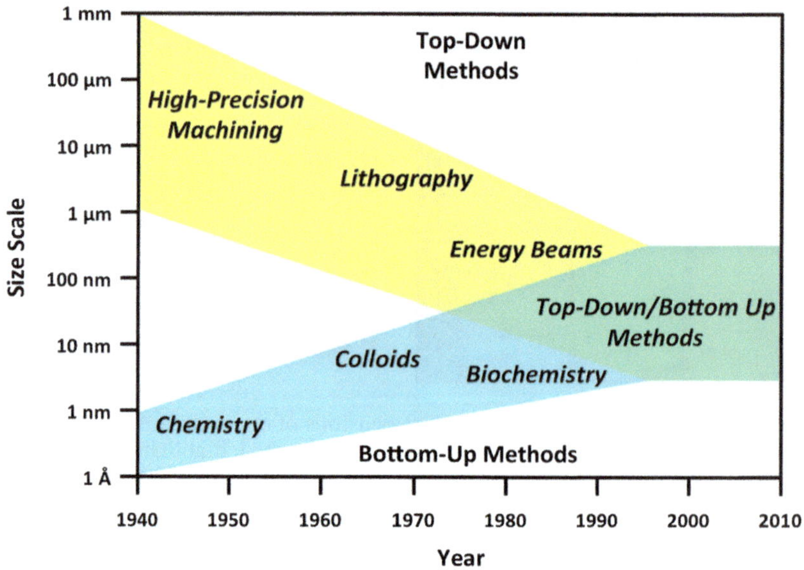

Figure 12.5 Development of top-down and bottom-up methods over recent decades.

These methods are often very convenient for production at the industrial scale because they result from the progress of fabrication methods that were already in use. Nevertheless, devices such as planetary mills, beam-assisted etching machines, or advanced extruders are currently used in many research laboratories to produce small amounts of extremely sophisticated nanostructured materials.

An example is represented by nanolithography, the technology most commonly used to manufacture computer chips. In this process, the material required in the final product is protected by a mask, while the material to be etched away remains exposed. The etching process is carried out either chemically or by means of UV light or X-rays, depending on the composition of the material to be removed. Electron and ionic beams are used as well, with the advantage of not requiring any mask. Milling is another example: powders are treated through repeated high-energy impacts, producing particles with smaller size and both structural and compositional changes. A variety of milling devices has been developed for different purposes, such as attrition mills, shaker mills, and planetary mills. All these milling approaches have a common drawback: the control of the dimensions of the particles is very poor. Nanopowder samples resulting from milling are generally polydisperse and often assembled into micrometre-sized clusters. Despite this, top-down methods are being considered, together with bottom-up approaches, for the production of nanomaterials aimed at the preservation of Cultural Heritage. Recently, $SrCO_3$ particles (with crystallites of 20–50 nm), obtained through ball-milling and then stably dispersed in propanol, were used in the deacidification of wood samples derived from Henry VIII's flagship the *Mary Rose* (see Section 16.5.1).[9]

12.2.2 Bottom-up Methods

Bottom-up methods deal with atoms that are assembled through appropriate experimental conditions into particles of defined size. Bottom-up fabrication resembles therefore the building of a brick house, where the bricks are replaced by the atoms or the molecules. Compared with top-down methods, bottom-up procedures are much better suited to provide control of the size, shape and structure of the generated nanoparticles. Two general methods are available to control the formation and growth of nanoparticles. Arrested precipitation is based on the depletion of the precursors during the formation of the nanomaterials and/or the presence of chemicals that are able to block the growth of the particles or to block the formation process. Another method is based on the restriction of the physical space available for the growth of individual nanoparticles through the use of templating agents. In both cases, building up a material from atomic or molecular species can be very time consuming. For this reason self-assembly techniques, where atoms and/or molecules are able to arrange themselves rationally, are often employed.

In the case of inorganic nanoparticles, bottom-up processes are commonly based on the reaction of inorganic or organometallic molecules as precursors. The sol–gel technique is among the most commonly used procedures for the production of inorganic oxide nanoparticles at the industrial level. The main advantages of the sol–gel technique are its cost-effectiveness, due to the low temperature processing, and the flexibility in terms of rheology, allowing easy shaping and embedding. The most commonly used precursors in the sol–gel production of oxide nanoparticles are alkoxides (*i.e.* molecules consisting of an organic group bound to a negatively charged oxygen), both because of their commercial availability and the chemical versatility of the M–OR bond, which provide a simple route to the synthesis of composite organic–inorganic materials. The sol–gel technique can be applied to the conservation of archaeological and historical wood. In fact, it enables the silicification of wood either by coating or impregnation with nanoparticulate silica sols, granting the mechanical consolidation of wood (see Section 16.4.1). Moreover, mesoporous silica nanoparticles with well-defined and controllable morphology and porosity, produced through a sol–gel synthesis, can potentially be used as catalysts to convert pollutant gases or to oxidize organic materials in order to minimize the formation of black crusts on artistic surfaces (see Section 14.2.1).

Spray pyrolysis is another method for the industrial production of nanoparticles. During the process, the reagents are introduced in the form of a spray (gas, liquids, aerosols or mixtures of them) into a heat source, such as plasma, flame or laser. The reagents decompose and particles start forming through homogeneous nucleation and growth. Rapid cooling results in the formation of nanoscale particles. In aerosol-based processes, solid or liquid particles are introduced at high pressure in a gas flow. The size of the solid particles or liquid droplets ranges from one molecule up to 100 μm. Spraying of the precursors into a hot atmosphere or onto a heated surface finally results in the drying/pyrolysis of the wet materials or in the deposition of coatings, respectively.

Nanostructured coatings are often produced through chemical vapour deposition (CVD), where a gaseous precursor is reacted with the surface of the substrate. The reaction is activated either thermally (thermal CVD) or by a plasma atmosphere (PECVD, plasma enhanced chemical vapour deposition). A similar approach is pursued in the preparation of nanopowders by the gas-phase condensation method. A bulk material is introduced in a chamber and is heated at a temperature far above the melting point, but lower than the boiling point, to produce a significant vapour pressure. A stream of gas (most commonly an inert gas) continuously moves the vapour phase away from the heating stage, inducing the steady evaporation of the material. As a consequence of their collisions with the gas molecules, the atoms in the vapour phase are rapidly cooled, and condense in small droplets. During the initial stages of the process the droplets are too hot to solidify; in fact, while still in the liquid phase, they collide and coalesce, leading to the formation of smooth and spherical droplets whose size is controlled by the environmental conditions. As the liquid particles are further cooled, solid nanopowders are finally obtained. This method is suitable not only for metal nanoparticles, but also for oxides (in this case the gas is oxygen).

Supercritical fluids (*i.e.* fluids taken to their supercritical state by controlling temperature and pressure) have been extensively used for the synthesis of nanoparticles. Given that they can dissolve a wide range of chemical precursors, these fluids are used to form nanoparticles through the rapid expansion of a supercritical solution.

An elegant method for the production of nanoparticles is the use of templating architectures that confine and control the growth of individual particles. Any material made of nanometric voids can be used to prepare inorganic nanoparticles, as far as these cavities are accessible to the chemical precursors. A bright example of this strategy is provided by Nature. Ferritin is a native protein responsible for the storage of iron through its mineralization as ferrihydrite within the spherical protein cage. Ferrihydrite can be dissolved without destroying the cage, leading to the formation of apoferritin, which can then be used as a template for the synthesis of inorganic nanoparticles with very well controlled size.

A similar approach is used in the preparation of nanoparticles in microemulsions. A microemulsion is a thermodynamically stable dispersion of two immiscible liquids, consisting of nanosized droplets of one liquid into the other, stabilized by a surfactant (see Chapter 8). The use of microemulsions in the production of nanostructured materials is sketched in Figure 12.6.

In particular, inverse microemulsions (w/o microemulsions: *i.e.* nanodroplets of water dispersed in oil) provide nanocompartments that can be used for the reduction or precipitation *in situ* of metal salts or ceramic materials.[10] Three properties of microemulsions are crucial for the preparation of nanoparticles. First, microemulsions are thermodynamically stable systems: once they are formed, they will be indefinitely stable as long as the thermodynamic parameters (namely the temperature) are not changed. This makes their preparation and storage extremely simple: in fact, they can be prepared by

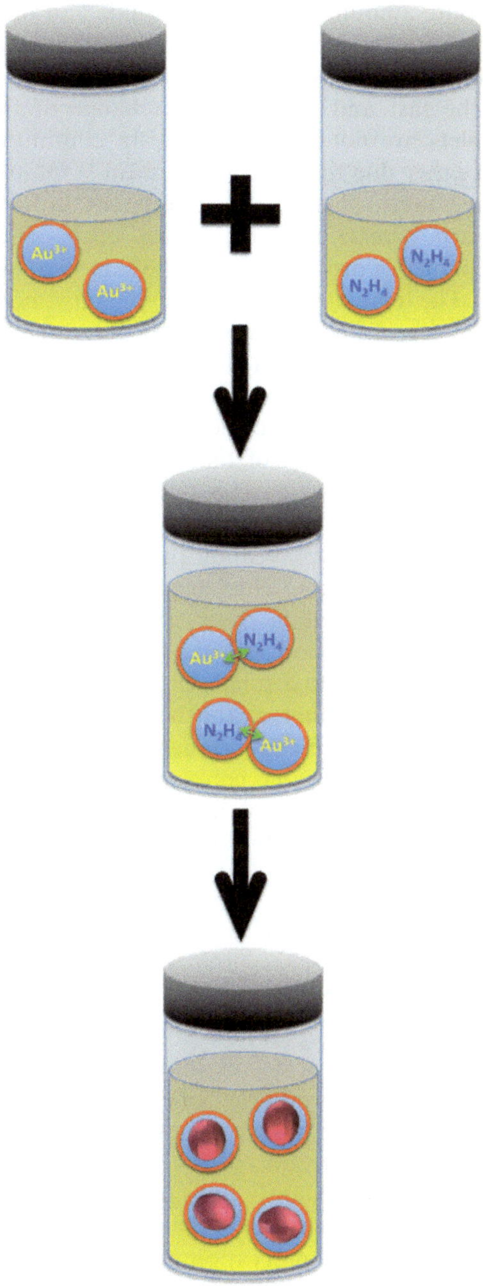

Figure 12.6 Schematic representation of the preparation of gold nanoparticles in inverse microemulsions through the reduction of Au(III) with hydrazine.

mixing the appropriate amounts of the two liquids and of the surfactant, the resulting system not being influenced by the preparation procedure (contrary to the case of emulsions). Second, one of the most important features of microemulsions lies in the fast and continuous exchange of the contents of the droplets. The droplets are not static, but are in continuous movement and collision with each other due to Brownian motion.[11] On each collision, some material interchange takes place between different droplets. It is therefore sufficient to mix two inverse microemulsions, each containing a different reactant within the aqueous droplets, to run the same reaction that would take place in bulk water. The templating effect of these nanosized droplets represents another crucial property of microemulsions in the synthesis of nanoparticles. In fact, when the reaction consists in the reduction of a metal salt or in the precipitation of an inorganic oxide, the use of microemulsions allows reduction of the growth of the particles down to the nanoscale. In particular, in the case of inverse microemulsions, the size of the droplets is mainly dictated by the ratio between the amounts of water and surfactant. Once the particles have reached the size of the droplets, they are stabilized through the adsorption of the surfactant on their surface, preventing further growth and leading to the formation of a stable suspension. This suspension can be eventually destabilized with a flocculating agent in order to separate and collect the solid content. However, the separation of nanopowders is often a long and complicated process, and leads to a small amount of powder. This is the main limitation of microemulsion technology for the synthesis of nanopowders at the industrial scale. Nonetheless, microemulsions represent an optimal tool for the synthesis of nanoparticles with various compositions and sizes at the laboratory scale. For instance, water-in-oil (w/o) microemulsions are used to produce nanoparticles of $Ca(OH)_2$ and $Mg(OH)_2$ (see Section 12.4), which are important materials for several applications in the preservation of Cultural Heritage.

12.3 Nucleation and Growth

Bottom-up methods for the preparation of nanoparticles are based on the assembly of a limited number of atoms into a solid phase. In the previous section, we introduced the synthesis of nanoparticles with templating structures. In this paragraph the basic concepts that control the early stages of nanoparticle formation are introduced, with a particular emphasis on the control of their dimensions, which is crucial for the effectiveness of several applications that involve nanoparticles. For instance, $Mg(OH)_2$ nanoparticles are successfully applied to the conservation of paper. In this case, the reduced dimensions of the nanoparticles lead to unmatched performance in the deacidification treatment, due to their high reactivity and to their easy conversion into carbonates.[12]

Nucleation is the process of forming a solid phase from a supersaturated, homogeneous liquid phase. This process is of fundamental importance for the control of the formation of inorganic crystals. An illustrative example is the synthesis of magnetic nanoparticles. This class of nanoparticles is of great interest for bio-medicine and magnetic recording, mainly because of their magnetic properties. In fact, because of their size, magnetic nanoparticles show a *superparamagnetic* behaviour.[13] Briefly, each nanoparticle consists of a single magnetic domain, behaving as a single magnetic dipole. This property is responsible for inducing motion and heat generation through the application of a static or oscillating magnetic field, respectively.

The magnetic responsiveness of these particles can be exploited also for the conservation of works of art. In particular, magnetic nanoparticles have been chemically embedded within hydrophilic gels, so as to produce magnetic gels (or "magnetic sponges").[14,15] These gels can be loaded with suitable micro-emulsions for the removal of detrimental polymer coatings from paints. The presence of magnetic nanoparticles allows the magnetic control of the application and the removalof the gel after the cleaning intervention, simply by applying a magnetic field. In this way the mechanical action and stress on the surface during the removal of the gel are minimized or completely avoided (see Sections 11.4 and 12.5).

During the synthesis of magnetic nanoparticles (*e.g.* magnetite, Fe_3O_4), the monomers are initially formed through the chemical reactions of a precursor. In particular, the formation of Fe_3O_4 molecules is achieved by rapidly increasing the pH of an aqueous solution of Fe(II) and Fe(III) ions upon addition of a strong base. The diagram in Figure 12.7, known as the La Mer diagram, shows the variation in the concentration of the monomer in solution as a function of time.

Three distinct phases (supersaturation, nucleation and growth) are shown, and the two different pathways that lead to particles with either a small or a high polydispersity. During the progress of the reaction, the concentration

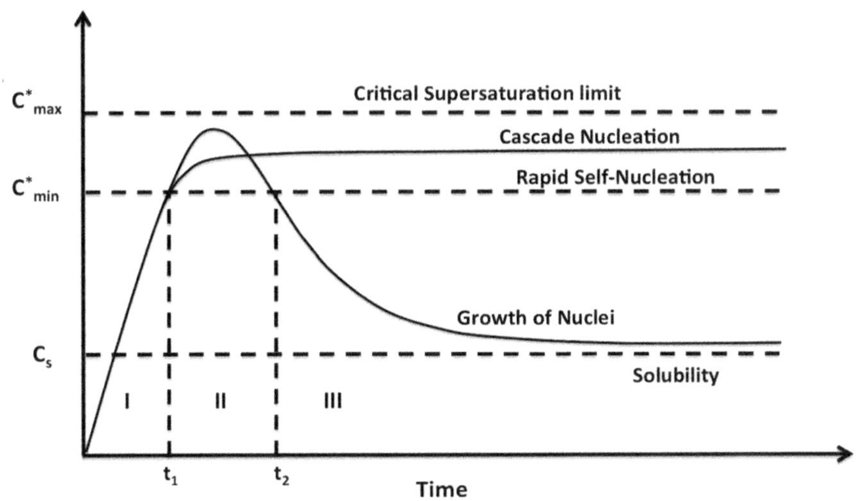

Figure 12.7 La Mer diagram. The supersaturation phase (phase I), the nucleation phase (phase II) and the growth phase (phase III) are indicated, as well as two different pathways leading to particles with different polydispersity.

of the monomer continuously increases before reaching a critical concentration threshold where the nucleation starts. This critical concentration, indicated as C^*_{min} in Figure 12.7, sets the border between the supersaturation phase (phase I) and the nucleation phase (phase II). During the nucleation, the phase separation begins, and the stable clusters separate from the monomer solution that is formed during the reaction. If the nucleation is not fast enough, a stationary concentration between the critical concentration of nucleation and the maximum critical concentration (C^*_{max}, which defines the upper limit of supersaturation) is established. Under these conditions, the monomer separates into a solid phase as quickly as it is produced by the reaction. New nuclei are continuously formed, while those already formed continue to grow, leading to the production of nanoparticles with many different sizes and a highly polydisperse sample. On the other hand, if the nucleation is extremely fast, it can overcome the rate of formation of new monomers. In this case the concentration of monomer in solution decreases again until it falls below the critical concentration of nucleation. From this moment on, which sets the border between the nucleation phase and the growth phase (phase III in Figure 12.7), the monomers formed in solution grow the nuclei that are already present, as new nuclei are no longer formed. In this way, the monomer concentration decreases until it reaches a stationary value that corresponds to the solubility concentration of the monomer, C_s. Because of the homogeneous growth of the nuclei, this process leads to the production of particles with low polydispersity.

For several applications it is desirable to use homogeneous nanoparticles, in terms of shape and size. This is the case, for instance, in applications where metal nanoparticles are used as pigments on the basis of their plasmonic properties.[16] In fact the size and its distribution determines the optical absorption and therefore the final colour. Another example is related to the consolidation of a porous matrix (*e.g.* wood or wall-paintings). In order to consolidate the surface and the bulk of a low-porosity artistic substrate, specific nanoparticles must be synthesized to match the porosity of the artefact. The use of monodispersed nanoparticles with small size (smaller than the average pore size) allows good penetration and, consequently, maximizes the beneficial effects. If, for example, the artistic substrate (*e.g.* the wall-painting) exhibits two main pore distributions (around two different average values), then a bidispersed nanoparticle population would grant good penetration and the interaction of particles with the walls of both the smaller and larger pores that need to be consolidated.

To obtain a monodispersed sample, the stage of nucleation must be as short as possible (*i.e.* $t_2 - t_1$ should be minimized; see Figure 12.7). In fact, the nuclei should all be formed at the same time and their growth should take place at the same time. Furthermore, the stages of nucleation and growth should be well separated, because the growth of particles during the nucleation process should be avoided. In other words, to obtain monodispersed nanoparticles it is fundamental to control both the nucleation and the growth rate, as the diagram in Figure 12.8 shows.

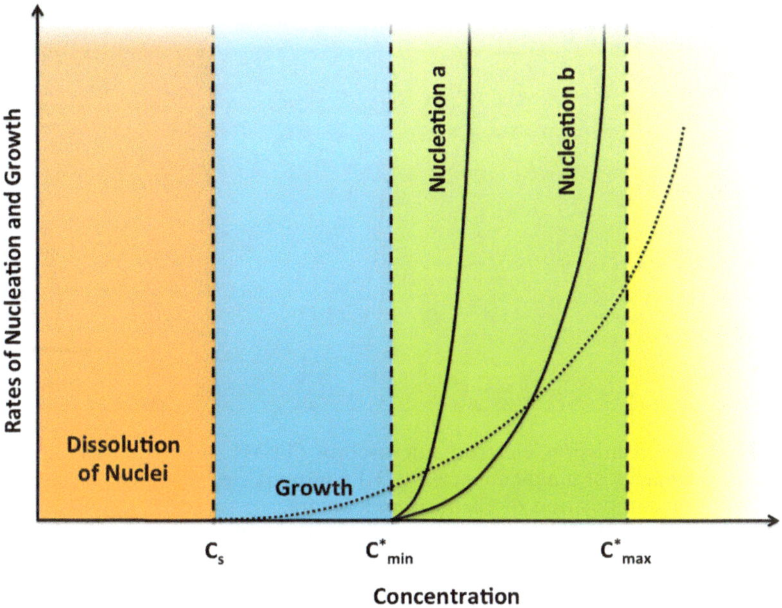

Figure 12.8 Rates of nucleation and growth displayed *vs.* the concentration of the monomer for two hypothetical preparations of nanoparticles (nucleations a and b).

Below the solubility limit of the monomers, the dissolution of the nuclei may occur. When the solubility concentration is reached, and if the nuclei are already present, then the growth starts. To produce new nuclei one has to reach at least the critical concentration, C^*_{min}. At this concentration the growth is already active, an event that makes it difficult to keep the nucleation and growth stages separated.

In practice, in order to obtain a sample as monodispersed as possible, the nucleation curve should be as steep as possible and, at the same time, the slope of the tangent to the curve of growth in the range of concentrations in which the nucleation takes place should be as low as possible. These two conditions provide a short nucleation phase and a reduction in growth of the nuclei in this time. Temperature represents another parameter that can be used to tune nucleation and growth, because it often affects the solubility of both the monomers and the final products.

To attain effective size control, different strategies are commonly pursued. Referring to the case of magnetic nanoparticles, two approaches are effective. In the first, iron precursors are dissolved in water at high concentration. To increase the solubility, the solution is heated to the boiling point. A boiling solution of a strong base, such as sodium hydroxide, is then mixed together with the Fe(II)–Fe(III) solution under strong stirring, leading to the formation of Fe_3O_4, which is poorly soluble in water. This procedure allows the rapid formation of a huge amount of nuclei and the fast consumption of all the metal

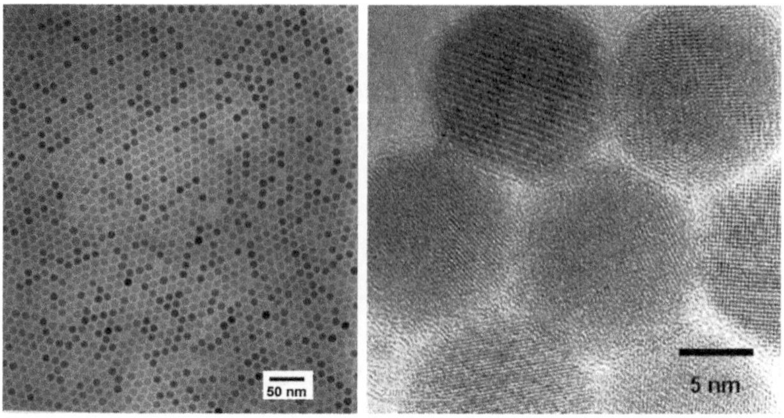

Figure 12.9 Transmission electron microscopy (TEM) and high resolution TEM images of maghemite (γ-Fe_2O_3) nanoparticles prepared through thermal decomposition of chemical precursors.
(Reprinted with permission from Ref. 19. Copyright 2001 American Chemical Society.)

ions, and results in the formation of small and quite monodispersed particles.[17,18] If the reaction is allowed to proceed for few hours, the dissolution of the less stable particles takes place, with an increase in the size of the nanoparticles, and a decrease in the polydispersity. When a higher monodispersity is needed, an effective procedure consists of the synthesis through thermal degradation of organometallic precursors in solvents with high boiling points. The use of such solvents brings the temperature up to about 200 °C, at which the organometallic precursors can be decomposed in the presence of stabilizing agents. This method was used for the preparation of highly monodispersed magnetic nanoparticles (see Figure 12.9),[19] with peculiar self-assembly properties driven by magnetic interactions.[20,21] Unfortunately, this method produces limited amounts of nanoparticles, because low concentrations of precursors are needed to achieve a high monodispersity.

12.4 Nanomaterials from Alkaline Earth Metals: Oxides and Hydroxides

In the last 20 years, alkaline earth metal hydroxides [*e.g.* $Ca(OH)_2$, $Mg(OH)_2$ and $Ba(OH)_2$] have proven to be vital materials for several restoration applications, especially when they are prepared in the form of nanosized particles, to be steadily dispersed in specific solvents (see Chapters 13–16). Alkaline earth metal oxide and hydroxide nanoparticles can be synthesized *via* several routes, but some of them are more suited than others to the production at large scale of materials to be used for the conservation of works of art. For instance, CaO nanoparticles were prepared by using an NaOH aqueous solution to precipitate $Ca(OH)_2$ nanocrystals from an ethylene glycol solution of calcium nitrate.[22] The crystals were then calcinated at 500 °C to obtain nanoparticles of few tens of nanometres. Similarly, CaO nanopods were synthesized through the precipitation of $CaCO_3$ nanoparticles from a saturated aqueous solution of $Ca(OH)_2$ by bubbling CO_2 in the presence of a polymer.[23] CaO nanoparticles were finally obtained by calcination of $CaCO_3$ at 700 °C. These two examples have a common drawback in the framework of conservation and consolidation of works of art, due the calcination process. In fact, during this treatment particles tend to fuse and grow, making the preparation of stable dispersions very difficult. Methods based on a homogeneous phase reaction in water and other organic solvents, without the need of the calcination step, have also been developed. By tuning the critical reaction parameters, such as the temperature, the concentrations and the ratios of the reactants, $Ca(OH)_2$ nanoparticles of different sizes and shapes can be produced. For example, it was shown that at temperatures above 100 °C the formation of nanoscaled particles in nonaqueous media is promoted.[24] The use of water[25] or glycols[4] as solvent leads to different results in terms of size, but not necessarily in terms of shape. In fact nanosized crystals of calcium hydroxide with a plate-like shape were obtained by both methods, with dimensions in the 30–60 nm range for glycols and in the 200–300 nm range for water. A drawback of these methods arises directly from the precursors commonly used for the synthesis (calcium chloride and sodium hydroxide), because sodium chloride is produced as a side-product. As described in Section 12.6, the presence of a high concentration of electrolytes (such as NaCl) leads to the destabilization of the nanoparticle dispersions, and requires the purification of the reaction products. Moreover, in this case the treatment of several types of works of art (*e.g.* consolidation of wall-paintings or stone) with the dispersions would lead to the contamination of the substrate with NaCl, which may produce detrimental effects such as salt efflorescence (see Section 2.2.2) or pigment alteration (it is well known that the famous azurite blue pigment can be degraded into green paratacamite, a basic copper chloride, when there is a source of chloride ions near the painting[26]). Thus, after the synthesis of nanoparticles in water, sodium chloride is usually removed by dialysis. The purification is slightly more difficult when glycols are used, and involves peptization and sonication in propanol. Glycols are adsorbed on the surface of nanoparticles, promoting their aggregation and the formation of micrometric clusters. The ultrasonic treatment with propanol promotes the physisorption of this alcohol on the nanoparticles' surfaces in place of the glycol molecules, and yields nanosized units that aggregate further into mass-fractal superstructures.

Nanoparticles of calcium hydroxide can also be prepared through a heterogeneous phase reaction of CaO with water. Under specific conditions, this reaction produces lime putty, *i.e.* a mixture of calcium hydroxide and water in equal amounts, usually containing a rather small amount of unreacted CaO within the core of the particles. The reaction can be forced to the full hydration of calcium oxide by operating at high temperature and pressure in an autoclave.

During this process, the conversion of oxide particles to hydroxide is accompanied by a change in volume, which leads to a fragmentation of the particles and a reduction in their size. Finally, these particles can be effectively dispersed in propanol through sonication. Moreover, no side-products are involved in this preparation method, so no further purification step is required. Recently, this method was used to produce dispersions of $Ca(OH)_2$ nanoparticles for the *in situ* consolidation of Mayan wall paintings in Mexican archaeological areas.[27]

Magnesium oxide and hydroxide nanoparticles were prepared through several different methods. Nanoparticles with different morphological structures can be prepared by precipitation in the homogeneous phase, depending on the concentration of the alkali solution. Needles and rods can be prepared at low alkali concentration, while platelet particles are obtained in highly alkaline conditions.[28]

In addition, $Mg(OH)_2$ nanoparticles with a platelet morphology have been prepared in w/o microemulsions.[29] The size of the particles obtained showed that smaller particles were first produced within micelles. Owing to the coalescence of these early products, the final particles were clearly larger than the size of the micelles, but still in the nanometre range. Calcium hydroxide nanoparticles were also prepared from w/o microemulsions, obtaining particles in the 2–10 nm size.[30] However, it is important to note that such small sizes imply enhanced particle reactivity, for example towards atmospheric CO_2 in the hydroxide carbonation process. This aspect may limit the application of this synthetic method, when the particles are used for the consolidation of carbonate-based substrates. In fact in this case the consolidating power on the degraded works of art would come from the progressive transformation, within the substrate's pores, of $Ca(OH)_2$ into a crystalline network of $CaCO_3$ that produces the strengthening of the matrix (see Chapter 13). If the highly reactive nanoparticles underwent carbonation in their final preparation steps, or during storage (even as dispersions), their consolidating power would be reduced. On the other hand, dispersions of partially carbonated nanoparticles could still be used for deacidification purposes (*e.g.* on paper or wood artefacts), the carbonate phases granting a milder deacidifying agent for aged, oxidised cellulose, which is sensitive to high levels of alkalinity (see Chapters 15 and 16).

Magnesium oxide and hydroxide nanoparticles have also been obtained by sonochemical methods.[31] Magnesium hydroxide nanostructures were prepared by reacting magnesium acetate and sodium hydroxide under ultrasonic radiation. Hydroxide nanoparticles were calcinated at 400 °C to produce MgO nanoparticles with a size of a few tens of nanometres.

Recently, magnesium hydroxide nanoparticles (with an average size of 90–150 nm) were synthesized through a coprecipitation reaction at 90 °C, starting from magnesium chloride and sodium hydroxide as precursors.[3] The precipitated $Mg(OH)_2$ nanostructures were purified by dialysis against distilled water. Finally, the dispersions of the nanoparticles in propanol were prepared by means of high-energy homogenization, followed by sonication.

The synthesis of barium hydroxide nanoparticles through a homogeneous phase reaction in water is hampered by the fact that $Ba(OH)_2$ is fairly soluble in water ($K_{sp} = 2.55 \times 10^{-4}$). Thus the attainment of a high degree of supersaturation, which is a critical factor in obtaining nanosized particles, is hindered. Instead, Giorgi *et al.* proposed a break-down approach for the preparation of fairly stable dispersions of $Ba(OH)_2$ nanoparticles in 1-propanol.[27] Barium hydroxide nanoparticles can be applied together with calcium hydroxide as mixed dispersions in propanol, for the effective conservation of wall-paintings in cases of sulfate pollution (see also Section 13.6).[27]

12.5 Surface Functionalization: Compatibilization and Functional Nanocomposites

Surface functionalization of nanoparticles is as important as the synthesis of the nanoparticles themselves. To exploit their potential fully, nanoparticles often have to be dispersed or, at least, compatibilized with the surrounding medium: in fact nanoparticles are very rarely used as bare powders. Surface functionalization is the process of modification of the nanoparticles' surfaces through the introduction of specific chemical functionalities. This process can efficiently promote the dispersion of the particles in a solvent, as described in detail in the following paragraph.

The surface of inorganic nanoparticles can be chemically modified by using the very same chemical reactions as used in bulk systems. As an example we will discuss the case of gold. Because of its affinity for sulfur, the functionalization of gold nanoparticles is as easy as mixing the particles with any thiol. The resulting product consists of gold nanoparticles covalently bound to the thiol molecule through an Au–S bond. The Brust and Schiffrin method[32] for the preparation of hydrophobic gold nanoparticles is based on such process.

When the nanoparticle is made up of an inorganic oxide, such as a transition metal oxide, the functionalization procedure is not as immediate as in the case of gold. However, one could take advantage of the silicon chemistry. In fact an almost endless variety of silane derivatives are commercially available and they can be easily coupled to the oxide surface through a pH-catalysed condensation reaction.[33] Silanes can also be used to produce an inert silica coating on the surface of inorganic nanoparticles to improve their stability. In the case of magnetic nanoparticles, the silica coating provides bio-compatibility and chemical stability. Furthermore, the coating stabilizes magnetic nanoparticle dispersions by shielding the magnetic dipole interaction with the silica shell and, because silica is negatively charged, the coating enhances the electrostatic repulsion between the nanoparticles.

Similar to silica, gold is another inorganic coating that has been used to functionalize the surface of inorganic nanoparticles. The Au shell provides chemical stability and offers a platform for further organic functionalization.

Surface functionalization does not necessarily take place through the formation of a covalent bond, but can be obtained through a physisorption process as well. For instance, silica nanoparticle sols can be modified either chemically or physically, and then applied to wood substrates for different applicative properties, such as conferring resistance to fire, water, insects or microorganisms (see Sections 16.4.1 and 16.4.2). In general, in the case of particle functionalization with a polymer, both the *in situ* and the *ex situ* approach can be followed. In the *ex situ* approach, the inorganic nanoparticles are first synthesized and then dispersed in a solution of the polymer, where the physisorption process takes place. The polymer-coated inorganic nanoparticles are finally precipitated and eventually dispersed in a good solvent. In some cases, the polymer coating and the preparation of nanoparticles can be combined within a single process or in a series of consecutive processes within the same reactor. This is the *in situ* approach. The *ex situ* approach is more general, because there are fewer limitations on the types of nanoparticles and polymers that can be used.

Both approaches allow the preparation of nanocomposites, *i.e.* a multiphase solid material where one of the phases has one, two or three dimensions of less than 100 nm, or structures having nanoscale repeat distances between the different phases that make up the material. Natural materials such as bones and shells are examples of composite materials with enhanced mechanical properties such as toughness, strength and hardness. Research on polymer nanocomposites has led to several products being available on the market, which represents one of the few examples where the potential of nanotechnology has been at least partly achieved. Advanced technologies where nanocomposites have emerged include magnetic media, bone cement, filter membranes, and solar cells. It is not surprising therefore that the interest in nanocomposite materials is mainly due to the vast range of properties that can arise from a combination of nanoparticles and the features of polymers. In fact the peculiar size-dependent properties of the particles, together with the processability and the chemical and morphological structure of polymers, result in innovative materials with unique characteristics. In addition, research on nanocomposites could also lead to innovative properties that do not derive directly from the properties of the single components. Nanocomposites are dealt with to a wider extent in Section 14.3, where several applications in conservation of Cultural Heritage are reported.

Here, we mention an example of a nanocomposite, namely the magnetic nanosponges loaded with microemulsions for the cleaning of Paraloid®-treated works of art (see also Section 11.4).[15,34] Magnetic nanoparticles functionalized with a polyethylene glycol derivative were chemically incorporated into a polyacrylamide matrix to obtain a chemical sponge that can be loaded with common solvents or with more sophisticated dispersed systems, such as microemulsions or micellar systems (see Figure 12.10).

This composite possesses the mechanical properties of the acrylamide matrix. In fact it can be easily manipulated, cut with a knife to the desired shape, and can load a large amount of water (about ten times its dry weight). At the same time, the presence of the magnetic nanoparticles makes the composite magnetically responsive, so that it can be magnetically handled. The system is

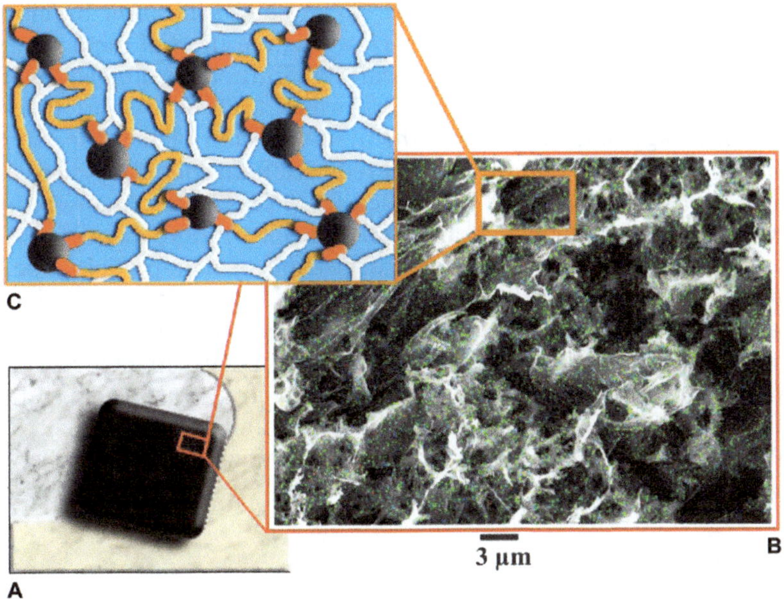

Figure 12.10 (A) A magnetic nanocomposite chemical gel (2×2×2 cm cube) used for the selective cleaning of artistic substrates, laid over a marble surface. (B) SEM-EDX image of the gel matrix, and mapping of iron. The green dots indicate ferrite magnetic nanoparticles incorporated into the gel polyacrylamide matrix. (C) Detail showing the magnetic nanoparticles (black spheres) chemically linked to polyethylene glycol (PEG) molecules (orange chains) through carboxylate functional groups (red chains). The functionalized magnetic nanoparticles are cross-linked to the polyacrylamide gel chains (white chains). (Image by Michele Baglioni.) The resulting "nanosponge" can be loaded with common solvents or with more sophisticated dispersed cleaning systems, such as microemulsions or micellar systems.

particularly efficient in the uptake and release of the material contained in the loaded phase, especially in the case of a microemulsion. This is very important for the conservation of Cultural Heritage, where the loaded magnetic nanosponges have been shown to be effective in the dissolution and removal of detrimental polymer coatings from the surfaces of works of art.

12.6 Stabilization of Dispersions

In the previous sections we have listed several different methods for the production of inorganic nanoparticles. However nanoparticles often need to be dispersed in a liquid or solid medium for further treatment. Different approaches can be used for this purpose; some of them provide a short-term solution (*i.e*, the dispersions do not show precipitation for up to a few days), while others lead to the production of dispersions stable for months or even years. Among the former, mechanical stirring and sonication by ultrasound represent the most common methods. However, the effect of these approaches, if not combined with proper chemical stabilization, is limited by the re-aggregation of individual nanoparticles, which leads in a short time to the formation of agglomerates and to the destabilization of the dispersion. It is important here to note that the stability required for a colloidal system strongly depends on the application it is intended for. For instance, many restoration interventions need systems that are stable for one or more days.

To achieve persistent stability of a dispersion of nanoparticles one has to counteract and/or take advantage of different interactions: van der Waals interactions, electrical double-layer interactions, steric interactions, hydrophobic forces and solvation forces.[35] Van der Waals interactions (see also Chapters 1 and 6) between two nanoparticles are always attractive and depend on the size of the nanoparticles, on their distance, and on the materials. The van der Waals force, F_{vdw}, between two identical spherical particles in the vacuum could be approximated as:[36]

$$F_{vdW} \cong \frac{AR}{12D^6} \tag{12.1}$$

where A is the Hamaker constant (depending on the materials that the particles and the solvent are made of), R is the radius of the particles and D is their distance. Van der Waals forces promote the aggregation of particles: the bigger the particles, the stronger the interaction. Therefore, to obtain a stable dispersion, one must modify the surface of the particles to keep them sufficiently distant and/or change the solvent to modify the Hamaker constant.

When ionic species are present in the dispersion (which is almost always the case for nanoparticle dispersions used in the conservation of Cultural Heritage), one has to take into account the electrical double-layer interactions. Ionic species interact attractively with oppositely charged interfaces, forming a double layer in the proximity of the surface. The electrostatic potential ψ as a function of the separation h from a flat surface can be approximated as:

$$\psi \cong \psi_0 e^{-h/k_d} \tag{12.2}$$

The decay length is given by the inverse of the Debye–Hückel parameter k_d, which corresponds to the thickness of the electrical double layer. The value of k_d is proportional to the charge of the counterions and to their concentration, reflecting that the ions' charge and concentration dramatically affect the stability of a nanoparticle dispersion. In fact the presence of ions dissolved in the solvent produces a screening of the electrostatic charges on the surface of the nanoparticles. In this case, the electrostatic stabilization is reduced, leading eventually to the flocculation (*i.e.* formation of large aggregates) of the particles.

Steric interactions are very important when particles stabilization is carried out using polymers. Particles coated by a polymer shell are considerably more stable against aggregation owing to a strong decrease of their surface energy, and also to the onset of steric repulsions. When a polymer is adsorbed on the surface of a nanoparticle, some parts of the molecule are adsorbed, while others extend into the solvent, forming a more or less compact layer. When two nanoparticles approach each other, the polymer layers partially overlap, leading to steric repulsion. The stronger steric stabilization is provided by copolymers: in fact copolymers bear some groups that can strongly adsorb at the surface of the particle, while other residues do not adsorb at all and can extend into the solvent.

Solvation forces are often important in nanoparticle dispersions and involve the molecules of solvent close to the nanoparticle surface. Finally, hydrophobic forces are important when non-polar molecules, or their non-polar groups, are adsorbed on nanoparticles, as in the case of surfactants.

Taking into account all these interactions, the formulation of stable nanoparticle dispersions requires an energy barrier between the nanoparticles to prevent or inhibit their close approach when attractive van der Waals interactions are likely to dominate the energetic scenario. Two approaches are commonly followed for this purpose. In the first the particles are stabilized electrostatically through surface charges exposed to a polar solvent. The uneven distribution of counterions leads to an effective charge at the level of the electrical double layer. When two nanoparticles approach each other, repulsion occurs, with a magnitude that depends on the surface potential and on the concentration of electrolytes. At low electrolyte concentrations, the repulsive energy is strong enough to balance the van der Waals attraction, and produces an energy barrier that prevents the approach of the particles and results in the stabilization of the nanoparticle dispersions. In the second approach non-ionic surfactants and/or polymers are adsorbed on the surface of nanoparticles. This second mechanism of stabilization is usually referred to as steric stabilization, because the repulsion between nanoparticles is provided by the steric hindrance of the surfactant/polymer. Charges have no role, making this method especially appealing in systems that contain electrolytes at a large concentration. The most effective stabilization is obtained when polymeric surfactants are used, which adsorb very strongly onto the particle surface.

In the framework of conservation of Cultural Heritage, very often nanoparticle dispersions of oxides and hydroxides are stabilized through electrostatic stabilization in a proper solvent. Nanoparticles are first charged during their preparation and an excess of charge is generated through extensive dialysis, followed by particle dispersion in an appropriate solvent. Propanol is generally the best candidate: in fact this alcohol is environmentally friendly, is quite volatile and has a low surface tension. Furthermore, propanol displays high affinity for the surface of alkaline earth metal oxide and hydroxide nanoparticles, acting not only as a solvent, but also as a stabilizing agent against aggregation.

Given that the stabilization of nanoparticle dispersions is one of the crucial aspects for the application of nanomaterials to the preservation of Cultural Heritage, it is worth adding here a few details. When brought in contact with a polar medium, nanoparticles might acquire surface electric charges. These surface charges commonly arise from ions derived from the solid (*e.g.* hydroxyl groups or calcium ions as in the case of calcium hydroxide nanoparticles commonly used for wall-painting consolidation, see Chapter 13, and paper/wood deacidification, see Chapters 15 and 16), which then influence the distribution of nearby ions in the polar medium (electrical double layer, see Figure 12.11). A detailed description of the theoretical aspects of the electrical double layer goes beyond the scope of this chapter. Briefly, a widely accepted model for the double layer predicts that some of the counterions are located close to the particle surface (Stern layer), while the remainder are broadly distributed in the diffuse double layer.[37] The Stern layer represents therefore the closest distance to the surface approachable by counterions. A schematic representation of the charge distribution and the corresponding potential energy is shown in Figure 12.11. When two nanoparticles approach each other in a dispersion, the diffuse layers partially overlap. In particular, the potential at the boundary between the Stern and the diffuse layer is the most important in determining the interactions among the nanoparticles. This potential difference between the Stern plane and the diffuse layer is called the zeta potential (ξ).[38]

The thickness of the double layer and the zeta potential are the crucial parameters for the formulation of a stable dispersion of particles. In practice, to

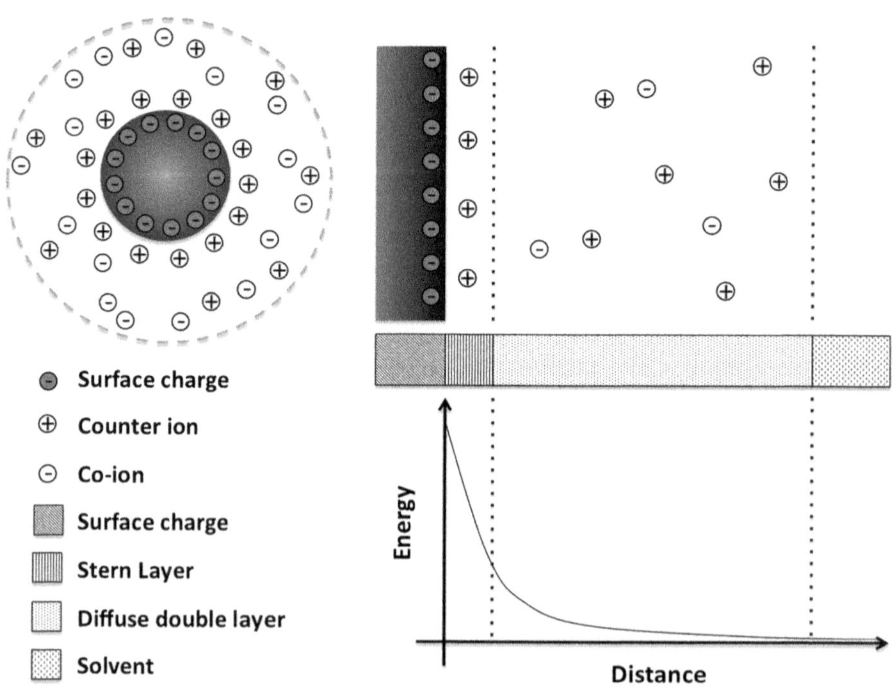

Figure 12.11 Sketch of a negatively charged nanoparticle in a polar solvent. Surface charges, counter ions and co-ions are indicated, as well as the Stern layer and the diffuse double layer.

ensure a sufficient repulsion between the particles and to avoid their aggregation, it is necessary that both the zeta potential and the double layer thickness are not too small. In particular, the double layer thickness is a function of the concentration and the valence of ions in solution. When the concentration of ions in the solvent is high (*i.e.* when the electrolyte concentration is large), the thickness decreases and so does the potential, leading to unstable dispersions.

Finally, it is important to note that, in the framework of the preservation of Cultural Heritage, the stabilization of particles through the use of electrolytes or additives (*e.g.* adsorbed polymers, surfactants, polyelectrolytes) could also imply additional drawbacks. The presence of electrolytes such as soluble salts on treated porous works of art (stones, wall-paintings) could cause crystallization and mechanical stress (salt efflorescence, see Sections 2.2.2 and 12.4), while surfactants, polymers and polyelectrolytes tend to remain on the treated artefacts and in the long term could exhibit undesirable properties (*e.g.* acidity), or release degradation products that could be potentially harmful to the artistic substrate (*e.g.* acids on paper matrices). Considering these drawbacks, nanoparticle formulation without stabilizers should be preferred to the "more stable" stabilized dispersions.

12.7 Consolidation and Protection

As stated in the previous sections, nanomaterials have been explored in recent decades as powerful tools for the conservation of cultural heritage. Inorganic nanoparticles, in particular, have been investigated and used to provide different functions. Two classic applications involve their use for the consolidation and for the protection of works of art.

The need for compatible consolidation of artefacts, such as wall-paintings and stones, arises from the degradation induced by both weathering and the use of detrimental restoration materials (*e.g.* incompatible synthetic polymer coatings), as explained in Chapters 2 and 3. For instance, in the last 20 years, nanoplatelets of $Ca(OH)_2$ have been successfully developed and applied to several conservation case studies involving immovable works of art, while mixtures of calcium and barium hydroxide nanoparticles, synthesized and dispersed in propanol as described in Section 12.4, have been recently used for the consolidation of wall-paintings in Mayan archaeological sites.[27] The main advantage is due to the high compatibility of these materials with the treated substrates, which limits the drawbacks and ensures a long-lasting beneficial effect. Moreover, this technology is simple to use and relatively inexpensive.

The use of inorganic nanomaterials for the consolidation of wall-paintings and stones will be widely discussed in Chapter 13, while nanocomposite materials (formed by nanoparticles included in a polymeric matrix) for the consolidation of stone will be treated in Chapter 14. Essentially, the rationale for the use of nanocomposites is to combine the main beneficial features of both nanomaterials and polymers (coatings, adhesives), while limiting in some cases the amount of polymer applied, hence reducing any possible detrimental effects.

Together with consolidation, the protection of works of art represents a crucial step for a comprehensive approach to the conservation of Cultural Heritage. There are many factors that can potentially damage works of art, including pollution and storage conditions. A factor that is often overlooked is the effect of environmental conditions: in fact biofouling represents a real menace in tropical climates, such as in the case of Mexican and Indian architecture. To inhibit the growth of microorganisms, innovative approaches have been recently investigated, including the application of nanoparticles with antibacterial properties. For this reason silver nanoparticles have already been used in fabrics and on stones, showing extremely promising results.[16]

Another example is related to the use of titanium dioxide (TiO_2) for construction materials, as a photocatalyst for self-cleaning, de-polluting and self-sterilizing surfaces. An important example is constituted by the self-cleaning "Dives in Misericordia" church in Rome, designed by the American architect Richard Meier, which is coated with a white colour containing the active photocatalytic titanium dioxide (TX Active® products by Italcementi) (see Figure 12.12).

Recently, the effects of nanostructured TiO_2 coatings have been studied on historic architectural stone surfaces (*e.g.* travertine) in order to obtain a self-cleaning effect through the combination of their superhydrophilic and photocatalytic photo-induced properties, while preserving the original artefact's aspect.[39] In practice, superhydrophilicity allows the formation of a uniform water film over the substrate, which prevents the adhesion of soil, while photocatalysis causes the UV light-promoted decomposition of organic and inorganic pollutants. Important parameters such as water absorption through capillarity and the static contact angle (see Chapter 1) have been assessed before and after the TiO_2 treatments, and it was found that, without UV light

Figure 12.12 The self-cleaning "Dives in Misericordia" church in Rome, designed by the American architect Richard Meier, and coated with the active photocatalytic titanium dioxide (TX Active® products by Italcementi).

exposure, the coating does not affect these properties. Moreover, the photo-induced hydrophilicity does not seem to produce higher water absorption (a possible drawback), making these applications a promising option for further testing.

Finally, it is important to recall here the use of alkaline inorganic nanoparticles [*e.g.* $Ca(OH)_2$ and $Mg(OH)_2$] for the pH adjustment of artistic and historical objects, such as paper and wood. This application has a high potential impact on the conservation of the vast historical and documentary cultural heritage worldwide, which is threatened by degradation mechanisms such as acid-catalyzed hydrolysis and oxidation.[40] This subject represents thus one of the main examples of the use of nanotools in conservation science, and will be discussed in detail in Chapters 15 and 16.

Further Suggested Reading

In this chapter we have briefly reviewed the properties of nanomaterials and the most common synthetic methods, especially focusing on the materials and the methods often used in the preservation of Cultural Heritage. The reader interested in specific aspects that are only briefly introduced in this chapter is recommended to refer to the following reading material.

The peculiar properties of materials at the nanoscale are described in detail in the book *Nanostructures and Nanomaterial* by Cao and Wang,[41] where, together with the methods for the synthesis and the fabrication, the properties and the applications of nanostructures and nanomaterials (particularly inorganic nanomaterials) are comprehensively presented.

The connection between nanomaterials and chemistry is well described in the book *Nanoscale Materials in Chemistry* by Klabunde, in which the preparation, the application, the characterization, and the environmental and toxicological aspects of nanoscale materials are reviewed from the viewpoint of a chemist.[1]

There is no doubt that, among the experimental techniques used to characterize nanomaterials, techniques of microscopy have played a major role in developing nanotechnologies over the last decades. The *Handbook of Microscopy for Nanotechnology* by Yao and Wang provides a broad overview of the various microscopy techniques used in the investigation of nanomaterials.[42] Scattering techniques have been also widely used in the characterization of nanoparticles and, in particular, of their liquid dispersions. In particular, light scattering techniques have become a standard method in many laboratories. The book *Particle Characterization: Light Scattering Methods* by Xu systematically describes the principles, instrumentation, data interpretation, and applications of light scattering methods for the analysis of nanoparticle dispersions.[43]

Finally, because this book should be of interest also to readers that are not very familiar with chemistry and physics, but who possess a strong background in architecture and restoration, the book *Nanomaterials, Nanotechnologies and Design: An Introduction for Engineers and Architects* by Ashby, Ferreira and Schodek deserves a special citation.[44] The authors explore the new approaches to design that the nanotechnologies offer, with a particular emphasis on the opportunities for designers, engineers and architects resulting from the use of innovative materials.

Review and Questions

The main points discussed in the present chapter may be summarized as follows:

- The reduced dimensions of nanomaterials lead to innovative properties that cannot be achieved with the common bulk materials.
- Inorganic nanomaterials can be prepared either by top-down or bottom-up methods.
- Among the bottom-up methods, the two crucial steps for the preparation of nanoparticles are nucleation and growth.
- The most commonly used nanomaterials in the conservation of Cultural Heritage are oxides and hydroxides of alkaline earth metals.
- For their application in the fields of protection and consolidation of works of art, nanoparticles often have to be formulated in stable dispersions in a suitable solvent.
- Nanocomposites have great potential in many applications because they combine the properties of the polymeric matrix with those of the nanoparticles embedded within.

Questions

1) Nanoparticles of oxides and hydroxides of alkaline earth metals have found widespread use in the preservation of Cultural Heritage. Explain why.
2) What are the most common methods employed in the preparation of nanoparticles of oxides and hydroxides of alkaline earth metals?
3) What are the interactions affecting the stability of nanoparticle dispersions?

Answers

1) Nanoparticles of oxides and hydroxides of alkaline earth metals have found widespread use in the preservation of Cultural Heritage especially because of their deacidifying properties, combined with their ability to penetrate into works of art and consolidate them. As an example, calcium hydroxide nanoparticles are very good deacidifying agents (*e.g.* for the deacidification of wood, paper and canvas) and, after their transformation into calcium carbonate, they provide an alkaline reservoir. Moreover, owing to their physico-chemical compatibility with carbonate-based substrates (*e.g.* limestone and frescoes), dispersions of calcium hydroxide nanoparticles have been widely used for the restoration of lifted and flaking paint layers in wall-paintings.

2) The most common methods employed in the preparation of nanoparticles of oxides and hydroxides of alkaline earth metals are based on wet chemistry. In particular, these methods are often based on the precipitation of oxides and hydroxides through the alkalinization of water solutions, in conditions of controlled nucleation and growth.
3) The stability of dispersions of nanoparticles is affected by different interactions: van der Waals, electrical double-layer, steric, hydrophobic and solvation forces. These interactions could either lead to the stabilization or to the destabilization of the dispersions. For this reason it is important to investigate their effect and extent in the formulation of stable dispersions to be used in the conservation of Cultural Heritage.

References

1. K. J. Klabunde (ed.), *Nanoscale Materials in Chemistry*, Wiley-Interscience, 1st edn, 2001.
2. http://www.ptlp.com/bookkeeper.html (accessed June 11 2012).
3. G. Poggi, R. Giorgi, N. Toccafondi, V. Katzur and P. Baglioni, *Langmuir*, 2010, **26**, 19084.
4. M. Ambrosi, L. Dei, R. Giorgi, C. Neto and P. Baglioni, *Langmuir*, 2001, **17**, 4251.
5. R. Giorgi, D. Chelazzi and P. Baglioni, *Langmuir*, 2005, **21**, 10743.
6. R. Giorgi, L. Dei, M. Ceccato, C. Schettino and P. Baglioni, *Langmuir*, 2002, **18**, 8198.
7. S.-M. Shih, C.-S. Ho, Y.-S. Song and J.-P. Lin, *Ind. Eng. Chem. Res.*, 1999, **38**, 1316.
8. P. C. Hiemenz and R. Rajagopalan, *Principles of Colloid and Surface Chemistry*, Marcel Dekker Inc, New York, 1997.
9. E. J. Schofield, R. Sarangi, A. Mehta, A. M. Jones, F. J. W. Mosselmans and A. V. Chadwick, *Mater. Today*, 2011, **14**, 354.
10. M. A. López-Quintela, *Curr. Opin. Colloid Int.*, 2003, **8**, 137.
11. M. A. López-Quintela, C. Tojo, M. C. Blanco, L. García Rio and J. R. Leis, *Curr. Opin. Colloid Int.*, 2004, **9**, 264.
12. R. Giorgi, C. Bozzi, L. Dei, C. Gabbiani, B. W. Ninham and P. Baglioni, *Langmuir*, 2005, **21**, 8495.
13. R. Kodama, *J. Magn. Magn. Mater.*, 1999, **200**, 359.
14. M. Baglioni, D. Rengstl, D. Berti, M. Bonini, R. Giorgi and P. Baglioni, *Nanoscale*, 2010, **2**, 1723.
15. M. Bonini, S. Lenz, R. Giorgi and P. Baglioni, *Langmuir*, 2007, **23**, 8681.
16. E. Falletta, M. Bonini, E. Fratini, A. L. Nostro, G. Pesavento, A. Becheri, P. L. Nostro, P. Canton and P. Baglioni, *J. Phys. Chem. C*, 2008, **112**, 11758.
17. M. Bonini, A. Wiedenmann and P. Baglioni, *J. Appl. Crystallogr.*, 2007, **40**, s254.
18. M. Bonini, A. Wiedenmann and P. Baglioni, *J. Phys. Chem. B*, 2004, **108**, 14901.

19. T. Hyeon, S. S. Lee, J. Park, Y. Chung and H. B. Na, *J. Am. Chem. Soc.*, 2001, **123**, 12798.
20. C. Neto, M. Bonini and P. Baglioni, *Colloids Surface A*, 2005, **269**, 96.
21. M. Bonini, E. Fratini and P. Baglioni, *Mat. Sci. Eng. C-Bio S*, 2007, **27**, 1377.
22. Z.-X. Tang, D. Claveau, R. Corcuff, K. Belkacemi and J. Arul, *Mater. Lett.*, 2008, **62**, 2096.
23. F.-C. Yu, N. Phalak, Z. Sun and L.-S. Fan, *Ind. Eng. Chem. Res.*, 2011, **51**, 2133.
24. K. Yura, K. C. Fredrikson and E. Matijević, *Colloid Surface*, 1990, **50**, 281.
25. P. Baglioni and R. Giorgi, *Soft Matter*, 2006, **2**, 293.
26. L. Dei, A. Ahle, P. Baglioni, D. Dini and E. Ferroni, *Stud. Conserv.*, 1998, **43**, 80.
27. R. Giorgi, M. Ambrosi, N. Toccafondi and P. Baglioni, *Chem–Eur J.*, 2010, **16**, 9374.
28. J. Lv, L. Qiu and B. Qu, *J. Cryst. Growth*, 2004, **267**, 676.
29. J. Wu, H. Yan, X. Zhang, L. Wei, X. Liu and B. Xu, *J. Colloid Interf. Sci.*, 2008, **324**, 167.
30. A. Nanni and L. Dei, *Langmuir*, 2003, **19**, 933.
31. M. A. Alavi and A. Morsali, *Ultrason. Sonochem.*, 2010, **17**, 441.
32. M. Brust, M. Walker, D. Bethell, D. J. Schiffrin and R. Whyman, *J. Chem. Soc., Chem. Comm.*, 1994, 801.
33. S. Laurent, D. Forge, M. Port, A. Roch, C. Robic, L. Vander Elst and R. N. Muller, *Chem. Rev.*, 2008, **108**, 2064.
34. M. Bonini, S. Lenz, E. Falletta, F. Ridi, E. Carretti, E. Fratini, A. Wiedenmann and P. Baglioni, *Langmuir*, 2008, **24**, 12644.
35. P. F. Luckham, *Adv. Colloid Interfac.*, 2004, **111**, 29.
36. V. A. Parsegian, *Van Der Waals Forces: A Handbook for Biologists, Chemists, Engineers, and Physicists*, Cambridge University Press, Cambridge, 2006.
37. O. Z. Stern, *Elektrochem.*, 1924, **30**, 508.
38. R. J. Hunter, *Zeta Potential in Colloid Science*, Academic Press, London, 1981.
39. E. Quagliarini, F. Bondioli, G. B. Goffredo, A. Licciulli and P. Munafò, *J. Cult. Herit.*, 2012, doi:10.1016/j.culher.2012.02.006.
40. J. Wouters, *Science*, 2008, **322**, 1196–1198.
41. G. Cao and Y. Wang, *Nanostructures and Nanomaterials: Synthesis, Properties, and Applications*, World Scientific Publishing Company, Singapore, 2nd edn, 2011.
42. N. Yao and Z. L. Wang (ed.), *Handbook of Microscopy for Nanotechnology*, Kluwer Academic Publishers, New York, 2005.
43. R. Xu, *Particle Characterization: Light Scattering Methods*, Kluwer Academic Publishers, Dordrecht, 2002.
44. M. F. Ashby, P. J. S. G. Ferreira and D. L. Schodek, *Nanomaterials, Nanotechnologies and Design: An Introduction for Engineers and Architects*, Butterworth-Heinemann, Burlington, 2009.

CHAPTER 13

Inorganic Nanomaterials for the Consolidation of Wall Paintings and Stones

P. BAGLIONI* AND R. GIORGI

University of Florence, Italy
*Email: baglioni@csgi.unifi.it; giorgi@csgi.unifi.it

13.1 Introduction

This chapter focuses on the pioneering application of nanoparticles to the conservation of immovable cultural heritage (wall paintings and stones). As reported in previous chapters, the chemical "corrosion" that induces the powdering of paintings, and the mechanical action of rain, wind, dust particles, *etc.*, are the main factors responsible for the weakening of the porous structure (particularly of surface layers) of materials used in cultural heritage. When in contact with the environment, even the most durable material undergoes a natural, progressive and irreversible process of degradation (see Chapter 2). All the alteration processes of materials are spontaneous, because they lead to the formation of compounds that possess a lower energy content and therefore are thermodynamically more stable than the initial materials. This means that counteracting or minimizing the effects of degradation processes, or even trying to restore the original status of a work of art, is a very ambitious goal that requires a very thorough knowledge of the materials' properties and of the physico-chemical mechanisms they are subjected to over time. In fact, a restoration intervention is rarely limited to the removal of surface material, *e.g.* dirt, grease, repainting or

RSC Nanoscience & Nanotechnology No. 28
Nanoscience for the Conservation of Works of Art
Edited by Piero Baglioni and David Chelazzi

Published by the Royal Society of Chemistry, www.rsc.org

salts. More often, the aim of the intervention includes the addition of substances that might provide support to the substrate, improving its mechanical resistance, possibly without altering its original physico-chemical properties. In some cases, the irreversible addition or replacement of various components to the substrate is necessary. A restoration is therefore to be considered an invasive operation.

This chapter reports on the consolidation of wall paintings, but the main concepts and materials can also be extended to stone surfaces. In particular, the application of nanomaterials will be illustrated. The topic of wall painting consolidation will be discussed by considering various criteria such as compatibility, minimal intervention and reversibility.

At the end of the 1980s, the consolidation (and protection) of wall paintings and carbonate stones was mainly carried out using synthetic polymers (also known as resins).[1] In fact, from the beginning of the 1960s, the use of polymers was strongly encouraged by important international institutions, and this technology was spread all over the world.[2] At that time, workers at a few academic institutions (a key role was played by Enzo Ferroni, Professor of Physical Chemistry at the Department of Chemistry of the University of Florence) commented on the misuse of polymers, but only a few conservators shared this conviction. It is worth noting that in Italy, for example, the two National Schools for Conservation *i.e.* the OPD ("*Opificio delle Pietre Dure*" a public institute of the Italian Ministry for Cultural Heritage in Florence) and the ISCR ("*Istituto Centrale per il Restauro*, Institute for the Conservation and Restoration") have proposed for decades two different and antithetical conceptual approaches to this matter. The former encouraged the use of inorganic materials, as recommended by Ferroni, while the latter supported the application of organic products.

With time, the negative effects of the application of polymers due to their physico-chemical incompatibility with the treated substrates have become evident (see Figure 13.1), and the number of case studies showing great alterations due to the presence of these materials has dramatically grown.[3–6]

On the other hand, it must be noted that few alternative options were available to conservators. In fact, the main inorganic methods used from the first half of the 20th century until the 1960s had shown several limitations and poor results. Examples include fluosilicates, sodium (and potassium) silicate, and calcium or barium aluminate solutions used to consolidate and protect stones.[7–9]

However, the combination of conservators' empirical methods and a diffuse lack of knowledge of the chemistry of the materials involved contributed to an enthusiastic and uncritical assent to the theoretical principles of polymers application.

In this framework, the first steps were taken towards the synthesis of innovative compatible inorganic materials with enhanced properties due to the reduction of the particle size, as an alternative to the use of polymers. This is the case for calcium hydroxide, the first nanomaterial used in restoration, which is almost useless if used as a water solution or aqueous dispersion, but it is one of the best consolidants for carbonatic materials if applied in the form of nanoparticles dispersed in specific non-aqueous solvents.

Before discussing, in the following sections, the application of hydroxide nanoparticles for the consolidation of wall paintings, it is worth recalling

Figure 13.1 (A–C) The progressing degradation of a wall painting in Cholula (Mexico) treated with a combination of acrylic and vinyl polymers. Polymers led to the complete disruption of the painting in a few years after the application. (Reprinted from M. Baglioni, D. Rengstl, D. Berti, M. Bonini, R. Giorgi and P. Baglioni, *Nanoscale*, 2010, **2**, 1723, © Royal Society of Chemistry.) (D) Detail from the *Templo Rojo Cacaxtla* (Puebla, Mexico) that was restored using an inorganic consolidation treatment.

briefly the ideal properties of an agent that will achieve long-lasting restoration effects, and the general principles to be followed in modern conservation (see Chapter 3).

In general, the proper conservation of degraded materials requires:

- Increasing the material's cohesion, reduced as a result of the loss of binder (powdering of the surface) or by the abrasive action of external agents such as wind or water;

- Reinforcement of the adhesion of the paint layer to the support in order to halt so-called surface flaking (exfoliation of the paint layer);
- Restoring the "smoothness" of surfaces, to minimize the effects of light scattering (opacity of the colour);
- Consolidation of the work of art materials to depths of up to few centimetres, obtaining a good impregnation of the porous matrix.

As a matter of fact, an estimate of the restored materials' lifetime is very difficult whenever the consolidating materials introduced within the porous structure of the works of art lead to strong modification of the original physical and chemical characteristics of the objects. Much has been done in recent years, and numerous experimental techniques have been developed to make a significant contribution to the knowledge on consolidating systems. Unfortunately, however, some structural limits remain.

In this context we can debate whether the restored materials will be more resistant to the same environmental conditions that led to the degradation of the original masterpieces, or if the restoration treatments will induce an even more serious degradation process. In the current scenario, such fundamental questions cannot be answered definitely. Restored materials, in spite of all intents and purposes, are "new" materials, with their own properties that are not necessarily similar to those of the original materials. The methods of accelerated ageing, adopted in the laboratory to simulate the effects of a lifetime, may consistently differ from the real environmental conditions to which the materials have been subjected for years and centuries. It is therefore very difficult to provide clear and reliable evaluations of the performance of restoration products and treatments over time, unless compatible materials are used to restore the original properties of the artefacts.

It is clear then that the conservation of cultural heritage is entrusted in the first instance with the task of deepening the knowledge of material properties and the dynamics of degradation processes. Motivated by this awareness, many researchers continue to work strenuously for the development of new and more effective restoration methods, in order to save many works of art otherwise condemned to inexorable disfigurement and deterioration.

13.2 Consolidation Treatments: a Historical Perspective

Since the earliest times, humankind has always shown a great interest in the care of historical and artistic heritage. Vitruvius and Pliny provided valuable information on the construction techniques and conservation of Greek and Roman artefacts of earlier times. Pliny reports that natural resins were used to protect Carthage residential buildings, built with tufa. Vitruvius advises that plaster concrete structures helped "to keep [the stone structures] elegant and durable, without defects, for a long period time", and "had to be made up of three layers of mortar and two of stucco".[10]

Archival research has documented a variety of restoration works carried out in different times and areas. It is well known, for example, that in ancient times marble surfaces were provided with a finish based on melted wax and oil, a subsequent heating of the layer and a smoothing treatment with linen cloths. Wax glue was used in the Middle Ages, by Andrea Pisano, on the marble facade of the cathedral of Orvieto (Italy). Archival documents of the San Petronio church in Bologna (Italy) highlight the application of linseed oil on marbles.[10] The use of plaster and stucco in the Renaissance, for the protection of stone materials, is widely documented in the literature. Inorganic materials were added to organic substances such as egg white, wax and oil.[10]

Nowadays, the restoration community is becoming growingly aware of the necessity to overcome the empirical approach used in the past, although some treatments did provide successful results.

Chemistry, and science in general, can contribute greatly to the preservation of works of art because it addresses the laws that rule the degradation, stability and reactivity of materials. The preservation of a valuable object is achieved by inhibiting the mechanisms that produce alteration. This is possible through a deep knowledge of the chemical composition and physico-chemical behaviour of the material in equilibrium with the environment.

13.3 Physico-chemical Compatibility and Durability

In the most recent decades experimental sciences have made a significant contribution to the centuries-long history of cultural heritage conservation, and to the formulation of two major operating principles for the selection of the appropriate methods of intervention.

1. The products should be applied to ensure maximum durability, as well as chemical and thermodynamic stability.
2. The treatment should be reversible when applying a substance of different composition from the original one, *i.e.* the removal of any applied products should be possible if deemed necessary.

However, complete reversibility cannot be achieved (see Chapter 3).

For this reason, in recent times many efforts have been carried out to produce and adopt restoration materials that mostly behave in the same way as those that make up the works of art. In one word: compatible materials. Compatibility means that the restoration materials exhibit very similar physico-chemical characteristics to those of the artefacts' materials, suggesting that the effects of the environmental parameters and the consequent degradation processes will be similar to those experienced by the original materials. Degradation will thus be homogeneously distributed within the materials, without any localized stress, as would occur, for instance, at the interfacial discontinuities formed between the painted layer of a wall painting and a polymeric organic coating film.

Compatibility, however, is not always an easily achievable feature. An example of the aforementioned problematic issues relates to the conservation of carbonate wall paintings (including, but not limited to, fresco paintings). It is well known that the degradation of these works of art is mainly due to corrosion and disaggregation of crystalline calcium carbonate (formed from lime, *i.e.* calcium hydroxide, see Chapter 2). However, the use of calcium carbonate and hydroxide for restoration, according to the compatibility criterion, has been historically limited owing to their low solubility in water. As a consequence, scarcely compatible materials, such as polymers, have been widely used since the middle of the 20th century. Time and natural ageing have shown that the use of many formulations of polymers results in a non-durable consolidation of both wall paintings and stone materials, and may indeed prove detrimental.

In Mexico, for example, several restoration projects have been performed using materials and recipes directly imported from Europe.[11] Unfortunately these products have led to devastating and unexpected degradation processes. Synthetic polymers, such as Paraloid® (an acrylate–methacrylate copolymer also known as Acryloid®), Mowilith®/Mowital® (acrylate–vinyl acetate copolymers), and Primal® (acrylate–methacrylate copolymer), were used in large amounts, with different results. In controlled environments such as colonial buildings the results were often acceptable. However, in the archaeological sites of Palenque, Bonampak, Teothihuacan, Cacaxtla and Kohunlich, where many types of polymers were applied as fixatives, secondary reactions were observed, *e.g.* the detachment and flaking of surfaces and a consistent acceleration of the chemical reactions involved in the degradation of paintings.[12–16] This was particularly evident in Cacaxtla, where wall paintings were treated some decades ago using the two consolidation methods available at that time: the Ferroni method (*vide infra*) in the *Templo Rojo* (Red Temple) and the application of synthetic polymers (mainly Paraloid®) in the *Murales de la Batalla* (Battle Paintings). The paintings treated with the Ferroni method (inorganic method) are still "in good health" while those treated with Paraloid® are now (about 25 years later) in a severe state of degradation. Moreover, the removal of synthetic polymers is not easy, and in some cases even impossible, as we have discussed in previous chapters of this book.

Similar degradation processes have also affected several European artefacts, but the degradation kinetics is slower, owing to less extreme environmental conditions, and apparently the artefacts are in a better state.

This prompted a revision of the protocols used for restoration, leading conservators to search for new materials and methods. The question that rises at this point is: what objectives should be achieved after a restoration intervention?

The answer cannot be restricted purely to aesthetic factors. Restoration is inherently invasive, and therefore it is necessary to know the effects on the interface properties and porous structure of materials. Given that the natural ageing process usually results in an increase in total porosity and the percentage of large pores, low values of porosity correspond to a high durability of the material. However it is difficult to translate the "low porosity" into quantitative terms. A good approach may be that based on the physical parameters typical of intact (*i.e.* not degraded) materials, and their comparison with the values measured on the degraded ones. This method can lead to the correct setting of the intervention and to the control of the appropriate preventive conservation conditions (such as the control of the environmental parameters). In general, the more compact the material, the greater its mechanical strength and the lower the possibility of attack by degradation agents (in particular gas or liquid phase water, saline solutions or air pollutants).

However, the combined action of several degradation agents can produce noticeable changes even on materials that originally were very compact. Marble, for instance, has a very low porosity (Italian marble has a porosity lower than 10%) but very often it shows severe degradation. Crystals of calcite, the main component of marble, have a thermal expansion coefficient that varies depending on their orientation within the crystalline network; under the effect of thermal stress, crystals shrink and expand anisotropically. This mechanism is particularly harmful when the material presents a high percentage of small pores (<1 μm), such as in marble. Travertine, with an open porosity similar to marble but with a higher percentage of large pores (about 82% of the pores are >1 μm), exhibits higher resistance for example to frost damage.

In principle, appropriate treatment of reinforcement/consolidation should reduce the porosity of the material and modify the pore size distribution. Unfortunately, the appropriate amount of consolidation products is not easy to determine, because it is not possible to monitor in real time the induced variation of porosity in the porous medium, and sometimes the remedy has proven worse than the disease. A significant decrease in the average size of pores can be, in fact, extremely harmful. Some treatments, for example, reduce the proportion of the larger pores, while in other cases this percentage increases as a result of the filling of the smaller pores. Obviously, this effect changes the vapour permeability of the wall and therefore the "breathing" ability of the material. Methodologically, treatments should be considered to be effective when they induce structural changes that are "consistent" with the physical and chemical features of the intact (not degraded) material.

The best conservation approach consists thus, as stated above, in the application of chemical products that are very similar to the original materials. This strongly reduces the risks of low compatibility, *i.e.* undesired secondary effects, especially in the long-term conservation perspective. Accurate physico-chemical investigations provide essential information for scientists, conservators, archaeologists and historians, for a correct planning of the conservation procedures.

This awareness started to stand out at the end of the 1980s, even if only gradually and with controversial opinions. Nowadays, the debate is enriched by a large amount of scientific data, which contribute to clarify the importance of compatibility, and this achievement has recently produced, in some cases, a relevant improvement of the conservation strategy for cultural heritage.

13.4 Ferroni (or Barium or Florentine) Method

The Ferroni (or barium) method, also known as the Florentine method, is a milestone in the pioneering use of compatible consolidation methodologies for the conservation of wall paintings.

The method was developed by Enzo Ferroni at the end of the 1960s, with the twofold purpose of solving the problem of sulfation and achieving the consolidation of flaking painted surfaces. A successful application of this method was carried out for the first time during the restoration of the wall paintings by Beato Angelico in Florence, performed by the conservator Dino Dini (see Figure 13.2).[17,18]

Figure 13.2 Wall paintings by Beato Angelico (15th century) in the San Marco convent (Florence, Italy). On the left, the effects of sulfation are highlighted, under glazing light; on the right, the paintings after the application of Ferroni method. (Credits by Piero Baglioni.)

The method (see also Section 3.4) has been reviewed in the literature,[19–21] but it will be briefly outlined here because it represents the starting point for the development of nanomaterials for consolidation of wall paintings. It consists of a two-step procedure that involves the application of ammonium carbonate and barium hydroxide aqueous solutions. Depending on the amount of sulfates to be removed, a 3% to 10% ammonium carbonate solution is applied with a cellulose poultice on the wall painting's surface, to transform gypsum into soluble ammonium sulfate, according to the following scheme:

$$CaSO_4{\cdot}2H_2O + (NH_4)_2CO_3 \rightarrow (NH_4)_2SO_4 + CaCO_3 + 2H_2O$$

Reaction Scheme 13.1

Ammonium sulfate (a very soluble salt) is mostly absorbed by the poultice; in this step, calcium carbonate forms as powdery filler with poor mechanical properties.

The application of an excess saturated aqueous solution of barium hydroxide results in a stable consolidation effect, due to the regeneration *in situ* of calcium hydroxide, through a double-exchange reaction with calcium carbonate, according to the following reaction scheme:

$$Ba(OH)_2 + CaCO_3 \rightarrow Ca(OH)_2 + BaCO_3$$

Reaction Scheme 13.2

The saturated solution of barium hydroxide (5% w/w) is usually obtained by dissolving crystalline barium hydroxide (as a powder) in water. Barium hydroxide also transforms the ammonium sulfate not removed after washing into barium sulfate, an insoluble salt that does not produce solubilization–recrystallization cycles that would harm the wall painting's porous matrix:

$$Ba(OH)_2 + (NH_4)_2SO_4 \rightarrow 2NH_3 + 2H_2O + BaSO_4$$

Reaction Scheme 13.3

The carbonation of barium hydroxide could also be responsible for the consolidation effect, because the crystalline structure of barium carbonate has some similarities with that of calcite. Although this process might have some importance it does not account for the improvements that the restoration treatment provides over time, which are definitely related to the slow transformation of the regenerated calcium hydroxide into crystalline calcium carbonate (calcite) that causes, as a long-term effect, the further mechanical strengthening (*i.e.* the consolidation) of the paintings. This phenomenon was first observed by the conservator Dini, a few years after the intervention on Beato Angelico's paintings in Florence.

13.5 Calcium Hydroxide Nanoparticles

In principle, calcium hydroxide (lime) is the ideal material to be used for the consolidation of plaster and wall paintings, because the maximum physico-chemical compatibility can be achieved. Lime is in fact the best material for wall painting reinforcement, because it is the "original" binder used by artists. It is important to note that even when pigments are laid onto a calcium carbonate substrate using organic or mixed organic–inorganic binders (the '*secco*' painting technique), calcium hydroxide still represents a suitable and compatible material to strengthen powdering and flaking paint layers. Indeed, upon carbonation, $Ca(OH)_2$ turns into $CaCO_3$, providing a crystalline network that is cohesive with the carbonate substrate, and mechanically reinforces the degraded painted surface.

Limewater, *i.e.* a saturated aqueous solution of calcium hydroxide, could be used to consolidate wall paintings and carbonate stones. Unfortunately, its efficacy is quite limited, mainly owing to the poor solubility of calcium hydroxide in water ($1.73\,g\,L^{-1}$ at 20 °C) and the high solubility of CO_2 in water, which reduces the amount of available binder. On the other hand, more concentrated systems, based on lime particles dispersed in water, cannot be used for two reasons: 1) the dispersions are made by large particle clusters that are not stable and the sedimentation is very fast; 2) as a direct consequence of point 1, the formation of white glazing over the painted surfaces is unavoidable.

Some improvements can be achieved by dispersing lime particles in non-aqueous solvents. However commercially available dispersions of earth alkaline hydroxides exhibit a broad size distribution and the mean dimensions are larger than several micrometres, the smallest particles being around 0.5 μm while 80% are larger than 1 μm. Similarly to aqueous dispersions, such formulations do not provide good results in terms of the consolidation effect.[22] On the other hand, lime acquires its unique properties when prepared at the nanosize scale.[23,24]

Researchers at CSGI (*Consorzio Interuniversitario per lo Sviluppo dei Sistemi a Grande Interfase*; Center for Colloid and Surface Science) at the University of Florence, originally developed, at the end of the 1990s, the first formulation based on calcium hydroxide nanoparticles in propanol,[25] which has been further developed and is today commercially available under the trademark Nanorestore®.

Although the synthetic methodologies allow the preparation of particles with ideal size and size distributions in aqueous or non-aqueous solvents for application on wall paintings, it is of paramount importance to avoid their application as aqueous dispersions, because they are not stable, owing to the tendency of the particles to form aggregates in water. As explained in Section 12.6, a substantial improvement was achieved by using short-chain alcohols as dispersing media that produce kinetically stable dispersions. In particular, 1-propanol and 2-propanol promote the de-aggregation and stabilization of the nanoparticles, as a result of the physical absorption of alcohols onto the surface of the nanoparticles.

The use of some organic solvents as carriers for nanoparticles implies several advantages.[25] Alcohols are environmentally friendly, volatile and, when compared with other organic solvents, have low toxicity. Furthermore, their rheological properties make them very easy to use (their density and viscosity are similar to those of water). For these reasons alcoholic dispersions of nanoparticles can be applied using simple techniques, such as brushing and spraying, as highlighted by tests on several porous materials. Moreover, the surface tension of alcohols is low enough to ensure optimal wetting of the wall paintings' matrix.

As a matter of fact, stable alcoholic dispersions of calcium hydroxide nanoparticles have been successfully applied since 1996, instead of polymers, as consolidants and fixatives to re-adhere lifted paint layers during many restoration workshops in Italy and elsewhere in Europe (see also Section 13.7).[23–27] In terms of chemical reactions, the dispersions of nanoparticles in organic solvents is conceptually similar to the application of a concentrated solution of limewater, well above the physico-chemical concentration limit imposed by the low solubility of calcium hydroxide in water. The consolidating effect is thus greatly enhanced.

The average size of particles is a key factor in obtaining good results for penetration and consolidation,[26] but their performance depends also on other important factors that determine their chemical reactivity and the final features of the calcium carbonate network that is formed upon carbonation. Particle size distribution, crystallinity, shape (crystal habitus), specific surface area, crystalline domains size, surface fractal structure and the presence of crystal defects all contribute to defining the kinetics of lime carbonation and the main characteristics of the final product. Moreover, the crystalline structure of the calcium carbonate formed is strictly related to the mechanical properties acquired by the materials onto which the nanoparticles are applied. Therefore much attention and great experience are required to choose the ideal characteristics of the nanoparticles in order to achieve a good consolidation result. Finally, the carbonation reaction is also affected by the diffusion of CO_2 within the porosity of the materials to be consolidated.

It is important to highlight here how the $Ca(OH)_2$ nanoparticles possess the original features of one of the oldest nanomaterials used by humankind, *i.e.* aged lime putty, which has been thus developed by modern colloid science into a new system, produced through specific synthetic routes and stably dispersed in an appropriate solvent, to be fully applicable to the preservation of wall paintings and limestone.

In fact, if we track the hydroxide preparation process back in time, we find that Cennino Cennini's Renaissance treatise and the more ancient classic sources (by Pliny and Vitruvius) report on the hydration process of calcium oxide (also known as "slaking of quicklime").[28] They also provide accurate recommendations for the improvement of the quality of slaked lime: a complete ripening of lime, which may require from several months to perhaps 2–3 years, could only be achieved upon prolonged storage under pure water. The best procedures for lime preparation were defined to improve the plasticity and workability of lime putties. Lime putty is a paste containing about 30–40% free water that was traditionally prepared by the artists through a slaking reaction, performed by using a slight excess of water with respect to the stoichiometric ratio; lime putty produces finer particles and higher specific surface areas, which results in higher reactivity and plasticity of the putty with respect to the dry hydrate.

Few scientific investigations have tried to understand the mechanisms underlying the process. The effect of prolonged storage of lime under water has been deeply analysed by Rodriguez-Navarro *et al.*[29,30] The authors recently contributed greatly to this subject and provided a clear general view of the process, supported by several forms of experimental evidence. Fresh and aged lime putties were investigated in terms of their crystal morphology and the size of the particles in order to study their changes upon ageing. The authors showed that prismatic crystals, which are present in fresh lime putty, underwent ripening upon ageing with significant size reduction and morphological modifications to a sub-micrometre plate-like habitus. This process was explained with the assumption that the solubility of the prism faces is higher than that of the basal faces of hexagonal portlandite. In fact, this difference may explain the start of a prismatic-to-plate-like transformation of particles; however, it was necessary further to demonstrate that secondary nucleation of plate-like hydroxide was taking place. By using electron microscopy and X-ray diffraction the authors provided clear evidence to support their conclusions (see Figure 13.3).

The higher surface area and the changes in morphology of lime particles account for the water retention and workability of plasters prepared by using ripened lime putty.

Interestingly, the carbonation process of lime is also greatly affected by the properties of calcium hydroxide particles. In particular, Cazalla *et al.*[31,32] demonstrated that, upon carbonation, the change in calcite: portlandite ratio *vs.* time (calculated over 200 days) is four times higher for aged putty than for commercially available freshly hydrated lime. As a matter of fact, all the appealing features of aged lime putty are enhanced when calcium hydroxide is stabilized in the form of nanosized platelets dispersed in propanol, which provides the ideal consolidating agent to reinforce wall paintings and limestone.

It must be pointed out that, owing to the number of interlinked factors, neither a unique universal formulation of nanoparticles nor a general application protocol can be defined for all possible conservation case studies. However some advice, based on the experience gathered in the last decade, can be highlighted and used to guide chemists and conservators towards the best solution to fit their conservation concerns. Preferably, the particle dimension should range from 50–250 nm with an average size around 150 nm, and they should possess, as already mentioned, a plate-like shape and a high crystallinity.

Particle sizes in the 50–250 nm range (see Figure 13.4) avoid the risks of substrate whitening caused by reduced penetration and deposition of particles over the surface, while the slow carbonation ensures the regular and well-organized building up of the calcium carbonate crystalline network, conferring a good consolidation effect on the treated material.

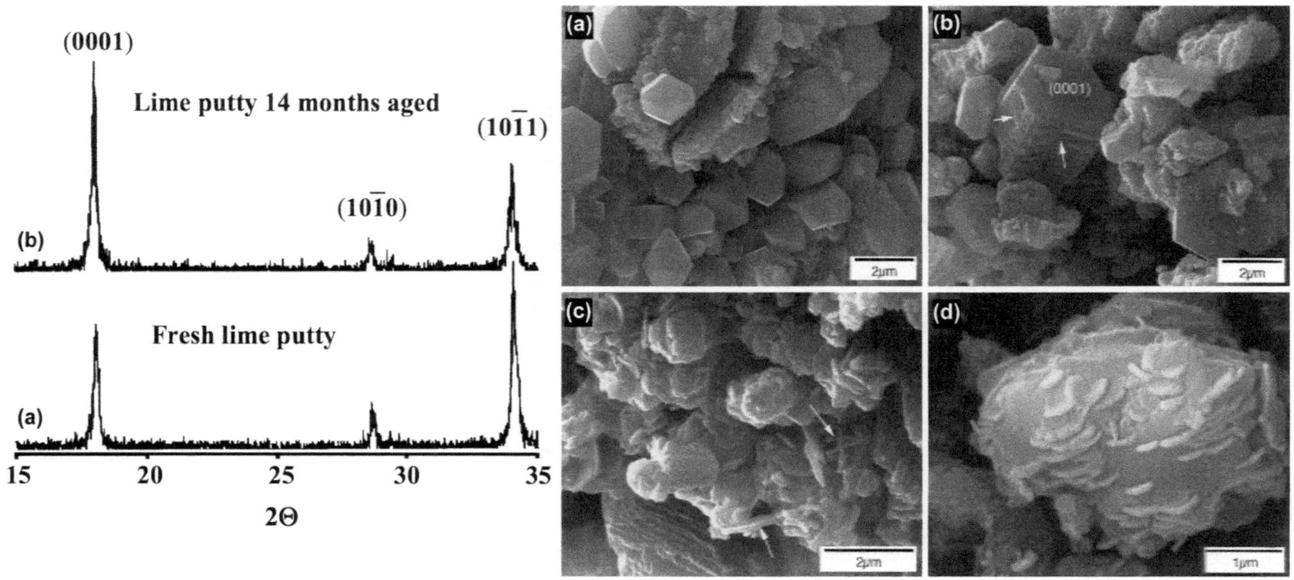

Figure 13.3 On the left, oriented aggregates XRD diffractograms: (a) fresh lime putty and, (b) 14-month-old aged lime putty. Cu*K*α radiation. On the right, scanning electron microscopy (SEM) micrographs: (a) portlandite crystals in a fresh slaked lime putty (sample A); (b) evolution of portlandite crystals showing features of corrosion (arrows) parallel to {0001} faces in 2-month-old aged lime putty (sample A); (c) 1-year-old lime putty showing submicrometre, platelike aggregates of CH (arrows) growing on preexisting portlandite crystals (sample B); (d) detail of submicrometre platelets on portlandite crystals in 1-year-old lime putty (sample B). The orientation of the crystals growing in this {0001} face seems to follow preexisting dissolution steps.
[Copyright (2005) Wiley-VCH. Used with permission from C. Rodriguez-Navarro, E. Hansen and W.S. Ginell, Calcium hydroxide crystal evolution upon aging of lime putty, *J. Am. Ceram. Soc.*, 1998.]

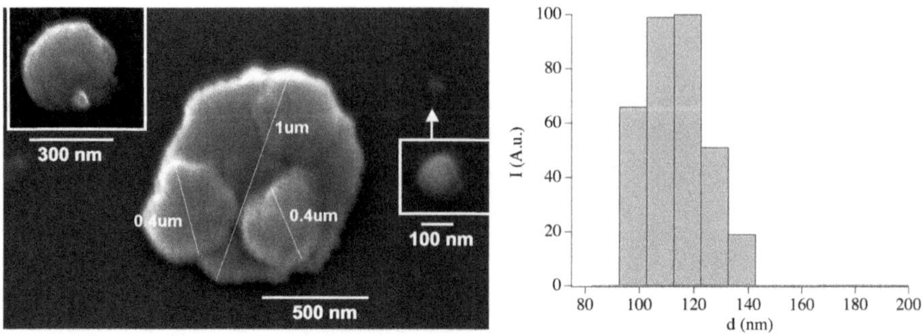

Figure 13.4 On the left, scanning electron microscopy (SEM) image of a cluster of calcium hydroxide nanoparticles prepared through a homogeneous phase reaction in water at high temperature. On the right, particle size distribution of calcium hydroxide nanoparticles in 2-propanol, as obtained by dynamic light scattering (DLS). These particles were obtained by a breakdown process starting from micrometric calcium hydroxide (lime putty) through a heterogeneous phase reaction (see Section 12.4). [Copyright (2010) Wiley-VCH. Used with permission from R. Giorgi, M. Ambrosi, N. Toccafondi and P. Baglioni, Nanoparticles for cultural heritage conservation: calcium and barium hydroxide nanoparticles for wall painting consolidation, *Chem. Eur. J.*, 2010.]

As described in Chapter 12, chemistry offers several pathways for tuning the final characteristics of nanoparticles (*e.g.* size, shape, polydispersity) in order to fit the requirements of a successful consolidation treatment. Currently, several synthetic routes have been studied and improved to obtain calcium hydroxide nanoparticles with good characteristics for the consolidation of degraded wall paintings and carbonate stones with porosity in the range 10–40%.

However, more efforts are still needed in order to understand how the physico-chemical properties and the environmental conditions interplay during the consolidation process. In fact, a huge amount of processes occur at the same time, leading to the formation and disruption of interfaces, to modifications of the physico-chemical characteristics of materials, and to the exchange of fluids within the pores of materials, such as the solvents used as nanoparticle carriers and gases (vapour and CO_2) from the environment.

The general guidelines for the application procedures and the selection of nanomaterials are determined on the basis of the acquired expertise, after more than 10 years of experimentation and collaborations between scientists and restorers (see also Section 13.7). Many research efforts are being pursued in order to understand fully the interacting processes that occur at the nanoscale interfaces; this may definitively pave the way to a full control of the consolidation/restoration of materials.

13.6 Magnesium, Strontium and Barium Hydroxides

An important practical limitation to the use of calcium hydroxide nanoparticles for the restoration of wall paintings and stones is the presence of soluble sulfate salts, *e.g.* sodium or magnesium sulfate, as possible contaminants. In fact sulfate ions react with calcium hydroxide through a double-exchange reaction (see Scheme 13.4), and produce gypsum and alkali metal hydroxides that are not effective for consolidation. The solubility product of calcium sulphate is 4.93×10^{-5} (at 25 °C), much lower than that of sodium or magnesium sulfates (typical contaminants of plaster and stones).[33] This reaction is not thermodynamically favoured in the solid state (ΔG^0 is $+86.2\,kJ\,mol^{-1}$), but it occurs to a good extent because water is usually present inside the walls and, therefore, the reactions take place in the solution phase.

Effective and durable consolidation is thus hindered, owing to the partial consumption of calcium hydroxide and to the formation of detrimental gypsum, which produces a white glaze over the work of art surface, and threatens the long-term stability of the wall painting, as discussed in Chapter 2.

A solution to this issue is the use of barium hydroxide nanoparticles, either alternatively or as a complement to calcium hydroxide. Barium hydroxide in fact favours the transformation of gypsum into insoluble barium sulfate (see also Section 13.5), as shown in Scheme 13.5.

When the first step in the Ferroni method cannot be applied, mixed formulations of $Ca(OH)_2$ and $Ba(OH)_2$ are particularly effective in the pre-consolidation of surfaces that are largely contaminated by sulfates.[34] Such substrates in fact often exhibit severe brittleness and fragility, and their direct cleaning would imply a consistent loss of material (*e.g.* pigments). Stable dispersions of mixed calcium and barium hydroxides can be obtained in 1-propanol, as explained in Section 12.4. Such formulations represent innovative tools, alternatives to the traditional pre-cleaning consolidation treatments that are carried out with caseinate, glues or acrylic polymers. The use of alkaline earth nanoparticles grants higher compatibility, long-term stability and effectiveness of the treatment.

A similar approach was proposed by Ciliberto *et al.*[33] for the synthesis and application of strontium hydroxide nanoparticles. These were prepared through a homogeneous phase reaction in water, starting from low cost raw materials and using milder synthetic conditions (lower temperature). The $Sr(OH)_2$ nanoparticles are a possible alternative to calcium hydroxide when large amounts of sulfates are present in wall paintings and plasters and cannot be removed with the first step of the Ferroni method.

Recently, mixed formulations based on magnesium and calcium hydroxide nanoparticles were successfully used for the consolidation of Angera stone, a

$$Ca(OH)_2 + Na_2SO_4 + 2H_2O \rightarrow 2NaOH + CaSO_4{\cdot}2H_2O$$

Reaction Scheme 13.4

$$Ba(OH)_2 + CaSO_4{\cdot}2H_2O \rightarrow BaSO_4 + Ca(OH)_2 + 2H_2O$$

Reaction Scheme 13.5

Figure 13.5 Collegio dei Ghisleri in Pavia (Italy). Architectonic and decorative elements in Angera dolomite stone were treated with mixed formulations based on calcium and magnesium hydroxide to build a chemically compatible mineral network.

dolomite stone (sedimentary rock) largely used for historical buildings in the north of Italy, and composed of calcium–magnesium carbonate (see Figure 13.5).[35] The rationale was that the use of both calcium and magnesium hydroxides can enhance the physico-chemical compatibility with the original stone, owing to the concomitant formation of calcium and magnesium carbonates.

13.7 Case Studies

The accounts of recent investigations and experimental tests carried out in several different sites with specific conservation issues will provide the reader with a more complete view of the potential impact of nanotechnology for the consolidation of mural paintings.

The formulations presented in this section are versatile and their application to a variety of conservation issues needs care in order to define parameters such as the ideal size and distribution of the particles to be used, the best concentration, the application time and method, and the ideal chemical composition of solvents and consolidating materials.

13.7.1 Wall Paintings from the Italian Renaissance and Mesoamerica

As already mentioned in Section 13.5, alcoholic dispersions of calcium hydroxide nanoparticles have been successfully applied in the last decade as consolidants and fixatives to re-adhere lifted paint layers.[23–27] In this section we report two significant case studies, concerning the preservation of wall paintings belonging to different cultural heritage traditions and characterized by exposure to diverse environmental conditions.

The first case study is the conservation of some 16th century *fresco* wall paintings by Santi di Tito, located on the Counterfaçade of the "Basilica di Santa Maria del Fiore" in Florence, the city Cathedral or "Duomo". The *frescoes* exhibited detached and flaking paint layers that were in great need of consolidation (see Figure 13.6).

Dispersions of calcium hydroxide nanoparticles in propanol were selected as a fully compatible material for the re-adhesion of the flaking parts. The particles were synthesized from a homogeneous phase, starting from solutions of NaOH and $CaCl_2$.[26] A purification step was carried out to eliminate the side-products (NaCl), and the hexagonal nanosized $Ca(OH)_2$ platelets (see Figure 13.6) were stably dispersed in propanol and applied on the wall substrate.

The definition of a unique application protocol that would take into account all the possible experimental conditions is beyond the scopes of this section. However some standard rules can be highlighted to illustrate the feasibility of this methodology.

Usually, the ratio between the amount of particles and the dispersion volume ranges roughly between 1 and 5 $g\,L^{-1}$, to grant a full consolidation effect while minimizing the formation of white veils on the treated surface. As a general rule, two to three applications at a low concentration are preferred to a single application using more concentrated dispersions. Typically, the dispersions are brushed onto the substrate using a veil of Japanese paper to protect the surface, and to distribute the particles homogeneously through the surface layers. The nanosized platelets possess a great capacity for water absorption, a key factor in the carbonation process. Eventually, the treated surface can be moistened with a cellulose pulp poultice soaked with water, placed on Japanese paper. Upon completion of the reaction with CO_2, a crystalline network of calcium carbonate is created, granting the re-adhesion of the flaking parts.

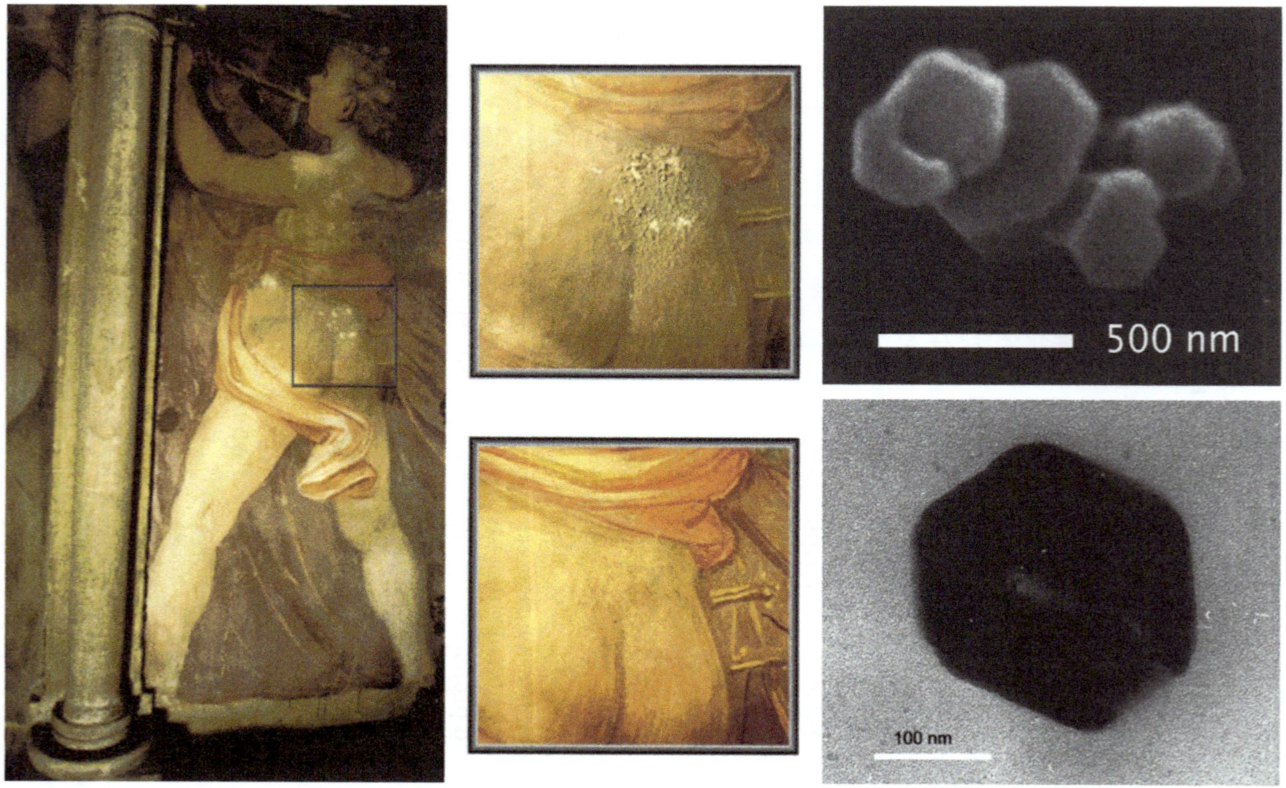

Figure 13.6 On the left, example of application of $Ca(OH)_2$ plate-like nanoparticles dispersed in 1-propanol for the consolidation of the wall paintings by Santi di Tito (16th century) *Gli Angeli Musicanti* on the Counterfaçade of the "Santa Maria del Fiore" Cathedral in Florence. The region labelled with the box was treated with the particles: (top) before the restoration and (bottom) after the restoration. On the right, SEM (top) and TEM (bottom) micrographs of $Ca(OH)_2$ particles with hexagonal habitus, obtained from homogeneous phase reaction at 90 °C.
[Reprinted with permission from M. Ambrosi, L. Dei, R. Giorgi, C. Neto and P. Baglioni, P. Colloidal particles of $Ca(OH)_2$: properties and application to restoration of frescoes, *Langmuir*, **17**, 2001, 4251–4255. Copyright 2001 American Chemical Society.]

In the case of the Santi di Tito's *frescoes*, the pre-consolidation treatment with the nanoparticle dispersions led to excellent results, providing both surface strengthening and re-adhesion that allowed the safe cleaning of the painting, to restore the work of art to its original aspect (see Figure 13.6). This case study is highly representative of the conservation issues that are usually met by conservators of wall paintings across Europe.

The second reported case concerns the conservation of Mesoamerican wall paintings, belonging to the Mayan site of Calakmul ("*La Antigua Ciudad Maya de Calakmul*"). The site is located in Campeche state, in the southeast part of Mexico, an area characterized by a sub-tropical climate. The Calakmul ruins, discovered in 1931, represent an exceptionally important historical record of the Classic Mayan period (250–800 AD). The site was included in the World Heritage list of the United Nations Educational, Scientific and Cultural Organization (UNESCO) in 2002. The "Proyecto Arqueologico Calakmul" was established in 1993 and involves archaeologists, architects, engineers, conservators and epigraphists, besides other specialists. The numerous excavation campaigns carried out in recent decades have identified about 6000 remains, including buildings, stelae, tombs, altars and mural paintings. At the end of 2004, some extraordinary mural paintings were discovered inside Structure I of the *Acropolis Chik Naab* (see Figure 13.7).[36] These paintings represent one of the most important documents of the pre-Columbian era and a precious documentation of Mayan daily life. In fact, a study accounting for the recent epigraphic investigations has highlighted the great importance of the discovery for the knowledge of this ancient civilization.[37]

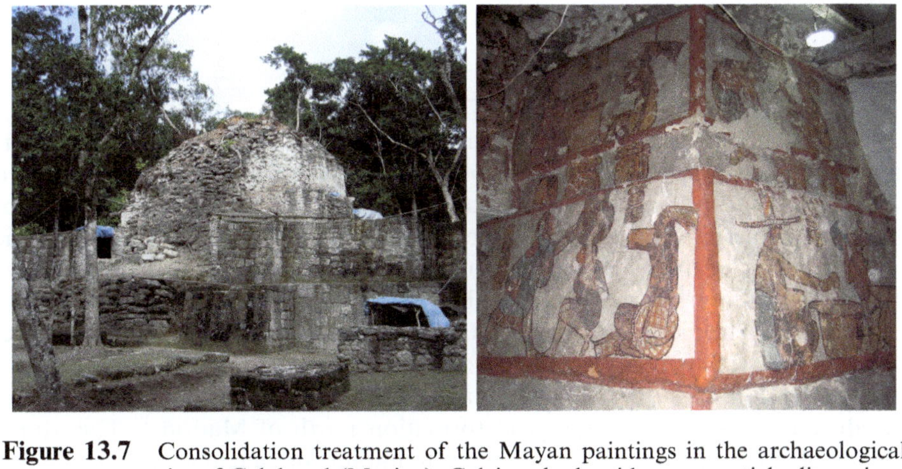

Figure 13.7 Consolidation treatment of the Mayan paintings in the archaeological site of Calakmul (Mexico). Calcium hydroxide nanoparticle dispersions were used to counteract the powdering of the murals. The paintings were discovered in 2004 and restored *in situ* using highly compatible consolidants.

It is worth noting that Mayan people intentionally entombed the painted surfaces. The filling materials (including stone and ceramic fragments) were taken away during the excavation, recovering thus the thick layer of plaster, made of fine powder of stucco mortar and limestone, and painted with a lime-based technique.[38] The conservation status of the newly exposed paintings necessitated immediate intervention to stabilize the painting layer and to reinforce the plaster.

The conservation issues were particularly challenging because of the sub-tropical climate of the Calakmul area. Indeed, the temperature is about 30 °C all the year round, and the relative humidity (RH) is very high for a large part of the year, ranging between 75 and 95%. Inside the building of Structure I, RH is often very close to saturation conditions, so that the painting surfaces are usually slightly wet.

In these cases, especially in the past, detachment of paintings ("*stacco*") was usually foreseen. In fact, the above-mentioned environmental conditions prevented application of the restoration materials commonly used in Mexico until the end of 1980s, *i.e.* polymers. Under these environmental conditions polymers degrade in a few years and promote further degradation because they alter the main physico-chemical properties of the original materials.

Thus, an innovative approach based on calcium hydroxide nanoparticles dispersed in isopropanol was followed.[38] The entire scene was treated with a calcium hydroxide dispersion using the brushing technique. The paint layer was protected with a Japanese paper tissue to minimize the mechanical action of the brush on top of the surface. The painting showed extensive powdering that was completely counteracted upon nanoparticle application and the complete carbonation of the particles. The concentration used was $5\,g\,L^{-1}$ and the full consolidation process required about 40 days.

13.7.2 The Consolidation of Limestone

Slaked lime [$Ca(OH)_2$] nanoparticles have also been investigated as consolidants for limestone and dolomitic stone. Calcium carbonate is in principle a durable and compatible material for the filling and consolidation of several porous carbonate-based stones. Some authors have expanded this basic idea by studying the kinetics of the carbonation of lime nanoparticles that have been used to improve the mechanical properties of stones, and by determining the physico-chemical characteristics of the newly formed crystalline phase. Interesting studies were conducted on the application of nanoparticles to a Spanish dolomitic stone, frequently used in Madrid (Spain) and obtained from Redueña, a Cretaceous geological formation north of Madrid.[39] This stone, made of dolomite (90–95%) and a small amount of calcite (5–10%), exhibits high open porosity. The consolidation intervention was performed using Nanorestore®; the treated samples were exposed, during carbonation, at RH values of 33% and 75%. The use of several non-destructive techniques provided information about the chemical, morphological and physical properties of the stone samples, upon completion of carbonation. Environmental scanning electron microscopy (ESEM-EDS), spectrophotometry, capillarity, water absorption under vacuum, ultrasound velocity, nuclear magnetic resonance (imaging, MRI, and relaxometry) and optical surface roughness (OSR) analyses were used to obtain a full description of the on-going processes and an evaluation of the changes caused by the treatment.

Under both the RH conditions, the stone samples showed a decrease in open porosity and saturation values; ultrasound velocity significantly increased in the three spatial directions, especially in the higher porosity sample exposed to RH 75%. This was explained in terms of a better and more homogeneous distribution of the consolidating product. Other evidence, provided by ESEM, X-ray diffraction (XRD) and transmission electron microscopy (TEM) analyses, suggested that transformation of portlandite [$Ca(OH)_2$] into vaterite ($CaCO_3$), monohydrocalcite ($CaCO_3 \cdot H_2O$) and calcite ($CaCO_3$) was faster at 75% RH (see Figure 13.8). Thus, the physical

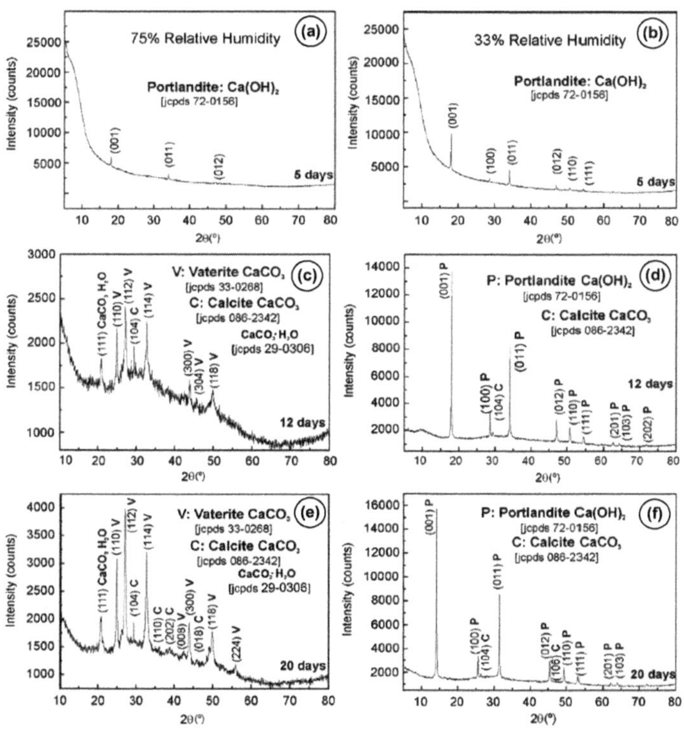

Figure 13.8 (Top) X-ray Diffraction (XRD) of the calcium hydroxide nanoparticle product on different days of consolidation in humid (75% RH) and dry environments (33% RH) showing the changes in the mineral phases during the carbonation process. a) 75% RH after 5 days; b) 33% RH after 5 days; c) 75% RH after 12 days; d) 33% RH after 12 days; e) 75% RH after 20 days; and f) 33% RH after 20 days.

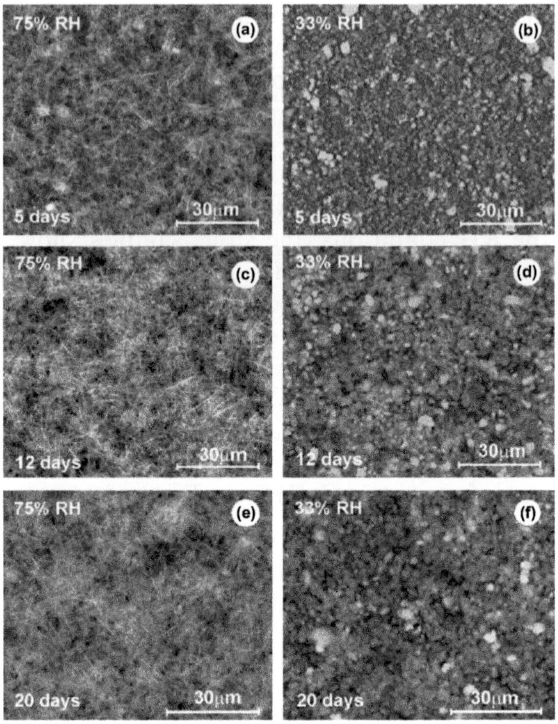

Figure 13.8 (Bottom) Environmental scanning electron microscopy (ESEM) images of the calcium hydroxide nanoparticle products on different days of consolidation in humid (75% RH) and dry environments (33% RH) showing the morphology of the crystals during the carbonation process. a) 75% RH after 5 days; b) 33% RH after 5 days; c) 75% RH after 12 days; d) 33% RH after 12 days; e) 75% RH after 20 days; and f) 33% RH after 20 days.
[Figure 13.8 has been reprinted from *Materials Characterization*, **61**, P. López-Arce, L. S. Gomez-Villalba, L. Pinho, M. E. Fernández-Valle, M. Álvarez de Buergo and R. Fort, Influence of porosity and relative humidity on consolidation of dolostone with calcium hydroxide nanoparticles: Effectiveness assessment with nondestructive techniques, pp. 168–184, Copyright (2010), with permission from Elsevier.]

and mechanical properties of the stones were significantly improved. Three-dimensional (3D) topographic maps obtained by OSR, which describe the average surface roughness, showed only small modifications so that no change in the surface appearance (colour, luminosity, reflectivity) could be detected. On the other hand, NMR relaxometry showed a lowering in T1 and T2 relaxation times. Such effect can be related to the decrease in the pore sizes, as a result of the repair of cracks and crystallization of newly formed calcite.

Some other applications of calcium hydroxide nanoparticles for stone consolidation have also been reported.[40] They include the case of the biocalcarenite stone known as Globigerina, typically derived from Malta, which exhibits similarities to several other lithotypes in the Mediterranean area (*e.g.* Lecce stone or Noto stone from the south of Italy). Ammonium oxalate and barium hydroxide solutions (well-established inorganic materials used for the consolidation of stone surfaces, due to the formation of calcium oxalate and barium carbonate, respectively) were compared with the calcium hydroxide nanoparticle treatment. Several parameters were selected for the physico-chemical characterization of stone samples, before and after the consolidation treatment. Surface analysis by electron microscopy combined with elemental analysis by EDS (energy dispersive X-ray spectroscopy), capillarity, pore size distribution evaluation by gas-porosimetry, colorimetric analysis and drilling force measurements were performed to evaluated the effects of the treatments. Calcium hydroxide nanoparticles provided the best results in terms of reinforcement of the stone surface, with negligible effects on the colour appearance of stone and small changes in the pore structure. The latter point is particularly important because these stones are mostly used in a seaside environment, and are thus exposed to large amounts of marine aerosol. Therefore, the permeability of the stone is essential to reduce the mechanical stresses due to salts.

Further Suggested Reading

L. Dei, P. Baglioni and M. Mauro, Materials for wall paintings conservation: changes of physicochemical properties, aging effects, and reversibility. Preprints of the Conference *Reversibility: Does it Exist*?, ed. A. Oddy and S. Carroll, London, 8–10 September 1999, p. 73.

L. S. Gomez-Villalba, P. López-Arce and R. Fort, Nucleation of $CaCO_3$ polymorphs from a colloidal alcoholic solution of $Ca(OH)_2$ nanocrystals exposed to low humidity conditions, *Appl. Phys. A-Mater.*, 2012, **106**, 213.

L. S. Gomez-Villalba, P. López-Arce, M. Alvarez de Buergo and R. Fort, Structural stability of a colloidal solution of $Ca(OH)_2$ nanocrystals exposed to high relative humidity conditions, *Appl. Phys. A-Mater.*, 2011, **104**, 1249.

P. López-Arce, L. S. Gómez-Villalba, S. Martínez-Ramírez, M. Álvarez de Buergo and R. Fort, Influence of relative humidity on the carbonation of calcium hydroxide nanoparticles and the formation of calcium carbonate polymorphs, *Powder Technol.*, 2011, **205**, 263.

M. Matteini, An assessment of Florentine methods of wall painting conservation based on the use of mineral treatments, in *The Conservation of Wall Paintings*, Proceedings of the symposium organized by the Courtauld Institute of Art and the Getty Conservation Institute, London, July 13–16, 1987, ed. S. Cather, GCI, Los Angeles, p. 137.

P. Mora, L. Mora and P. Philippot, in *Conservation of Wall Painting*, Butterworths, London, 1984.

P. Baglioni and R. Giorgi, Soft and hard nanomaterials for restoration and conservation of cultural heritage, *Soft Matter*, 2006, **2**, 293.

Review and Questions

The main points discussed in the chapter may be summarized as follows:

- Calcium hydroxide is the original binder of wall paintings and plaster, and crystalline carbonate is one of the main components of stone artefacts. Unfortunately, the use of $Ca(OH)_2$ solutions or aqueous suspensions is hindered by its low solubility and by the high tendency to form clusters when used in the form of particle dispersions in water.
- Nanoscience provides new tools for the consolidation of wall paintings, plasters and stones, based on highly compatible inorganic materials. These represent an alternative to the use of synthetic polymers that have been proved to be detrimental under diverse environmental conditions.
- Short-chain aliphatic alcohols are excellent candidates as carriers for nanoparticles; the solvents' low surface tension grants easy and quick wetting of the substrates' matrix. Nanoparticle dispersions exhibit high kinetic stability (no stabilizers are needed), which is perfectly compatible with the average application times.
- Upon carbonation, the hydroxide builds a crystalline network of calcium carbonate that grants the reinforcement of wall paintings, plasters and limestone.
- The synthesis of earth-alkaline metal hydroxides provides several tools for the consolidation of salt contaminated plasters and other carbonate-based stone. Barium, strontium, magnesium and calcium hydroxides, especially in mixed dispersions, behave as poly-functional composite materials, owing to their different chemical reactivity and to the physico-chemical features of formed carbonates.

Questions

1) What are the typical degradation processes that can be counteracted using calcium hydroxide nanoparticles?
2) What is the best application procedure to apply nanoparticle dispersions?
3) How is the best concentration of nanoparticles chosen?
4) What are the main advantages related to the use of nanoparticle dispersions?
5) Can hydroxide nanoparticles be used for the consolidation of paintings realized with the '*secco*' technique?
6) If white veils form upon nanoparticle application, how can they be removed? What are the conditions that favour the veiling process?

Answers

1) This treatment is effective at the surface level, *i.e.* up to few hundreds of microns. It is not addressed to deeper, bulk structural consolidation. The

application of calcium hydroxide aims to recreate or reinforce the original paint layer. Degradation implies the corrosion of calcium carbonate and mainly results in the powdering of the colours. This is the typical degradation phenomenon that is counteracted by the application of nanoparticles. The flaking of the paint layer can also be fixed when the detachment is limited to a few millimetres and the painted layer is thin. In this case, the application of nanoparticles dispersions causes the re-adhesion of the flaking parts, which is effective after the complete carbonation of calcium hydroxide.

2) Nanoparticle dispersions can be applied by brushing and/or by low-pressure spraying with appropriate protection of the surface with Japanese paper. The first option should be preferred because the penetration of the particles is favoured, and larger amounts of consolidant can be used. In the case of paint flaking, the mechanical action of the brush helps the re-adhesion process. Spraying is a good option when the degradation status is so severe that the action of brushing could cause damage. Thus, spraying is typically useful when a pre-consolidation treatment is necessary.
3) The concentration mostly depends on the porosity of the material to be treated. Typically, a concentration of about $5\,g\,L^{-1}$ is optimal for degraded wall paintings and porous limestones. More diluted dispersions (dilution ratios from 1 : 1 to 1 : 10) can also be used. Usually a two-step treatment using a $2.5\,g\,L^{-1}$ dispersion is preferred rather than a single treatment with a $5\,g\,L^{-1}$ dispersion.
4) The main advantage relies on the very high physico-chemical compatibility of inorganic nanoparticles with the substrate. The use of solid particles also allows their linear distribution through the substrate section, from the surface (where most of the particles deposit) to the inner layers.
5) Calcium hydroxide nanoparticles can also be used on most of the paintings carried out with a '*secco*' technique. Upon natural weathering, the original organic binders become much less sensitive to pH variations such as those induced by the application of hydroxides onto the paintings' surface. In fact, "mineralization" of binders occurs with time.
6) Veil formation mostly depends on the amount of particles that were applied. It is of fundamental importance to stop the application when the porous substrates reach the solvent saturation. It is also important to avoid rapid evaporation of the alcohol, a situation that is enhanced by dry environmental conditions. In case of veil formation, which appears a few minutes after application, it is convenient to apply some pure alcohol by brushing. If the veiling persists, it is necessary to apply to the painted surface a cellulose-pulp compress, impregnated with demineralized water, for a few hours. A light white veil usually disappears, in 12–24 hours, when the treated materials are exposed to conditions of high relative humidity (>75%).

References

1. C. Horie, in *Materials for Conservation: Organic Consolidants, Adhesives and Coatings*, Butterworth-Heinemann, London, 1987.
2. P. Mora, L. Mora and P. Philipot, *Conservation of Wall Painting*, Butterworths, London, 1984.
3. R. L. Feller, in *Accelerated Agings – Photochemical and Thermal Aspects*, The Getty Conservation Institute, Los Angeles, CA, 1994, p. 63.
4. M. Lazzari and O. Chiantore, *Polymer*, 2000, **41**, 6447.
5. O. Chiantore and M. Lazzari, *Polymer*, 2001, **42**, 17.
6. M. Favaro, R. Mendichi, F. Ossola, U. Rosso, S. Simon, P. Tomasin and P. A. Vigato, *Polym. Degrad. Stabil.*, 2006, **91**, 3083.
7. G. Spoto, *Acc. Chem. Res.*, 2002, **35**, 652.
8. R. Franchi, G. Galli and Manganelli Del Fa, *Stud. Conserv.*, 1978, **23**, 23.
9. R. Rossi-Manaresi, *Proceedings of the meeting of the Joint Committee for the Conservation of Stone, Bologna, October 1–3, 1971. (ICOM, ICOMOS, International Centre for Conservation – Working group on the treatment of stone)*, ed. R. Rossi-Manaresi and G. Torraca, Centro per la Conservazione delle Sculture all'Aperto, Bologna, 1972, p. 83.
10. M. Matteini and A. Moles, *ICOM Committee for Conservation. 5th Triennial Meeting, Zagreb, 1–8 Oct. 1978*. Preprints, International Council of Museums, Paris, 1978, 78/15/5/1-10.
11. H. Orea amd V. Magar, *Preprints of the 13th Triennial Meeting ICOM Committee for Conservation, ICOM-CC, Rio de Janeiro, 22–27 September 2002*, ed. R. Vontobel, James and James, London, p. 176.
12. A. Espinosa, in *Proceedings of in situ Archaeological Conservation, 6-13 April 1986, Mexico City*, ed. H. W. M. Hodges, INAH, Mexico City, 1987, p. 84.
13. J. Riederer, in *Committee for Conservation, 7th Triennial Meeting, Copenhagen*, ICOM, Paris, 1984, p. 21.
14. D. Magaloni and T. Falcon, *Conservacion del Templo Rojo de Cacaxtla, Temas y problemas*, 1 Coloquio del Seminario de Estudio del Patrimonio Artistico, ed. A. Torres Michua and E. X. De Anda Alanis, UNAM-IIE, Mexico, 1997, p. 107.
15. I. Torres and G. Pelaez, *Conservacion y restauration de la pintura mural en Cacaxtla, in Cacaxtla, proyecto de investigation y conservacion*, Gobierno del Estado de Tlaxcala-INAH, Mexico, 1990, p. 77.
16. T. López and V. M. Dávila, *Proceedings of the 33rd International Symposium on Archaeometry, 22–26 April 2002, Geoarchaeological and Bioarchaeological Studies*, ed. H. Kars and E. Burke, Vrije Universiteit, Amsterdam, p. 189.
17. E. Ferroni, V. Malaguzzi and V. Rovida, *Proceedings of the ICOM-CC Plenary Meeting*, The International Council of Museums-Committee for Conservation, Amsterdam, Netherlands, 1969.
18. E. Ferroni and P. Baglioni, *Scientific methodologies applied to works of art. Proceedings of the symposium, Florence, Italy 2–5 May 1984*, ed. P. L. Parrini, Montedison progetto cultura, Milan, 1986, p. 108.

19. C. Giovannone, M. Ioele, P. Santopadre and U. Santamaria, *Bollettino ICR*, 2003, **6–7**, 59.
20. M. Matteini and A. Moles, *Proceedings of the ICOM Committee for Conservation 7th triennial meeting: Copenhagen, 10–14 September 1984: preprints*, Paris, 1984, pp. 84.15.15–84.15.19.
21. M. Matteini, *Proceedings of the symposium organized by the Courtauld Institute of Art and the Getty Conservation Institute, London, July 13–16 1987*, ed. S. Cather, Getty Conservation Institute, Marina del Rey, 1991, p. 137.
22. I. Brajer and N. Kalsbeek, *Stud. Conserv.*, 1999, **44**, 144.
23. P. Baglioni, E. Carretti, L. Dei and R. Giorgi, in *Self-Assembly*, ed. B. H. Robinson, IOS Press, 2003, p. 32.
24. P. Baglioni and R. Giorgi, *Soft Matter*, 2006, **2**, 293.
25. P. Baglioni, L. Dei and R. Giorgi, *Stud. Conserv.*, 2000, **45**, 154.
26. M. Ambrosi, L. Dei, R. Giorgi, C. Neto and P. Baglioni, *Langmuir*, 2001, **17**, 4251.
27. R. Giorgi, M. Baglioni, D. Berti and P. Baglioni, *Acc. Chem. Res.*, 2010, **43**, 695.
28. D. V. Thompson and C. Cennini, *The Craftsman's Handbook. "Il Libro dell' Arte"*, Dover Publications, NewYork, 1978.
29. C. Rodriguez-Navarro, E. Hansen and W. S. Ginell, *J. Am. Ceram. Soc.*, 1998, **81**, 3032.
30. C. Rodriguez-Navarro, E. Ruiz-Agudo, M. Ortega-Huertas and E. Hansen, *Langmuir*, 2005, **21**, 10948.
31. O. Cazalla, C. Rodriguez-Navarro, E. Sebastian, G. Cultrone and M. J. De La Torre, *J. Am. Ceram. Soc.*, 2000, **83**, 1070.
32. K. Elert, C. Rodriguez-Navarro, E. S. Pardo, E. Hansen and O. Cazalla, *Stud. Conserv.*, 2002, **47**, 62.
33. E. Ciliberto, G. G. Condorelli, S. La Delfa and S. Viscuso, *Appl. Phys. A-Mater.*, 2008, **92**, 137.
34. R. Giorgi, M. Ambrosi, N. Toccafondi and P. Baglioni, *Chem. Eur. J.*, 2010, **16**, 9374.
35. D. Chelazzi, G. Poggi, Y. Jaidar, N. Toccafondi, R. Giorgi and P. Baglioni, *J. Colloid Interf. Sci.*, 2012, http://dx.doi.org/10.1016/j.jcis.2012.09.069.
36. R. Carrasco, in *Maya*, ed. P. Schmidt, M. de la Garza and E. Nalda, Rizzoli International Publications, New York, 1998, p. 372.
37. R. Carrasco, V. A. Vazquez Lopez and S. Martin, *PNAS*, 2009, **106**, 19245.
38. R. Giorgi, D. Chelazzi, R. Carrasco, M. Colon, A. Desprat and P. Baglioni, *Proceedings of the Munich IIC Congress 2006 "The Object in Context: Crossing Conservation Boundaries"*, ed. D. Saunders, J. H. Townsend and S. Woodcock, 2006, p. 162.
39. P. López-Arce', L. S. Gomez-Villalba, L. Pinho, M. E. Fernández-Valle, M. Álvarez de Buergo and R. Fort, *Mater. Charact.*, 2010, **61**, 168.
40. P. Croveri, L. Dei, R. Giorgi and B. Salvadori, *Proceedings of the 10th International Congress on Deterioration and Conservation of Stone*, ed. D. Kwiatkowski and R. Lofvendahl, Stockholm, 2004, vol. 1.

CHAPTER 14

Biomineralization, Geopolymers and Hybrid Nanocomposites

L. DEI* AND D. CHELAZZI

University of Florence, Italy
*Email: luigi.dei@unifi.it; chelazzi@csgi.unifi.it

14.1 Biomineralization

Biomineralization is the process through which minerals are formed by organisms. Biomineralization has a vast range of applications in several fields, including the controlled precipitation of $CaCO_3$, the study of extraterrestrial biological processes, tooth and bone tissue formation, and the formation of pathological concretions such as renal and salivary stones.[1–3] Calcium carbonate, calcium phosphate and silica (glass), are the most common minerals found in skeletal and supporting bio-inorganic structures produced by animals. Biogenic mineral structures have a composite inorganic–organic nature, where an organic matrix is combined with nano- or microscale minerals in the amorphous or crystalline state.[4] The study of these materials aims at the characterization of the mineral layers, of the organic phases, and of the organic–inorganic interface.[5] In fact, organisms produce a variety of composite biominerals with different structural levels, well organized into a complex hierarchy from the nanometric to the macroscopic scale. The hierarchical arrangement seems to play a fundamental role in preventing the mechanical rupture of the mineral phase, through the production of interfaces that deflect the cracks, hampering their propagation across the substrate. Moreover in several cases, at the lowest structural level, biomineral crystals do not reach the necessary dimensions to allow the cracks to spread (the so-called "Griffin length"), and are arranged in a precise way leading to a very

RSC Nanoscience & Nanotechnology No. 28
Nanoscience for the Conservation of Works of Art
Edited by Piero Baglioni and David Chelazzi

Published by the Royal Society of Chemistry, www.rsc.org

effective adaptation to loads.[6] As a result, biominerals exhibit unique properties that are different from those of the corresponding purely inorganic phases. It has been shown, for example, that the presence of organic nanolayers at the interface between microtablets of calcium carbonate imparts toughness to molluscs' nacre, compensating for the brittleness of the mineral phase,[7,8] while spicules in siliceous sponges are greatly more flexible and tough than synthetic glass rods.[9,10]

Research efforts have focused on the mechanisms of mineral formation, and particularly on the presence of precursor phases that exhibit a short-range order with respect to the mature phases into which they crystallize.[11] Many organisms, thus, produce first a colloidal phase from which the crystals are then nucleated and grown, as opposed to the regular nucleation and growth that occur in a saturated solution (as in many inorganic synthetic pathways, see Section 12.3).[12] Proteins seem to have an important role during the formation of the precursor and of the mature composite organic–inorganic matrix, in fact mineralized tissues have been shown to contain several different macromolecules that are also found in other tissues of the producer organisms.[13,14]

This concise introduction to such a vastly studied topic highlights the importance of two key possibilities for useful applications in the field of cultural heritage conservation, i.e. the use of (i) biogenic minerals and (ii) composite organic–inorganic materials, for the consolidation of inorganic artistic substrates such as stones and wall-paintings. Both these aspects will be addressed in the following sections. For several obvious reasons (the general conditions and the spatial scales related to dealing with porous artistic substrates), bacteria rather than larger organisms will be considered for the deposition of consolidating biominerals on stone. In this case, the production of minerals is considered as "induced" rather than "controlled", owing to the dependence of the mineral type produced on the environmental conditions.[15]

14.1.1 Bacterial Biomineralization of Calcium Carbonate

The aim of this section is to overview some of the most recent and important applications of calcium carbonate biomineralization for the consolidation of immovable works of art or artistic substrates, such as architectural and sculptural stone, whose degradation has been discussed in Chapter 2. As previously reported, consolidation with synthetic products (acrylic polymers, epoxy resins) commonly leads to the formation of incompatible films on the treated surfaces, and inorganic consolidants with a higher physico-chemical affinity for the works of art materials should be preferred (see also Sections 3.3–3.5). Indeed, $CaCO_3$ is the most compatible material for the consolidation of carbonate stones, and its precipitation *in situ* is ruled by four factors. Of these, the first three are related to the precursor system that initiates the precipitation:[16]

1) concentration of Ca^{2+} ions;
2) concentration of dissolved inorganic carbon;
3) pH;
4) availability of nucleation sites.

When the product of the activities of calcium and carbonate ions exceeds the $CaCO_3$ solubility constant ($K_{so\ calcite} = 4.8 \times 10^{-9}$ at 25 °C), the system is supersaturated and calcium carbonate is likely to precipitate.[17]

In the last two decades, carbonate mineralization induced by bacteria has been highlighted as an environmentally friendly method to consolidate artistic carbonate stone. As a matter of fact, several carbonate rocks have been cemented in natural processes through carbonate precipitation induced by microorganisms.[18] The precipitation of carbonates has been reported for many different bacterial species, and has been studied in saline and freshwater aquatic environments, and in soil.[19–23]

One strategy involves the hydrolysis of urea in order to obtain a protective calcite layer on degraded limestone substrates (see Figure 14.1). In this case,

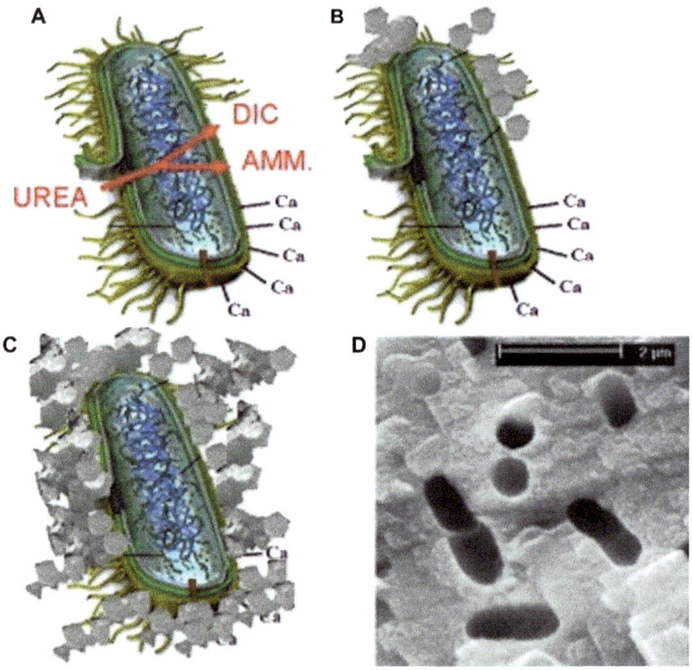

Figure 14.1 Simplified representation of carbonate precipitation induced by hydrolysis of urea. (A) In the solution, calcium ions are attracted to the bacterial cell wall, which has a negative charge. Both dissolved inorganic carbon (DIC) and ammonium (AMM) are released, as a consequence of the addition of urea. (B) In the presence of calcium ions, supersaturation can occur, leading to the precipitation of calcium carbonate on the bacterial cell wall. (C) When the cell becomes fully embedded, the nutrient transfer is hampered, causing the death of the bacterial cell. (D) Imprints of bacterial cells involved in the precipitation of calcium carbonate (bar is 2 µm).
[Reprinted from *Ecological Engineering*, **36**, W. De Muynck, N. De Belie and W. Verstraete, "Microbial carbonate precipitation in construction materials: A review", Pages No. 118–136, Copyright (2010), with permission from Elsevier.]

bacteria with a high initial specific urea degradation rate will allow a high substrate turnover for a limited number of bacteria. Another important factor to be considered for the selection of species is the ζ-potential, which measures the potential of the electric layer at the surface of the bacteria. In fact, microbes with a high negative ζ-potential will effectively adhere to limestone (which has a positive ζ-potential). The advantages involved in the use of this technique are the easier control and the potentially rapid production of large amounts of carbonate.

The complex metabolic processes involved in the production of biominerals have been investigated in several studies, and a comprehensive outline of the essential findings was provided by Zamarreño *et al.* in a recent publication on the subject, where freshwater bacteria were investigated as a limestone consolidant.[24] A fundamental step in the production of biominerals involves the interaction between Ca^{2+} ions (which must be present in the bacterial growing medium) and the bacteria walls, which are negatively charged, leading to a change in the ionic charge and to the aggregation of the bacteria, which thus become the nuclei of the minerals produced and increase the size of crystals.[25,26] The carbonate ions, which combine with calcium ions, can either have a biotic (bacterial metabolism) or an abiotic origin.[27]

During the process of mineral formation, the bacteria produce extracellular polymeric substances (EPS) that work as nucleation sites but also aggregate crystals into bigger organic–inorganic composite clusters. Consequently, different compositions of EPS result in different composite structures.[28–30] It is thus evident that the selection of the bacterial strain, the conditions and the chemical media used for their incubation and growth are all crucial factors in preparing the proper consolidation procedure. The characterization of the biominerals and assessment of the treatment on stone is typically performed through X-ray diffractometry (XRD), scanning electron microscopy (SEM), and staining techniques (for the detection of both bacteria and EPS), coupled with optical and confocal scanning laser microscopy.

In their work,[24] Zamarreño *et al.* focused on several important aspects, concerning the disposal of bacteria (genera *Pseudomonas* and *Acinetobacter*) in the organic–mineral network produced, the cells' viability after the formation of biominerals, and the quantification of the beneficial consolidant effect on limestone. Histological staining showed that the carbonate crystals are formed around polysaccharides, the latter also serving as bridges between smaller crystals, to form sheet-like structures.[24] Bacteria are present both within the polysaccharide cores and as an outer coating on the carbonate crystals, and act as an integrating part of the composite structures, rather than remaining as isolated cells or clusters.[24] It was observed that bacteria deposit crystals on the surface of limestone samples, and also penetrate into the porous matrix, producing an effective reduction of the stone's porosity (by around 40–45%). Such a reduction is compatible with a strengthening of the substrate without causing the obstruction of the pores. The "breathing" of the surface (*i.e.* the vapour transfer) is thus granted, while other detrimental effects in real applications are avoided, such as the formation of salts within the blocked pores, which would lead to micro-pressure and cracks. The same study[24] also showed that the collection and overnight incubation of bacteria, after the mechanical breaking of the crystals, resulted in only scarce viability of the released microorganisms, which remain alive for no more than two weeks. These data confirmed the applicative potentiality of the method, because uncontrolled bacterial growth could be detrimental for the artistic substrate.

In previous research work, Rodriguez-Navarro *et al.* tested different bacterial genera, obtaining interesting results.[31] Their work considered the limitations and drawbacks involved in the use of a soil bacterium (*Bacillus cereus*) for the consolidation of ornamental stone.[32,33] In fact, despite the formation of a compatible and highly coherent carbonate cement on the stone surface, the treatment lacked in-depth penetration, and led to the formation of a surface film (consisting of EPS) that clogged the stone's porosity. Moreover, the *Bacillus* bacteria may form endospores that could lead to uncontrolled growth. Consequently, they focused on *Myxococcus xanthus*, a non-pathogenic aerobic soil bacterium, to induce carbonate precipitation on samples (slabs) of a porous calcarenite limestone widely used in Spanish historical buildings. Their results showed that the carbonates formed by bacteria adhere strongly to the stone substrate, owing to the epitaxial growth on the pre-existing calcite grains.[31] The strong adhesion and the organic–inorganic nature of the newly formed calcite enhanced resistance against ultrasound. Bacteria-induced cementation was reported up to depths of more than 500 μm, and the most significant carbonate precipitation occurred in the first 5–10 days after the treatment, without any clogging of the substrate's pores. Moreover, viability tests showed that the drying of treated stone samples led to the death of the bacteria.

Interestingly, as shown by Urzi *et al.*, bacterial genera that are commonly found in building stones, such as *Geodermatophilus* and *Micrococcus*, are also able to induce carbonate precipitation under appropriate laboratory conditions, supporting the hypothesis that autochthonous bacteria could interact with the newly introduced bacteria to provide biomineralization.[34]

Several authors have debated the penetration depth of bacteria and of their consolidating action, because the treatment could be intended either as a surface coating (granting consolidation and waterproofing properties), or for providing binding action through the cross-section of the substrate. De Muynck *et al.* considered *Bacillus sphaericus* for the surface treatment of cementatious materials, and demonstrated that the treatment led to the formation of a layer of crystals with dimensions ranging between 10–40 μm and 100 μm, and with a greater resistance against freeze–thaw cycles and penetration by chloride. Similar effects are obtained with some well-established coating materials, such as acrylic polymers, organic–inorganic materials and other silicon-based products (silanes, siloxanes, silicates).[35] The same authors also provided an extensive review on the same subject (see the Further Suggested Reading section of this chapter).[36]

A very interesting application involves the use of bacteria for the removal of salts (*e.g.* sulfates) and crusts from artistic substrates, with a twofold positive effect since the removal of gypsum can also result in its conversion to calcite. Alfano *et al.* used Carbogel to deliver *Desulfovibrio vulgaris* in order to remove black crusts from stone.[37] This technique was effective in removing sulfate deposits (up to 85%) in a relatively short time (24 hours) from the altered tuff stone surfaces of the 12th-century Matera Cathedral (Italy) (see Figure 14.2). The method had already proved to be effective in the removal of 98% of the sulfates from 2–3 mm thick crusts on a marble fragment from the cathedral of Milan, after a 45-hour treatment. Carbogel allowed easy application of the bacteria, maintaining a good level of microbial activity, and simple removal from the substrate, providing altogether better performance than other carriers, such as sepiolite and Hydrobiogel-97. The authors concluded that the feasibility of this very promising method should be evaluated case by case, owing to

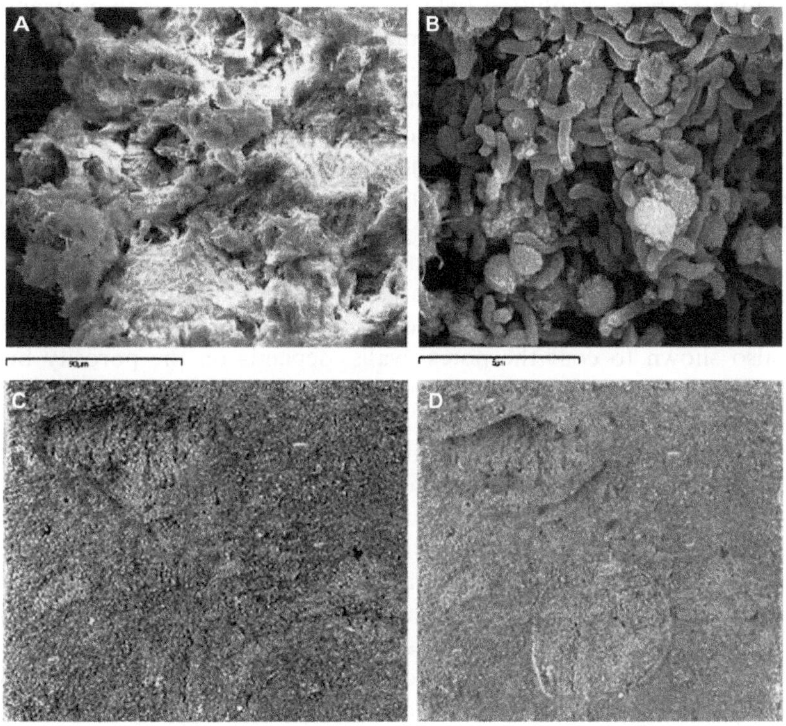

Figure 14.2 Scanning electron microscope (SEM) images of: (A) Carbogel (bar is 90 μm); (B) Carbogel loaded with *Desulfovibrio vulgaris* cells (bar is 5 μm). (C) Tuff stone specimen under stereoscopic observation (10×), before treatment with *D. vulgaris*. (D) The specimen after 48 hours of treatment with *D. vulgaris* in Carbogel (bioremoval of sulfates).
[Reprinted from *International Biodeterioration & Biodegradation*, **65**, G. Alfano, G. Lustrato, C. Belli, E. Zanardini, F. Cappitelli, E. Melo, C. Sorlini and G. Ranalli, "The bioremoval of nitrate and sulfate alterations on artistic stonework: The case-study of Matera Cathedral after six years from the treatment", Pages No. 1004–1011, Copyright (2011), with permission from Elsevier.]

differences in factors such as the crusts' chemical nature, thickness and uniformity.

A different approach was reported by Tiano *et al.*, based on the assisted precipitation of carbonate crystals within the pores of calcareous stones. In this case, however, the bioremediation was obtained directly using natural and synthetic polypeptides, rather than living microorganisms.[38]

Organic matrix molecules (OMM), such as proteins rich in aspartic acid, are involved in the biomineralization of calcite by *Mytilus edulis* (a shellfish). Both OMM and, alternatively, poly(aspartic acid), which is commercially available and does not need expensive and time-consuming extraction processes, have been used to obtain the controlled precipitation of carbonate crystals. For the treatment of stone samples in closed environment, these bio-inducing macromolecules (BIMs) were dissolved in a $CaCl_2$ solution, and added to an ammonium carbonate solution (which develops CO_2). In open environments, BIMs are dissolved in a supersaturated calcium bicarbonate solution, after fluxing CO_2, and sprayed over stone substrates. The addition of commercial $CaCO_3$ nanoparticles to the solution was considered to maintain supersaturation in the stone pores. It was found that the size of the crystals increases with the concentration of BIMs, hence providing a tunable tool for the reinforcement of stones with different porosity; in fact the size can be controlled to bridge the pores in the substrate. For example, on a marble substrate, poly(aspartic acid) induces the formation of larger and regularly oriented crystals, in comparison to purely inorganic precipitation from supersaturated solutions (see Figure 14.3). The penetration of the organic molecules, which were also shown to coat the pores' walls, depends on the porosity but was evaluated to be around 1 cm. Laboratory tests showed that treatments with a low concentration of BIMs resulted in some reinforcement of the calcareous stone samples, without consistent alteration of physico-chemical properties such as the colour and the water uptake.

To conclude this section, it is important to mention the research developed on biological mortar, *i.e.* a mixture of finely ground limestone, bacteria, and a feeding medium containing a calcium salt. In this case, cementation results from nucleation and growth of microbiologically produced carbonate crystals on the surface of carbonate aggregates. These systems were set and optimized in order to avoid problems due to physico-chemical incompatibility of traditional repair mortars with degraded substrates.[39,40]

14.2 Geopolymers

Geopolymers are a class of synthetic materials made up of aluminosilicates with potential use in a number of areas, especially as additives or as alternatives to common cements, or for ceramic applications. The name geopolymers was coined by Davidovitz,[41] but the scientific literature reported something similar 25 years earlier.[42] Indeed, the more appropriate term for such material should be "inorganic polymers",[43] because they derive from a reaction between a solid aluminosilicate and a highly concentrated aqueous solution of alkali hydroxide

Figure 14.3 (Top) SEM image of calcium carbonate crystals grown on a marble substrate using the ammonium carbonate method. Bar is 5 μm. (Bottom) SEM micrograph of a marble sample coated using poly(aspartic acid) in $CaCl_2$. Bar is 20 μm. The ammonium carbonate method was used to grow the crystals.
[Reprinted from *Journal of Cultural Heritage*, **7**, P. Tiano, E. Cantisani, I. Sutherland and J. M. Paget, "Biomediated reinforcement of weathered calcareous stones", Pages No. 49–55, Copyright (2006), with permission from Elsevier.]

or silicate. These materials tend to be used as substitutes of traditional cements in a very broad range of applications, with a significant reduction in the emission of greenhouse gases (CO_2) with respect to regular cement.[44] Several crystalline and amorphous compounds are described as geopolymers, but sometimes they are referred to as:[45] "low temperature aluminosilicate glasses",[46] "alkali-activated cement",[47] "geocement",[48] "alkali-bonded ceramic",[49] "inorganic polymer concrete",[50] and "hydroceramic".[51] All these materials deal with almost the same basic chemistry, the result of coupled alkali-mediated dissolution and precipitation reactions in an aqueous medium substrate.[45] Despite the common basic chemistry behind these materials, their microstructure and their physical, mechanical, chemical and thermal properties strictly depend on the raw materials from which they derive. A conceptual model for geopolymerization was proposed in the 1950s,[42] and it consists of three stages: (i) destruction–coagulation, (ii) coagulation–condensation, and (iii) condensation–crystallization. Indeed, a recent paper[45] explains in more detail this conceptual model with at least five steps, consisting of dissolution of the materials coming from the aluminosilicate source (the so-called raw materials), speciation equilibrium, gelation, reorganization and polymerization followed by hardening. In particular, the last part of the mechanism (structural reorganization) determines the microstructure and pore distribution of the material, which is the crucial topic for the desired application. For the characterization of the geopolymers the most suitable techniques are XRD, transmission and scanning electron microscopy (SEM), nuclear magnetic resonance (NMR), infrared and Raman spectroscopy, and small angle neutron scattering (SANS).

Research work on geopolymers has focused on applications and related technologies, for example as cements, ceramics and zeolites to impart high mechanical strength, fire resistance, chemical stability, and enhancement of adhesion or cohesion properties. Although the application of geopolymers in the area of conservation of cultural heritage is still dawning, it is a very promising field for contemporary art and architecture, where Portland cement plays a very important role. In particular, the potential high compatibility of several classes of geopolymers with Portland cement and, more in general, with cementitious materials, is a very stimulating starting point for the development of more environmentally friendly, economical, durable and high performance materials for the consolidation and reinforcement of bulk and surface cementitious artefacts.

14.2.1 Surfactant-Template Sol–Gel Synthesis of Mesoporous Silica

Mesoporous silica nanoparticles are defined as chemically and thermally stable nanomaterials with well-defined and controllable morphology and porosity.[52] These particles possess external and internal surfaces that can be selectively functionalized with multiple organic and/or inorganic groups. One of the most common methods to prepare these particles is surfactant-template sol–gel synthesis. The basic idea is the formation of liquid-crystalline mesophases of amphiphilic molecules (surfactants) that behave as templates for the *in situ* polymerization of orthosilicic acid. In this way the polymerization is topologically driven by the pre-formed ordered template made up of the surfactant molecules. The synthesis can be performed either in acidic or basic conditions, and the source of silica can be fumed silica, sodium silicate, or a tetra-alkyl oxide of silane.

The mechanism of the template-driven formation of mesoporous silica is quite simple and can be depicted in the following way. At a certain concentration surfactants form ordered structures made up of geometrical arrays of the micelles (*i.e.* hexagonal rods) that can be seen as true liquid crystals. The addition of an inorganic precursor to the polar solvent (for example water) that surrounds the hydrophilic surface of the micelles results in a hydrolysis reaction that leads to the formation of a regular array of silica cylinders over the pre-existing surfactant template. The last step of the surfactant-template sol–gel synthesis is the removal of the surfactant and solvent molecules. This is achieved by washing the sample with suitable solvents or by temperature changes (calcination).[53] The factors that affect the synthesis and the properties of the final silica include the molecular structure of the surfactant and its concentration, the surfactant type (cationic, anionic or non-ionic), the presence of counterions, the polarity of the reaction medium, the ordering of the template structure, pH and temperature. Electrostatic interactions drive the formation of the silica mesoporous structures. In fact, cationic surfactants strongly interact *via* electrostatic forces with anionic inorganic precursors, whereas anionic surfactants attract cationic inorganic complexes. The sol–gel synthesis mediated by non-ionic surfactants is normally driven by hydrogen bonds. The most important application of mesoporous silica is in the formation of heterogeneous catalysts with large pore size and high surface area.[54] As such, they could be used in the conservation of cultural heritage as catalysts to convert pollutant gases or for the rapid oxidation of organic materials in order to prevent and/or minimize the formation of black crusts. However, much work has to be done to assess whether their use can be really suitable and feasible.

14.3 Organic–Inorganic Hybrid Nanocomposites

Composite materials are generally defined as three-dimensional combinations of at least two chemically different components separated by an interface. Traditionally, they include a binding matrix, often polymeric, and a filler material, such as inorganic particles. If one or more components exhibit nanometric size or structuration, the term "nanocomposite" is adopted.

Nanocomposite materials have been widely explored in recent decades for many different technological applications. By adding inorganic nanoparticles to a polymer, multifunctional systems can be obtained that exhibit better performance than either the pure polymer or the traditional filled polymeric materials. Typical features of nanocomposites are:

- Improved mechanical and rheological properties;
- Thermal and flame resistance;
- Chemical resistance (*e.g.* to acidic conditions);
- Decreased permeability and increased moisture resistance;
- Protection against ultraviolet (UV) radiation.

Nanoscale inorganic materials, used for the synthesis of composites, usually include metal oxides (*e.g.* zinc, aluminium and titanium oxides), carbonates (*e.g.* $CaCO_3$), and silicon-containing compounds (silica, silicates, *etc.*). Nanocomposites are involved in many different fields, from medicine and pharmacy to papermaking, textiles, plastics and ceramics, military and aerospace applications. A review of the different synthetic methods and applications is beyond the scope of this book, and inclusive literature is provided in the Further Suggested Reading section. As a general statement, one relatively simple way to obtain nanocomposites is through the blending of the polymer particles with inorganic particles in a common medium (water or an organic solvent), avoiding the separation of the two components. Wang *et al.* (see Further Suggested Reading) provided a comprehensive review on the fabrication of nanocomposites, considering colloidal particle assembly methods, and the formation of composites from particle blends and colloidal hybrids.

In this chapter we will report some of the most recent applications of nanocomposites in the field of cultural heritage preservation, and provide the rationale for using composite systems in different case studies.

The first application concerns the use of nanocomposites for the consolidation of stone and mural paintings. In spite of the main disadvantage attributable to their physico-chemical incompatibility (see also Sections 3.3–3.5), synthetic polymers have been and are still extensively used mainly because of their optical and adhesive effects on substrates such as wall paintings, *i.e.* the saturation of colours and the re-adhesion of local detachments, and owing to the general belief that they could act as a stable and reversible protective coating on the artistic surface. On the other hand, as highlighted in Section 14.1.1 and in previous chapters (see Chapters 12 and 13), compatible and effective treatments can be performed using materials such as calcium hydroxide and carbonate, either in the form of nanoparticle dispersions in non-aqueous solvents, or through bio-induced deposition. For example, dispersions of $Ca(OH)_2$ nanoparticles in propanol may grant the homogeneous consolidation of porous degraded carbonate matrices such as *frescoes* or mural paintings, at a depth that depends on the particle size and on the substrate's porosity, re-establishing the cohesion of the damaged work of art through the formation of a crystalline three-dimensional $CaCO_3$ texture within the substrate's pores (see Chapter 13). The treatment results in surface strengthening, good compatibility, long-term stability, and re-adhesion of local detachments. The nanoparticles behave thus as a consolidant, rather than as an adhesive.

However, in the presence of large detachments the use of organic adhesives cannot be avoided, and a possible alternative is the combination of the two classes of material (polymers and inorganic nanoparticles) to provide the re-adhesion of the detached portions and the consolidation of the surface with the minimum amount of polymers, minimizing the detrimental effects due to their physico-chemical incompatibility with the treated substrates. Nuccio *et al.* reported the synthesis and characterization of $Ca(OH)_2$ nanoparticles/Paraloid B72 [poly(EMA/MA) 70:30] nanocomposites and their application for the conservation of stone materials.[55] The authors followed two different synthetic strategies to obtain stable systems, and used optical microscopy to show that the morphology of the composite films and the size of $Ca(OH)_2$ nanoparticle agglomerates change depending on the different synthetic pathways and nanoparticle:polymer ratios. The first synthetic strategy consisted in mixing a dispersion of nanoparticles (in 2-propanol) with acetone, followed by the addition of the appropriate amount of Paraloid. The second synthesis consisted in mixing limewater (*i.e.* a saturated calcium hydroxide solution) with a solution of Paraloid in a 2-propanol–acetone blend, at 70 °C. The latter method is less easily controlled, owing to the separation of a precipitate formed by the polymer and the nanoparticles. The first method was thus preferred, and concentrations of about 1% w/w Paraloid and 0.5% w/w $Ca(OH)_2$ were selected for the tests. It is worth noting that, in restoration practice, the pure polymer solution concentrations range from 5 to 10% w/w (*e.g.* in acetone). Figure 14.4 shows the optical microscopy image of a film cast from the first composite system on a microscope glass. Within the polymer film, nanoparticles form aggregates whose dimensions vary greatly, from 10 μm (or less) up to large clusters of about 100 μm.

The measurement of the contact angle (see Section 1.4.1) on the films showed that their wettability is similar to that of pure polymer films, suggesting that these coatings provide strong protection against bulk water absorption. Moreover, the mechanical tests (Scotch tape™ test) on degraded plaster model samples indicated that the composite system provides a satisfactory consolidation of the treated substrates.

Recently some composite liquid systems were obtained by mixing a solution of Paraloid B72 and a dispersion of $Ca(OH)_2$ nanoparticles in an alcohol–acetone blend. Similar to the previous example the concentration of

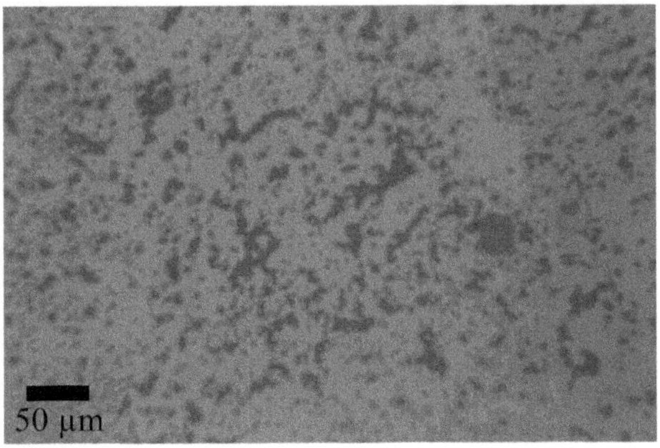

Figure 14.4 Optical microscope image of a film cast on a microscope glass from a nanocomposite system obtained by mixing a dispersion of nanoparticles (in 2-propanol) with acetone, then adding Paraloid B72.
(Reprinted from ref. 55 by the kind permission of Arcadia Ricerche Copyright owner.)

the polymer in the systems was considerably decreased with respect to the standard values of the pure polymer solutions used for restoration purposes, but was still sufficient to form the adhesive films (see Figure 14.5).

Within the films, $Ca(OH)_2$ nanoparticles aggregate to form clusters whose dimensions span from less than 1 μm up to 5–10 μm. The results showed that the inclusion within the film maintains a satisfactory degree of hydroxide carbonation. Calcite has a unit cell volume of about 368 Å^3, while that of portlandite is 55 Å^3, thus the carbonation induces a mechanical stress within the plastic film. In fact, the $CaCO_3$ crystals formed can grow up to break the

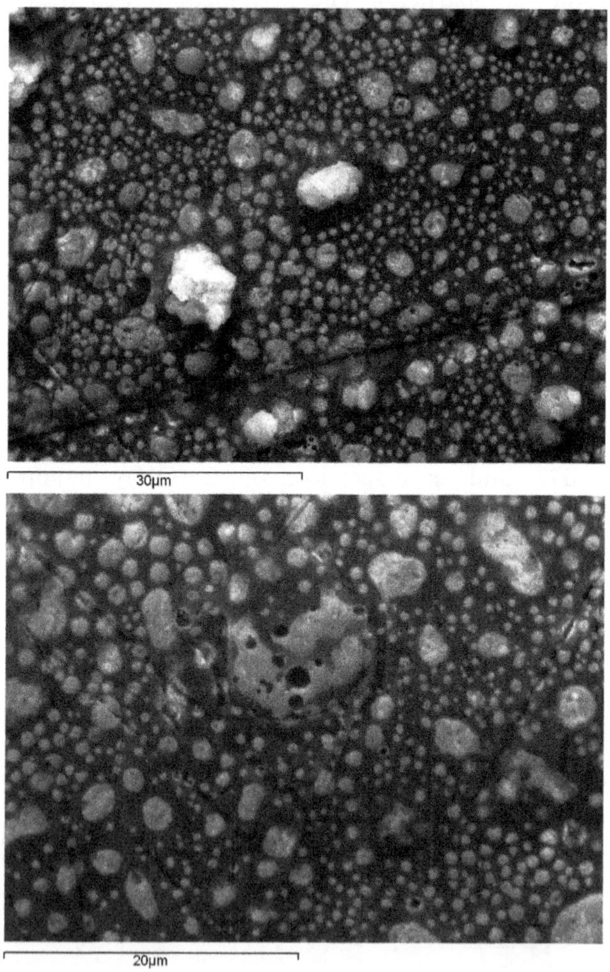

Figure 14.5 SEM images of a composite film of Paraloid B72 and $Ca(OH)_2$ particles, laid on a golden platelet, after carbonation (the linear cracks are due to scratches on the platelet). (Top) $CaCO_3$ crystals protruding from the film surface; bar is 30 μm; (Bottom) detail of holes formed upon drying of the film; bar is 20 μm.

film's surface (see, for instance, the larger crystals in Figure 14.5, top panel). Moreover, the presence of particles decreases the plasticity of the film, presumably leading to the formation of holes upon drying (see Figure 14.5, bottom panel).

Interestingly, preliminary tests indicated that treatment with these composite systems can decrease the loss of permeability to water vapour (permeability is typically reduced to a large extent by pure polymeric coatings), and can increase the substrate's contact angle (hence decreasing the wettability of the surface by bulk water). Altogether these two effects are particularly appealing for real applications, because they would allow the surface to "breathe" (see also Section 14.1.1) and grant impermeability. Moreover, the Scotch tape™ test showed that the application of composite systems led to surface consolidation and re-adhesion of powdery substrates, thus opening perspectives for the use of these materials in real case studies.

A preliminary study on the synthesis of $Ca(OH)_2$ nanoparticle–tetraethyl orthosilicate nanocomposites was carried out by Naldini *et al.*[56] Tetraethyl orthosilicate (TEOS) can be applied as a monomer in alcoholic solution, and is absorbed through capillarity by porous matrices. At a given temperature, relative humidity and pH, the monomers form covalent bonds with silicate crystals and create a three-dimensional network of amorphous silica. In fact, TEOS has been elected as a compatible compound for the consolidation of silicate stones, and has been used on limestone and carbonate substrates.[57] On a calcium carbonate matrix, secondary bonds (hydrogen bonds, London and van der Waals forces; see Chapter 4) will be formed instead, so the consolidating effect will be strongly reduced. The $Ca(OH)_2$ nanoparticle–TEOS nanocomposites were thus considered for tests on carbonate substrates, in order to achieve increased consolidation and reduce the relative amount of TEOS to enhance the physico-chemical compatibility of the treatment. The authors prepared a composite system by mixing the two separate components (see Figure 14.6), *i.e.* a dispersion of calcium hydroxide nanoparticles and a solution of TEOS, both in isopropanol.

Composite films were cast on microscope slides and investigated through optical and SEM-EDX experiments. The results showed that the cross-linking of TEOS produces a polymeric network that reduces the clustering of $Ca(OH)_2$ nanoparticles, with respect to films cast from pure $Ca(OH)_2$ nanoparticle dispersions (see Figure 14.7).

The $Ca(OH)_2$–TEOS and $Ca(OH)_2$–Paraloid B72 nanocomposites represent two promising options when the degree of substrate degradation and lack of cohesion are so severe that the use of polymeric materials (*e.g.* adhesives) is necessary.

In the last 15 years, organo-modified silicates have been widely used to impart higher chemical, thermal and mechanical resistance to organic polymers. For applications in the field of conservation science these nanocomposites are considered as consolidating and protective agents on porous substrates. D'Arienzo *et al.* investigated composite systems obtained by dispersing organoclay nanoparticles (Cloisite 30B, *i.e.* sodium montmorillonite modified by *N*,methyl-*N*,tallow-*N,N'*,2-hydroxyethyl-ammonium chloride)

Figure 14.6 SEM image of a composite $Ca(OH)_2$ nanoparticle–TEOS film. The white boxes on the right highlight silica particles (based on EDX analysis). Bar is 1 μm. The inset on the left shows the liquid $Ca(OH)_2$ nanoparticle–TEOS nanocomposite system.
(Reprinted from ref. 56 by the kind permission of Arcadia Ricerche Copyright owner.)

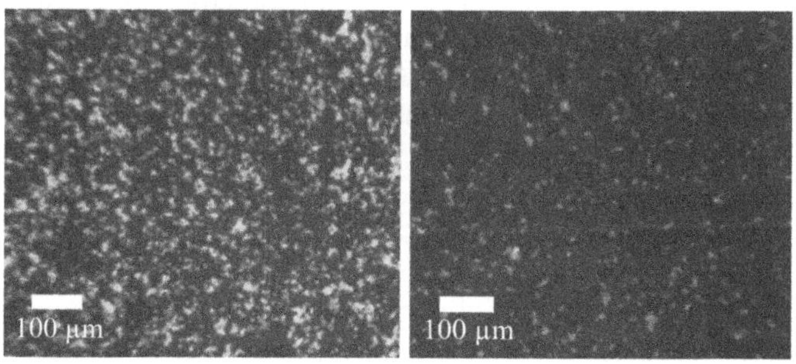

Figure 14.7 Optical microscope image (44×) of a film cast from a dispersion of $Ca(OH)_2$ nanoparticles (left) and from a composite $Ca(OH)_2$ nanoparticle–TEOS system.
(Reprinted from ref. 56 by the kind permission of Arcadia Ricerche Copyright owner.)

upon sonication into a solution of a low-viscosity commercial acrylic and vinylidenfluoride-based polymer (Fluormet CP). The latter product is widely used for stone restoration because it combines the adhesive properties of acrylic polymers with the chemical stability of fluorinated elastomers.[58] The nanocomposite system was evaluated for the consolidation of tuff stone, which has a high content of both amorphous silicates and zeolite phases. An extended physical investigation of treated stone samples highlighted that even a low content (1–4% w/w) of nanoparticles in the polymer can provide a consolidating effect. Optical assessments and SEM analyses showed that, upon brushing the tuff samples, a polymeric layer was formed on the stone surface, down to a depth of 3 mm, coating the inner walls of the pores and linking the disaggregated parts. Mechanical tests (uniaxial compression and abrasion) showed that the mechanical strength of the samples was increased by the treatment, with a reduction in stone pulverization during the abrasion tests. The presence of the organoclay nanoparticles enhanced the consolidating effect of the polymer, and increased the system's compatibility with silicate substrates. Moreover, colorimetric measurements did not show any significant chromatic alteration in the treated stone. However, the treatment led to a stronger reduction of permeability in comparison to the application of the pure polymer, probably due to a superficial partial clogging of the pores by the montmorillonite layers. Even though complete occlusion of the stone's pores did not occur, this issue should be considered in future work and *in situ* applications.

De Ferri *et al.* reported the application of coatings obtained from two siloxane products charged with different quantities of organo-modified silica nanoparticles, functionalized with 1,1,1-trimethyl-N-(trimethylsilyl)silanamine.[59] The composite systems (solutions) were brushed onto samples of traditional monumental and architectural stone, such as limestone, sandstone and granite. The treatment led to a strong increase of the stones' hydrophobicity (static contact angles of more than 140°), due to the chemical properties of the particles' functionality (silanamine) and to the morphology of the nanoparticle aggregates. Interestingly, capillary absorption tests showed that the water absorption of marble and sandstone increased for several days after the treatment, while for granite samples the water absorption was reduced. Moreover, the treatment produced only a slight colour alteration (on sandstone) or no significant colour change at all (marble, granite).

In a previous study, Manoudis *et al.* investigated the effect of artificially induced roughness on the water repellency of mineral substrates coated with polymer–silica nanoparticles composite films.[60] Calcium carbonate tablets were used as marble samples, and treated with a composite system obtained by dispersing the nanoparticles into a poly(methyl methacrylate) solution in toluene. The results showed that even at low concentration (1%) the silica particles maximized the hydrophobicity of the coated samples. Atomic force microscopy (AFM) highlighted that the nanocomposite coating produced a strong increase in the surface roughness, resulting in an increase of the surface contact angle (see the Wenzel and Cassie–Baxter models in Section 1.4.1.2).

Besides consolidation, nanocomposite systems have been considered for the cleaning of works of art. A significant example is the use of magnetic nanosponges loaded with microemlusions for the removal of detrimental coatings from the surface of works of art. These systems (see Sections 11.4 and 12.5) are prepared by functionalizing the magnetic nanoparticles followed by their chemical crosslinking to a polyacrylamide gel matrix. The final composite exhibits the mechanical properties of the acrylamide matrix and is magnetically responsive, thus providing great advantages in handling, application and removal.

Nanocomposites have also been investigated as protective coatings against ultraviolet (UV) radiation, pollution gases, and moulds or bacteria. Several microorganisms, in fact, may cause physico-chemical changes on stone substrates, producing physical and chemical damage. As a result, both the aesthetic appearance and the mechanical resistance of the substrates are severely affected.

As a final example of the application of nanocomposites in the conservation of works of art, we report two studies on titanium- and copper-based composite nanocoatings for providing resistance to UV or biodegradation.

Owing to its photocatalytic activity, titanium dioxide can be applied for environmental purification, self-cleaning applications, and for surface functionalization of cellulose to provide bactericidal properties.[61–63] A recent study by Afsharpour *et al.* considered TiO_2 nanocomposites for the protection of paper.[64] Coatings on model and real paper samples were carried out by either a direct application of the nanoparticles or by laying Klucel G (a physical gel made up of a hydroxypropyl cellulose mixed with ethanol) on paper and then spraying a dispersion of nanoparticles in ethanol ($2\,g\,L^{-1}$) onto the paper surface, to obtain a nanocomposite layer made of TiO_2 particles and cellulosic polymer (Klucel G; see Figure 14.8).

Fourier transform infrared (FTIR) and UV measurements supported the idea that the nanoparticles are bound to cellulose (either the substrate or the Klucel gel) through physical interactions. The FTIR measurements also indicated that, compared with untreated paper, the application of the nanocomposite protective coating inhibited the formation of cellulose oxidation products upon accelerated thermal ageing[65] (15 days, roughly equivalent to 100 years of natural ageing). On the contrary, paper treated with pure TiO_2 nanoparticles exhibited degradation. Mechanical tests showed that the Klucel and TiO_2 composite coating maintains a good resistance of paper to tensile strength after ageing, while untreated paper exhibits a strong decrease in mechanical properties. The anti-fading property of the composite coating was

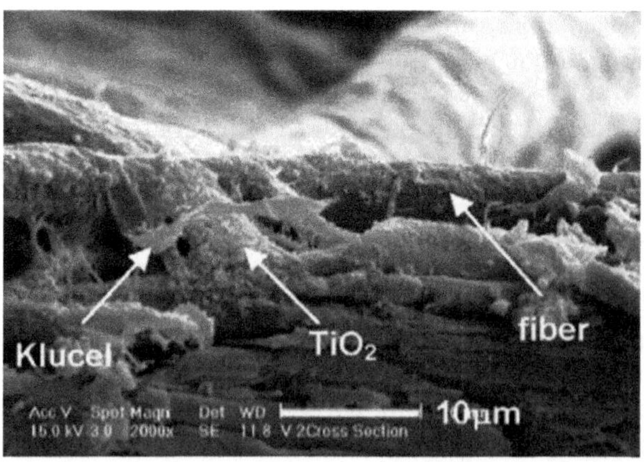

Figure 14.8 SEM image of a nanocomposite TiO_2–Klucel film on cellulose fibres. [Reprinted from *Journal of Cultural Heritage*, **12**, M. Afsharpour, T. Rad Fereshteh and H. Malekian, "New cellulosic titanium dioxide nanocomposite as a protective coating for preserving paper-art-works", Pages No. 380–383, Copyright (2011), with permission from Elsevier.]

also investigated by ageing paper samples, coloured with the traditional colorant alizarin lake, with a UV lamp at 270 nm. The results showed that the colour fading is slower in samples treated with the nanocomposite, while it is accelerated on samples simply treated with TiO_2 nanoparticles.

The better performance of composite coatings is due to the behaviour of TiO_2 as a photocatalyst. In fact, while this chemical absorbs UV light, acting as a screening agent, it will produce radicals that oxidize organic compounds, such as dyes or cellulose. The TiO_2–Kulcel composite layer can thus be used as a UV screening agent, using the beneficial properties of TiO_2, and avoiding its direct interaction with the dyes and the cellulose of the paper. The authors reported that the coating could also be used to prevent the formation of biofilms on the treated paper.

Ditaranto *et al.* reported the application of copper-based nanocoatings to prevent and remedy microbial proliferation on stone.[66] Core–shell copper nanoparticles were synthesized through an electrochemical procedure called the "sacrificial anode" technique,[67] and mixed with a commercial consolidant and water-repellent product, based on ethyl silicate and polysiloxane oligomers in a white spirit solution (a mixture of hydrocarbons distilled from oil, with a boiling point of 150–200 °C). The synthetic method allows great control of the particle size dispersion. In particular, spherical nanoparticles with a mean core of 2–4 nm are obtained.[67] However, once embedded in the composite film, the particles will form clusters with larger dimensions (*e.g.* from tens to about one hundred nanometres). The use of nanocomposites was justified by the benefit of a single-step treatment, since in some cases the separate application of the water-repellent materials and biocides has resulted in mutual interference (*e.g.* biocides such as quaternary ammonium salts may decrease the water repellency if applied before the hydrophobic coatings).[68] An additional advantage is that the gradual release of copper by the composite film, when exposed to aqueous solutions, could minimize the number of treatments required for maintenance. The particle concentration in the composite dispersed system ranged from 0.02 to 0.28% w/w. Both glass slides and calcareous stones with different porosity were treated and analysed to assess the effectiveness of the method. Surface analysis of the composite films through XPS (X-ray photoelectron spectroscopy) showed that the surface concentration of copper increases with time, as a result of copper surfacing over the modification of the silicate–polysiloxane product (*i.e.* hydrolysis of ethyl silicate, evaporation of ethanol and polymerization of siloxane moieties), reaching a plateau 40 days from the application. The XPS data also highlighted that greater stone porosities imply lower surface concentrations of copper, since copper particles penetrate into the inner layers. Accordingly, electro-thermal atomic absorption spectroscopy (ETAAS) analysis showed that the amount of metal ions released into aqueous solutions, in contact with the films, increased over time to reach a plateau after a few (1–3) hours. The preventive action against the formation of biofilms, even at the lower particle concentrations, was confirmed by preliminary *in vitro* tests, which showed complete inhibition of the growth of *Arthrobacter histidinolovorans*. Cioffi *et al.* had previously demonstrated the biological activity of these nanomaterials against a wide range of microorganisms, highlighting that the bioactivity could be tuned by changing the concentration of nanoparticles.[67] One possible drawback, however, is due to colour alteration of stone upon treatment, as indicated by colorimetric data. The chromatic changes are similar to those caused by the application of the pure commercial product, and are enhanced by both high concentrations of particles in the composite system, and by low porosity of the stone. However, the colour changes were shown to decrease upon ageing of the treated substrate (*e.g.* after two months). This application can thus be considered to be promising in the cultural heritage field.

Further Suggested Reading

On the bacterial biomineralization of calcium carbonate:

W. De Muynck, N. De Belie and W. Verstraete, Microbial carbonate precipitation in construction materials: A review, *Ecol. Eng.*, 2010, **36**, 118.

On geopolymers:

T. Bakharev, Durability of geopolymer materials in sodium and magnesium sulfate solutions, *Cem. Concr. Res.*, 2005, **35**, 1233.
H. F. W. Taylor, *The Chemistry of Cement*, Academic, London, 1964.
C. K. Yip, G. C. Lukey and J. S. J. ven Deventer, The coexistence of geopolymeric gel and calcium silicate hydrate at the early stage of alkaline activation, *Cem. Concr. Res.*, 2005, **35**, 1688.

On polymeric nanocomposites:

S. N. Bhattacharya, M. R. Kamal and R. K. Gupta, *Polymeric nanocomposites: theory and practice*, Hanser Verlag, Munich, 2008.
J. H. Koo, *Polymer nanocomposites: processing, characterization and applications*, McGraw-Hill, New York, 2006.
T. Wang and J. L. Keddie, Design and fabrication of colloidal polymer nanocomposites, *Adv. Colloid Interfac.*, 2009, **147–148**, 319.

Review and Questions

This chapter deals with the application of biogenic minerals, geopolymers and composite organic-inorganic nanocomposites for the consolidation and protection of works of art. The main theoretical and applicative factors are:

- Calcium carbonate is one of the most compatible consolidating materials for carbonate-based stone. Both biominerals and synthetic organic–inorganic phases exhibit unique properties that are different from those of the corresponding purely inorganic phases.

- The use of bacteria for the deposition of biominerals is a promising consolidating tool, and several studies have been reported. Important factors to be considered are the possible alterations of the substrate upon treatment (*e.g.* avoiding the complete clogging of the stone's pores), and the control of cell viability on the substrate.
- The fundamental properties of geopolymers and their possible future applications for cultural heritage conservation are described, including some remarks on the technique of surfactant-template sol–gel synthesis for obtaining mesoporous silica.
- Synthetic organic–inorganic hybrid nanocomposites can be used for consolidation or protection, as UV-shielding materials or for inhibiting the growth of undesirable microorganisms. The organic part (polymers) usually provides hydrophobic and adhesive properties to the film. The inorganic portion grants a higher physico-chemical compatibility with carbonate and silicate stones, and provides consolidating or protective properties. Alteration of the porosity and colour of the treated stone are possible drawbacks to be evaluated in practical applications.

Questions

1) Name and define two mechanical properties that are typically affected by the presence of organic phases in hybrid organic–inorganic materials such as biominerals. How are these properties altered?
2) Recall two main functions of bio-induced mineral layers on stone substrates. Provide a reasonable threshold value for the filling of pores by biominerals (including the effect of any growing media).
3) Why can geopolymers reduce the emissions of greenhouse gases?
4) What are the expected applications of geopolymers and mesoporous silica?
5) Evaluate the potential reversibility of a treatment with a hybrid acrylate–$Ca(OH)_2$ nanocomposite on limestone.

Answers

1) Two mechanical properties typically affected by the presence of organic phases are toughness and brittleness. Toughness can be expressed as the amount of energy per unit volume that a material can absorb before any rupture occurs. Qualitatively, it can be defined as the ability of a material to absorb energy and deform if mechanically stressed, without undergoing rupture. Brittle materials, on the other hand, tend to break without deforming when a mechanical stress is applied. Biominerals can exhibit an increased toughness and a reduced brittleness, compared with the corresponding purely inorganic phases.
2) The treatment can be intended either as a surface coating, granting consolidation and waterproofing properties, or for providing a deeper binding action through the cross-section of degraded substrates. Some

authors have reported pore filling of about 40–45% as a good value for granting consolidation while maintaining vapour transfer through the stone layers.

3) The geopolymers can reduce the emissions of greenhouse gases because they can substitute for cementatious materials in some applications. Indeed, the production of cements involves a great deal of energy and produces large amounts of CO_2 during the stage of heating the raw materials. Geopolymers can be obtained at room temperature, avoiding high temperature heating steps, and therefore reducing the use of energy and decreasing emissions relative to those associated with the production of cements.
4) Although they need to be further assessed, geopolymers are expected to be good compatible materials for conserving and treating contemporary art/architecture "objects" made of cement; on the other hand, mesoporous silica could have potential use to catalyse some processes dealing with pollution.
5) Provided that the ageing of the polymer network (acrylate) does not prevent its solubilization by solvents or other cleaning tools (*e.g.* microemulsions), the organic component of the nanocomposite could be at least partially removed. On the other hand, the cleaning of the acrylate network might cause the partial removal of residual $Ca(OH)_2$. Furthermore, the removal of polymer domains linking clusters of $CaCO_3$ (formed upon carbonation of the particles) could partially turn the consolidating nanocomposite into $CaCO_3$ filler, with reduced binding ability within the stone's pores.

References

1. S. Mann, B. R. Heywood, S. Rajam and J. D. Birchall, *Nature*, 2002, **334**, 692.
2. K. L. Thomas-Keprta, D. S. McKay, S. J. Wentworth, T. O. Stevens, A. E. Taunton, C. C. Allen, A. Colemann, E. K. Gibson and C. S. Romanek, *Geology*, 1998, **26**, 1031.
3. G. H. Nancollas and W. Wu, *J. Cryst. Growth*, 2000, **211**, 137.
4. P. Gilbert, M. Abrecht and B. H. Frazer, *Rev. Mineral Geochem.*, 2005, **59**(1), 157.
5. S. Weiner, *J. Struct. Biol.*, 2008, **163**, 229.
6. J. D. Currey, *Science*, 2005, **309**, 253.
7. S. Kamat, X. Su, R. Ballarini and A. H. Heuer, *Nature*, 2000, **405**, 1036.
8. A. G. Evans, R. Z. Wang, Z. Suo, I. A. Aksay, M. Y. He and J. W. Hutchinson, *J. Mater. Res.*, 2001, **16**(9), 2475.
9. C. Levi, J. L. Barton, C. Guillemet, E. Le Bras and P. Lehuede, *J. Mater. Sci. Lett.*, 1989, **8**, 337.
10. M. Sarikaya, H. Fong, N. Sunderland, B. D. Flinn, G. Mayer, A. Mescher and E. Gaino, *J. Mater. Res.*, 2001, **16**, 1420.

11. Y. Politi, Y. Levi-Kalisman, S. Raz, F. Wilt, L. Addadi, S. Weiner and I. Sagi, *Adv. Funct. Mater.*, 2006, **16**, 1289.
12. S. Weiner, I. Sagi and L. Addadi, *Science*, 2005, **309**, 1027.
13. L. Addadi, D. Joester, F. Nudelman and S. Weiner, *Chem. Eur. J.*, 2006, **12**, 980.
14. J. L. Arias and M. S. Fernández, *Biomineralization: From Paleontology to Materials Science*, Editorial Universitaria, Santiago, 2007.
15. M. A. Rivadeneyra, R. Delgado, A. del Moral, M. R. Ferrer and A. Ramos-Cormenzana, *Microbiol. Ecol.*, 1994, **13**(3), 197.
16. F. Hammes and W. Verstraete, *Rev. Environ. Sci. Biotechnol.*, 2002, **1**, 3.
17. J. W. Morse, in *Carbonates: Mineralogy and Chemistry*, ed. R. J. Reeder, Mineralogic Society of America, Washington, DC, 1983, vol. 11, p. 227.
18. C. Buczynski and H. S. Chafetz, *J. Sediment. Petrol.*, 1991, **61**, 226.
19. E. Boquet, A. Boronat and A. Ramos-Cormenzana, *Nature*, 1973, **248**, 21.
20. H. M. C. Danielli and M. A. Edington, *Geomicrobiol. J.*, 1983, **3**, 1.
21. C. B. Lipman, *Carnegie Inst.*, 1929, **391**, 231.
22. J. Parraga, M. A. Rivadeneyra, R. Delgado, J. Iniguez, M. Soriano and G. Delgado, *React. Funct. Polym.*, 1998, **36**, 265.
23. M. A. Rivadeneyra, J. Parraga, R. Delgado, A. Ramonos-Cormenzana and G. Delgado, *FEMS Microbiol. Ecol.*, 2004, **48**, 39.
24. D. V. Zamarreño, R. Inkpen and E. May, *Appl. Environ. Microb.*, 2009, **75**(18), 5981.
25. T. J. Beveridge and W. S. Fyfe, *Can. J. Earth Sci.*, 1985, **22**, 1893.
26. M. A. Rivadeneyra, G. Delgado, A. Ramos-Cormenzana and R. Delgado, *Res. Microbiol.*, 1998, **149**, 277.
27. H. L. Ehrlich, *Earth Sci. Rev.*, 1998, **45**(1–2), 45.
28. C. Buczynski and H. S. Chaftez, *J. Sediment. Petrol.*, 1991, **61**, 226.
29. Y. Van Lith, R. Warthmann, C. Vasconcelos and J. A. Mckenzie, *Geomicrobiol. J.*, 2003, **1**, 71.
30. O. Braissant, G. Cailleau, C. Dupraz and E. P. Verrecchia, *J. Sediment. Res.*, 2003, **73**, 485.
31. C. Rodriguez-Navarro, M. Rodrigues-Gallego, K. B. Chekroun and M. T. Gonzalez-Muñoz, *Appl. Environ. Microb.*, 2003, **69**(4), 2182.
32. P. Tiano, L. Biagiotti and G. Mastromei, *J. Microbiol. Methods*, 1999, **36**, 139.
33. S. Castanier, G. Le Métayer-Levrel, G. Orial, Loubière and J. P. Perthuisot, in *Of Microbes and Art: the Role of Microbial Communities in the Degradation and Protection of Cultural Heritage*, ed. O. Ciferri, P. Tiano and G. Mastromei, Plenum, New York, NY, 2000, p. 201.
34. C. Urzi, M. Garcia-Valles, M. Vendrell and A. Pernice, *Geomicrobiol. J.*, 1999, **16**, 39.
35. W. De Muynck, K. Cox, N. De Belie and W. Verstraete, *Constr. Build. Mater.*, 2008, **22**(5), 875.
36. W. De Muynck, N. De Belie and W. Verstraete, *Ecol. Eng.*, 2010, **36**, 118.
37. G. Alfano, G. Lustrato, C. Belli, E. Zanardini, F. Cappitelli, E. Melo, C. Sorlini and G. Ranalli, *Int. Biodeter. Biodegr.*, 2011, **65**, 1004.

38. P. Tiano, E. Cantisani, I. Sutherland and J. M. Paget, *J. Cult. Herit.*, 2006, **7**, 49.
39. G. Le Metayer-Levrel, S. Castanier, G. Orial, J. F. Lubiere and J. P. Perthuisot, *Sediment. Geol.*, 1999, **126**(1–4), 25.
40. G. Orial, T. Vieweger and J. F. Loubiere, *Les mortiers biologiques: une solution pour la conservation de la sculpture monumentale en pierre*, Art Biology and Conservation, Metropolitan Museum, New York, 2002.
41. J. Davidovitz, *J. Therm. Anal.*, 1991, **37**, 1633.
42. V. D. Glukhovsky, *Soil Silicate Articles and Constructions* (Gruntosilikatnye virobi i konstruktsiii), Budivelnik Publisher, Kiev, 1967.
43. J. R. Van Wazer, *Inorg. Macromol. Rev.*, 1970, **1**, 89.
44. E. Gartner, *Cem. Conc. Res.*, 2004, **34**, 1489.
45. P. Duxson, A. Fernández-Jiménez, J. L. Provis, G. C. Lukey, A. Palomo and J. S. van Deventer, *J. Mater. Sci.*, 2007, **42**, 2917.
46. H. Rahier, B. van Mele, M. Biesemans, J. Wastiels and X. Wu., *J. Mater. Sci.*, 1996, **31**, 71.
47. A. Palomo and J. I. L. De La Fuente, *Cem. Conc. Res.*, 2003, **33**, 281.
48. P. V. Krivenko, in *Proceedings of the First International Conference on Alkaline Cements, Concretes*, ed. P. V. Krivenko, VIPOL Stock Company, Kiev, Ukraine, 1994, p. 11.
49. S. Mallicoat, P. Sarin and W. M. Kriven, *Ceram. Eng. Sci. Proc.*, 2005, **26**, 37.
50. M. Sofi, J. S. J. van Deventer, P. A. Mendis and G. C. Lukey, *J. Mater. Sci.*, 2007, **42**, 3107.
51. Y. Bao, M. W. Grutzeck and C. M. Jantzen, *J. Amer. Ceram. Soc.*, 2005, **88**, 3287.
52. B. G. Trewyn, I. I. Slowing, S. Giri, H.-T. Chen and W. S.-Y. Lin, *Acc. Chem. Res.*, 2007, **40**, 846.
53. A. C. Pierre, *Introduction to Sol-Gel Processing*, Kluwer Academic Publishers, Boston-Dordrecht-London, 1998.
54. Y. Ren, B. Yue, M. Gu and H. He, *Materials*, 2010, **3**, 764.
55. A. Nuccio, E. Carretti, P. Baglioni and L. Dei, *Atti del Convegno di Studi "Scienza e Beni Culturali XXVII. 2011. Governare l'Innovazione – Processi, strutture, materiali e tecnologie tra passato e futuro", Bressanone 21–24 June 2011*, ed. G. Biscontin and G. Driussi, Arcadia Ricerche S.r.l., Marghera-Venezia, 2011, p. 521.
56. M. Naldini, E. Carretti, P. Baglioni and L. Dei, *Atti del Convegno di Studi "Scienza e Beni Culturali XXVII. 2011. Governare l'Innovazione – Processi, strutture, materiali e tecnologie tra passato e futuro", Bressanone 21–24 June 2011*, ed. G. Biscontin and G. Driussi, Arcadia Ricerche S.r.l., Marghera-Venezia, 2011, p. 29.
57. A. E. Charola, *APT Bulletin*, 1995, **26**, 10.
58. L. D'Arienzo, P. Scarfato and L. Incarnato, *J. Cult. Herit.*, 2008, **9**, 253.
59. L. De Ferri, P. P. Lottici, A. Lorenzi, A. Montenero and E. Salvioli-Mariani, *J. Cult. Herit.*, 2011, **12**, 356.

60. P. Manoudis, S. Papadopoulou, I. Karapanagiotis, A. Tsakalof, I. Zuburtikudis and C. Panayiotou, *JPCS*, 2007, **61**, 1361.
61. A. Fujishima, K. Hashimoto and T. Watanabe, *Photocatalysis: Fundamentals and Applications*, Best Knowledge Center (BKC), Tokyo, Japan, 1999.
62. M. Charmon and H. Yasuyoshi, *Int. J. Food Microbiol.*, 2006, **107**, 99.
63. W. A. Daoud, J. H. Xin and Y.-H. Zhang, *Surf. Sci.*, 2005, **599**, 69.
64. M. Afsharpour, F. T. Rad and H. Malekian, *J. Cult. Herit.*, 2011, **12**, 380.
65. ASTM D6819-02e3, *Standard Test Method for Accelerated Temperature Aging of Printing and Writing Paper by Dry Oven Exposure Apparatus.*
66. N. Ditaranto, S. Loperfido, I. van der Werf, A. Mangone, N. Cioffi and L. Sabbatini, *Anal. Bioanal. Chem.*, 2011, **399**, 473.
67. N. Cioffi, L. Torsi, N. Ditaranto, G. Tantillo, L. Ghibelli, L. Sabbatini, T. Bleve-Zacheo, M. D'Alessio, P. G. Zambonin and E. Traversa, *Chem. Mater.*, 2005, **17**, 5255.
68. M. Malagodi, M. P. Nugari, A. Altieri and G. Lonati, *Proceedings of the Ninth International Congress on Deterioration and Conservation of Stone*, ed. V. Fassina, Elsevier, Amsterdam, The Netherlands, 2000, vol. 2, p. 225.

CHAPTER 15
Inorganic Nanomaterials for the Deacidification of Paper

R. GIORGI

University of Florence, Italy
Email: giorgi@csgi.unifi.it

15.1 Introduction

This chapter deals with the subject of conservation of books, manuscripts and archives, which represent, in terms of the number of items, the most abundant historically valuable heritage.[1] According to their physico-chemical properties, these artefacts are among the most delicate, being typically based on the superposition of colouring matter over a thin layer of cellulose-based support.

The ageing of these objects depends, as will be illustrated in the following sections, on several factors; some of them are typically environmental, others are strictly connected with the papermaking technique. This contribution will highlight the main conservation issues of manuscripts realized according to the classical European tradition, which since the early Middle Ages was based on the use of iron-gall inks applied over handmade paper. The latter was typically produced from rags and was characterized by very well preserved cellulose substrates; starting from the end of 19th century, wood replaced rags as a cellulose source, and low quality paper started to be used.

In the first part of this chapter, the chemical composition of paper materials and inks will be reviewed together with a short history of papermaking processes. This information is essential to understand the degradation pathways that threaten the preservation of documents and contribute to the loss of the mechanical properties of paper. In particular, the strict correlation

RSC Nanoscience & Nanotechnology No. 28
Nanoscience for the Conservation of Works of Art
Edited by Piero Baglioni and David Chelazzi

Published by the Royal Society of Chemistry, www.rsc.org

between hydrolysis, which is a pH-dependent process, and oxidation, which mainly follows a radical mechanism, will be described, with special emphasis on the role of acidity in the natural ageing of paper.

In this conceptual framework, some of the most recent and significant contributions of nanotechnology to manuscript conservation will be introduced and discussed.

15.2 Cellulose-based Materials in Art

Paper is constituted of a network of natural cellulose chains held together by entanglement and hydrogen bonds, and is typically obtained after the deposition of an aqueous suspension of cellulose fibres on a board.

Cellulose is a carbohydrate produced by plants to construct the highly tensile tissue in cell walls. Plant fibres such as cotton, jute, flax and hemp are more or less pure cellulose. Highly vascular plants, such as trees, also contain lignin. Wood is a natural composite material made of roughly 40–50% cellulose, 15–25% hemicelluloses and 20–30% lignin (see Chapter 16).[2]

Cellulose is a linear semi-crystalline polysaccharide whose degree of crystallinity ranges between 50 and 90%. The carbon atoms in the pyranose ring are numbered from 1 to 5, and the carbon in the attached methanolic group is numbered 6. The glucose units are linked through a β(1–4)-glycosidic bond, giving long chains that form crystallites.

The β position of the –OH on C1 determines a 180° rotation of the following glucose unit around the 1–4 axis of the pyranoside ring. The steric encumbrance between two units is thus reduced. For this reason, the repeating monomeric unit of cellulose is usually given as a disaccharide, formed by two glucose molecules, formally named cellobiose.

The molecular weight of cellulose ranges from 0.5×10^6 to $2.5 \times 10^6\, g\, mol^{-1}$, with the average molar mass reaching $1.5 \times 10^6\, g\, mol^{-1}$, or even higher in native cellulose.

The average number of monomer units present in a cellulose chain is expressed by the degree of polymerization (DP), defined as the ratio between the molecular weight of the cellulose molecule and the weight of a glucose unit, *i.e.* the number of units in a chain. Depending upon the source, the degree of polymerization of native cellulose, from plant material, may vary between 9000 and 15 000.

Polydispersity greatly depends on the source of cellulose; cotton and linen cellulose is less polydisperse than wood cellulose. A high DP is related to a high mechanical strength of the fibres. The cleavage of bonds can lead to a significant reduction in molecular weight. For cellulose in cotton, scission of less than 1% of the total chain links (*e.g.* due to acidity or oxidation) can cause the loss of fibre strength. The pulping process, which is a fundamental step in the paper industry, reduces the DP of wood cellulose to about 5000.[3]

Cellulose is scarcely soluble in water, but is highly hygroscopic because –OH groups give hydrogen bonds with the H_2O molecules. Owing to its hygroscopicity, cellulose absorbs or emits water depending on the temperature and on relative humidity. In general, cellulose absorbs water at ambient conditions up to 6–7% of its weight.

The hydroxyl groups of cellulose can form intramolecular hydrogen (H-)bonds between glucose units or intermolecular hydrogen bonds between different chains. These bonds are responsible for the supramolecular structure of cellulose (see Figure 15.1).

Intermolecular H-bonds, in fact, are thought to occur between $O_{(3)}H$ and $O_{(6)}H$. Assisted by dipole and Van der Waals interactions (in orthogonal directions with respect to the hydrogen intermolecular bonds), H-bonding favours the alignment in parallel strands, resulting in a highly regular sequence and rigid molecular chains (the average distance between cellulose chains is 0.54 nm). The chains partly form highly crystalline domains. In cellulose obtained from cotton, the crystalline regions typically represent about 70–80% of the cellulose, while the rest exhibits amorphous structure.

The elementary fibrils are 2–4 nm in cross-section and 100 nm in length, and are composed of successions of elementary crystallites. The internal cohesion between crystallites is achieved through polymer molecules extending from less ordered interlinking and non-crystalline regions. This "fringe-fibrillar" model for the structure of fibres was proposed by Hearle in 1958[4] and experimentally confirmed by Pionteck *et al.*[5] In 1984, Vanderhaart and Atalla established that native cellulose (also named I-cellulose) was composed of two distinct crystalline phases that they called I–α and I–β, which are present in variable proportions depending on the cellulose source.[6]

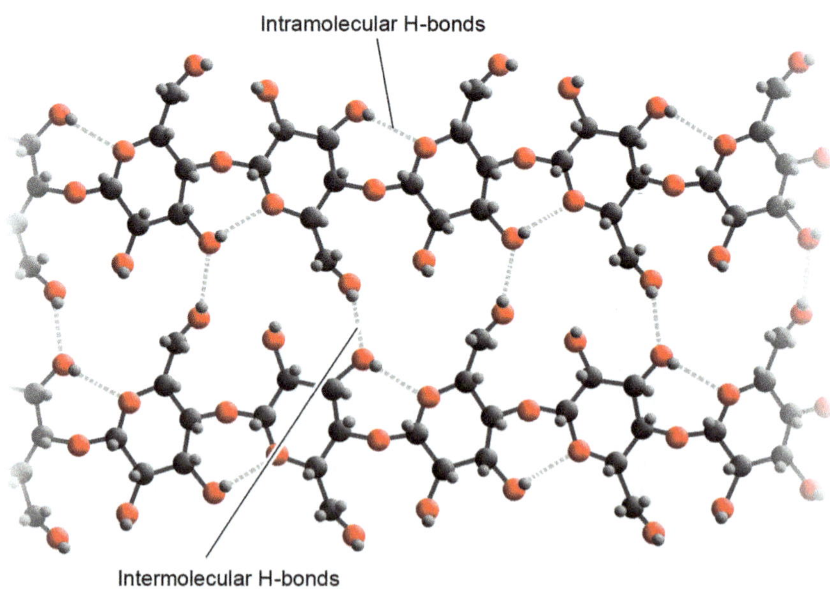

Figure 15.1 Intramolecular and intermolecular hydrogen (H-)bonds in a cellulose molecule. (Credits by Michele Baglioni.)

The ultrastructure of native cellulose is, thus, the more or less random aggregation of elementary fibrils into microfibrils of 10–30 nm width, grouped into macrofibrils 100–400 nm wide, which are structured in different cell wall layers.[2] The cell-layered structure of cotton fibres and wood implies the presence of interfaces, pores and channels ranging from 1 to 5 nm in width, that determine the accessibility and swelling properties of the fibres.

As a consequence of its physical properties, cellulose is a fundamental component of several support materials used for works of art, including paper documents and drawings, and easel paintings (wood panel and linen canvas).

15.3 Papermaking Processes over the Centuries

Cellulose is the major component of a paper document, but several substances have been commonly used in papermaking processes to provide a sheet with its ideal properties.[7,8] The chemical composition of paper mainly depends on: (i) the origin of the fibres; (ii) the chemical treatment which the raw material has been subjected to during pulping; (iii) possible additional treatments to refine the quality and improve the properties of the paper. The presence of fillers and additives for the sizing of paper sheets is particularly important, because they greatly affect the characteristics of paper.

In the Middle Ages cotton, linen or other textile fibres obtained from old rags were mainly used as a fibre source. However at the end of 15th century, as a consequence of the large production of books following the invention of printing machines, the demand grew so that the textile industry (and the recycling of rags) could not supply enough waste rags to the paper industry. In the absence of an alternative source of cellulose, many efforts were made to accelerate the mechanical (and even chemical) treatment of textile rags in order to provide large amounts of pure cellulose without time-consuming processes. Facilities such as the Hollander beater (1680), which was used to produce paper pulp from cellulose-containing plant fibres, or the continuous paper-making machine invented by Louis-Nicolas Robert (1799) tackled this task. However, it was only in the second half of the 19th century that an industrial procedure was developed to obtain cellulose, of different purity, by using wood as the raw material.

Cellulose is structured into fibres, which show different morphologies depending on their origin. Cotton fibres are isolated long lint and linter fibres. Therefore paper cellulose obtained from cotton fibres is always relatively pure. Wood pulp, on the other hand, even when highly processed, contains various amounts of other constituents, because wood fibres are bound together by a polymeric complex composed of lignin linked to hemicellulose. This complex is found in the cell walls and the middle lamella (see Section 16.2). Both hemicellulose and lignin have to be removed in order to produce high quality paper because lignin discolours in sunlight. In fact, delignification of wood-pulp is not considered in the production of paper with low life expectancy (*e.g.* newspaper, wrapping paper). The delignification Kraft method, developed by

Carl Dahl in 1884 and adopted worldwide ever since, is chemically aggressive so that cellulose depolymerization occurs. It is based on the action of alkaline compounds (sodium hydroxide) working at temperatures ranging from 150 to 180 °C. As outlined in the following sections, alkaline degradation of cellulose is an important factor that may contribute to reducing the quality of paper and its stability.

Paper production processes also changed with time in terms of the additives used for the sizing of paper. Sizing is performed to reduce the hydrophilic nature of cellulose and retard the penetration rate of aqueous fluids such as inks. This was achieved in older Arabian tradition by using compatible materials such as starch. Later, in the European Middle Ages, sizing involved the use of animal glue (this procedure was invented in Fabriano, Italy), another excellent material in terms of chemical and physical stability. In addition, at the beginning of the 19th century, the classical sizing method, which consisted in the application of glue on an already formed sheet, was replaced by adding aluminium sulfate to pulp during paper production (internal sizing) and by the addition of amphiphilic rosin (colophony). The resin acids (*e.g.* abietic acid) form salt bridges between their anionic hydrophilic head and the aluminium ions linked to the hydroxyl groups in the paper fibres. The hydrophobic part of the acid molecules points away from the paper surface, creating a hydrophobic layer. This sizing procedure provided great economic advantages and it was used until the beginning of the 20_{th} century when, unfortunately, it was found to be one of the main causes of paper degradation. In fact resin acids are responsible for inducing acid-catalysed hydrolysis. In modern times, acid sizing has been replaced by alkaline sizing paper technology.

15.4 Degradation of Books and Manuscripts

Paper is based on a natural organic material and as such is susceptible to several degradation phenomena. Chemical, physical and biological processes may cause irreversible structural changes in the constituting fibres, resulting in discoloration and embrittlement. Among the chemical processes, hydrolysis and oxidation are the main routes responsible for reducing the mechanical strength of paper.

Both paper composition and environmental conditions are decisive factors for the stability and durability of paper. Chemical composition may affect stability, which depends on the:

- Fibre type and length;
- Crystallinity of cellulose;
- Presence of fillers (*e.g.* calcium carbonate, clay and others);
- Presence of functional groups (aldheydes, ketones or carboxylic groups) formed during papermaking;
- Presence of acidic components (*e.g.* sizing agents);
- Presence of metal ion impurities (*e.g.* iron and copper ions).

The following environmental factors favour degradation, both in a direct and an indirect way:

- Temperature;
- Relative humidity (RH);
- Oxygen and light exposition;
- Air pollutants (*e.g.* O_3, NO_x, SO_2);
- Microorganisms (fungi and bacteria).

As a general criterion, the stability of paper may be expressed by the degree of polymerization (DP) of cellulose. The DP of new, undamaged modern paper cellulose typically ranges between 1200 and 1500. Upon ageing, a decrease below 300 DP units is possible. At such a low DP value the tensile strength of the original paper drops to 20% of its initial value, and paper crumbles into dust. In conservation practice, the DP is an important parameter that can be empirically measured with different techniques, one classic methodology being determination *via* viscosimetry.[9]

15.4.1 Reactions in Acidic Medium

Acid-catalysed hydrolysis is the essential and most typical degradation process of glycosidic bonds, which affects di-, oligo- and poly-saccharides. The molecular mechanism of acidic hydrolysis, confirmed by model and tracer studies,[10] is outlined in Figure 15.2. This process occurs in three steps: in the first, the proton of the catalysing acid interacts rapidly with the glycosidic oxygen linking two sugar units, forming a conjugate acid. A slow cleavage of the C–O bond follows, yielding an intermediate cyclic carbonium cation. Protonation may also occur at the ring oxygen, resulting in ring opening and the formation of a non-cyclic carbonium cation. It is uncertain which type of carbonium ion is most likely to be formed. Probably both modifications take place, with a greater likelihood of the cyclic cation in most cases. The tautomeric carbonium–oxonium ion occurs in a half-chair conformation. The carbonium cation finally initiates rapid addition of a water molecule, resulting in the stable end product and release of the proton.

The extent of depolymerization following acid-catalysed hydrolysis depends on the acid strength and concentration, as well as on temperature and duration of the reaction. The kinetics of the acid degradation of cellulose has been widely studied, one of the basic statements being:[11]

$$S = n^0(1 - e^{-kt}) \qquad (15.1)$$

Where n^0 is the number of polymer bonds available for degradation, and k is the rate constant of the hydrolysis reaction. S is defined as the number of scissions suffered by cellulose chains, and is related to DP_n:

$$S = (1/DP_n) - (1/DP_n^0) \qquad (15.2)$$

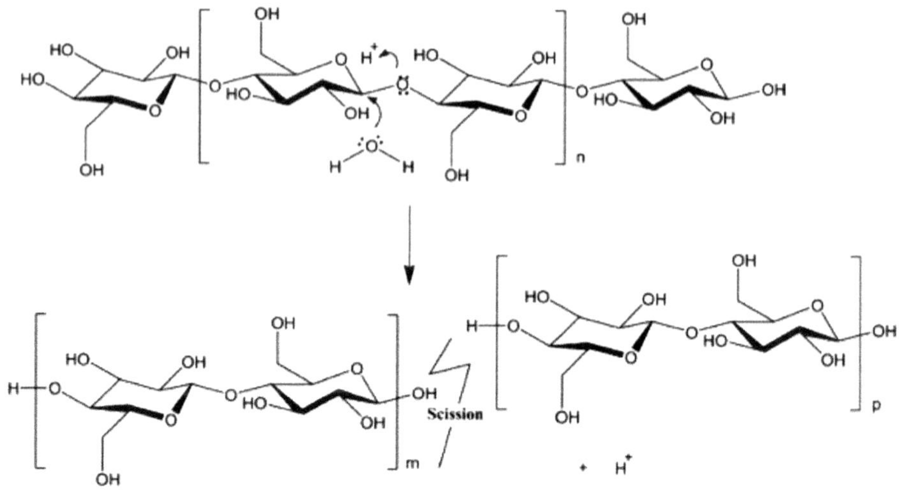

Figure 15.2 Acid-catalysed hydrolysis of β(1–4)-glycosidic bonds in cellulose. [This figure was published in S. Sequeira, C. Casanova and E. J. Cabrita, Deacidification of paper using dispersions of $Ca(OH)_2$ nanoparticles in isopropanol. Study of efficiency, *Journal of Cultural Heritage* 2006, **7**, 264–272, Copyright © 2006 Elsevier Masson SAS. All rights reserved.]

Where $DP_n{}^0$ is the initial degree of polymerization and DP_n is at time t.

As a matter of fact, S is widely used when studying the degradation status of cellulose-based art materials, since it accounts for the large variability of DP_n. The DP of paper depends upon the papermaking process and the condition of the fibres used, therefore it is not identical for paper of different origins, even when new and not aged. Evaluation of the number of scissions (S) is thus usually preferred when describing the behaviour of artefacts undergoing acidity, photo- or thermal oxidation, and other degradation processes.

The kinetic law for the acidic hydrolysis of cellulose can be simplified at low degrees of degradation, when the number of scissions is still lower than the number of polymer bonds available for degradation ($S \ll n^0$), giving:

$$S = n^0 kt \qquad (15.3)$$

The parameter n^0 would thus play the role of accessibility.[12] For many applications, however, the kinetic law is further simplified, being assumed to be:

$$S = kt \qquad (15.4)$$

This very simple, linear relationship has been experimentally verified many times, but there are exceptions in which a deviation from linearity was observed as a rapid decrease in the reaction rate in the first stage of degradation.[13,14] The behaviour was explained by simultaneous action of two kinds of splitting reaction: the classic β(1–4)-glycosidic bond cleavage and the cleavage of the so-called "weak links".[15] Weak links are β(1–4) bonds that are more sensitive than the normal ones, owing to the presence of other functional groups in place of hydroxyls in the chain. In fact, celluloses isolated from wood or other plant materials contain a greater or lesser number of carboxyl and aldehyde groups, randomly distributed along the chains. By inductive effect these modified units cause weaker linkages.[16] The difference between the two rate constants can be as high as 10^5.

It is also interesting to recall the study by Calvini *et al.*,[17] who introduced the concept of "levelling-off" *DP* (*LODP*), which is the asymptotic limit of cellulose degradation. The initial phase of paper degradation is generally well described by a pseudo-zero order kinetic, but after a certain time a downward curvature to lower degradation rates is observed and the graph tends towards a limit value. In fact, the classical model to describe cellulose depolymerization [leading to eqn (15.4)] is based on the assumption that all bonds in cellulose are equally available for scission, as they may be in a homogeneous system. However, considering that the ratio of crystalline bonds to amorphous linkages in cellulose is usually higher than 9 : 1, this may not be the case in paper degradation. Therefore, the *LODP* refers to the state of cellulose when only crystalline regions remain.

15.4.2 Reactions in Alkaline Medium

The alkaline degradation of cellulose has been studied because it is considered an essential factor in the Kraft process for delignification of wood pulp. The initial reaction is the solvation of hydroxyl groups by hydroxyl ions, causing a swollen state. At elevated temperatures the polysaccharides are attacked by strong alkali solutions, with a large number of reactions taking place, the most important being:[18,19]

- Dissolution of undegraded polysaccharides;
- Peeling of end-groups with formation of alkali-stable end-groups;
- Alkaline hydrolysis of glycosidic bonds and acetyl groups;
- Degradation and decomposition of dissolved polysaccharides, hydrolysed fragments and peeled monosaccharides.

The most important reactions responsible for the loss of polysaccharides and reduction of the cellulose chain length are peeling and hydrolysis reactions. At temperatures of about 100 °C, which occur during the heating period in pulping, the degradation of polysaccharide chains starts from the reducing end-groups by peeling reaction (primary peeling). At temperatures above 150 °C, the chains are split by alkaline hydrolysis and new reducing groups are formed, which are also subjected to endwise degradation (secondary peeling). The peeling reaction of polysaccharides involves the elimination of reducing end-groups by a β-alkoxy elimination of various carboxylic acids, thus reducing the chains by one monomeric unit at the time.[20] Endwise peeling is expected to cleave about 50–60 glucose units until a competing reaction takes place, which stops degradation.[21,22]

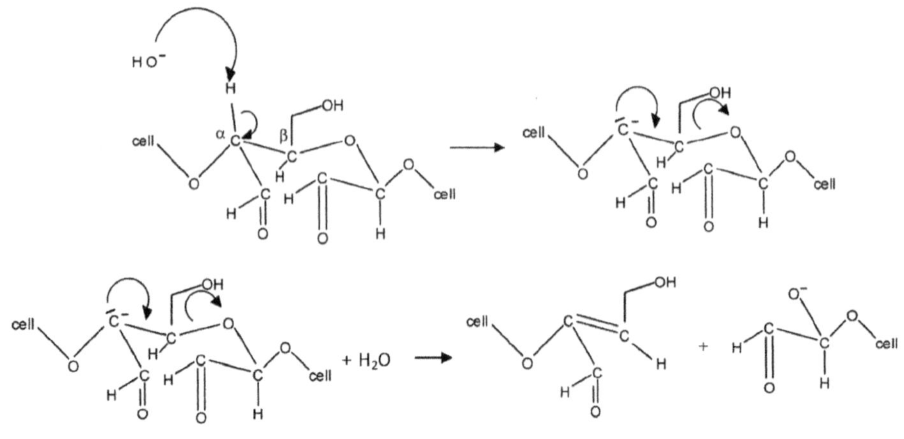

Figure 15.3 Alkaline depolymerization of oxidized cellulose through β-alkoxy elimination reaction.

It is worth noting that oxidized cellulose groups are known as "potential" sites of degradation, because oxidized chains can be easily degraded in alkaline medium even at mild concentrations and at room temperature.

The glycosidic bond cleavage proceeds by β-alkoxy elimination, as illustrated in Figure 15.3. Comparable to the cleavage at the C2-carbonyl group, cleavage may also take place at a C3-carbonyl group. Furthermore, oxidation at the C6-hydroxyl group can also lead to cleavage of the glycosidic bond.

These reactions must be considered when treating cellulose-based works of art and objects with alkaline solutions for restoration purposes (e.g. deacidification processes), or when dissolving cellulose in strongly alkaline solutions for analytical purposes. In fact, in the viscosimetric determination of the polymerisation degree, the strongly alkaline Cuen (Bis(ethylenediamine)copper(II)hydroxide) is used to prepare cellulose solutions, and the obtained DP_v is usually consequently underestimated.[9]

15.5 Iron-gall Ink Corrosion

The "ink corrosion" process describes the corrosive effect of iron-gall ink as it is applied onto a support such as paper (or parchment). Discolouration, embrittlement, and perforation of paper are the final manifestations of paper degradation. Iron-gall ink corrosion has been the subject of many studies, which have identified acid-catalysed hydrolysis and transition metal-catalysed (auto-) oxidation of cellulose as the main deterioration mechanisms.[23]

In order to better understand the conservation methods that will be described in the following sections, a more detailed description of both the composition and chemical degradation pathways of inks will be provided.

15.5.1 Iron-gall Ink Composition and Preparation

The first writing fluids used in the Western world were suspensions of carbon particles in media such as gum or oil. The difficulty of dispersing soot in an aqueous medium and the sensitivity of gum-based inks to moisture, as well as their low adhesiveness to skins and parchment, were most probably the reasons for the development of iron-gall inks.

The circumstances that led to the discovery of the reaction between gallotanninic acid and iron salts, responsible for the black–blue colour of iron-gall ink, are unknown. Application of carbon inks containing impurities of iron salts on tanned skins may have been decisive. In antiquity, iron-gall ink was already being used, among writing fluids made of soot. In *ca.* 200 BC, Philon of Byzantium mentioned the use of an invisible ink. The secret writing was executed with an extract of gall-nuts and made visible by the use of iron salts. Records of Pliny (23–79 AD) refer to shoemakers who treated freshly tanned boots with "atramentum" to dye them black. Atramentum is most likely to be iron vitriol (iron sulfate).[24]

Through the centuries, iron-gall ink became the most important writing fluid and was intensely used in the period between the Middle Ages and the 20th century, until, at the beginning of industrialization, achievements of the colour industry played a prominent role in the slow replacement of iron-gall ink by organic colorants and pigments.

A multitude of iron-gall ink recipes with varying components and compositions have been passed down to us, testifying its widespread use.[25] Artists and literates were fond of iron-gall ink because of its durability and richness in colour, and many valuable documents and artistic works have been produced using this dark coloured ink.

The expression "iron-gall" ink may be considered to be misleading, because only iron ions are thought to be present. In fact, it is known that even large amounts of copper ions can be found in some documents. Recent analytical investigations using in-air PIXE (proton induced X-ray emission), performed on several historical documents, have shown a great variation in copper concentration through the time-span of iron-gall ink use.[26,27] Molar ratios of copper to iron ions as high as 0.7 were found. "Ferrous sulfate", "green copperas" and "iron vitriol" are just a few names found in the literature for describing iron(II) sulfate heptahydrate. The same applies for copper(II) sulfate pentahydrate, which may also be called "blue copperas" or "copper vitriol". Both vitriols were already known in the Middle Ages, but separation of these two minerals after their extraction from mines could not be accomplished. Therefore, iron vitriol was often contaminated by copper vitriol and by vitriols of other metals.

Four essential constituents are required to prepare iron-gall inks: water and/or wine as the dispersing medium; Arabic gum as a binder; iron(II) sulphate salt; and tannic acid extracted from gall-nuts. The reaction of iron ions with this acid is responsible for the formation of the dark and insoluble metal–organic complex salt. Sulfuric acid is the side-product of this reaction.

Tannic acid (TA; penta-m-digalloyl-glucose) is a tanning agent found in gall-nuts. Gall-nuts develop when gall wasps (*Cynips tinctoria*) insert their eggs into the thin branches and leaves of young oak trees. Hosting the young wasps, it takes about 3–6 months for the nuts to become 8–15 mm in diameter. They may be light-brown or greenish-black in colour. Tannic acid, also referred to as gallotannic acid, is extracted from gall-nuts by boiling ground-up nuts in acidic aqueous solution (gall-nuts usually contain up to 55% w/w of tannins). This procedure leads to hydrolyzation of gallotannic acid to its basic components, which are gallic acid (depending upon pH, gallic acid or gallate) and glucose. Simple aqueous extraction from crushed nuts without using acids or boiling leads to a much smaller amount of hydrolysed product (*i.e.* gallic acid), giving a larger amount of tannic acid.

Pure water (ancient recipes suggested rain-water) was usually used for ink preparation. Very often water was also mixed with (or sometimes replaced by) wine, beer or vinegar. The latter also acted as a preservative, preventing the ink from being attacked by mould.

The method of extraction is decisive in determining the tannic acid to gallic acid ratio in ink. This is of small significance for the ink preparation, because both substances are known to form complexes with iron ions, forming a coloured pigment. However it was found that a darker colour was obtained when mainly gallic acid is present in the formulation. Inks with a larger amount of tannic acid appeared brownish rather than black. Because of the natural origin of gall-nuts, the amount and ratio of the tanning agents (gallic acid : gallotannic acid) may vary, making it difficult to produce inks of constant quality. Therefore calling the ink 'iron-gall' ink or 'ferric gallate' ink is not fully correct. It would be more precise to refer to this ink as 'iron gallotannate' or 'ferric gallotannate' ink, but the term iron-gall ink is still the most widely used.

Iron(II) ions form only water-soluble, colourless complexes with gallic acid. As soon as iron ions are oxidized by oxygen, coloured and water-insoluble complexes are formed with gallic–tannic acid or gallate–tannate in various stoichiometric ratios depending upon pH, as shown in Table 15.1.

Two models have been proposed to describe the chemical formation and structure of iron-gall ink. The first was proposed by Wunderlich in 1991,[28] but the generally accepted one is by Krekel,[29] who described a two-step reaction, involving iron(II) oxidation by oxygen, to give an octahedral coordinate of two iron(III) ions by gallic acid and water molecules, as shown in Figure 15.4.

The last essential component of writing materials is the binder. The most frequently used was Arabic gum, a natural resin obtained from various acacia trees very common in the Sudan or other African countries. It can be dissolved

Table 15.1 Iron(III)-gall complexes at various pH values; LH_4 = gallic acid. (From S. Sequeira, C. Casanova and E. J. Cabrita, *Journal of Cultural Heritage*, 2006, **7**, 264.)

Complex	*pH*	*Colour*
Fe(LH)	2.5	Grey
$Fe(LH)/[Fe(LH)_2]^{3-}$	3.5–7.4	Blue–purple
$[Fe(LH)_2]^{3-}$	3–7.5	Blue–purple
$[Fe(LH)_2]^{3-}/[Fe(LH)_3]^{6-}$	7.5–10.4	Red–purple
$[Fe(LH)_3]^{6-}$	10.5	Red

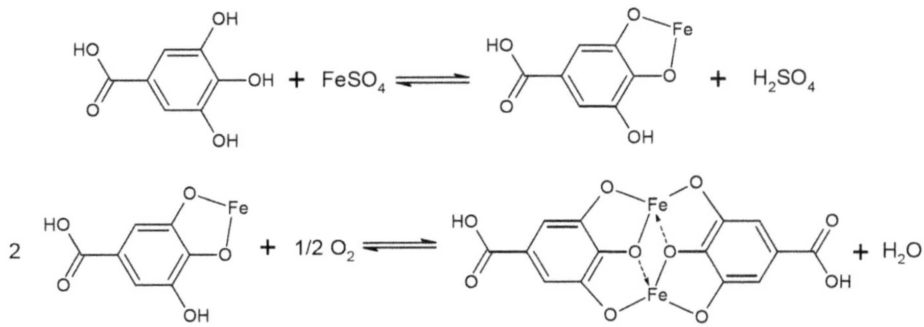

Figure 15.4 Reaction scheme for the formation of the colourless iron(II) gallate complex from gallic acid and iron sulfate, and further oxidation to the black iron(III) pyrogallate complex.

in water, giving slightly acidic solutions. The chemical structure of Arabic gum is very complex. Natural factors influence its molecular weight, which varies between 250 000 and 1 000 000 g mol^{-1}. About 70% of its weight is made up of branched heteropolysaccharides, the remaining weight being due to a glycoprotein. D-galactose, L-arabinose, L-rhamnose and D-glucuronic acid are the main constituents of the polysaccaride fraction, determining the hydrophilic character of this part. Hydrophobic proteins in the glycoprotein unit are responsible for the lipophilic character of this fraction. The chemical composition of Arabic gum makes the material applicable as a stabilizer, thickening agent, emulsifier, filler and antioxidant. In iron-gall ink preparation, Arabic gum serves as a binder by stabilizing the pigments in aqueous solution. Furthermore, it works as a glue by sticking the pigments onto the paper surface. The thickening property of Arabic gum prevents the writing fluid penetrating too quickly into the paper, thus giving clearer and sharper ink lines. Recent studies have shown that Arabic gum has an influence on the extent of iron-gall ink corrosion of paper. Scanning electron microscope (SEM) images of well-preserved areas on original manuscripts have revealed a thick covering layer of gum. It has been suggested that the protective effect is of a physical nature, *i.e.* Arabic gum protects paper by surrounding the corrosive components. Eventually Arabic gum will also be degraded by the corrosive substances, consequently damaging the paper support.

15.5.2 Iron-gall Ink Corrosion Mechanisms

Depending on the ingredients used for preparation of the ink, the pH of iron-gall ink may vary between 1.0 and 3.0. Consequently, acid-catalysed degradation affects glycosidic bonds, leading to a decrease *DP* as a consequence of random chain cleavage. It is worth noting that hydrolysis takes place in a heterogeneous reaction system (cellulose: solid state; adsorbed water: liquid phase), and lower reaction rates are expected than for homogeneous systems, depending on cellulose crystallinity, which influences the accessibility of potential reaction centres (higher crystallinity implies lower accessibility).[30]

Besides acid-catalysed hydrolysis, transition metal ions catalyse the oxidation of cellulose. Owing to the low gallotannic acid content of ink, some of the iron ions do not participate in the formation of the iron(III)–gallic complex, and are thus "free" to participate in other reactions.

Both iron and copper can easily change their oxidation status and participate in redox reactions. In particular, iron and copper ions promote radical formation, which causes extreme oxidative damage to cellulose. Throughout the process, oxygen and moisture play a fundamental role in allowing these reactions to take place. Transition metal ions support the formation of hydrogen peroxide and hydroperoxide radicals. Several authors have investigated both the mechanism and the factors that lead to the formation of hydrogen peroxide in acidic medium during the reaction of iron(II) ions with atmospheric oxygen (see Reaction Schemes 15.1 and 15.2).[31,32]

Both reactions are favoured in acidic medium and may be suppressed in a neutral environment. The Russell test (inspired by the homonymous effect used in photography) is based on the reaction of hydrogen peroxide with a sensitized Ilford Orthochromatic Film kept in contact with paper,[33,34] and provides a simple way to monitor the presence of hydrogen peroxide within the cellulose fibres.

The presence of peroxides has been also reported in alkaline environments; in this case, the mechanism of formation is neither fully understood nor widely agreed.

The key factor in determining cellulose oxidation is the further decomposition of peroxides caused by the presence of iron(II), which results in the formation of hydroxyl ions and highly reactive hydroxyl radicals, and in the oxidation of iron(II) to iron(III).

This reaction is the well-known "Fenton reaction" (Reaction Scheme 15.3). Other transition metals (*e.g.* copper) react in the same way with hydrogen

$$Fe^{2+} + O_2 + H^+ \rightarrow Fe^{3+} + HOO^{\bullet}$$

Scheme 15.1

$$Fe^{2+} + HOO^{\bullet} + H^+ \rightarrow Fe^{3+} + H_2O_2$$

Scheme 15.2

$$Fe^{2+} + H_2O_2 \rightarrow Fe^{3+} + HO^- + OH^{\bullet} \text{ (Fenton reaction)}$$

Scheme 15.3

$$Cu^{+} + H_2O_2 \rightarrow Cu^{2+} + HO^{-} + OH^{\bullet} \text{ (Fenton-related reaction)}$$

Scheme 15.4

$$OH^{\bullet} + R'H \rightarrow H_2O + R'^{\bullet}$$

Scheme 15.5

$$R'^{\bullet} + O_2 \rightarrow R'\text{–}OO^{\bullet}$$

Scheme 15.6

$$R'\text{-}OO^{\bullet} + R''\text{–}H \rightarrow R'\text{–}OOH + R''^{\bullet}$$

Scheme 15.7

peroxide; in this case, we refer to "Fenton-related reaction" (or "Fenton-like reaction"; see Reaction Scheme 15.4).

The origin of these terms is historical, since in 1876 H. J. H. Fenton used mixtures of iron(II) salts and hydrogen peroxide, referred to as "Fenton's reagent", for the oxidation of tartaric acid.

The highly electrophilic hydroxyl radical ($HO^{\bullet}$) (in acidic solution: $E^0 = +2.8$ V) easily takes away hydrogen from R–H bonds in cellulose (Reaction Scheme 15.5), forming organic radicals ($R'^{\bullet}$). These radicals may participate in various reactions, which finally result in chain cleavage. Indeed, organic radicals react easily with oxygen and interact with other undamaged cellulose molecules (R″–H) to form a new radical ($R''^{\bullet}$) and cellulose hydroperoxide (R′–OOH) (see Reaction Schemes 15.6 and 15.7).

Oxidation sites are randomly distributed in the cellulose chains and the process only stops when two radicals match together. Obviously, these reactions affect the mechanical properties of paper, because cellulose undergoes several chain scissions, especially due to complete oxidation of C2 and C3 positions, as well as cross-linking reactions. It is very common to observe perforation of paper in correspondence with inks (see Figure 15.5), because the mechanical resistance of paper is locally so reduced that simple handling (even if gentle) is enough to induce powdering of paper.

All the oxidized transition metals ions produced after "Fenton" or "Fenton-like" reactions should be brought back to their reduced state in order to be catalytically active towards further cellulose autoxidation. This is what usually occurs. In fact, paper contains some reducing components (*i.e.* oxalic acid, aldehydes, superoxide anion and others) capable of reducing transition metal ions, initiating a further cycle of paper degradation. A few possible regenerating reactions are reported in Reaction Schemes 15.8–15.10.

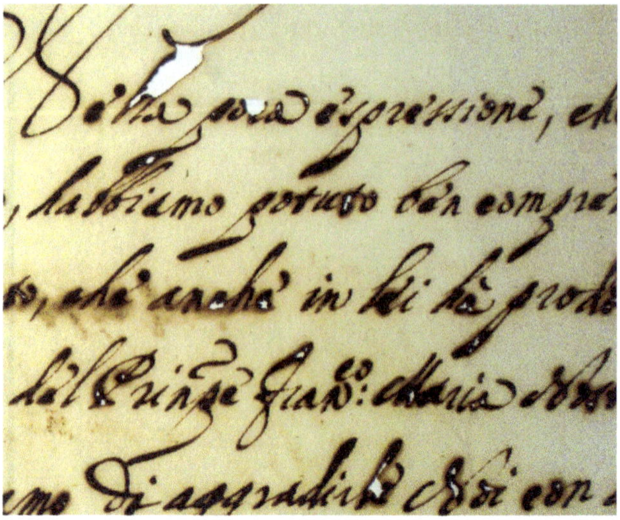

Figure 15.5 Typical example of metal–gall ink corrosion on an 18th-century document.
[Reprinted with permission from *Langmuir*, **26**, G. Poggi, R. Giorgi, N. Toccafondi, V. Katzur and P. Baglioni, Hydroxide nanoparticles for deacidification and concomitant inhibition of iron-gall ink corrosion of paper, 19084–19090. Copyright (2010) American Chemical Society.]

$$2Fe^{3+} + H_2C_2O_4 \rightarrow 2Fe^{2+} + 2CO_2 + 2H^+$$

Scheme 15.8

$$Fe^{3+} + O_2^- \rightarrow Fe^{2+} + O_2$$

Scheme 15.9

$$2Fe^{3+} + R\text{–}CHO + 2OH^- \rightarrow 2Fe^{2+} + R\text{–}COOH + H_2O$$

Scheme 15.10

Ink corrosion can be observed even under strong alkaline conditions. Alkalinity, in fact, favours the rapid elimination of the superoxide anion (O_2^-) from the hydroperoxide radical ($HOO^•$), additionally releasing a proton. Superoxide ion, as already reported, may then act as a reducing agent for Fe(III) and Cu(II) ions. In this type of reaction, the reduced form of the transition metal is regenerated by transformation of the superoxide anion to oxygen (see Reaction Scheme 15.9).[35]

Both acidic and alkaline conditions, thus, play a role in the production of hydroxyl radicals in aqueous solution. Recently, radical formation in Fe^{3+}–Fe^{2+} and Cu^{2+}–Cu^{+} aqueous solutions has been investigated in order to obtain reliable data to be used for modelling the ink-promoted degradation reaction on paper. In particular, Figure 15.6 shows the hydroxyl radical production in the pH range typical of historical paper documents, even after conservation treatment.

The catalytic activity of copper ions increases with pH. On the other hand, the activity of iron ions exhibits a maximum at pH 8.5, then decreases rapidly; this may be attributed to the precipitation of hydrous ferric(III) oxides. Taking a closer look at the hydroxyl radical production rates, it is clear that copper ions show higher catalytic activity over the entire observed pH range. The high reactivity of copper ions in alkaline oxygen systems was also reported by D. P. Graves.[36] Jian-Hui Sun *et al.* investigated the Fenton oxidation process in a pH range of 2–6 using *p*-nitroaniline (PNA) as a radical scavenger. The maximum degradation rate constant of PNA was found to be at pH 3, as represented in Figure 15.7.[37]

In summary, it appears evident that pH control of paper is a very urgent task, because acidity favours hydrolysis of cellulose while also taking part in the process of radical formation. On the other hand, even high pH may contribute to oxidative degradation. Therefore, preventive conservation requires treatments able to neutralize acidity (deacidification), and the final target of the treatment should be to reach a neutral pH so as to minimize the oxidative degradation process catalysed by transition metal ions. As a matter of fact, high alkalinity could hinder the long-term preservation of paper.

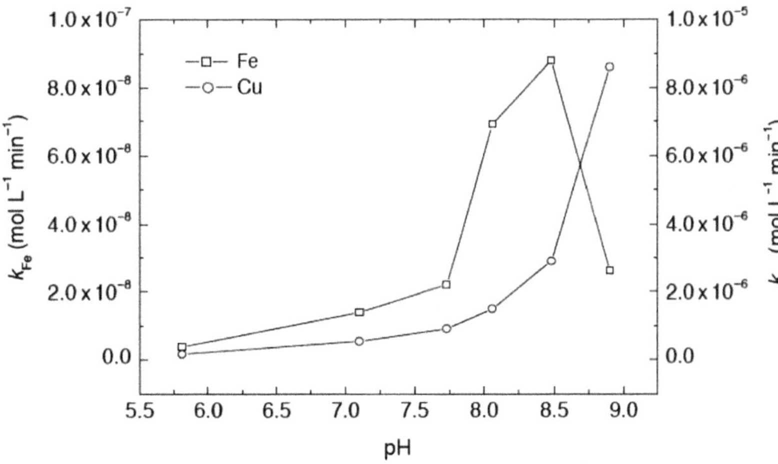

Figure 15.6 Reaction rate constants of oxidizing species, at 25 °C, for systems containing iron and copper.
[Reprinted from *Acta Chim. Slov.* **50**, M. Strlič, J. Kolar, V. S. Šelih, D. Kocar and B. Pihlar, A comparative study of several transition metals in Fenton-like reaction systems at circum-neutral pH, 619–632, Copyright (2003), with permission from the authors.]

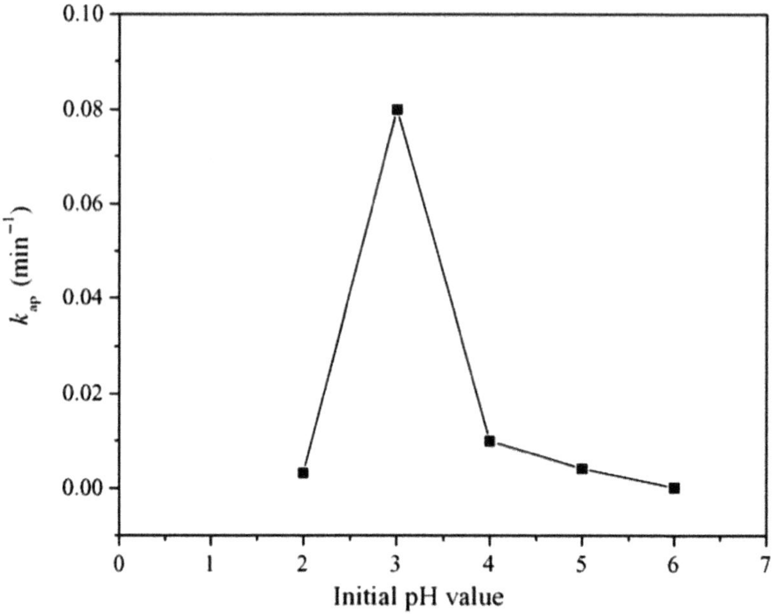

Figure 15.7 Degradation rate constant of PNA ($[PNA]_0 = 0.181$ mM; $[H_2O_2]_0 =$ 10 mM, $[Fe^{2+}]_0 = 0.05$ mM, temperature $= 20 \pm 1$ °C) as a function of pH. [Reprinted from *Journal of Hazardous Materials*, **148**, J.-H. Sun, S.-P. Sun, M.-H. Fan, H.-Q. Guo, L.-P. Qiao and R.-X. Sun, A kinetic study on the degradation of p-nitroaniline by Fenton oxidation process, 172–177, Copyright (2007), with permission from Elsevier.]

15.6 Mass-Deacidification and Single-Item Treatment

Deacidification of paper has a long tradition, and several different methods are today available to counteract acid hydrolysis.[38,39] Usually, the application of alkaline materials is carried out to provide paper with an alkaline reserve, in order to protect it from recurring acidity, which may be due to acidic environmental pollution or to chemical reactions (for example, oxidation initiates the development of carboxylic acid groups on cellulose). The alkaline reserve is realized by applying an excess of deacidifying agents. Solutions or dispersions of alkaline earth hydroxides are typically used to achieve the almost instantaneous neutralization of free hydrogen ions, with formation of water. The hydroxide excess is slowly converted into carbonate upon reaction with atmospheric carbon dioxide, forming an alkaline buffer.

Commonly used alkaline materials contain calcium and magnesium ions. These cations may participate in the formation of salts as they react with carboxyl anions present in cellulose as a result of oxidation processes.

Most of the deacidification treatments can be carried out manually. However, owing to the tremendous number of cellulose-based documents threatened by low pH values, individual treatments for each document are inconceivable. In fact, interventions on single items are time-consuming and expensive. Books under restoration are usually disassembled in order to separate the cover (often made in parchment or leather) and treat single pages. On the other hand, single-item treatments allow conservators to follow all the intervention steps constantly.

As a consequence, some methods have been developed to treat several documents at once, possibly without time-consuming preparatory procedures (*e.g.* dismantling). These methods are referred to as mass-deacidification treatments.

It is clear that the urgent need for interventions on a huge number of items (this case is quite common if we consider historical libraries) poses several conditions in terms of economic sustainability. In other words, the net costs largely depend on the time required for the intervention and the materials to be used. The search for new methods of intervention must consider these factors in order to develop technological solutions that might be fully exploitable. The Library of congress in Washington (USA) has commissioned the Preservation Technologies, Limited Partnership (PTLP) company to provide book preservation services to the library using PTLP's Bookkeeper mass-deacidification process, a well-established method. On the library website (www.loc.gov/preservation) it is reported that "the fifth contract, awarded in January 2011, will enable the Library to treat 1,250,000 books and at least 5,000,000 sheets of manuscripts by the end of October 2015. The long-term objective is to treat at least 250,000 books and 1,000,000 manuscript sheets annually for the next 25 years (the remaining 25 years of the 35-year plan). By the end of 2010, the Library had extended the useful life of 3 million books and 8 million sheets of manuscript materials from the national collections."

It is clear that any innovative technology must fulfil similar requirements in order to be competitive with the existing procedures. Mass-deacidification aims to neutralise acidity selectively, without any risk of damaging inks and other soluble or insoluble materials. Many research efforts have been dedicated to this purpose since the beginning of the 1970s.

In the following sections, a survey of several methodologies will be provided in order to clarify the importance and potential impact of the recent achievements obtained in this field thanks to nanotechnology. Deacidification processes are classically subdivided into aqueous, non-aqueous and gaseous treatments. Non-aqueous treatments generally exhibit fewer side effects than aqueous ones, thus they are typically selected for mass-deacidification.

15.6.1 Aqueous Treatments

Aqueous deacidification methods are usually applied by immersion, and in some cases by spraying. The deacidification system is generally a solution of bicarbonates or hydroxides of alkaline earth metals. In particular, a semi-saturated water solution of calcium hydroxide ($pH \sim 10.5$) or bicarbonate solution ($pH \sim 8.5$) is traditionally used.[40] Deacidification with calcium compounds usually leads to higher paper pH (pH > 9) than with magnesium compounds, because of the higher water solubility of calcium hydroxide than magnesium hydroxide.

The neutralization reaction involving aqueous alkaline substances is very effective owing to the high mobility of the reactive components. Immersion of paper in solutions also favours the washing out of acidic components and soluble metal salts, which may contribute to oxidation processes. Additionally, aqueous treatments generally improve the visual appearance of treated papers, because a large amount of hydro-soluble dirt is removed.

Nonetheless, aqueous treatments also exhibit severe disadvantages. Permanent changes in the mechanical strength of paper have been observed, as well as the bleeding of ink components. Furthermore, the application of aqueous saturated calcium or magnesium hydroxide solutions results in a very high pH that can damage cellulose and promote ink discolouration by decomposition of the ink complexes.

Despite these negative effects, aqueous treatments are still widely used by conservators. Restoration practice recommends the application, after deacidification and drying, of an aqueous dispersion (2–5% w/w) of hydroxyethylcellulose (tylose), to provide paper with re-sizing. This is usually necessary because aqueous treatments, especially at high pH, partially remove the original sizing.

15.6.2 Gaseous Treatments

In principle, gaseous deacidification represents the ideal method to treat paper supports that exhibit solvent sensitive inks, dyes and adhesives.[41–42] On this basis, the mass deacidification technique based on the use of gaseous diethylzinc (DEZ), developed by G. B. Kelly and J. C. Williams and co-workers at the Library of Congress in Washington, was enthusiastically welcomed.

Diethyl-zinc reacts readily with acids and water, resulting in zinc ions and zinc oxide as shown in Reaction Scheme 15.11 (acid neutralization) and 15.12 (formation of an alkaline reserve).

Unfortunately, the reaction of diethyl zinc with oxygen is explosive, and the technical management of the facility was complex and not convenient, so this method was abandoned after few years.

$$Zn(C_2H_5)_2 + 2H^+ \rightarrow Zn^{2+} + 2C_2H_6$$

Scheme 15.11

$$Zn(C_2H_5)_2 + H_2O \rightarrow ZnO + 2C_2H_6$$

Scheme 15.12

15.6.3 Non-aqueous Treatments

Several non-aqueous deacidification methods have been proposed and used during the last three decades, some of them being specifically designed as mass-deacidification processes.[41–48] In this section, three methods will be described in more detail; this choice is related to the fact that they introduced some innovative features, which have guided further research efforts performed in this field. These systems, still used worldwide, are known as the Wei T'o, Battelle, and Bookkeeper processes, and are based on the use of different organic solvents and deacidifying agents. However, they are all designed to lead to the formation of magnesium hydroxide within the cellulose fibres.

Most of the organic solvents proposed for paper deacidification involve health risks and/or are highly flammable. Nonetheless, these methods are of outmost importance for paper conservation because non-aqueous solvents reduce the risk of ink solubilization, migration and bleeding. Thus, treatment of documents can be carried out without previous verification of the water solubility of the inks, enhancing the feasibility of mass deacidification.

Among the above-mentioned processes, the oldest method is the Wei T'o, which was invented by R. Smith in the 1970s and was one of the first processes to be used for mass-deacidification.[41–43] The method is based on the use of magnesium alkoxides/carbonates (magnesium methoxymethyl-carbonate, MMMC) in a mixture (1 : 9 v/v) of methyl alcohol and chloro-fluorocarbons (CFC). The basic idea was to find a chemical compound soluble in organic solvents and which is able to form alkaline material upon hydrolysis. In fact, paper in equilibrium at 50% relative humidity exhibits a water content of about 5–7%, and hydrolysis starts after volatilization of the solvent (see Reaction Schemes 15.13 and 15.14).

After neutralization of acidity, the excess of magnesium hydroxide reacts with carbon dioxide, leading to the formation of magnesium carbonate, which acts as an alkaline reserve, as shown in Reaction Scheme 15.15.

The Wei T'o process provided the conceptual basis for the development of other methods based on the hydrolysis of Mg–organic salts, such as the Sable (France), the CSC Book Saver (Spain), and the Papersave (Germany).[44,45] The latter, also known as the "Battelle process" because it was developed by

$$CH_3OMgOCOOCH_3 + 2H_2O \rightarrow Mg(OH)_2 + 2CH_3OH + CO_2$$

Scheme 15.13

$$CH_3OMgOCOOCH_3 + H_2O \rightarrow MgCO_3 + 2CH_3OH$$

Scheme 15.14

$$Mg(OH)_2 + CO_2 \rightarrow MgCO_3$$

Scheme 15.15

$$Mg(OC_2H_5)_2 + 2H_2O \rightarrow Mg(OH)_2 + 2C_2H_5OH$$

Scheme 15.16

$$Ti(OC_2H_5)_4 + 4H_2O \rightarrow Ti(OH)_4 + 4C_2H_5OH$$

Scheme 15.17

Battelle Ingenieurtechnik GmbH, is based on the use of a complex of magnesium and titanium alkoxides in hexadimethyldisiloxane (HMDO). During the treatment, magnesium alkoxide neutralizes the free acids in the paper, and the excess is converted into magnesium hydroxide upon reconditioning (see Reaction Scheme 15.16). Magnesium hydroxide is then converted into a carbonate reserve as explained above. Titanium alkoxide reacts with moisture, forming titanium hydroxide (see Reaction Scheme 15.17), which then decomposes into TiO_2 and water.

The main difference from the Wei T'o method is the choice of solvent. HMDO offers several advantages in terms of lower eco-toxicological impact. It is worth noting that CFCs have been banned and the original formulation of Wei T'o can no longer be used.

As a matter of fact, the search for new ways of handling Mg–organic salts, with potential application to paper deacidification, is limited by stringent rules; costs, ecotoxicology issues, chemical and physical inertness all represent important aspects that must be considered in order to formulate efficient and reliable conservation methods. The Battelle process can be considered as a representative example of the fulfilment of these requirements.

The Bookkeeper process introduced relevant differences with respect to methods that involve the use of organic solutions. This process is based on the application of a dispersion of solid magnesium oxide (MgO) particles, which according to the producers have sub-micron size. Perfluorinated solvents are used as the dispersing medium, to ensure complete inertness towards inks and paper materials. These solvents are highly apolar and not compatible with MgO, and the oxide particle size is not small enough to confer stability, thus the kinetic stability of the dispersion is granted by the presence of a stabilizer that remains on the paper after treatment, with unknown effects in the long term. Magnesium oxide particles react with acids in paper (typically reported as sulfuric acid equivalents, see Reaction Scheme 15.18) and, when used in excess, form a reserve of magnesium carbonate to protect the paper from recurring acidity.

The application of solid particles implies possible drawbacks due to the penetration depth of particles within cellulose fibres. In particular, a veiling effect is usually detected on low porosity paper, so that in some cases the

$$MgO + H_2SO_4 \rightarrow MgSO_4 + H_2O$$

Scheme 15.18

application of the Bookkeeper method is not recommended. Moreover, the presence of stabilizers (whose exact composition is protected by patents pending) with fluorinated alkyl chains increases the surface hydrophobicity. Despite these limitations, the application of the Bookkeeper method has provided very good results in most cases, so it is probably the most commonly used deacidification process worldwide at the present time.[46–48]

15.6.4 Application of 'Anti-oxidants'

As discussed in Section 15.5.2, metal-gall ink corrosion depends on the *in situ* production of hydrogen peroxide, which reacts with iron(II) ions to give highly reactive hydroxyl radicals. Minimizing the concentration of hydroxyl radicals either by trapping or transforming them into less harmful substances should therefore slow down the degradation process.

Different materials can be used for this purpose:

- Peroxide stabilizers, to prevent the decomposition of hydrogen peroxide formed *in situ*;
- Peroxide decomposers, to favour the decomposition of hydrogen peroxide to stable non-radical products;
- Radical scavengers that act as a "sacrificial anode" for cellulose, scavenging hydroxyl radicals and forming stable products;
- Chelating substances that stop the catalytic activity of transition metal ions by blocking all active sites, so as to counteract oxidative degradation.

In contrast to metal chelating agents, both radical scavengers and peroxide decomposers/stabilizers exhibit a non-specific anti-oxidative action, and can be used in different conditions.

J. G. Neevel proposed the use of phytate (myo-inositol hexaphosphate) to inhibit the catalytic activity of iron by chelation.[31] This method proved to be very effective and further improvements were made, making it a suitable and popular conservation treatment for restorers. However, after the discovery of substantial amounts of copper ions in inks, phytate treatment was questioned. J. Kolar *et al.* proposed the use of halides, which behave as radical scavengers in acidic medium, blocking the oxidative degradation pathway of cellulose.[49,50] In the following paragraph a further description of these two conservation treatments will be provided.

Phytate is an abbreviation for myo-inositol hexaphosphate, the salt of myo-inositol hexa phosphoric acid (IP6) (see Figure 15.8), a substance found in several foods of plant origin, such as cereals and legumes. It is of great importance in food chemistry, because it influences the bioavailability of mineral elements by the formation of stable complexes. Indeed, phytate is a chelating agent and forms complexes with divalent metal ions, which are poorly soluble at neutral pH. It has to be emphasized, at this stage, that phytate inhibits transition metal-catalysed oxidation but is not effective against the natural oxidation of cellulose.

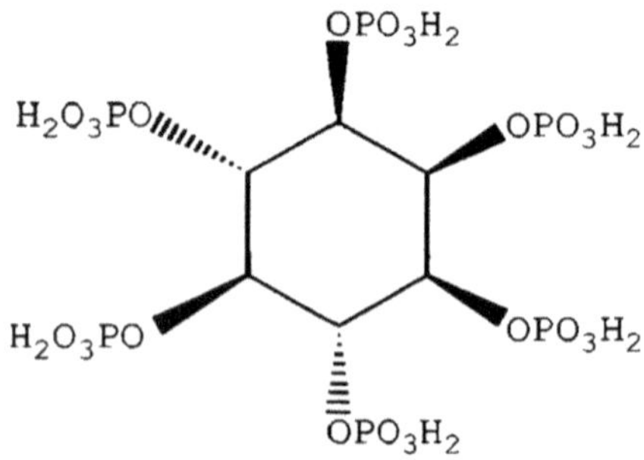

Figure 15.8 Structure of phytic acid (myo-inositol hexaphosphoric acid).

Phytate effectively blocks all the coordination sites of the iron(II) ion, preventing it from interacting with peroxides and suppressing thus the formation of hydroxyl radicals. After the iron(II)–phytate complex is formed, iron(II) is oxidized to iron(III), leading to an iron(III)–phytate complex. Application of phytate salts is typically followed by a deacidification treatment to prevent the cellulose fibres and the phosphorylated groups of phytate from undergoing hydrolysis. Neveel obtained the best results by combining either Ca–phytate (1.7 mmol L^{-1}) or Mg–phytate (2.2 mmol L^{-1}) with an aqueous calcium bicarbonate (11 mmol L^{-1}) treatment.[51,52] The pH is a critical factor in the phytate treatment. Whereas tri-[iron(III)-phytate] is insoluble, mono- and di- [iron(III)-phytates] are partially soluble depending on the pH value, leading to a partial removal of iron ions. Neevel stated that the iron(III)–phytate complexes are unreactive and white and do not change the visual appearance of the treated object. Unfortunately, phytates are scarcely soluble in organic solvents, and are thus indicated for aqueous conservation treatments. Efforts have been made to modify phytates to allow application in organic solvents. Sala *et al.*[53] evaluated the effectiveness of myo-inositol 1,2,3-tris- and 1,2,3,5-tetrakis(di-hydrogenphosphate) for the inhibition of iron-gall ink corrosion, and they found these substances to be almost as effective as phytate. In fact, inositol molecules with a lesser degree of phosphorylation exhibit some hydroxyl groups that can be provided with suitable functionalization, so as to allow the use of phytate in less polar solvents.

A further method to influence the solubility of phytate is the use of different cations. Neevel reported promising results using lithium aluminium phytate in ethanol.[54]

It has been reported that phytate causes the specific inhibition of iron(II) catalysed oxidation, while copper(I) ions maintain their catalytic ability. This would rule out the further use of phytate on documents containing copper(I) ions. Nonetheless, further experiments showed that phytate stabilizes iron-gall inks containing different transition metals, in agreement with a study of Persson *et al.*, who considered several phytate–metal complexes, investigating their solubility and relative stability at pH 7.4. The following stability order was found:[55]

$$Cu^{2+} > Zn^{2+} > Ni^{2+} > Co^{2+} > Mn^{2+} > Fe^{2+} > Ca^{2+}.$$

The high affinity of phytate for copper (II), with pronounced binding capacity between pH 5 and 7, might justify its use on documents containing only Cu(II) ions, because the reduction of the copper (II) complex to copper (I) is inhibited.

Malešič *et al.* proposed the use of quaternary ammonium halides (and pseudo-halides) to inhibit iron-gall ink corrosion.[50,56] It has been stated that iodide, bromide and the pseudo-halogen thiocyanate exhibit anti-oxidant properties because they react with hydroxyl radicals in a wide temperature and pH range.[50,56] The term "anti-oxidant" is used in this context according to what is usually meant by conservators, *i.e.* referring to every chemical able to counteract the typical final effects of oxidation, independently of its chemical behaviour.

It has also been reported that in acidic solution halides catalyse the heterolytic decomposition of peroxides.[57] However, the application of iodide salts leads to paper discolouration. Further experiments by Malešič, using bromide salts, highlighted the pronounced effect of the bromide counterion. Investigation of several quaternary ammonium and phosphonium bromides pointed towards an outstanding anti-oxidant effect as the alkyl-chain length of the cation is increased.[58–60] The advantage of quaternary alkyl ammonium halides resides in their high solubility in organic solvents. Moreover, halides are effective independently of the transition metal content, as suggested by the authors. A disadvantage may be the necessity of repeating the conservation treatment in order to maintain the stabilizing effect. It must also be noted that, according to the literature, a further deacidification treatment must be applied on paper in order to obtain satisfactory results from bromide application.[61]

Quaternary alkyl–ammonium halides were also found to be effective in inhibiting the corrosion of copper objects in acidic medium. Cetyltrimethylammonium bromide (CTAB) reduced corrosion by blocking the metal surface through electrostatic adsorption. A bromide layer is adsorbed on the surface, providing the anchor for the cetyltrimethylammonium ions (as illustrated in Figure 15.9).

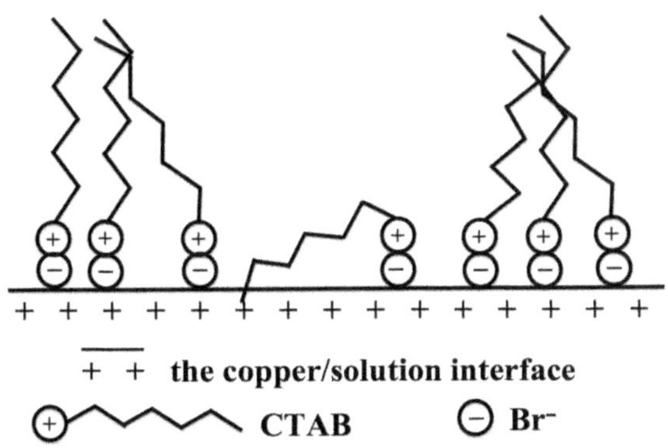

Figure 15.9 Schematic illustration of alkyl-ammonium halides adsorbing to a positively charged surface.
[Reprinted from *Corrosion Science*, **45**, H. Ma, S. Chen, B. Yin, S. Zhao and X. Liu, Impedance spectroscopic study of corrosion inhibition of copper by surfactants in the acidic solutions, 867–882, Copyright (2003), with permission from Elsevier.]

This leads to the formation of a hydrophobic surface that prevents oxygen or other substances from interacting with the copper surface, thus inhibiting metal oxidation.[62] Similar effects might occur on paper following the application of quaternary alkyl–ammonium halides.

15.6.5 New Perspectives: Nanoparticles

Significant developments in non-aqueous deacidification have been achieved since 2002 at the Centre for Colloid and Surface Science (CSGI) in Florence. Kinetically stable dispersions of calcium hydroxide [$Ca(OH)_2$] or magnesium hydroxide [$Mg(OH)_2$] nanoparticles in isopropanol or propanol have been successfully applied for paper deacidification.[63–71]

An important feature of dispersions in alcohol (or other organic solvents) is their easy application onto substrates (*e.g.* by spraying, brushing and immersion), which makes mass deacidification highly feasible. Additionally, the method involves the use of environmentally friendly solvents with low toxicity.

The size of nanoparticles (see Figure 15.10) allows kinetically stable dispersions in medium polar solvents to be obtained without adding stabilizers. This feature also minimizes the risk of white veil formation, which is sometimes caused by treatments performed with the Bookkeeper method. The amount of alkaline material (buffer) delivered onto the paper substrates can be measured, allowing the control of the final pH.

Most of the hydroxide nanoparticles transform into carbonates in a few days, and this quick reaction reduces the potential effects of cellulose alkaline degradation when the deacidification treatment is carried out on oxidized paper. However, mixed hydroxide–carbonate formulations can also be used in order to provide an alkaline buffer while reducing the amount of hydroxide to the minimum necessary to neutralize acidity.

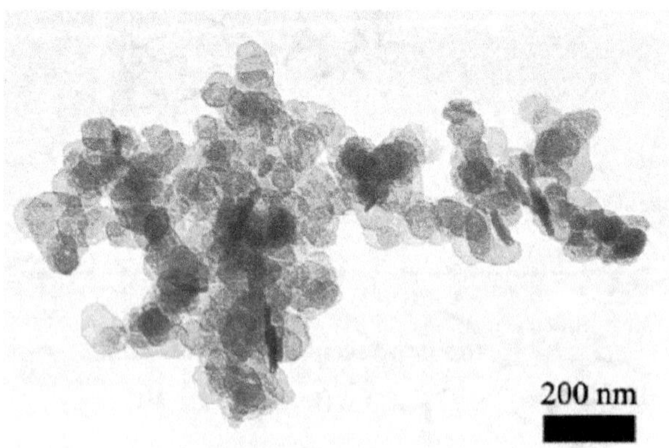

Figure 15.10 Transmission electron microscopy (TEM) picture of $Mg(OH)_2$ nanoparticles synthesized through a homogeneous phase reaction at 90 °C. [Reprinted from *Soft Matter*, **2**, P. Baglioni and R. Giorgi, Soft and hard nanomaterials for restoration and conservation of cultural heritage, 293–303, Copyright (2006), with permission from RSC.]

As highlighted in Section 15.5.2, it is evident that pH plays a fundamental role in the radical formation and in the catalytic activity of transition-metal ions. A pH range from 7 to 8 is the ideal condition to inhibit the above-mentioned processes, avoiding the hydrolysis of cellulose. Thus, a single-step treatment that allows fine pH adjustment seems to be the best approach to be followed in order to prevent paper documents from ageing.[69,70]

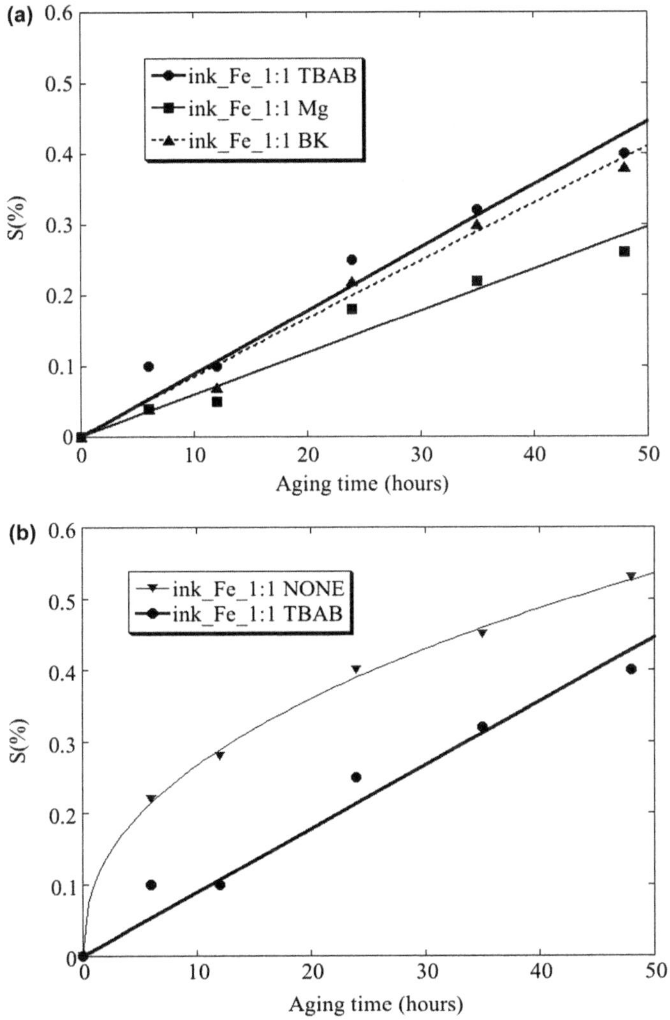

Figure 15.11 Cellulose broken bonds [$S(\%)$] *vs.* ageing-time plots of Whatman® paper painted with a 1:1 (w/w) iron sulphate : gall-nut ink (ink_Fe_1:1). (a) Comparison between samples treated with anti-oxidant TBAB treatment and paper preventively deacidified with magnesium hydroxide nanoparticle dispersions (Mg) and the Bookkeeper method (BK). (b) Comparison between paper samples treated with anti-oxidant TBAB treatment and the unprotected control (NONE).

In a recent study,[70] the application of nanoparticle dispersions has been compared with other methods for the stabilization of inked paper. Inked paper model samples were treated with different methodologies before undergoing accelerated ageing based on temperature and relative humidity. Untreated inked samples were also aged. The percentage of broken bonds (S) was plotted *vs.* ageing time (t) to highlight the effectiveness of each treatment. The most effective treatments were expected to reduce the increase of S upon ageing.

Figure 15.11(a) shows the S *vs.* t plot for samples treated with, respectively, an aqueous solution of halides, a dispersion of magnesium hydroxide nanoparticles in propanol (up to a final pH of 7.5), and the Bookkeeper method. Figure 15.11(b) shows the same plot for untreated samples and samples treated with the halide solution. The anti-oxidant treatment based on halides is clearly effective (see Figure 15.11b), and the positive effects are comparable to those granted by treatment with $Mg(OH)_2$ nanoparticles (see Figure 15.11a); however, the application of nanoparticles grants more stable pH conditions due to the alkaline buffer that is left on the substrate, while the treatment with halides would require a further deacidification process. Thus, a single step non-aqueous deacidification method appears to be an ideal candidate process for the stabilization of manuscripts.

It is worth noting that the accelerated ageing conditions are so strong that after 48 hours the untreated paper crumbles into dust, and even the treated samples lose some of their mechanical properties; however, in the 0–36 h range, which corresponds to more realistic ageing conditions, the difference between treated and untreated sample is significantly high.

15.7 The Conservation of Canvas and Textiles

Linen and cotton canvas used as oil painting supports may be degraded by acidity that affects cellulose.[72] Some of the materials used in the painting technique contain or develop acid compounds; this is the case for varnishes, used as protective coatings applied over the paint layer,[73] and a contribution may also come from the drying-oil medium used as binding material.[74] Upon natural ageing, these acids may migrate through the canvas texture and catalyse acidic hydrolysis of cellulose. Further amount of acids come from the polluted environment, which is also relevant in indoor conditions (museums, galleries and exhibitions).[75] The final effect is the progressive depolymerization of cellulose due to the cleavage of glycosidic bonds and the loss of the mechanical properties of canvas. Linen fabric degradation can be evaluated by pH measurements and it is not rare to observe pH values of 3.5–4.0.[76] This process, already described in the context of paper conservation, is dramatically enhanced in the case of canvas, because textiles are tensioned into a frame and are subjected to relatively high tensile strengths. In fact, it is quite common to observe cracks and tears, which are usually counteracted simply by the application of re-lining to reinforce the degraded original support.

The application of alkaline materials on the back of paintings, directly onto the linen or cotton texture, has been proposed to protect canvas from acidic degradation. This method was also considered as a preventive conservation treatment, because of the possibility of conferring an alkaline reserve (*i.e.* carbonates) on textiles. Nevertheless, only a few studies have been addressed to this topic and few applications of deacidification techniques have been documented. This is mostly due to the intrinsic difficulty in applying a liquid phase (such as a solution or dispersion) to canvas paintings. The stratigraphy of oil paintings is complex, and materials with different affinities for polar and apolar solvents are alternately layered one over the other. This may lead to the swelling of the preparation layer (made with mixed organic–inorganic material such as animal glue and lead carbonate or calcium sulfate) when polar solvents are used, and of the oil binding media when apolar solvents are used. The application of liquid phases to canvas may also cause the swelling of textile fibres, with a consequent alteration of the texture, under the mechanical stress due to tensioning. The final effect is the detachment of the paint layer.

A limited number of deacidification techniques have been investigated as potential candidates for the protection of canvas paintings from acid-catalysed cellulose depolymerization. Among others, the application of magnesium methoxymethylcarbonate as a methanol and trifluorochloroethane solution (Wei T'o) was investigated by Hackney *et al.*[77] The application can be performed by brushing, to achieve better penetration, or by spraying, which is less invasive and risky. The proposed blend of solvents can be safely used on the painting, but has been forbidden because of the environmental impact of chlorofluorocarbon solvents. In spite of these drawbacks, magnesium methoxymethylcarbonate is the only chemical used so far to fight the acidity of canvases.[78]

Some potentially good alternatives may come from paper conservation. The main limitation is due to the fact that most of the non-aqueous deacidification methods require dry conditions that cannot be granted for canvas paintings. In this framework, nanoparticles represent a really intriguing new research area. Alkaline nanoparticles, opportunely dispersed in pure or blended liquids with adequate polarities, represent a simple way to provide textiles with carbonates that neutralize acidity and confer long-term pH stability on canvas.

Further Suggested Reading

Anon, TAPPI (Technical Association of the Pulp and Paper Industries), *T 435 om-02: Hydrogen Ion Concentration (pH) of Paper Extracts (Hot Extraction Method)*. TAPPI Press, Atlanta, GA, 2002.

Anon, TAPPI (Technical Association of the Pulp and Paper Industries), *T 509 om-02: Hydrogen Ion Concentration (pH) of paper extracts (cold extraction method)*. TAPPI Press, Atlanta, GA, 2002.

Anon, TAPPI (Technical Association of the Pulp and Paper Industries), *T 529 om-02: Surface pH measurements of paper*. TAPPI Press, Atlanta, GA, 2002.

J. W. Baty, C. L. Maitland, W. Minter, M. A. Hubbe and S. K. Jordan-Mowery, Deacidification for the conservation and preservation of paper-based works: a review, *Bioresources*, 2010, **5**, 1955.

V. Daniels, A. Donnithorne and P. Smith (ed.), *Works of Art on Paper, Books, Documents and Photographs – Techniques and Conservation,* preprints to International Institute for Conservation Congress – Baltimore 2002, The International Institute for Conservation of Historic and Artistic Works, London, 2002.

Review and Questions

Paper conservation is a highly challenging issue because it depends on the chemical stability of an organic biopolymer such as cellulose. Paper is a thin 3D molecular network of complex chemical composition that is threatened by both environmental conditions (*e.g.* air pollution) and endogenous agents that induce auto-catalytic degradation pathways. The main points discussed in the chapter may be summarized as follows:

- The evolution of papermaking techniques has produced, in the last two centuries, poor quality paper. This is mostly due to the change in raw material (from rags to wood) to obtain cellulose, and to the introduction of acidic sizing, based on alum and rosin.
- Acidity is the main factor that favours paper degradation, because it catalyses, even at room temperature, the hydrolysis of cellulose, which causes the loss of the typical mechanical properties of paper (elasticity, tensile strength resistance, and folding endurance).
- Iron-gall inks have been used for more than 1000 years in Europe as excellent writing materials in terms of chemical stability and chromatic yield. However, the study of a great number of recipes from ancient treatises has shown that very often the right proportion of reactants to be used for making inks was not respected. The excess of iron salts was demonstrated to be the cause of further cellulose degradation due to oxidation reactions *via* a radical mechanism.
- Several deacidification methods have been developed in the past decades. Most of the efforts were addressed to the formulation of non-aqueous systems that prevent solubilization of labile materials applied on paper.
- Nanotechnology is providing innovative materials and formulations that overcome the main limitations presented by some of the methods most commonly used worldwide. The main advantage is represented by the small dimension of particles that prevent veiling, and the possibility to apply the alkaline agent to the paper directly, instead of using a precursor that requires a drying pre-treatment and further hydrolysis to make the deacidification possible.

Questions

1) Why is acidity so detrimental to paper's mechanical stability?
2) How can paper acidity be estimated?

3) Why is it important to adjust pH to around neutrality in order to limit ink corrosion processes?
4) What are the main risks involved in the use of mass-deacidification methods?

Answers

1) The mechanical properties of paper depend on its degree of polymerization (*DP*). Acid-catalysed hydrolysis of cellulose causes a rapid decrease in *DP*. The non-crystalline regions of cellulose are easily attacked in acidic environments to reach the levelling-off *DP*, where only crystalline regions remain. This value, usually around 300 *DP* units, corresponds to paper crumbling into dust.
2) Paper pH is traditionally measured by conservators using a glass-electrode with a flat bulb that is put in contact with a drop of distilled water deposited on the paper surface. The measurement is usually performed a few tens of seconds after the application of water. Alternatively, pH-indicator strips based on single colorants can be used to indicate pH exceeding threshold values. For instance, chlorophenol red gradually turns from yellow to violet when the pH passes from 4.7 to 6.7. However, the most reliable method consists in using a glass-electrode to measure the pH of a solution obtained after acidity extraction in water (either cold or hot) of paper. The main limitation is the amount of paper required (at least 200 mg), which is not acceptable for valuable documents. Recently applications have been developed for the non-destructive and non-invasive determination of pH using near infrared (NIR) spectroscopy.[79]
3) The pH is one of the factors that promote the formation of radicals within paper fibres. In the presence of free iron(II) ions, low pH favours the formation of hydroperoxide radicals. It has also been demonstrated that pH higher than 8 may enhance the catalytic activity of both iron(II) and copper(I) ions. In fact, the activity of these ions is minimized in the pH 6–8 range, which thus represents the best condition to inhibit the ink corrosion process.
4) Mass-deacidification aims to optimize the ratio between the benefits and the costs of deacidification treatments. The basic idea is to reduce the operation time as much as possible. In fact, deacidification is commonly performed through immersion of single sheets, which implies the need to dismantle the book. This operation is time consuming. On the other hand, the best option would be the treatment of several items at the same time, which requires the use of chemically inert solvents (*e.g.* exhibiting low polarity) to deliver the alkaline agents within paper fibres, to avoid solubilization of inks.

   Detractors highlight the importance of visually inspecting each page of a book to predict possible treatment drawbacks. For instance, a single page might include inks belonging to different historical periods (*e.g.*

scholars' or artists' notes beside the main text), which have different chemical composition and exhibit different affinity for solvents. Moreover, there are limitations to the homogeneous penetration of alkaline agents through the book bulk.

References

1. J. Wouters, *Science*, 2008, **322**, 1196.
2. J. C. Roberts, *The Chemistry of Paper*, RSC paperbacks, London, 1996.
3. W. J. Barrow, *Permanence/Durability of the Book – VII. Physical and Chemical Properties of the Book Papers, 1507–1949*, W. J. Barrow Research Laboratory Inc., Richmond, VA, 1974.
4. J. W. S. Hearle, *J. Polym. Sci.*, 1958, **28**, 432.
5. H. Pionteck, W. Berger, B. Morgenstern and D. Fengel, *Cellulose*, 1996, **3**, 127.
6. D. L. Vanderhaart and R. H. Atalla, *Macromolecules*, 1984, **17**, 1465.
7. P. Bower, *Proceedings of the 4th International Conference of the Institute of Paper Conservation, London 1997*, ed. J. Eagan, The Institute of Paper Conservation, Haddon Leigh, NJ, 1998, p. 1.
8. D. Hunter, *Papermaking – The history and Technique of an Ancient Craft*, Dover Publications Inc., New York, 1978.
9. UNI 8282 – equivalent to the ISO standard 5351/1.
10. J. F. Harris, *Appl. Polym. Symp.*, 1975, **28**, 131.
11. E. Montroll, *J. Am. Chem. Soc.*, 1941, **63**, 1215.
12. A. Sharples, *J. Polym. Sci.*, 1954, **13**, 393.
13. M. Marx-Firgini, *J. Appl. Polym. Sci.*, 1987, **33**, 2097.
14. A. M. Emsley, M. Ali and C. M. Eley, *Cellulose*, 1997, **4**, 1.
15. R. L. Feller, S. B. Lee and J. Bogaard, *Historic Textile and Paper Materials. Conservation and Characterization. Advances in Chemistry Series, 212*, ed. H. Needles and S. H. Zeronian, American Chemical Society, Philadelphia, PA, 1986, p. 329.
16. E. H. Immergut, B. G. Ranby and H. Mark, *Ricerca Sci.*, 1995, **25**, 308.
17. P. Calvini, *Cellulose*, 2005, **12**, 485.
18. S. A. Rydholm, *Pulping Processes*, Interscience Publishing, New York, 1965.
19. E. Sjöström, *Tappi*, 1977, **60**, 151.
20. D. Fengel and G. Wegener, *Wood: Chemistry, Ultrastructure, Reactions*, Walter de Gruyter, Berlin, 1983, p. 296.
21. O. Franzon and O. Samuelson, *Svensk Papperstid.*, 1957, **60**, 872.
22. B. Alfredsson, L. Gedda and O. Samuelson, *Svensk Papperstid.*, 1961, **64**, 694.
23. J. A. Emery and H. A. Schroëder, *Wood Sci Technol.*, 1974, **8**, 123.
24. M. P. Merryfield, *Original Treatises on the Arts of Painting*, Dover Publications Inc, New York, 1967, p. 70.
25. M. Zerdoun Bat-Yehouda, *Les Encres Noires au Moyen Âge*, CNRS, Parigi, 1983, p. 152.

26. P. Del Carmine, L. Giuntini, W. Hooper, F. Lucarelli and P. A. Mandò, *Nucl. Instr. Meth. B*, 1996, **113**, 354.
27. F. Lucarelli and P. A. Mandò, *Nucl. Instr. Meth. B*, 1996, **109/110**, 644.
28. C. H. Wunderlich, R. Weber and G. Bergerhoff, *Zeitschrift für anorganische und allgemeine Chemie*, 1991, **598/599**, 371.
29. C. Krekel, in *Tintenfraß Schäden und ihre Behandlung*, ed. G. Banik and H. Weber, Werkhefte der Staatlichen Archivverwaltung Baden-Württemberg, Stuttgart, 1999, p. 25.
30. H. A. Krässig, *Cellulose: Structure, Accessibility and Reactivity*, Polymer Monographs, Gordon and Breach Science Publishers, New York, 1993, **11**, p. 167.
31. J. G. Neevel, *Restaurator*, 1995, **16**, 143.
32. J. G. Neevel and C. T. J. Mensch, *Preprints of the ICOM Committee for Conservation 12th Triennal Meeting, Lyon*, ed. J. Bridgland, James and James, London, 1999, p. 528.
33. V. Daniels, *Non-destructive Testing of Fibre-reinforced Plastics Composites*, ed. J. Summerscales, Elsevier Applied Science Publisher Ltd., 1990, vol. 2, p. 413.
34. V. Daniels, *Br. J. Non-Destruct. Test.*, 1993, **35**, 375.
35. A. L. Rose and T. D. Waite, *Environ. Sci. Technol.*, 2005, **39**, 2645.
36. D. P. Graves, *The influence of cobalt ion concentration on the degradation of methyl β-D-glucopyranoside in oxygen-alkali*, PhD thesis, The Institute of Paper Chemistry, Lawrence University, Appleton, WI, January 1981.
37. J. H. Sun, S. P. Sun, M. H. Fan, H. Q. Guo, L. P. Qiao and R. X. Sun, *J. Hazard. Mater.*, 2007, **148**, 172.
38. H. J. Porck, *Mass Deacidification, An update of Possibilities and Limitations*, ECPA reports, ECPA– European Commission on Preservation and Access, Amsterdam, 1996.
39. P. Sparks, *Technical Consideration in Choosing Mass Deacidification Processes*, Council on Library & Information Resources, Washington, 1990.
40. W. J. Barrow and R. C. Sproull, *Science*, 1959, **129**, 1075.
41. H. D. Burgess, E. M. Kaminska and A. Boronyak-Szaplonczay, *Evaluation of Commercial Mass Deacidification Processes: Akzo-DEZ, Wei T'o and fmc-mg3. Phase I: Naturally Aged Papers,* Canadian Conservation Institute, Ottawa, 1992.
42. E. M. Kaminska and H. D. Burgess, *Evaluation of Commercial Mass Deacidification Processes: Akzo-DEZ, Wei T'o and fmc-mg3. Phase II: New and Artificially Aged Modern Papers*, Canadian Conservation Institute, Ottawa, 1994.
43. R. D. Smith, *Conservation of Library and Archive Materials and the Graphic Arts*, ed. G. Petherbridge, Butterworth, London, 1987.
44. J. Havermans, R. Van Deventer, S. Pauk and H. Porck, *Deacidification of Books and Archival Materials with the Battelle Process*, CNC-Publikaties, 9, Den Haag, 1996.

45. S. Pauk and H. Porck, *Oriënterend onderzoek naar de effecten van de Battelle en Diethylzink massaontzuringsmethoden/Pilot research into the effects of the Battelle and diethylzinc massdeacidification methods*, CNC-Publikaties, 8, Den Haag, 1994.
46. Preservation Technologies, Bookkeeper Process, Cranberry Township, PA, *http://www.ptlp.com* (accessed 07/12/2012).
47. S. Buchanan, *An Evaluation of the Bookkeeper Mass Deacidification Process. Technical Evaluation Team Report for the Preservation Directorate, Library of Congress*. Preservation Directorate, Library of Congress, Washington, DC, 1994.
48. S. Zumbühl and S. Wuelfert, *Stud. Conserv.*, 2001, **46**, 169.
49. J. Kolar, A. Možir, M. Strlič, G. de Bruin, B. Pihlar and T. Steemers, *e-Pres. Sci.*, 2007, **4**, 19.
50. J. Malešič, J. Kolar, M. Strlič and S. Polanc, *e-Pres. Sci.*, 2005, **2**, 13.
51. U. Henniges, R. Reibke, G. Banik, E. Huhsmann, U. Hähner, T. Prohaska and A. Potthast, *Cellulose*, 2008, **15**, 861.
52. A. Potthast, U. Henniges and G. Banik, *Cellulose*, 2008, **15**, 849.
53. M. Šala, J. Kolar, M. Strlič and M. Kočevar, *Carbohyd. Res.*, 2006, **341**, 897.
54. J. G. Neevel, *The Iron Gall Ink Meeting, 4th & 5th September 2000, The University of Northumbria, Newcastle upon Tyne*, ed. J. E. Brown, The University of Northumbria, Newcastle upon Tyne, 2000, p. 127.
55. H. Persson, M. Türk, M. Nyman and A. S. Sandberg, *J. Agric. Food Chem.*, 1998, **46**, 3194.
56. J. Malešic, M. Strlič, J. Kolar and S. Polanc, *J. Mol. Catal. A-Chem.*, 2005, **241**, 126.
57. W. C. Bray and R. S. Livingston, *J. Am. Chem. Soc.*, 1923, **45**, 1251.
58. J. Kolar, A. Možir, A. Balažic, M. Strlič, G. Ceres, V. Conte, V. Mirruzzo, T. Steemers and G. de Bruin, *Restaurator*, 2008, **29**, 184.
59. M. Strlič, J. Kolar, V. S. Šelih, D. Kocar and B. Pihlar, *Acta Chim. Slov.*, 2003, **50**, 619.
60. J. Kolar, M. Strlic, M. Budnar, J. Malesic, V. S. Selih and J. Simcic, *Acta Chim. Slov.*, 2003, **50**, 763.
61. E. Huhsmann and U. Hähner, *Restaurator*, 2008, **28**, 274.
62. H. Ma, S. Chen, B. Yin, S. Zhao and X. Liu, *Corros. Sci.*, 2003, **45**, 867.
63. R. Giorgi, L. Dei, M. Ceccato, C. V. Schettino and P. Baglioni, *Langmuir*, 2002, **18**, 8189.
64. R. Giorgi, C. Bozzi, L. Dei, C. Gabbiani, B. W. Ninham and P. Baglioni, *Langmuir*, 2005, **21**, 8495.
65. R. Giorgi, L. Dei, C. V. Schettino and P. Baglioni, *Works of Art on Paper, Books, Documents and Photographs – Techniques and Conservation*, preprints to International Institute for Conservation Congress – Baltimore 2002, ed. V. Daniels, A. Donnithorne and P. Smith, The International Institute for Conservation of Historic and Artistic Works, London, 2002, p. 69.
66. S. Sequeira, C. Casanova and E. J. Cabrita, *J. Cult. Herit.*, 2006, **7**, 264.

67. E. Stefanis and C. Panayiotou, *Restaurator*, 2007, **28**, 185.
68. E. Stefanis and C. Panayiotou, *Restaurator*, 2008, **29**, 125.
69. G. Poggi, P. Baglioni and R. Giorgi, *Restaurator*, 2011, **32**, 247.
70. G. Poggi, R. Giorgi, N. Toccafondi, V. Katzur and P. Baglioni, *Langmuir*, 2010, **26**, 19084.
71. D. Chelazzi, G. Poggi, Y. Jaidar, N. Toccafondi, R. Giorgi and P. Baglioni, *J. Colloid Interf. Sci.*, 2012, http://dx.doi.org/10.1016/j.jcis.2012.09.069.
72. N. Ryder, *The Conservator*, 1986, **10**, 31.
73. S. Brewis and T. G. Halsall, *J. Chem. Soc.*, 1961, 646.
74. M. P. Colombini, F. Modugno, M. Giacomelli and S. Francesconi, *J. Chromatogr. A*, 1999, **846**, 113.
75. S. Hackney, *Stud. Conserv.*, 1984, **29**, 105.
76. G. F. Davidson and T. P. Nevell, *J. Textile Inst.*, 1965, **47**, 439.
77. S. J. Hackney, J. Townsend and N. Wyplosz, *Preprints of the ICOM Committee for Conservation 11th Triennal Meeting, Edinburgh*, ed. J. Bridgland, James and James, London, 1996, p. 271.
78. S. J. Hackney and G. A. Hedley, in *Preprints of the ICOM Committee for Conservation 7th Triennal Meeting, Copenhagen*, ed. D. de Froment, ICOM in association with the J. Paul Getty Trust, Paris, 1984, p. 16.
79. EU FP6 project SurveNIR (*Near Infrared Tool for Collection Surveying*), http://www.science4heritage.org/survenir/ (accessed on December 28, 2012); T. Trafela, M. Strlic, J. Kolar, D. A. Lichtblau, M. Aanders, D. Pucko Mencigar and B. Pihlar, *Anal. Chem.*, 2007, **79**, 6319.

CHAPTER 16

Inorganic Nanomaterials for the Deacidification of Waterlogged Wood

D. CHELAZZI,* R. GIORGI AND P. BAGLIONI

University of Florence, Italy
*Email: chelazzi@csgi.unifi.it; giorgi@csgi.unifi.it; baglioni@csgi.unifi.it

16.1 Introduction

Humankind have used wood since their earliest origins, for defence and heating, as well as for the creation of tools, musical instruments, weapons, shelters, furniture, transport and artefacts, to mention only the main applications. The main advantages in the use of wood for several purposes lie in its availability and relatively easy working, compared with other materials such as metals and stone. The various wood species exhibit different mechanical and chemical properties, making it a very versatile material. The combination of flexibility and hardness, in fact, makes wood appealing for a wide number of applications if compared with stone and ceramics, and its low thermal and electrical conductivity makes it a good insulating material with respect to the metal parts with which it is usually mated in construction.

Nevertheless, wood is sensitive to physical and chemical transformation, and is attacked by several micro- or macro-organisms, leading to a wide range of degradation paths. The role of science in wood conservation has traditionally dealt with the understanding of the anatomical structure, chemical composition, physical characteristics, metabolic processes, interaction with

RSC Nanoscience & Nanotechnology No. 28
Nanoscience for the Conservation of Works of Art
Edited by Piero Baglioni and David Chelazzi

Published by the Royal Society of Chemistry, www.rsc.org

other living species, as well as with the production of a number of compounds and materials, such as coatings and consolidants meant for the preservation of the wooden tissue from degradation agents.

Recently, the important contribution of nanotechnology to the preservation of works of art has involved the conservation of wooden objects and artefacts, leading to promising results and opening new potentialities in conservation science.

This chapter will be focused on one of the main chemical issues in wood preservation: acidity. In particular, the presence of high acidity levels in archaeological, waterlogged wood will be discussed, and strategies for its neutralization, using compatible nanotechnology, will be presented. Waterlogged wood includes different kinds of remains, ranging from small objects to majestic warships, and is an important source of anthropological/historical information and aesthetic references. The preservation of such objects has consequently acquired great importance for intellectual, social and economic reasons.

A preliminary discussion of the structure, chemical composition and degradation of wood will be covered, as an essential background for all end-users dealing with the preservation of such a complex material. The state-of-the-art methods for conservation of archaeological wood will also be introduced, in order to provide a complete pattern of solutions to degradation issues. Innovative materials for coating and consolidation will be thus discussed, which also involve nanotechnology. Finally, two important case studies will be discussed, regarding the application of inorganic nanomaterials for the conservation of historical waterlogged shipwrecks (the *Vasa* and the *Mary Rose*), and the preservation of beautiful church organs belonging to the European cultural heritage.

16.2 Wood Composition and Structure

Knowledge of the chemical composition and anatomical structure of wood is a necessary step prior to its conservation. Wood is defined as the main conductive and mechanical tissue of the tree: it is a complex, three-dimensional composite solid material, whose structure may vary extensively according to different species.

As a first fundamental classification, trees are divided into two botanical groups: Gymnosperms (softwoods) and Angiosperms (hardwoods). Differences between these two groups are mainly due to the macro- and microanatomic structure and chemical composition. Softwoods generally have needlelike or scalelike leaves, and are hence referred to as conifers, while hardwoods have broad leaves. It is important to note that the terms "hardwoods" and "softwoods" have no reference to the actual hardness of the wood species: some softwood may be harder than some hardwood. Some of the most common species of hard and softwoods are reported in Table 16.1.

Different wood species exhibit a wide variation in properties related to their appearance, such as texture, grain pattern and colour, which affect their use for decorative and artistic purposes. Moreover, physical properties such as density, thermal conductivity and diffusivity, heat capacity and electrical conductivity, and mechanical properties such as elasticity and hardness, all vary according to different wood species.

Table 16.1 The main hardwoods and softwoods.

Softwoods	*Hardwoods*
Araucaria	Alder, Applewood, Ash, Aspen
Cedar, Cypress	Basswood, Beech, Birch
European Yew	Cherry, Chestnut, Cottonwood
Fir	Ebony, Elm, Eucalyptus
Hemlock	Ironwood
Kaya	Mahogany, Maple
Larch	Oak, Olive
Pine	Poplar
Redwood	Rosewood
Spruce	Teak
	Walnut, Willow

Table 16.2 Average chemical composition (weight %) of hardwoods and softwoods. Adapted with permission from R.C. Pettersen, in *The Chemistry of Solid Wood*, ed. R. Rowell, *Adv. Chem. Ser.*, 207, American Chemical Society, Washington DC, 1984, Chapter 2, p. 76. Copyright 1984 American Chemical Society.

Component name	*% in Softwoods*	*% in Hardwoods*
Cellulose	45–53	40–50
Hemicellulose	14–20	23–31
Lignin	25–30	20–24
Extractives	4–9	2–7

Before discussing the anatomical structure of wood, its chemical composition will be introduced. The main chemical components of wood are cellulose, hemicellulose, lignin and both organic and inorganic (ash) extractives. A precise chemical composition of wood, however, cannot be defined even for a given tree species. In fact, it varies with geographical location, climate and soil conditions, and among tree parts. However, average values can be defined according to wood groups (softwoods and hardwoods) and species. Table 16.2 shows the average values for the chemical composition of hard and softwoods.

16.2.1 Cellulose, Hemicellulose and Pectin

Cellulose has already been discussed in Chapter 15, and only some specifics related to wood cellulose will be provided as complementary information.

As for paper cellulose, the degree of polymerization (DP; see Section 15.2) is a fundamental parameter to be considered in order to determine the degradation status of wood and to predict its durability and physical performance. It has been determined that the average DP for molecular chains of native cellulose from several sources of plant material may vary from 9–10 000 up to 15 000.[1] These molecular weight determinations are done using light-scattering experiments, and the DP obtained, referred to as DP_w, is overestimated owing to the exponential increase of scattering with molecule size.[2] The number average degree of polymerization, DP_n, is obtained from osmometry measurements, which are linear with respect to molecular size. The calculated $DP_w : DP_n$ ratio is thus a measure of the molecular weight distribution, and has a value of nearly 1 for native cellulose in the secondary walls of plant cells, indicating that cellulose polymer chains in secondary walls are monodisperse, *i.e.* their molecules have roughly the same size.[3] Cellulose in the primary walls, on the other hand, is polydisperse.

Cellulose crystallinity is another important parameter when dealing with conservation of wood artefacts and objects (see Section 15.2). Native cellulose is partially crystalline, but more amorphous and smaller complex sugars are also found in wood, the main ones being hemicelluloses, a heterogeneous group of low molecular weight polysaccharides.

A typical hemicellulose chain is formed by 50–300 sugar monomers (mannose, galactose, xylose, glucose, arabinose, while hardwoods may also contain traces of rhamnose) and, in some cases, by a smaller number of uronic acids (glucuronic, galacturonic), combined in different ways. The molecular weight of hemicellulose is, thus, usually much lower than that of cellulose. Hemicellulose is embedded in the cell walls of plants, bound to cellulose with pectin, another important wood constituent, to form a network of cross-linked fibres with structural function. The partial structure of a typical hardwood hemicellulose molecule (xylan) is shown in Figure 16.1.

It is important to highlight that hemicellulose has a random and amorphous structure, with little strength or resistance to chemical attack such as hydrolysis or oxidation. Accordingly, hemicellulose will be preferentially degraded, with respect to cellulose, during natural or accelerated ageing of wood, and a lower content of hemicellulose is reasonably expected for strongly degraded, archaeological wood. In fact, hemicellulose is easily hydrolysed by dilute acids, alkalis or by many hemicellulase enzymes.

Pectin is a heterosaccharide mainly composed of homogalacturonan, rhamnogalacturonan I and rhamnogalacturonan II, which vary in their chain lengths and complexity, but are shorter than cellulose and amorphous. Pectins

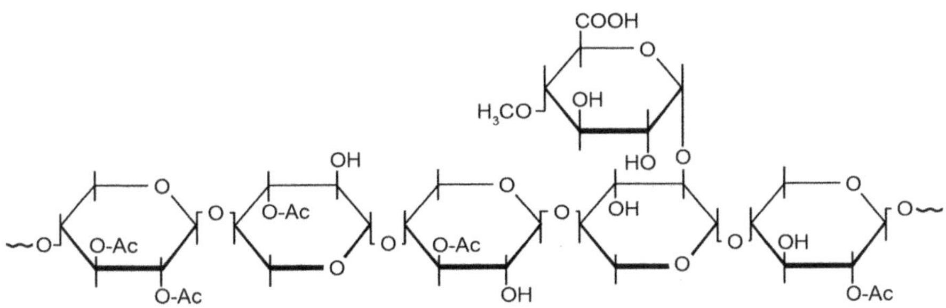

Figure 16.1 Partial formula of hardwood xylan. A single terminal chain is found, approximately every 10 xylose units, consisting of methylglucuronic acid attached to the C2 position of xylose.
[Readapted from *Macromolecular Symposia* **120**, (1997) pp. 183–196, Copyright 1997, with permission from Wiley-VCH.]

are more soluble in water than cellulose, and, under acidic conditions, can form gels. They are also susceptible to degradation in both neutral and alkaline conditions, following different mechanisms modulated by pH and temperature.[4] Pectins can form complex compounds with bivalent cations (Ca^{2+}, Mg^{2+}), and the resulting network is solvable by using chelating agents, such as ethylenediamine tetraacetic acid (EDTA). This factor must be taken in account when coupling wood deacidification methods based on magnesium or calcium compounds with iron extraction treatments based on chelating compounds (see also Section 16.5.1).

All wood polysaccharide components (globally referred to as holocellulose) can form carbohydrate–lignin complexes *via* covalent bonds.[5,6] Most commonly, alkali-liable benzyl ester bonds and fairly alkali-stable benzyl ether and gycosidic bonds are thought to be formed in wood between sugars and lignin.[7] Thus, strong alkalis should be used with care on wood for deacidification purposes, because not only could they directly damage cellulose or hemicellulose (see also Section 15.4 for alkaline degradation of holocellulose), but, at high temperature, they could also break bonds between lignin and polysaccharides, leading to structural damage.

16.2.2 Lignin

The second main component of wood, together with cellulose, is lignin, which is an aromatic polymer whose composition is complex, and varies according to different wood species. Essentially, lignin is constituted by a three-dimensional network of phenilpropane derivatives (coniferyl, sinapyl and *p*-coumaryl alcohol) that are variously bonded in an irregular array through ether and carbon–carbon bonds. It is a rigid, amorphous and thermoplastic polymer with adhesive properties. Some bonding patterns that are typically found in lignin are shown in Figure 16.2. The dominant types of linkage are aryl ether bonds.

It has been hypothesized that lignin forms a single macromolecule, whose growth is dependent only on the presence of phenolic radicals. This theory is still debated, and much about the structure of lignin remains to be clarified.[8] Lignin is deposited predominantly in the walls of secondarily thickened cells and in intercellular regions (middle lamellae). It confers rigidity on wall cells and makes them more resistant to microbial degradation.[9] As a consequence of having few hydrophilic groups, lignin is insoluble in water, but it is dissolved by the combined action of acid and ethanol at high temperature (see Section 16.3). Owing to its insolubility in water and to its resistance to degradation, lignin is expected to be a major component in naturally degraded and archaeological wood.

16.2.3 Extractives

As reported in Table 16.2, complex sugars and lignin can represent more than 90% of the chemical composition of wood. The remaining 2–9% (increasing up to 20% in tropical wood) is made up by extractives, *i.e.* metabolic products that

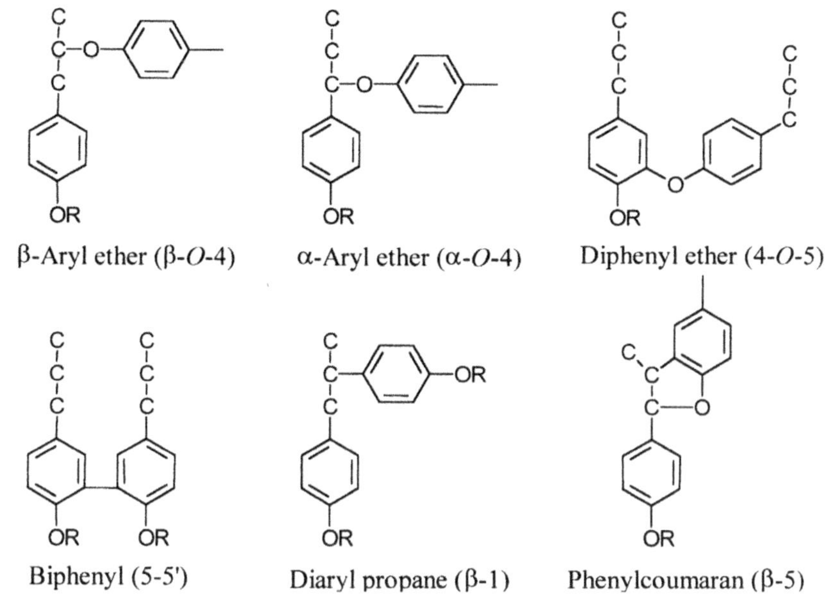

Figure 16.2 Typical bonding patterns in lignin.
[Reprinted from *Journal of Photochemistry and Photobiology B: Biology*, **56**, O. Lanzalunga and M. Bietti, Photo- and radiation chemical induced degradation of lignin model compounds, pp. 85–108, Copyright (2000), with permission from Elsevier.]

are not part of the wood structure, and can be extracted by organic solvents (ethanol, acetone, CH_2Cl_2, chloroform, ether, *etc.*), water (hot or cold) or with vapour flux. Extractives participate in tree metabolism and provide a defence against microbial attack.

They come in a great variety, some contained inside the cell lumen, others deposited on the cell walls, or inside the walls, embedded between the cellulose fibrils. They usually include alcohols, fats, waxes, alkaloids, proteins and amino acids, phenols (complex and low molecular weight derivatives), carbohydrates (starches, simple sugars, polysaccharides and gums), terpenes, glycosides (*e.g.* saponins), tannins (high molecular weight), and complex mixtures such as essential oils, mucilages and resins.

The concentration of these compounds varies in different species and in different regions of the same plant. A large amount of extractives causes an increase in the odour and the colour of wood (towards dark red, brown or green), and results in lower permeability and hygroscopicity, and higher acidity and decay resistance.[2]

Finally, the chemical composition of wood includes several mineral elements and compounds. Calcium, potassium and magnesium are the most abundant elements in wood ash, but traces of sulfur, phosphorus, silicon, boron, chlorine and heavy metals are also present.[10] Ash, *i.e.* the inorganic residue after ignition of wood at high temperature, is usually less than 1% of the total mass for wood from temperate zones, while it may be slightly higher for wood from tropical areas.[2]

Extractives must be considered when using water or organic solvents for wood treatments, as in determination of pH through water extraction, or in aqueous and non-aqueous deacidification treatments. Leaching of extractives will take place, depending on several factors that include the duration of treatment, temperature, the solvent's polarity and surface tension, porosity of wood, and, in the case of archaeological objects, the wood's residual content of extractives.

16.2.4 Anatomical Structure

A thorough description of wood parts and of its organizational macroscopic and microscopic structure is beyond the scopes of this work (see Further Suggested Reading), but an elemental anatomical and structural description will be provided before discussing conservation methods.

In describing some aspects of the cellular and anatomic structure of wood, reference will be made to the three-dimensional orientation of wood material. Wood anisotropy is defined by a radial, tangential and longitudinal direction with respect to the trunk axis (see Figure 16.3).

Given that wood is a cellular composite material, the description of its anatomy will start from the cell, and then extend to the three-dimensional cellular arrangement. Softwoods and hardwoods can be kept as the two main reference groups, even though, as for chemical composition, variability is found among tree species or even within a single tree.

Wood is composed of several distinct types of cell, all of which have elongated form and an external membrane (cell wall) with different thicknesses. During cell growth, the cell walls undergo some modifications such as lignification (increase of stiffness while losing permeability), and in their fully mature form they are mainly composed of varyingly oriented cellulose microfibrils, as well as of smaller sugars and lignin. Figure 16.4 shows the layered structure of wood cell walls. The primary and more external wall (P) is formed by randomly oriented microfibrils, lignin, hemicellulose and pectin. The secondary wall is divided into three layers (S_1, S_2, S_3), the middle one being the thickest (up to

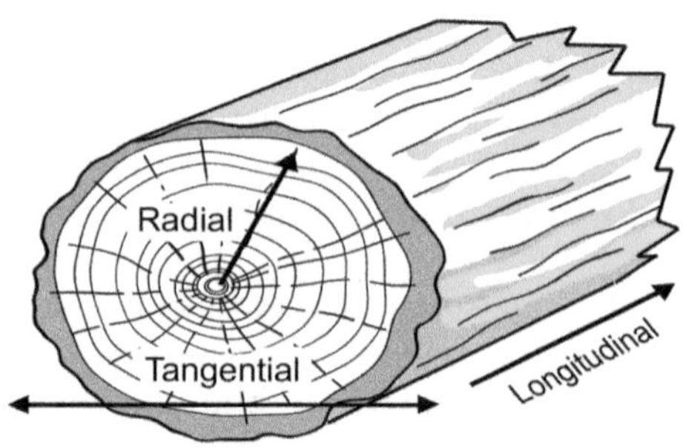

Figure 16.3 Radial, tangential and longitudinal directions in wood anisotropy. (Image by Michele Baglioni.)

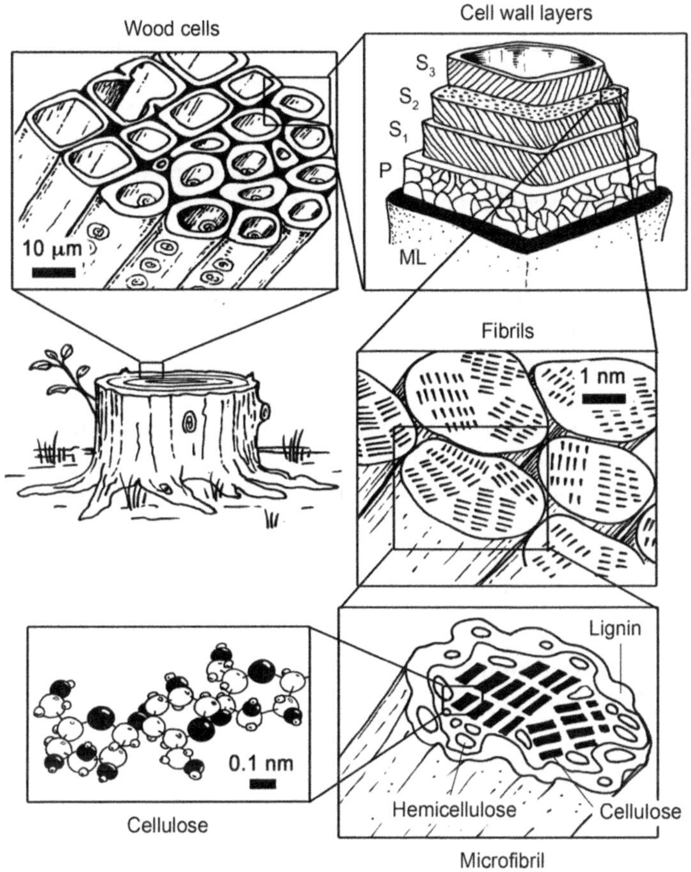

Figure 16.4 Layered structure of wood cell walls.
(Image by Michele Baglioni, adapted with permission from P. Hoffmann and M.A. Jones, in *Archaeological Wood. Properties, Chemistry, and Preservation*, ed. R.M. Rowell and R.J. Barbour, *Adv. Chem.* 225, American Chemical Society, Washington DC, 1990. Copyright 1989 American Chemical Society.)

5 μm). Within each layer, the microfibrils are parallel, but each layer has a different orientation. A network is thus formed, conferring great mechanical resistance and rigidity to the secondary wall. The walls have a nanoscale porosity, with radii of even less than 10 nm.

The middle lamella (ML) links the primary walls of different cells. It consists of a complex matrix of lignin (70–80%), pectin and hemicellulose, and has an isotropic structure.

The bulk of a living tree is constituted mostly of dead, lignified hollow cells whose internal section, called the "lumen", does not contain cytoplasm. Cells of different kinds are disposed to form a complex structure, which varies greatly with wood type. Cell dimensions, width of the lumen and the spatial arrangement of cells are all crucial factors in the evaluation of penetration of solvents, solutions and dispersions of particles (such as consolidants, de-acidifying agents, *etc.*) within the wood matrix, and must be considered when dealing with wood preservation and treatment. A strong modification of the cellular structure is expected for aged, archaeological wood, due to wall collapse and other degradation phenomena, and a full structural characterization with optical and electronic microscopy (*e.g.* scanning electron microscopy, SEM) should be carried out before selecting the appropriate conservation methods.

Softwood has a simpler structure, and is referred to as "omoxile" because it shows one main type of elementary cell, tracheids, used both for a structural function and lymph conduction. Tracheids are longitudinally oriented cells and have no openings on the top, their average width being 30–50 μm and their average length 3–5 mm.

A second type of small living cells, called parenchyma, composes the wood rays, *i.e.* radially arranged structures. Parenchyma contains storage cells and contain nutrimental substances and inorganic compounds. A type of specialized parenchyma, called epithelial cells, has very thin walls and a large lumen, and is located in the intercellular spaces (resin or gum canals). Their function is the secretion of different substances (resin, gum, latex, *etc.*).

Table 16.3 Softwood and hardwood cell types by their function. (Source: R.A. Parham and R.L. Gray, in *The Chemistry of Solid Wood*, ed. R. Rowell, *Adv. Chem.*, 207, American Chemical Society, Washington DC, 1984, Chapter 1. Copyright 1984 American Chemical Society. Reprinted with permission.)

	Function			
Wood group	*Structure*	*Conduction*	*Storage*	*Secretion*
Softwoods	Tracheids	Tracheids	Parenchymatic radial cells	Epithelial cells, Parenchymatic axial cells
Hardwoods	Fibres	Vessels	Parenchymatic radial cells	Epithelial cells, Parenchymatic axial cells

Table 16.4 Cell dimensions in different kinds of softwood and hardwood (Source: D. Fengel and D. Grosser, in *Ullmanns Encyklopadie der technischen Chemie*, 4th ed. Verlag Chemie, Weinheim, p. 669–679. 1976. Copyright Wiley-VCH Verlag GmbH & Co. KGaA. Reproduced with permission.)

		Tracheid/fibre dimensions		*Vessel dimensions*	
Wood species	*Wood density ($g\ cm^{-3}$)*	*Length (mm)*	*Diameter (μm)*	*Length (mm)*	*Diameter (μm)*
Red fir (softwood)	0.30–0.64	1.7–3.7	20–40	–	–
Pine (softwood)	0.30–0.86	1.4–4.4	10–50	–	–
Oak (hardwood)	0.39–0.93	0.6–1.6	10–30	100–400	10–400
Beech (hardwood)	0.49–0.88	0.6–1.3	15–20	300–700	5–100

Hardwood is called "eteroxile" wood, and shows two main different types of cell: fibres, which provide structure, and vessels, for conduction purposes. Fibres are elongated dead cells, with thick walls and small lumens, and without side openings. Their length is a maximum of 3–4 mm, and their lumen has an average radius of 15–20 μm. Vessels are dead cells shorter than tracheids, but with a larger lumen, and have lymph conduction functions only. Tables 16.3 and 16.4 summarize the main cell types, their functions and dimensions, for soft and hardwoods.

16.3 Degradation of Wood and Waterlogged Wood

Many factors cooperate in the degradation of wooden artefacts and objects. Wood structure and components can be degraded through physical changes, chemical reactions, insect attack and degradation by microorganisms.

Changes in temperature and fluctuations of relative humidity are among the most common physical causes of wood degradation. The loss of moisture due to water evaporation causes differential shrinkage along the three directions of wood (radial, tangential and longitudinal), eventually resulting in cracking or deformation of wooden objects. For fresh wood, air drying results in radial and tangential shrinkage of about 3–6% and 5–10% respectively, while longitudinal shrinkage is considerably lower (~0.5%).

Waterlogged wood, on the other hand, experiences more drastic shrinkage, due to degradation of the cell walls, and consequently to a higher moisture content. Loss and leaching of chemical components due to natural ageing, in fact, results in the formation of voids and cavities inside the walls and porosity is thus increased. Moreover, the loss of the crystalline parts of cellulose, which are less hygroscopic, favours the uptake of water by degraded wood. The cavities formed are then filled with water, and the shape of the wooden object is preserved by impregnation of the porous matrix. Once the object is recovered and exposed to air, however, evaporation of water causes the weakened and degraded walls to collapse, resulting in dramatic macroscopic damage, the so-called "house of cards" effect, due to water's relatively high surface tension ($\gamma \approx 72\,\text{mN m}^{-1}$ at room temperature). For oak, a typical material used in wooden artefacts and objects, radial shrinkage upon air drying increases from 4% in fresh wood up to 12% in waterlogged wood, while tangential shrinkage increases from 8% to 24%.[11] However, depending on the degree of leaching and weakening of cell walls, and consequently on the maximum moisture content, shrinkage may reach values as high as 60%. A strongly degraded small to medium sized archaeological wooden object measuring 30 cm along the tangential direction, for instance, could shrink down to a final size of only 12 cm, with severe cracking occurring consequently.

The maximum moisture content (M) of wood can indeed be used as a practical parameter for waterlogged wood degradation, and for the evaluation of the extent of shrinkage due to drying. Its mathematical expression is:

$$M = [(m_w - m_0)/m_0]\cdot 100 \tag{16.1}$$

where m_w is the water saturated wood weight and m_0 is the weight of the dry wood. Values of M of 185–200% may be considered as the lower limit for indicating wood degradation, but the threshold varies according to wood species. For heavily degraded waterlogged wood, M values of 500% or 1000% are not uncommon.[11]

Degradation of cell walls in waterlogged wood is mainly due to chemical or microbial action. The two main chemical degradation reactions that affect wood holocellulose are oxidation and hydrolysis, the latter being catalysed by either acidic or strongly alkaline environments. Both reactions have been described in Sections 15.4–5. It is well known that degradation of waterlogged wood usually results in a strong decrease of the cellulose content, leaving structures that are mainly composed of lignin.[12] Lignin itself may, however, undergo degradation due to several causes.

Treatment of wood with concentrated mineral acids does not usually cause dissolution of lignin but its condensation.[13–16] Solvolysis of lignin can instead be carried out with alcohol in the presence of hydrochloric acid. For instance, treatment with boiling ethanol containing 2% hydrochloric acid dissolves lignin. The main depolymerization reaction is the breaking of aryl ether bonds.

Alkalinity affects the various types of linkage and structural elements of lignin in different ways. As Fengel *et al.* observed: "aryl ether bonds are typically cleaved in alkaline medium, and aryl–alkyl or alkyl–alkyl carbon–carbon bonds are destroyed to some extent".[15] Both diaryl ether and carbon–carbon bonds, on the other hand, are "stable under the same conditions".[15] Moreover α- and β-aryl ether linkages are the dominant types of linkages in hardwood and softwood lignins, the alkaline cleavage of these bonds contributes essentially to lignin degradation.[17–20] The most easily cleaved linkages are α-aryl ether bonds in phenolic arylpropane units. The only prerequisite for this reaction is a sufficient ionization of the phenolic group in the alkaline medium".[15] In phenolic lignin units, "β-aryl ether bonds are also rapidly cleaved in a first order reaction, but only in the presence of hydrogen sulfide ions, which are sufficiently nucleophilic to cause cleavage".[15] In non-phenolic units, "β-aryl ether bonds are cleaved much more slowly. The reaction velocity depends on the hydroxide concentration, but is independent of the presence of hydrogen sulfide ions".[15]

Alkaline deacidification treatment of archaeological wood with a high content of reduced sulfur compounds should thus be performed with particular care. This situation is not uncommon for waterlogged wood, because under water sulfate-reducing bacteria may produce reduced compounds that penetrate inside the wood matrix (see Section 16.5.1). However, alkaline depolymerization of lignin is not straightforward, because it requires high temperature and condensation reactions compete with reactions that cause its solubilization.[18,19]

Metallic ions such as copper and iron are known to degrade cellulose (see Section 15.5) and lignin structures.[21] The presence of metallic ions, due either to inorganic compounds or to microbial activity, is thus a major threat for wood conservation. Even though lignin is resistant to chemical degradation by microorganisms, the lignocellulose composed matrix is attacked *via* oxidative cleavage by white rot fungi through the use of copper laccase, and by brown rot fungi *via* iron oxalate complexes.[22,23] Finally, the aromatic moieties in lignin are capable of absorbing ultraviolet–visible (UV–Vis) light, inducing oxidative degradation pathways by photochemical reaction.[24]

16.4 Conservation of Archaeological Wood

The following sections provide an overview of the standard and state-of-the-art materials and techniques that are usually suggested for the conservation of archaeological and waterlogged wood. These have been roughly divided into bulk treatments (consolidants) and surface treatments (coatings). We recommend consulting the Further Suggested Reading section, in order to acquire more specific information about the several conservation methods that will be considered.

Innovative materials for wood consolidation and preservation will also be introduced in Sections 16.4.1 and 16.4.2, to grant perspectives beyond the state-of-the-art methodologies. Nanotechnology for the treatment of wood, on the other hand, will be discussed in sections 16.4.1–2 (silica nanosols, nano-latexes) and 16.5.1 (dispersions of alkaline nanoparticles).

16.4.1 Consolidants

The state-of-the-art materials used in conservation of waterlogged wood have their main role in increasing the mechanical strength of wood during drying, and in consolidating its highly porous and degraded structure, making up for the loss of holocellulose, and in the worst cases even lignin, from the wood cell walls. In fact, together with controlled drying, the use of consolidants is typically advised to limit severe shrinkage damage. Several techniques have been developed and reviewed in recent decades.[11,25]

A gradual dehydration of waterlogged wood, prior to its consolidation, can be carried out using low surface tension liquids such as acetone ($\gamma \approx 25\,\mathrm{mN\,m^{-1}}$), 2-propanol ($\gamma \approx 23\,\mathrm{mN\,m^{-1}}$) and dimethyl ether ($\gamma \approx 12\,\mathrm{mN\,m^{-1}}$), in order to replace water and decrease or eliminate the risk of cell wall collapse. However, the risks inherent to the use of flammable solvents usually limit them to the treatment of small objects.

In the case of waterlogged wood recovered from burial under salt water, removal of salts is a necessary step before controlled drying, in order to avoid blooming. For major objects, however, the incomplete removal of salts can produce migration phenomena and surface crystallization, also depending on the storage conditions (see Section 16.5.1). The presence of salts can also be a severe issue if hygroscopic consolidants are selected.

Consolidation of wood is classically performed by impregnation of the wood matrix using different materials. One of the most widely used consolidants is polyethylene glycol (PEG), $H(OCH_2CH_2)_nOH$ with $10 < n < 90$. Depending on the molecular weight of the formulation used, PEG can be in the form of liquid, semi-liquid or waxy materials. Usually, PEG is applied either in alcohol or water solution, by spraying waterlogged wood or soaking it in a vat. The glycol molecules penetrate the wood matrix and displace water inside cells. A full treatment can cover a period of months or even years (on large objects) because the molecular size and percentage of PEG in solution must be increased in small increments in order to stabilize the wood structure fully. The treatment confers a dark colour and a waxy consistency, hence leading to a major alteration of the original appearance of the object. The PEG is also used as a pre-treatment step in freeze-drying of wood and leather, in order to inhibit the formation of ice crystals inside the object during the process and to prevent severe shrinkage.[26] Freeze-drying is then carried out at low temperatures (–40 °C), using vacuum in order to have the frozen ice crystals sublimate.

An important drawback in the use of PEG is that it is corrosive to metals (PEG degradation will be further discussed in Section 16.5.1). This limitation poses problems for the treatment of many waterlogged wooden objects, which very often include metals in both bulk and ionic state. Moreover, blends of small and medium molecular weight PEG show high hygroscopicity, hence particular care should be taken when treating objects with a high salt content.

Other possible classic consolidation treatments are the sucrose and the acetone–rosin methods. Apart from some aesthetic issues (surface line cracking and colour alteration) the main drawback of the sucrose method relies in the hygroscopicity of sugar.[11] In the acetone–rosin method replacement of water is carried out by soaking the wooden object in saturated acetone solutions of natural rosin. The method ensures a dimensional stability that almost approaches the performance of PEG blends, and is particularly suggested for wood with low porosity. For such substrates, in fact, penetration of PEG could be difficult when using high molecular weight glycols.[27] A great advantage of the method relies in the fact that rosin does not react with metals, while PEG causes metal corrosion. However, given the high cost of the materials and the risks inherent in handling acetone, only small objects are usually treated. A pre-treatment of the object in diluted hydrochloric acid is sometimes suggested, in order to increase the penetration of acetone into wood, but the acid could be detrimental to the residual cellulose.[25]

A very promising alternative to PEG is represented by consolidation treatments with silicon-based materials. Inspiration for the chemistry of the process comes directly from nature itself, being based on phenomena such as the silicification of buried wood, a long natural process. Accordingly, a wide number of silicon compounds, particularly silicofluorides, have been used for treating wood. Recently, a consolidation method based on silicone oil has been developed at Texas A&M University. Even though the method has been so far limited to relatively small objects, it has proved successful in stabilizing many different organic materials, including waterlogged wood.[11] Advantages rely in efficient dimensional stabilization, a good post-treatment aesthetic appearance and a decreased need for strict environmental control around the wood artefact.

In the last 25 years, sol–gel processes have been developed that allow the production of sol films of modified metal oxides (see Sections 12.2.2 and 14.2).

The sol–gel technique has important applications in the field of archaeological and historical wood conservation, because it enables a partial but permanent silicification of wood by either coating or impregnation with nanoparticulate silica sols, granting the mechanical consolidation of wood, and improving parameters such as the Brinell hardness and the bending strength.[28] The sols, also referred to as "nanosols", are made of inorganic particles, with dimensions below 50 nm, dispersed in water or organic solvents, and are applied pure or with a wide range of additives.

The treatment of wood can be carried out by spraying or impregnation, depending on the depths one wants to reach through the wood core. As far as silica nanogels are considered, adhesion to the wood substrate is due to the formation of hydrogen bonds between the Si–OH groups and the hydroxyl groups of wood's cellulose. By modifying the nanosols with specific organic groups, it is possible instead to have primary bonds formed between the hydroxyl groups and the nanoparticles (see Figure 16.5).[29,30]

Treatment with properly modified nanosols, with high solid content, can indeed provide dimensional stabilization, limiting both swelling and shrinkage of wood.[31] As an lternative to the application of sol particles, it is possible to impregnate wood with sol precursors, *e.g.* tetraethoxysilane (TEOS), so that the sol–gel process occurs in the presence of bound water in the wood material. This method leads to some increase in the strength and dimensional stabilization of wood.[32]

In fact, nanosol technology has great potential for the restoration of wood monuments or wooden parts of historical buildings, as proved by interventions already performed by restorers, also using commercial products. One limitation seems to be related to the penetration of the nanoparticles within the wood bulk: depths of about 4–5 mm have been reported in some applications, but this parameter is strictly connected with the wood's type and degradation status.[28]

Research is also investigating a new and fascinating consolidation methodology based on the regeneration of a network of bonds within the degraded wood by using a completely compatible material, *i.e.* cellulose. Recently, cross-linkable cellulose ethers were tested by Salvini *et al.* as highly compatible and non-toxic consolidating agents for waterlogged wood.[33] The concept behind

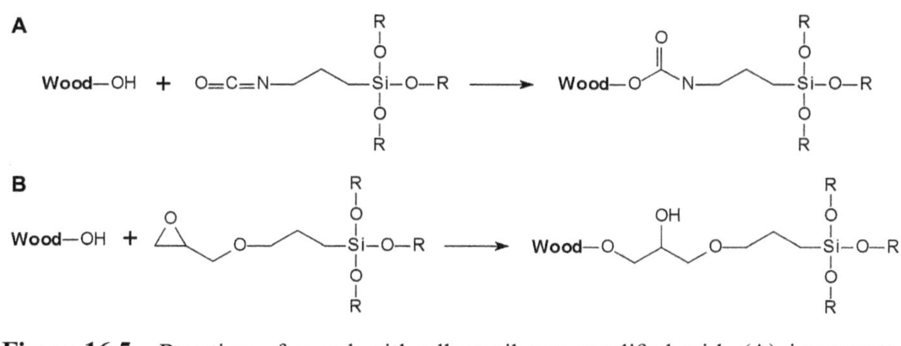

Figure 16.5 Reaction of wood with alkoxysilanes, modified with (A) isocyanatopropyl and (B) glycidyloxypropyl groups. (Reproduced from Ref. 28.)

this application is that molecules of chemically modified cellulose might penetrate inside the bulk of treated wood (*e.g.* by soaking) and then both cross-link and link with the substrate's cellulose or lignin, providing consolidation. A low DP cellulose derivative was used, because in real applications high molecular weight molecules could not penetrate the wood matrix, even if it were degraded and consequently more porous. Heavier molecules, however, could still be considered for surface treatments. Treatment tests on lignin flours showed, through Fourier transform infrared (FTIR) analysis, that the cellulose ethers exhibit high affinity for lignin, which is typically the main component in strongly degraded wood. However, the penetration of a colloidal dispersion of cellulose ethers through a cubic sample (1 cm^3) of archaeological wood was limited, possibly owing to the still excessive weight of the cellulose molecules, or to the presence of macromolecular agglomerates in the colloidal dispersion. The method is thus to be considered as a promising solution to be further investigated. The same authors also considered hydroxylated oligoamides as consolidating agents, to obtain water-soluble hydrophilic compounds with lower molecular weight than cellulose ethers (see Further Suggested Reading). Oligoamides with different molecular weight and hydrophilic : hydrophobic ratio were synthesized, and exhibited high affinity for lignin. Experimental data also showed that the oligoamides could penetrate through cubic samples of wood (see above), and the treatment enhanced the physical properties of degraded wood specimens. In fact, the specimens' density was increased, while the volumetric shrinkage was decreased with respect to untreated wood, making the treatment a good candidate for preservation methodologies.

Christensen *et al.* investigated the biomimetic use of cellulose and chitosan for wood preservation, based on the characteristic of biomaterials to self-arrange in light open or porous structures that exhibit good mechanical properties.[34] Open structures should avoid blockage of the wood pores, also allowing the future application of consolidation materials. Highly crystalline rods of cellulose can be prepared by treating fresh cellulose fibres with sulfuric acid. The rods can be collected and dispersed in water (using a tip sonicator), and the dispersion is used to treat archaeological wood samples. The rods form sheets within the wood matrix upon freeze-drying, when frozen with liquid nitrogen, but SEM evaluation could not distinguish the rods' sheets from the original cellulose, since their structure is very similar. Besides cellulose, chitosan was also considered, and the results showed that the polymer could be dissolved in dilute monovalent acids: even at low concentrations (2% w/w) the polymer solutions exhibit some viscosity, and upon freeze-drying they turn into a foam-like hard mattress. Moreover, chitosan solutions penetrated and stabilized wood to a greater extent than cellulose rod dispersions. Initial tests showed that mixed treatments with PEG increase the penetration of both systems, because PEG acts as a surfactant, limiting the flocculation of the crystals, which is instead favoured by release of ions from archaeological wood. Further study will aim at using hybrid consolidants with inorganic parts for increased mechanical strength or to provide acid-neutralizing properties (*e.g.* alkaline nanoparticles).

16.4.2 Coatings

There is a vast amount of literature on wood coatings; the term usually refers to a wide class of natural and synthetic compounds, used both for decorative purposes and for protection of wood surfaces.[35–42]

The main tasks that a coating has to achieve are to block or inhibit oxidation and hydrolysis caused by moisture, to shield UV radiation, and to limit physical damage (from abrasions, hits, *etc.*). Moreover, for application on artefacts, the aesthetic factor is particularly important: the coating has to provide the proper appearance, without causing discoloration or other undesired effects. Coatings are often applied as liquids, either solutions (aqueous or organic) or aqueous emulsions/dispersions.

There are several aspects to be considered in the use of coatings for conservation purposes, related to both their effectiveness and possible drawbacks. The emission of acidic volatile organic compounds (*e.g.* acetic acid) has been reported for several poly(vinyl acetate) wood adhesives and coatings used in museums for display cases or sealed enclosures,[43,44] raising concerns about the use of these polymers either on supporting materials (boxes, cabinets, frames) and directly on cellulose or protein-based artistic substrates such as wood, paper, canvas, parchment and leather. In fact, the lowering of VOC (volatile organic compound) release is also an ecotoxicological priority in the development of any conservation material. The UV curing of wood coatings, based on the formation of a three-dimensional network upon irradiation with UV light of monomers/oligomers (*e.g.* acrylic resin with a small amount of fluorinated monomer), is a popular and widely used alternative application with respect to solvent- or emulsion-based adhesives, because it grants lower VOC release, due to the absence of solvents and to the use of reactive diluents and non-volatile cross-linkable oligomers.[40,45] Recently, the photopolymerization of wood coatings with acceptable surface properties was obtained even under visible light, opening perspectives for applications where the use of UV curing is discouraged.[41] Curable acrylic coatings, however, can undergo yellowing upon any exposure to UV following curing, suggesting that screening agents such as light-absorbers could be coupled with the coatings, also to grant protection to wood from UV in the 280–400 nm range.[46]

New improvements in the development of high performance water based acrylic coatings are coming from research work on nanolatexes, *i.e.* dispersions of polymeric particles with diameters of about 100 nm or less.[47] A comprehensive overview on the interactions of nanolatexes with the wood matrix has been provided by Mader *et al.*[48] These coating formulations are advantageous as a consequence of several factors related to the small dimensions of the polymer particles. An easier film formation and a higher penetration within the fibres are the main points reported by the literature, both leading to increased consolidation and protection of the substrate. The penetration point, however, is still debated, because the first studies could not find a correlation with either particle size or the formulation's rheology, which some authors reported as a crucial aspect.[49]

Several methods can be used to obtain nanolatexes, resulting in different film properties. One practical concern is to limit the quantity of surfactant in the formulation. For example, very small polymer particles (about 5 nm) can be used as "seeds" for polymerization, or the use of surfactants can be combined with a steric–electrosteric stabilizing system. The latter technique resulted in a commercial formulation.[48] Another practical alternative that has been developed is to use non-conventional surfactants (*e.g.* polymeric surfactants), which has also led to the formulation of commercial products.[48] Stable transparent or translucent films can be thus produced on wood surfaces. The selection of the appropriate surfactant is also critical because the actual manufacturing processes deal with systems with limited concentration, whose penetration inside wood matrices seems to be strongly affected by surface tension. For example, strong interactions among the nanolatex particles result in decreased penetration rates; it has been reported that surface tensions below 45 $mN\,m^{-1}$ are typically necessary to provide easy wetting of wood. Research efforts are now focusing on the interaction mechanisms and on the use of surfactants to improve the performance of water-based systems, and to provide nanolatexes with particles of about 25 nm with improved penetration.

Finally, the silica-based nanomaterials described in the previous sections (nanosols) can be modified with specific chemical functions, to provide the treated wood (see Figure 16.6) with several features (protection from mechanical, chemical and photo-oxidative degradation), including an increased resistance to attack by insects and microorganisms. Suitable additives, such as polysiloxanes and fluorinated compounds, can also be used to confer the desired properties (*e.g.* hydrophobicity) to wood.[28,32]

In this regard, it must be considered that more work can and must be done in order to increase the long-term performance of inorganic wood composites that are exposed to outdoor weathering. The improvement of both nanolatexes and silicon-based nanosols could open an emerging market for safe and environmentally friendly wood products.

16.5 Case Studies

As discussed in the previous sections, there are a vast amount of techniques that can be considered for the conservation of archaeological and waterlogged wood,[11] for instance concerning the dimensional stabilization or surface treatments to provide wood with useful properties. In the following sections we will discuss two case studies, specifically dealing with the use of nanoparticles for the deacidification of archaeological wood samples coming from the Swedish *Vasa* warship, and with the decrease of acidic emissions of oak wood. Up to the present time, these applications constitute a small and specific part of all the possible treatments, yet we have highlighted them as promising cases that show the potential and effective impact of nanotechnology.

Before introducing the case studies, it is worth highlighting that an extended and comprehensive overview of some of the most recent scientific contributions

Figure 16.6 Wood beam ceiling in castle Hof (Riesa, Germany) partially damaged by brown rot fungi. In the box, a detail is provided of wood from the rear side, restored by nanosol application.
(Reproduced from Ref. 28.)

to the preservation of waterlogged wooden shipwrecks was provided by the Proceedings of the "Shipwrecks 2011" international conference, Stockholm, 2011 (see Further Suggested Reading). Besides consolidation methods, the Proceedings also deal with advanced wood characterization, the full discussion of which is beyond the scope of this chapter. A list of analytical methods, and their uses, is reported below to provide an essential core of the most interesting possibilities:

- X-ray spectroscopy, for the speciation of both sulfur and iron compounds in waterlogged wood;
- UV-Microspectrophotometry, to investigate the distribution of lignin at the sub-cellular level in wood;

- NMR (nuclear magnetic resonance), to determine and quantify functional groups and intermonomeric bonds in lignin, and thus characterize fungal degraded wood and evaluate lignin in organic matter;
- GPC (gel permeation chromatography), to determine the increase or decrease in the average molecular weight of extracted lignin;
- DE-MS (direct exposure mass spectroscopy), for rapid evaluation and comparison of archaeological wood (*e.g.* loss of polysaccharides);
- Py-GC-MS (analytical pyrolysis–gas chromatography–mass spectroscopy), to obtain qualitative and semi-quantitative information on the type of wood, on relative amounts of lignin and holocellulose, and transformations of lignin upon ageing;
- SEC (size exclusion chromatography), for the chemical analysis of wood;
- Tensile strength measurements, for the determination of the degree of degradation of wood;
- TRS (time-resolved diffuse optical spectroscopy), a non-invasive technique to obtain information on the composition, structure and water content of wood and waterlogged wood;
- GASMAS (gas in scattering media absorption spectroscopy), a non-invasive and non-destructive technique for the analysis of wood's gas diffusion properties.

As a matter of fact, a full characterization is the perfect match to the use of innovative materials for consolidation/deacidification, and the two approaches should be considered to be complementary in the preservation process. Naturally, as for the setting and use of innovative materials, the presence of advanced facilities and the related highly specialist skills imply that the involvement of scientific staff in restoration projects must be enhanced, and the collaboration between conservators and scientists further tightened.

16.5.1 The Deacidification of the *Vasa* Warship

The *Vasa* ship represents a unique case in archaeological wood restoration, and a challenge for restorers and conservation scientists. Its story traces back to 1626: the Hasburg Empire threatened the Swedish control of the Baltic Sea, and the Swedish King Gustav Adolf II ordered the enlargement of one of the ships under construction for the new fleet. The ship was completed in two years, but eventually sank during its maiden voyage in 1628.[50] The wreck then rested for 333 years on the seabed in Stockholm harbour, before successful rescue operations in 1961 made it the largest marine archaeological find of its kind ever salvaged.[51]

Large marine archaeological wooden artefacts, in fact, are relatively rare owing to the microbial and chemical degradation of wood occurring in salt- and oxygen-rich water. At its recovery, however, the wood of the *Vasa* appeared to be in good condition. This was due to several reasons: the low water temperature slowed down deterioration processes; the low salinity of the Baltic Sea discouraged *Xilofagus* organisms; the *Vasa* wood was quite fresh at its sinking, so that biological degradation of cellulose fibres by rot fungi was only at the initial phase. Moreover, the large amount of organic matter in the increasingly polluted water in Stockholm harbour consumed the dissolved oxygen. In such anoxic conditions, bacteria use sulfate ions as an oxygen source to metabolize organic compounds, producing elemental and reduced sulfur, and the environment around the *Vasa* was thus made inhospitable to several fungi and other wood-degrading microorganisms.

After the salvage, extensive treatment was started to stabilize and dry the ship's timbers. The *Vasa* is, in fact, the first major object for which aqueous solutions of PEG were used. The spraying treatment went on intermittently for 17 years; the ship was then slowly dried and eventually put on display in 1990 in the Vasa Museum, Stockholm, where it is kept under safe hydrothermal conditions ($T = 18–20\,°C$, $RH = 51–59\%$). Moreover, in order to inhibit all photo-induced or enhanced degradation processes, the light levels in the museum are kept below 100 lux (see Figure 16.7), UV filters have been placed on all the windows, and daylight is not allowed to fall on the ship.

About 240 tonnes of polymer were consumed during the whole treatment. In the 5–10 cm thick oak planks the content of PEG is about 30%, while in the much thicker heavy timbers it is about 10%. More than half of the *Vasa*'s outermost layer of wood consists of solid PEG. The treatment showed good results in size stabilization of the waterlogged wood, but could not prevent the development of salt formation and high acidity, which turned out to be a great menace for the *Vasa*.

In July 2000, salt formation (hydrated sulfates) and pH values below 2 were observed on *Vasa*'s wood in numerous areas (reported as more than 600 in 2002). A combination of advanced X-ray analyses showed that elemental sulfur had accumulated within the beams (0.2–4% by mass), up to some centimetres beneath the surface.[52] In fact, after the recovery of the wreck, a sulfur oxidation process had started, producing sulfuric acid (Scheme 16.1):

Figure 16.7 (Left) A view of the Vasa Museum in Stockholm during opening hours, in 2007, showing the low light levels maintained to inhibit photochemical degradation. (Photograph by Daria Noferini.) (Right) The *Vasa* warship. (Photograph by Karolina Kristensson.) (Both photographs are printed with the kind permission of the Swedish National Maritime Museums.)

$$S(s) + 3/2\ O_2 + H_2O \rightarrow 2H^+ + SO_4^{2-}$$

Scheme 16.1

The total potential yield of sulfuric acid resulting from oxidation of the initial amount of reduced sulfur compounds is estimated to be at least 5000 kg.[52]

One of the causes of underwater preservation of the ship turned out to be, thus, a severe conservation problem after recovery: sulfuric acid catalyses the hydrolysis of cellulose, leading to a decrease in the mechanical strength of wood, as described in Chapter 15. Moreover, the crystallization of sulfates (especially natrojarosite and gypsum) causes mechanical stress inside the wood pores.

Sulfuric acid, however, is not the only degradation agent present in the *Vasa* wood. During and after the salvage of the wreck, about 5500 new iron bolts, coated with epoxy and zinc, were inserted into the empty holes left by the rusted ones, and in the following 40 years the new bolts underwent serious degradation and corrosion.[53] Dissolved iron(II) compounds from the corroding metal penetrated the wood, oxidizing further with time to form particles of rust. The characteristic "black oak" colour of the *Vasa* is indeed mainly due to small particles of various iron(III) oxyhydroxides (rust) included in the wood and in the PEG. Moreover, iron and PEG are involved in interconnected reactions, leading to further degradation. While acidity dissolved the bolts' zinc coating, the corrosion of bulk metal iron was accelerated by contact with PEG inside the humid wood.[54]

Oxidation of reduced sulfur compounds to sulfuric acid is probably catalysed by the presence of iron ions, and iron compounds also catalyse the oxidative degradation of cellulose (see Chapter 15), resulting also in the production of oxalic acid from oak wood.[55] Finally, iron(III) ions interact with the oxygen atoms in PEG, weakening bonds within the PEG molecule and facilitating its degradation by oxidation, which leads to formation of a hemiacetal and finally to formic acid.

As an overall result, the *Vasa* wood shows low pH values, ranging from 4 (which is almost the natural pH for oak, mainly due to extractives such as tannins) to 0. Surface spots, where precipitation of sulfates and iron salts has occurred, exhibit very low pH, and acidity increases through the first centimetre in the bulk wood. The presence of acetic and oxalic acid, in fact, increases with depth, while other acids (formic, glycolic) show no spatial correlation. The concentration of sulfates decreases through the section up to the first centimetre.

Nowadays, the *Vasa*'s historical, cultural and economic importance is so crucial that many efforts are being made for its proper conservation. In preliminary tests, the Swedish curators applied soda poultices as an emergency cure on *Vasa*, but the deacidification effects were superficial and lasted only for short time periods (from a few weeks to three months). Several Swedish foundations and government agencies eventually decided to combine and make funds available to the National Maritime Museums (NMM) for conservation of the ship. Research tasks were stated within the project "Cure the *Vasa*", and focused on the study of oxidation and decomposition processes, on methods for the removal of acid and prevention of its development, on the study of decomposition of wood by iron and acid, on the removal of iron compounds from the *Vasa* wood, and on the stability and decomposition of PEG.

Collaboration between Swedish Maritime Museums authorities and CSGI (*Consorzio Interuniversitario per lo Sviluppo dei Sistemi a Grande Interfase*; Centre for Colloid and Surface Science) started in 2003, aimed at solving the problem of deacidification.[56–60] The focus was set on the use of non-aqueous dispersions of alkaline nanoparticles [*e.g.* $Ca(OH)_2$, $Mg(OH)_2$] for the treatment of the *Vasa* wood. The main reasons for testing these materials with respect to aqueous alkaline solutions were as follows.

1. The use of solid particles in a non-aqueous solvent allows the controlled release of OH^- ions within the wood matrix. On the other hand, a concentrated alkaline solution provides a sudden neutralization, delivering all at once a large amount of hydroxyl ions that could be harmful to the wood fibres, *e.g.* through β-alkoxy elimination (see Section 15.4). Moreover, in this case, the alkaline excess would be left in the form of free, mobile OH^- ions, which are aggressive to cellulose.
2. Carrier solvents with lower surface tension than water (*e.g.* alcohols) are able to wet the wood matrix quickly and effectively.

In fact, the treatment of *Vasa* timbers with sodium carbonate–bicarbonate aqueous poultices removed PEG and temporarily raised the surface pH, but the neutralization was effective only in the first 0.5 cm below the wood's surface, and did not provide long-lasting effects.[56]

Instead, it was expected that nanoparticles would penetrate within the wood's pores, neutralizing acidity and leaving a harmless alkaline reserve to prevent the recurring formation of acidity, which is due to all the interconnected degradation reactions within the *Vasa*'s timbers. Previous experience and results on cellulose treatment at CSGI had indicated that alkaline nanoparticles provide a long-lasting alkaline reserve to paper fibres, and that the long-term stabilization of cellulose at a pH of 6–7 prevents the formation of oxidizing radicals due to the catalytic activity of iron(II) in acidic environments (see also Sections 15.5 and 15.6).[61–63] However, no deacidification based on nanomaterials had been previously performed on wood.

Samples of the ship's oak and pine timbers (see Figure 16.8) were received and characterized at the CSGI laboratories, defining quantitative parameters in order to evaluate the degradation of wood and the effectiveness of restoration methods.

All the samples showed clear evidence of the treatment with PEG, visible as a black crust (where it enclosed iron compounds such as rust) or as white stains on the surface, or black layers at depths up to 0.5–1 cm beneath the surface. Stains of rust were also visible on some samples. Preliminary FTIR analysis of the black, white and orange crusts confirmed the presence of PEG and iron compounds. The consistency of the samples was softer on the surface and generally within the first centimetre through the cross-section, where the wood was impregnated with PEG. At higher depths the samples showed a harder

Figure 16.8 Samples (15–20 cm long) of PEG-treated waterlogged wood from the *Vasa* ship.

consistency and looked more similar to untreated wood. Measurements performed locally with indicator papers highlighted the fact that the pH varied along the surface and with depth, reaching very low values (0–2) in correspondence with rust stains and very acidic areas, and higher values (2–4) at greater depths, where PEG penetration was lower. In order to obtain average pH values, cubic blocks (3×3×3 cm) were cut from samples along the radial direction, which underwent milling and extraction in cold water. The pH of the extraction solution was measured, showing values of 2–3 for both pine and oak.

Thermal analysis of the samples revealed more about the conservation status of the cellulose. The maximum pyrolysis temperature of cellulose (also referred to as T_{max}) is a parameter observable through DTG (differential thermogravimetry) analysis of wood, paper and textiles (linen, cotton, canvas). This kind of micro-destructive analysis requires a few milligrams of sample and is carried out in less than an hour. The micro-samples are burnt (under N_2 or O_2 atmosphere), and the weight loss is measured as temperature is increased. The T_{max} is identified as a maximum in the first derivative of the weight loss *vs.* temperature curve, and represents the point at which the rate of thermal decomposition of cellulose has its highest value (see Figure 16.9). The T_{max} value varies according to the conservation status of cellulose, whether it is acidified, hydrolysed or oxidized. Higher pyrolysis temperatures are associated with a better degree of conservation of wood and cellulose.[64–66]

Pyrolysis temperatures as low as 325 °C were found for acidic *Vasa* oak cellulose (see Figure 16.9), and of 335 °C for pine. These are low values, in comparison to the ones associated with fresh wood (fresh oak: 345 °C; fresh pine: 367 °C), and are surely indicative of the degradation of the polymer chains. The shift in the pyrolysis temperature can be explained by different considerations.

First, acidification induces depolymerization of the cellulose chains, producing a large fraction of lower molecular weight oligomers, which require less energy in order to be thermally decomposed. Second, the acidic environment promotes the swelling and unpacking of the cellulose fibres, and consequently less protection of the individual bonds. Moreover cellulose dehydration, one of the first steps in the thermal degradation of wood, is an acid-catalysed process and, therefore, requires less energy than in a fresh, non-acidic sample. These considerations lead to the conclusion that T_{max} is

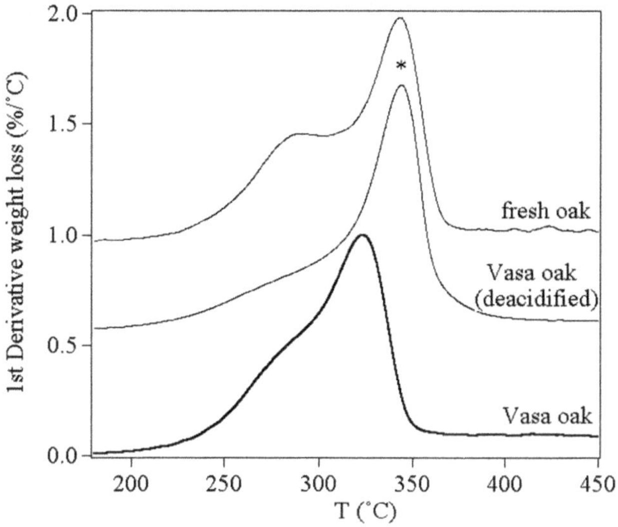

Figure 16.9 DTG (differential thermogravimetry) analysis of oak wood (fresh oak wood, *Vasa* wood before and after the deacidification treatment with alkaline nanoparticles). The maximum pyrolysis temperature (T_{max}) is marked with a "*" on the curve of the deacidified *Vasa* oak.

sensitive to the acidic environment of the samples. It is also interesting to note the strong decrease in the intensity of the peak shoulders at about 280 °C (due to hemicellulose) in the DTG curves of *Vasa* wood (see Figure 16.9), indicating that hemicellulose has been largely degraded in the 333 years on the seabed and during the following acid degradation.

In order to confirm these results and get an insight at the molecular level, research efforts focused on finding a relationship between the pyrolysis temperature of cellulose and its degree of polymerization (DP). Unfortunately, standard measurements of wood's cellulose DP cannot be performed, because wood is not soluble in the solvent used for the pre-treatment of the sample (owing to the presence of lignin).[67] Thus, for *Vasa*'s samples, DP had to be estimated in an indirect way. Whatman paper can be considered as pure cellulose, and was the ideal reference material for tests. Acidification of the paper with sulfuric acid (pH 1.2, 4 hours) resulted in a shift of T_{max} similar to that observed for the *Vasa* samples, and in a strong decrease in DP (typical of aged paper). These results supported the hypothesis, based on thermal analysis, that *Vasa* wood cellulose had undergone a strong depolymerization caused by the production of acids.

Preliminary deacidification tests were carried out on the same acidified paper sample models. Magnesium and calcium hydroxide nanoparticle dispersions in 2-propanol, in particular, were chosen for tests on acidified Whatman paper. The paper samples were soaked for 24 hours in a dispersion of $Mg(OH)_2$ nanoparticles. The treatment resulted in raising the pyrolysis temperature back to almost its original value. No increase in the DP of the paper was recorded after the deacidification treatment.[58]

These results indicated that the acidic environment had been efficiently neutralized, as was confirmed by pH measurements on the treated samples. As expected, no regeneration of the depolymerized cellulose occurred. However, no further decrease in DP was observed upon deacidification, leading to the conclusion that the application of nanoparticles can be tuned so that alkalinity is high enough to neutralize acidity, without damaging the fibres.

The increase in T_{max} after treatment with nanoparticles can also be explained by considering that earth alkaline metal ions (*e.g.* Ca^{2+}) might form a network of coordination bonds with oxidized sites in the degraded cellulose (*e.g.* carboxylic groups turn into carboxylate upon neutralization), so as to increase resistance to thermal degradation.[60]

Following these preliminary tests, the deacidification method was performed on the *Vasa* samples. Treatment of wood is associated with some problems in comparison with paper. Wood has a complex three-dimensional structure, making the penetration of particles inside the pores a key factor for a good deacidification treatment and for the creation of an alkaline reserve. Archaeological wood is usually more porous than fresh wood because bacterial and chemical degradation lead to a decrease in cell wall thickness, resulting in a larger cellular lumen. This is also true for *Vasa* wood, even though several wood degrading microorganisms had been discouraged by anoxic conditions. Figure 16.10 shows SEM images of *Vasa* wood cells from degraded areas. The walls are collapsed, and PEG can be noticed in the upper and lower corners of the picture, as a waxy material partially filling the cells. The dimensions of the lumens range from 5 to 20 μm.

The first step in the treatment of the *Vasa* samples was thus the partial removal of PEG from the wood fibres, in order to allow the penetration of

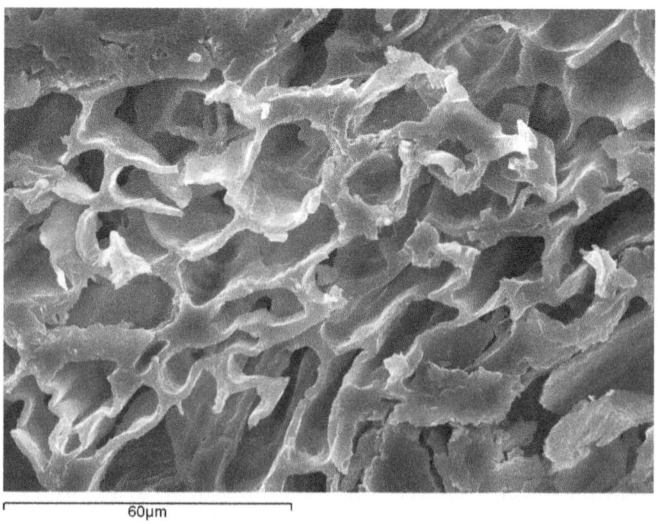

Figure 16.10 SEM image of cell walls from degraded areas of *Vasa* wood samples. Bar is 60 μm.

nanoparticles. Following the washing step, treatment on blocks was carried out using either $Ca(OH)_2$ (mean size of about 200–250 nm) or $Mg(OH)_2$ (120–150 nm) nanoparticle dispersions, in order to determine whether the difference in size and type of the particles would result in different performance. Small *Vasa* wood blocks (3×3×3 cm) were soaked into 0.07 M suspensions for 13 hours, and were allowed to dry until 2-propanol was completely evaporated. The average pH of the whole deacidified blocks, measured with the cold extraction method (as described above), was 4–5 for the $Ca(OH)_2$ treated oak sample and 6–7 for pine. These values are similar to those of fresh wood for oak and higher, but still safe, for pine (pH of fresh pine ≈5.5). The two main degradation parameters, pH and cellulose T_{max}, were then checked all along the cross-sections of the treated blocks, in order to monitor the local effects of deacidification.

The surface of the treated samples showed alkaline pH, ranging from 8 to 9 for $Ca(OH)_2$ treated wood and from 7 to 8 for $Mg(OH)_2$. Similar values were also found up to the first 0.5 cm beneath the surface, where T_{max} had reverted to values typical of fresh wood (see Figure 16.9), indicating that effective deacidification had occurred within the fibres.

However, the situation was different at depths of 0.5–1 cm beneath the surface. For $Ca(OH)_2$ treated blocks, the local pH decreased, reaching values of 4 for oak and of 5–6 for pine. Pyrolysis temperatures decreased too, gradually reaching lower values typical of acidified wood, indicating that 1 cm can be considered as the average penetration limit reached with the technique adopted for $Ca(OH)_2$ nanoparticle treatment. Figure 16.11 shows an SEM picture of calcium hydroxide nanoparticle clusters adhering to the wood fibres of *Vasa* oak wood, and Figure 16.12 shows two treated samples (pine and oak).

Neutralization of sulfuric acid within the fibres is followed by the formation of calcium sulfate crystals. However, the dimensions of the needle-like sulfate crystals formed from calcium hydroxide nanoparticles are smaller than the cell lumen, so no mechanical stress is produced (see Figure 16.11). Moreover, conversion of $Ca(OH)_2$ (portlandite) to $CaCO_3$ (calcite) was monitored through X-ray diffractometry (XRD) measurements, and was shown to occur four days after the application. This is an important aspect, because calcite provides a milder and less aggressive alkaline reserve on the cellulose fibres. The $Mg(OH)_2$ treated wood showed better features in terms of particle penetration; good values of pH (around 5–6) and T_{max} were found at depths up to 2 cm.

The results obtained by treating *Vasa* wood with nanoparticles were thus promising. However, in order to check the efficacy of treatment, accelerated ageing of the treated samples was carried out.[58,59,61] Blocks of *Vasa* oak and pine wood, deacidified with $Ca(OH)_2$ and $Mg(OH)_2$ suspensions, underwent accelerated ageing in an oven at a temperature of 90 °C and relative humidity of 78% (using a saturated KCl solution). In these conditions hydrolysis and oxidation of cellulose occur faster, leading to a decrease in pH, T_{max} and DP. After 120 hours of ageing, the pyrolysis temperature of the treated oak wood was still significantly higher than that of the untreated wood. For pine wood the situation was slightly different: after almost 60 hours of ageing, the T_{max} of

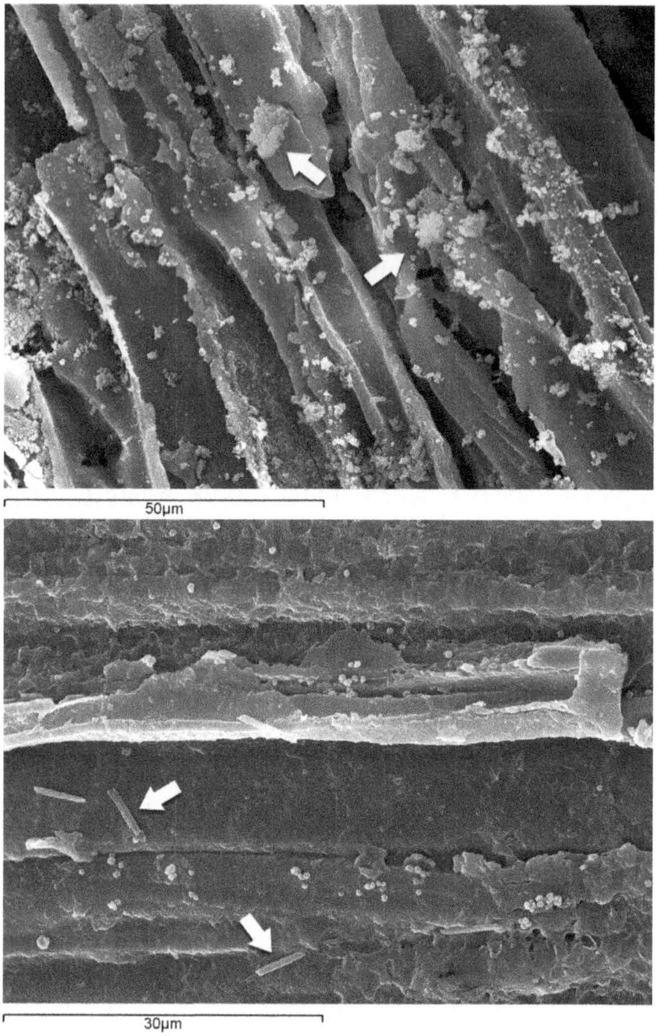

Figure 16.11 (Top) SEM pictures of $Ca(OH)_2$ nanoparticles clusters adhering to the fibres of a *Vasa* oak wood sample, following treatment by soaking in particle dispersions. Bar is 50 μm. (Bottom) Detail showing the formation of needle-like calcium sulfate crystals (confirmed by EDX) upon neutralization of sulfuric acid by the alkaline nanoparticles. Bar is 30 μm. The arrows highlight the presence of particles.

wood treated with $Ca(OH)_2$ decreased to values typical of acidic samples, while for wood treated with $Mg(OH)_2$ it remained closer to the value for fresh wood, even after 120 hours.

Conversion of accelerated ageing time into real ageing is not straightforward. Whatman paper was once again used as reference cellulose material. The same accelerated ageing procedure was carried out on acidified Whatman samples, and resulted in a strong decrease of T_{max} and DP to values typical of very

Figure 16.12 Samples of *Vasa* wood, following washing in water and treatment by soaking in dispersions of $Ca(OH)_2$ nanoparticles in 2-propanol. The treated samples are about 8 cm (pine) and 2 cm (oak) long, and have been cut from larger original *Vasa* samples, which are shown in the insert (see also Figure 16.8).

degraded historical paper, unless treatment with nanoparticles was provided. By matching the results obtained on wood and on the reference Whatman paper, it is possible to suggest that both calcium and magnesium hydroxide nanoparticles might provide long lasting resistance against the recurring acidity in *Vasa*'s wood. Magnesium hydroxide nanoparticles, however, are smaller and grant better penetration inside the wood matrix, hence allowing longer duration of resistance to recurring acidity.

Research efforts are now being focused on increasing the performance of the deacidification treatment. Deeper and more homogeneous penetration of the particles can be achieved in different ways. One possible method is to further reduce the particle size.[60] Another method relies on the use of different solvents than 2-propanol. The ideal solvent should be absorbed in large amounts and easily wet the wood matrix, fully evaporating after some time, but not too quickly so that nanoparticles can be carried deeply inside the wood bulk.

The type of solvent is also important when considering its interactions with the wood fibres. Very inert solvents could be very useful for treatments on waterlogged wood. Fluorinated solvents, such as perfluoropolyethers, show good features in this sense: they are chemically inert, non-toxic and environmentally safe, thermally and chemically resistant, and have low surface tension.

A stable suspension of $Mg(OH)_2$ nanoparticles was realized for tests on archaeological wood, using commercial perfluoropolyesthers (Galden®, Solvay-Solaxis) with a low surface tension similar to that of 2-propanol (about 20 mN m^{-1}), and a relatively low density and viscosity. In order to obtain a stable dispersion in Galden, dry solid nanoparticles were initially dispersed in a 10% aqueous solution of a Fluorolink® (Solvay-Solaxis) series surfactant, previously brought to pH 11 with ammonia in order to prevent the carboxyl groups of the surfactant neutralizing the alkaline nanoparticles. The surfactant, whose structure formula at alkaline pH is $Cl–(C_3F_6O)_n–CF_2–COO^-$, is adsorbed on the surface of the particles, which thus gain affinity for the fluorinated phase. Nanoparticles are finally extracted in Galden, and the suspension obtained is stable for many hours, which is sufficient for the application on archaeological wood.

The first application was done on a block (4×4×4 cm) of *Vasa* pine wood, previously washed and allowed to dry as described before. The block was the soaked for 24 hours in a suspension of $Mg(OH)_2$ nanoparticles in Galden, and was characterized after the treatment. An alkaline pH (around 8) was detected throughout the bulk, from the surface to the core of the sample, using an indicator paper. The DTG curves of samples derived from depths up to 2 cm showed the same T_{max} as fresh pine wood, confirming that deacidification had occurred in a homogeneous way throughout the wood bulk.

Energy-dispersive X-ray spectroscopy (EDS) analysis revealed the presence of fluorine (1–3% by weight) up to the first 1–3 mm beneath the surface, decreasing to 1% at 5 mm. Fluorine was not detected at greater depths. Hydrothermal ageing of the treated block was carried out in the same conditions as described before ($T = 90$ °C, RH = 78%), and the results were promising: after more than 300 hours of ageing (more than double the ageing time after applications with 2-propanol) the wood showed a pH of 7 all across its section, and a cellulose pyrolysis temperature typical of fresh pine at depths up to 1 cm. From 1 to 2 cm, T_{max} decreased to 357 °C, a value still much higher than that characteristic of strongly acidified wood.

Dispersions of magnesium hydroxide nanoparticles in fluorinated solvents could thus be efficiently used for the deacidification of the *Vasa* wood samples, granting a more homogeneous treatment and, consequently, a longer resistance to accelerated hydrothermal ageing than dispersions in 2-propanol. This better performance, however, is partially balanced by the introduction of fluorine in the wooden matrix. The stability of fluorinated surfactants inside wood is still a debated topic, and the possibility of surfactant degradation, leading to the production of acidic compounds, is still to be fully investigated and understood.[68]

Another stimulating issue is represented by the upscaling of the deacidification method (either with 2-propanol or fluorinated solvents) in order to treat large timbers. The partial removal of PEG from the acidic timbers, in order to grant the nanoparticles access to the wood fibres, is associated with several difficulties. Apart from the logistical problems due to washing and treating large objects, it must be considered that the drying step, following deacidification, should be performed in controlled conditions, and a new wood consolidant should be quickly applied in order to avoid shrinking and physical collapse of the wooden structure.

Research is also focusing on methods for the extraction of iron from wood, which could be coupled with deacidification treatments based on alkaline nanoparticles. So far, extraction treatments have been described using aqueous solutions of complexing agents such as DTPA (diethylenetriamine pentaacetic acid).[69,70] The effectiveness of extraction depends on several factors, including wood porosity, the pH of both wood and the extracting solution, concentration of the solution and frequency of treatments. It has been reported that, even under ideal conditions, extraction of iron to a large extent may take months or years to occur. New impregnation with PEG, followed by freeze-drying, has been performed after the extraction of iron compounds from wood.

Finally, it is worth reporting on an important case study involving another historical shipwreck, that of the *Mary Rose*. Sunk off the south coast of England in 1545, the ship was recovered in 1982 and has been stabilized with different methods in the last 30 years, including extensive PEG spraying treatment. Similarly to the *Vasa*, in fact, the timbers contain about 2 tonnes of sulfur species (originating from the reduction of sulfate ions by sulfate-reducing bacteria), which can oxidise to form sulfuric acid. Recently, particles of $SrCO_3$ have been tested on wood samples from the *Mary Rose*.[71,72] Strontium carbonate was selected because it is a mild alkaline agent that has affinity for sulfate ions, and particles (with crystallites of 20–50 nm) were produced using high energy ball milling, starting from commercial particles. Dispersions in 2-propanol were obtained by sonication. Small wood samples were soaked in the dispersions and underwent further sonication (3 days). Analysis showed that the treatment resulted in the penetration of particles through samples of only about 1 mm (see Figure 16.13),[71] with respect to the deeper penetration observed with calcium and magnesium hydroxide nanoparticles as reported in the previous sections. Neutralization of the sulfuric acid formed is expected to some extent, but an interesting feature is the effect on the reduced sulfur compounds within the wood.

Spectroscopic data obtained by XANES (X-ray absorption near edge structure) showed that the treatment caused a change in the proportions of the sulfur species: the quantity of reduced species decreased, and formation of both sulfoxide and $SrSO_4$ (which is stable and insoluble) occurred. Further investigation revealed that longer treatments (one week) on larger samples of oak wood (replicas of the *Mary Rose* wood, 20×20×25 mm blocks) resulted in the penetration of particles up to 6 mm through the wood if dispersions of aqueous particles were used, while a penetration of about 2 mm could be reached using dispersions in propanol.[72] A diffusion coefficient of $3\times10^{-12}\,m^2s^{-1}$ was estimated for aqueous dispersions along the grains of fresh oak, which means that to obtain a penetration of 10 mm would require about one year. For 2-propanol dispersions, the coefficient is about one order of magnitude lower. Naturally, the presence of PEG within the fibres slows the process, posing the conservation problems already discussed for the *Vasa* case. Finally, conversion of iron sulfates into both siderite ($FeCO_3$) and goethite [FeO(OH)] occurred, preventing the formation of sulfuric acid, but the reaction is not instantaneous and seems to occur only with a portion of the particles, while others diffuse through the sulphate-impregnated wood without reacting.

This method represents thus a very interesting application, even if more research is needed to evaluate its full practical feasibility.

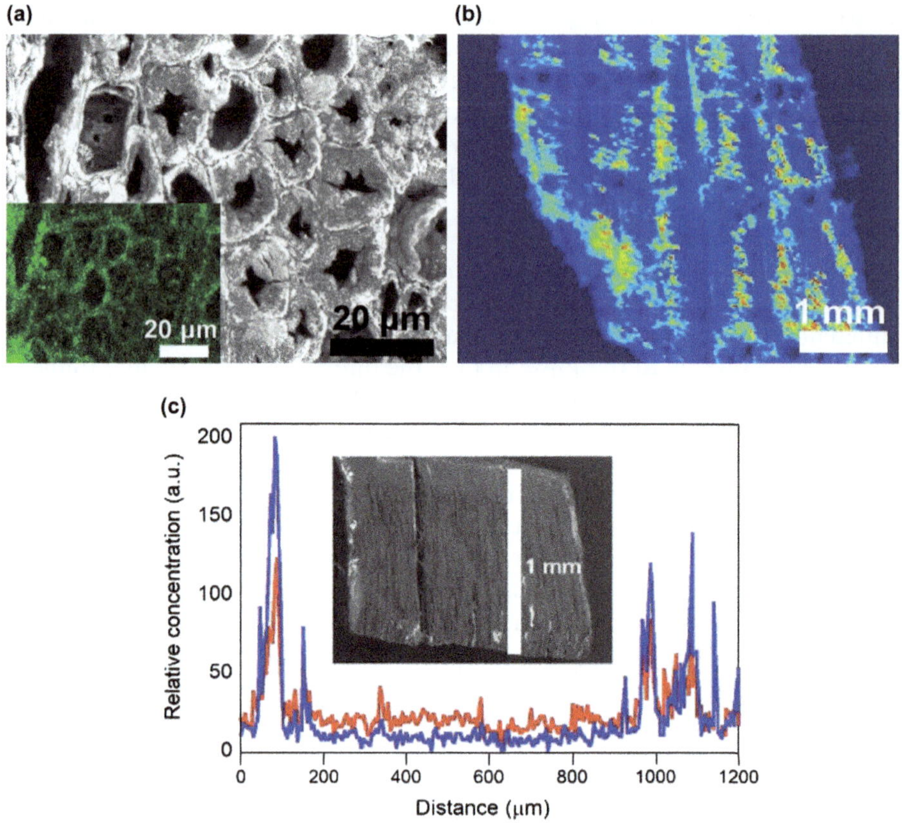

Figure 16.13 (a) SEM micrograph of *Mary Rose* timbers, treated with $SrCO_3$ particles. The inset shows the EDX strontium content map of same area. (b) X-ray fluorescence image showing the strontium distribution on the macroscale. (c) Sulfur and strontium profile as obtained using EDX analysis. The insert shows the line analysis throughout the sample cross-section.
[Reprinted from *Materials Today*, **14**/7-8, E. J. Schofield, R. Sarangi, A. Mehta, A. M. Jones, F. J. W. Mosselmans and A. V. Chadwick, Nanoparticle de-acidification of the Mary Rose, Pages No. 354–358, Copyright (2011), with permission from Elsevier.]

16.5.2 The Neutralization of Acidic VOC Emission from Wood

Acidic emissions from wood are a well known issue for conservators, mainly owing to the corrosion that emissions promote on materials in close contact or lying near the emitting substrate. Harmful situations occur when cellulose, protein, metals and other materials sensitive to acidity are stored inside wooden display cases and storage boxes, or directly put in contact with wooden frames or hardware, particularly if the wood is relatively fresh. Typical acidic volatile organic compounds (VOC) emitted by wooden substrates include acetic and formic acid, and aldheydes.[73,74]

The lowering of emissions is usually pursued by applying coatings onto the wooden surface, but it has been shown that the emission of acidic volatile organic compounds (*e.g.* acetic acid) occurs for several poly(vinyl acetate) adhesives and wood coatings used in museums for display cases, cabinets,

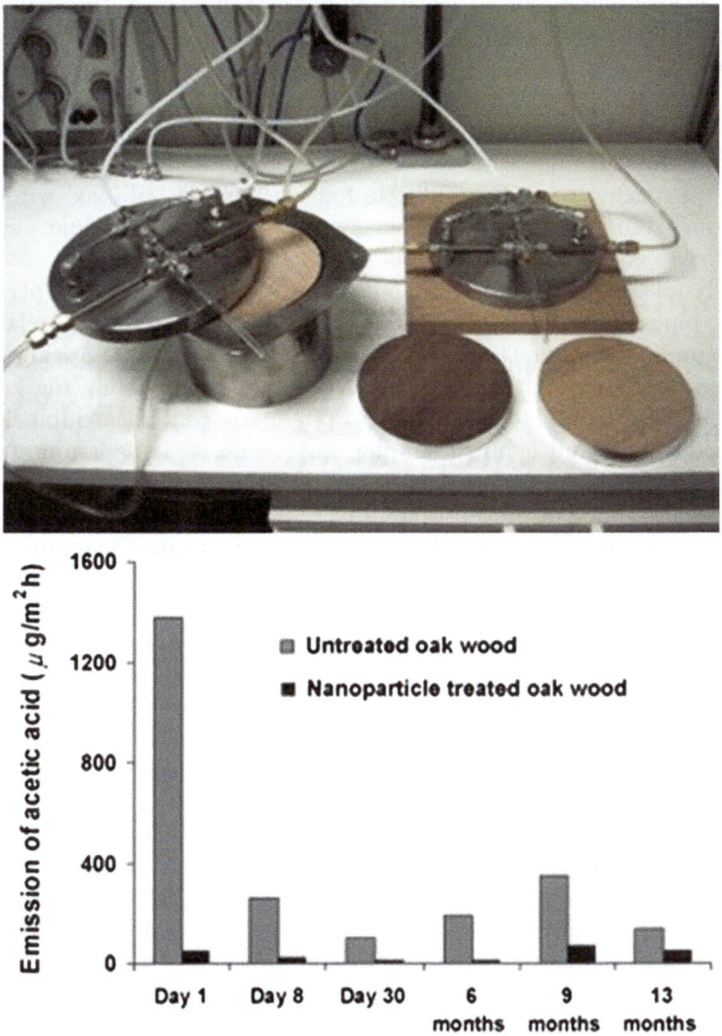

Figure 16.14 (Top) The field and laboratory emission cell (FLEC), used to determine the emission of VOC from oak wood disks. (Bottom) Emission of acetic acid vapour (μg $m^{-2}h^{-1}$) from oak, both untreated and treated with $Ca(OH)_2$ nanoparticle dispersions.
[Reprinted from *Journal of Cultural Heritage*, **10**, R. Giorgi, D. Chelazzi, E. Fratini, S. Langer, A. Niklasson, M. Rådemar, J.-E. Svensson and P. Baglioni, Nanoparticles of calcium hydroxide for wood deacidification: Decreasing the emissions of organic acid vapors in church organ environments, Pages No. 206–213, Copyright (2009), with permission from Elsevier.]

boxes or sealed enclosures, raising concerns about the use of these polymers either on supporting materials (boxes, cases, frames) and directly on cellulose or protein based artistic substrates such as wood, paper, canvas, parchment and leather.[43,44]

On the other hand, the treatment of boxes and cases with alkaline nanoparticles could be a simple, reliable and affordable solution to decrease any acidic emission. Wood hardware, in fact, could be easily provided with an alkaline buffer that penetrates the first layers, limiting the surface development of VOCs. Moreover, the treatment of valueless practical objects would be faster and easier than dealing with works of art or precious historical objects.

This approach was successfully tested on samples of oak wood, while considering solutions for the decreasing of emissions of acidic vapours in church organ environments.[75] The context was a fascinating case study: acids are slowly released from the wooden parts of organs, both during playing and when the pipes are silent. As a result, the pipes undergo corrosion and cracking, and eventually collapse. The problem has affected several historical organs all across Europe, and solutions had initially focused on coating the lead pipes with protective layers. However, several polymeric coatings exhibit the aforementioned issues, and a much simpler way to decrease emissions from oak wood (the main wood component in organs) was found. Calcium hydroxide nanoparticle dispersions in 2-propanol were used to treat model discs of fresh oak wood, using similar protocols to those adopted during the treatment of the *Vasa* samples.

The results (see Figure 16.14) showed that the particles formed an alkaline buffer on the surface and within the first millimetre of the dense fresh wood, and were effective in decreasing the emission of VOCs (particularly acetic acid, which is very corrosive to lead) to very low limits ($<70\ \mu g\ m^{-2}\ h^{-1}$) throughout the 13 months of monitoring following the application. Untreated wood emitted high concentrations of acetic acid ($200–400\ \mu g\ m^{-2}\ h^{-1}$).

Lower levels of emissions are expected from real, aged oak wood parts, so the method could be even more effective on real cases. Moreover, the internal wooden parts of the organs have no aesthetic value, so the possible white glaze due to surface deposits of nanoparticles, if used at high concentrations, does not pose any problems. These data, and the easy application mode (brush or spray), make the method extremely feasible for the treatment of any wood hardware, including cases or boxes in museums.

Further Suggested Reading

On the anatomical structure and composition of wood:

R. M. Rowell, *The Chemistry of Solid Wood, Advances in Chemistry 207*, ed. R. M. Rowell, American Chemical Society, Washington, D.C., 1984.

On the conservation methods of waterlogged archaeological materials:

D. L. Hamilton, *Methods of Conserving Archaeological Material from Underwater Sites*, Anthropology 605, Conservation of Archaeological

Resources I, Nautical Archaeology Program, Department of Anthropology, Texas A&M University, College Station, Texas, 1 January 1999.

On the advanced characterization and preservation of waterlogged wooden shipwrecks:

Proceedings of *Shipwrecks 2011, Chemistry and Preservation of waterlogged wooden shipwrecks – An International Conference*, ed. M. Ek, The Vasa Museum and KTH, Stockholm, 18–21 October 2011.

Review and Questions

This chapter deals with the description of the main issues in the conservation of archaeological and waterlogged wood, and the application of nanotechnology for the protection, consolidation and deacidification of wooden objects. The main theoretical and applicative concepts are:

- A fundamental knowledge of wood composition, structure, and main degradation paths, is the basis for deciding the appropriate restoration intervention. Essential information was provided in the chapter. Ideally, a wood conservation team could include consultant members such as agronomists or botanists, in order to provide support to conservators and materials scientists.
- Advanced diagnostic facilities (*e.g.* SEM-EDX, X-ray spectroscopy, thermal analysis) should be performed prior to any restoration intervention when dealing with archaeological wood whose anatomy, structure and composition might have been strongly altered through its conservation history.
- Amongst the traditional restoration materials, PEG is typically found as a consolidation agent. However, its use has shown significant drawbacks that can imply the raising of acidity within wooden fibres, or contribute to the problem if highly acidic or iron-containing wood is dealt with. Case studies concerning the conservation of two highly important shipwrecks, the *Vasa* and the *Mary Rose*, have been introduced and discussed. Dispersions of alkaline nanoparticles proved to be promising tools to neutralize acidity within the wood fibres. Innovative consolidants based on cellulose and other organic molecules have been discussed as possible applications, but further research efforts are necessary.
- Acidic emissions from wood are an issue of concern with respect to several aspects of conservation, *e.g.* storage. Traditional acrylate–vinyl based coatings are largely used, but develop acidity in their turn. The use of dispersions of alkaline nanoparticles to stop acidic emissions from the wood surface has been discussed. Innovative coatings obtained from silica and other inorganic nanosols are effective in functionalizing wood for different applications, such as protection from UV and microorganisms, or in providing surface consolidation.

Questions

1) Would you expect the chemical composition of archaeological, waterlogged wood to be richer in hemicellulose, cellulose or lignin? Why?
2) What kind of physico-chemical interaction is to be expected between silica-based nanosols and the treated wood substrates, in coating interventions? How does this affect the reversibility of the treatment?
3) Mention one possible advantage of carbonate-based nanoparticles with respect to hydroxides when adjusting the pH of a degraded, possibly oxidized, wood substrate.
4) If pectin is still present to some extent in the wood matrix, what risks could be involved in using both alkaline nanoparticles and chelating agents to deal with any acidic and iron-containing compounds affecting the substrate?
5) Briefly comment on the advantages of having waterlogged oak wood stabilized, respectively, at pH 5 and 7 through a deacidification treatment.

Answers

1) Holocellulose is less resistant than lignin to both chemical agents (acidity, alkalinity) and microorganisms. Hemicellulose, being less crystalline than cellulose, is degraded at a faster rate. So the composition is expected to be richer in lignin, rather than cellulose and hemicellulose.
2) If silica nanogels are considered, adhesion to the wood substrate is due to the formation of hydrogen bonds (between the Si–OH groups and the hydroxyl groups of wood's cellulose), so it is a weak interaction based on secondary bonds. By modifying the nanosols with organic groups, it is possible to have primary bonds formed between wood and the nanoparticles. Consequently, the reversibility of the treatment will be decreased, but further stability of positive effects could be granted.
3) Aged (oxidized) cellulose is sensitive to alkaline pH, so carbonates could provide a milder deacification, with reduced risks for the treated substrate. The setting (carbonation) of hydroxide into carbonate is another factor that decreases the risk of damage to the wood fibres.
4) Pectin can form complex compounds with bivalent cations (such as Ca^{2+}, Mg^{2+}), and the resulting network is soluble by using chelating agents (*e.g.* EDTA). Treatment with calcium or magnesium hydroxide nanoparticles provides neutralization of acidity, and might result in the formation of a network between sites in pectin chains and metal ions. However, EDTA could damage this network, with detrimental effects.
5) A pH of 5 is closer to that of fresh oak wood, and poses no risk of alkaline degradation of cellulose. However, stabilization at pH 7 is still safe for cellulose, and would also inhibit the oxidizing activity of any degradation compound containing iron ions (rusting bolts *etc.*).

References

1. D. A. I. Goring and T. E. Timel, *Tappi*, 1962, **45**(6), 454.
2. R. C. Pettersen, in *The Chemistry of Solid Wood, Advances in Chemistry 207*, ed. R. Rowell, American Chemical Society, Washington, D.C., 1984.
3. E. Sjöström, *Wood Chemistry: Fundamental and Applications*, Academic Press, New York, 1981, p. 56.
4. C. M. G. C. Renard and J. F. Thibault, *Carbohydr. Res.*, 1996, **286**, 139.
5. O. Erilkssen, D. A. I. Goring and B. O. Lindgren, *Wood Sci. Technol.*, 1980, **14**, 267.
6. T. Iversen, U. Westermark and B. Samuelsson, *Holzforschung*, 1987, **41**, 119.
7. O. Karlsson, T. Ikeda, T. Kishimoto, K. Magara, Y. Matsumoto and S. Hosoya, *J. Wood Sci.*, 2004, **50**, 142.
8. F. De Angelis and S. Reale, *La Chimica e L'Industria*, 2006, **88**(1), 58.
9. R. Vanholme, B. Demedts, K. Morreel, J. Ralph and W. Boerjan, *Plant Physiol.*, 2010, **153**, 895.
10. M. K. Misra, K. W. Ragland and A. J. Baker, *Biomass Bioenergy*, 1993, **4**(2), 103.
11. D. L. Hamilton, *Methods of Conserving Archaeological Material from Underwater Sites*, Anthropology 605, Conservation of Archaeological Resources I, Nautical Archaeology Program, Department of Anthropology, Texas A&M University, College Station, Texas, 1 January 1999.
12. G. Giachi, F. Bettazzi, S. Chimichi and G. Stacciali, *J. Cult. Herit.*, 2003, **4**, 75.
13. J. Gierer, *Sven. Papperstidn.*, 1970, **73**, 571.
14. J. Gierer and L. Sundholm, *Sven. Papperstidn.*, 1971, **74**, 345.
15. D. Fengel and G. Wegener, *Wood – chemistry, ultrastructure, reactions*, Walter de Gruyter, Berlin and New York, 1984.
16. J. Gierer, *Holzforschung*, 1982, **36**, 43.
17. S. A. Rydholm, *Pulping Processes*, Interscience Publishing, New York, 1965.
18. J. Gierer, *Wood Sci. Technol.*, 1980, **14**, 241.
19. J. Marton, in *Lignins. Occurrence, Formation, Structure and Reactions*, ed. K. V. Sarkanen and C. H. Ludwig, Wiley-Interscience, New York, 1971, p. 639.
20. J. Gierier and I. Norén, *Papier*, 1981, **18**, 35(10A).
21. E. Franceschi, I. Cascone and D. Nole, *J. Therm. Anal. Cal.*, 2008, **92**(1), 319.
22. H. Claus, *Micron*, 2004, **35**, 93.
23. S. R. Collinson and W. Thielemans, *Coord. Chem. Rev.*, 2010, **254**, 1854.
24. X. Q. Whang, B. H. Fei and H. Q. Ren, *Spectrosc. Spect. Anal.*, 2009, **29**(5), 1272.
25. D. W. Grattan, *Stud. Conserv.*, 1982, **27**, 124.
26. W. R. Ambrose, *Conservation of Stone and Wooden Objects*, 1970, **2**, 53, IIC, London.

27. H. McKerrell, E. Roger and A. Varsanyi, *Stud. Conserv.*, 1972, **17**, 111.
28. B. Mahltig, C. Swaboda, A. Roessler and H. Böttcher, *J. Mater. Chem.*, 2008, **18**, 3180.
29. P. Tingaut, O. Weigenand, H. Militz, B. de Jèso and G. Sèbe, *Holzforschung*, 2005, **59**, 397.
30. P. Tingaut, O. Weigenand, C. Mai, H. Militz and G. Sèbe, *Holzforschung*, 2006, **60**, 271.
31. K. Ogiso and S. Saka, *Mokuzai Gakkaishi*, 1994, **40**, 1100.
32. C. Mai and H. Militz, *Wood Sci. Tech.*, 2004, **37**, 339.
33. G. Cipriani, A. Salvini, P. Baglioni and E. Bucciarelli, *J. Appl. Polym. Sci.*, 2010, **118**, 2939.
34. M. Christensen, H. Kutzke and F. K. Hansen, in Proceedings of *Shipwrecks 2011, Chemistry and Preservation of Waterlogged Wooden Shipwrecks – An International Conference*, ed. M. Ek, The Vasa Museum and KTH, Stockholm, 18–21 October 2011.
35. D. J. Wigglesworth, *The Chemistry and Physics of Coatings*, RCS, Cambridge, 1994.
36. M. de Meijer, *Prog. Org. Coat.*, 2001, **43**, 217.
37. S. T. Eckersley and A. Rudin, *Prog. Org. Coat.*, 1994, **23**, 384.
38. B. G. Bufkin and G. C. Wildman, *For. Prod. J.*, 1980, **30**(10), 37.
39. A. S. Kimerling and S. R. Bhatia, *Prog. Org. Coat.*, 2004, **51**, 15.
40. R. Bongiovanni, F. Montefusco, A. Priola, N. Macchioni, S. Lazzeri, L. Sozzi and B. Ameduri, *Prog. Org. Coat.*, 2002, **45**, 359.
41. F. Mauguière-Guyonnet, D. Burget and J. P. Fouassier, *Prog. Org. Coat.*, 2006, **57**, 23.
42. S. Chang and P. Chou, *Polym. Degrad. Stab.*, 1999, **63**, 435.
43. J. L. Down, M. A. McDonald, J. Tétreault and R. S. Williams, *Stud. Conserv.*, 1996, **41**, 19.
44. J. Tétreault and E. Stamatopoulou, *Stud. Conserv.*, 1997, **42**, 141.
45. S. P. Pappas, *Radiation Curing: Science and Technology*, ed. S. P. Pappas, Plenum Press, New York, 1971.
46. S.-T. Chang and P.-L. Chou, *Polym. Degrad. Stabil.*, 1999, **63**, 435.
47. C.-S. Chern, *Principles and Applications of Emulsion Polymerization*, Wiley, Hoboken, New Jersey, 2008.
48. A. Mader, A. Schirò, M. Brischetto and B. Pizzo, *Prog. Org. Coat.*, 2011, **71**, 123.
49. M. de Meijer, B. van de Velde and H. Militzt, *J. Coat. Technol.*, 2001, **73**, 39.
50. B. Ohrelius, *Vasa, The King's Ship*, ed. Rabén & Sjrgen, Chilton Books, 1963.
51. M. Sandström, Y. Fors and I. Persson, *Vasa Stud.*, 2003, **19**, 1.
52. M. Sandström, F. Jalilehvand, I. Persson, U. Gelius, P. Frank and I. Hall-Roth, *Nature*, 2002, **415**, 893.
53. B. Håfors, *Vasa Stud.*, 2001, **18**, 185.
54. E. Guilminot, F. Dalard and C. Degigmy, *Eur. Fed. Corros. Publ.*, 2000, **28**, 300.

55. G. Almkvist and I. Persson, *Holzforschung*, 2008, **62**, 704.
56. R. Giorgi, D. Chelazzi and P. Baglioni, *Langmuir*, 2005, **21**, 10743.
57. R. Giorgi, D. Chelazzi and P. Baglioni, *App. Phys. A*, 2006, **83**, 567.
58. D. Chelazzi, R. Giorgi and P. Baglioni, *Proceedings of the International Heritage, Weathering and Conservation Conference (HWC – 2006)*, Madrid, Spain, June 21–24, 2006, ed. R. Fort, M. Alvarez de Buergo, M. Gomez-Heras and C. Vasquez-Calvo, Taylor & Francis, London, 2006, p. 797.
59. D. Chelazzi, R. Giorgi and P. Baglioni, *Macromol. Symp.*, 2006, **238**, 30.
60. P. Baglioni, D. Chelazzi, R. Giorgi, G. Poggi and N. Toccafondi, in *Proceedings of Shipwrecks 2011, Chemistry and Preservation of Waterlogged Wooden Shipwrecks – An International Conference*, ed. M. Ek, The Vasa Museum and KTH, Stockholm, 18–21 October 2011, p. 136.
61. R. Giorgi, C. Bozzi, L. Dei, C. Gabbiani, B. W. Ninham and P. Baglioni, *Langmuir*, 2005, **21**, 8495.
62. R. Giorgi, L. Dei, M. Ceccato, C. Schettino and P. Baglioni, *Langmuir*, 2002, **18**, 8198.
63. G. Poggi, R. Giorgi, N. Toccafondi, V. Katzur and P. Baglioni, *Langmuir*, 2010, **26**, 19084.
64. I. C. A. Sandu, M. Brebu, C. Luca, I. Sandu and C. Vasile, *Polym. Degrad. Stab.*, 2003, **80**, 83.
65. S. Vicini, E. Princi, G. Luciano, E. Franceschi, E. Pedemnte, D. Oldak, H. Kaczmaerk and A. Sionkowska, *Thermochim. Acta*, 2004, **418**, 123.
66. S. Soares, G. Camino and S. Levchik, *Polym. Degrad. Stab.*, 1995, **49**, 275.
67. ASTM D4243–99(2004)e1: Standard Test Method for Measurement of Average Viscometric Degree of Polymerization of New and Aged Electrical Papers and Boards, 2004, American Society for Testing and Materials – Edition: 2004.
68. S. Zumbühl and S. Wuelfert, *Stud. Conserv.*, 2001, **46**, 169.
69. Y. Fors and M. Sandström, *Chem. Soc. Rev.*, 2006, **35**, 399.
70. G. Almkvist and I. Persson, in *Heritage Microbiology and Science: Microbes, Monuments and Maritime Materials*, ed. E. May, M. Jones and J. Mitchell, RSC Publishing, Cambridge (UK), 2008.
71. E. J. Schofield, R. Sarangi, A. Mehta, A. M. Jones, F. J. W. Mosselmans and A. V. Chadwick, *Mater. Today*, 2011, **14**, 354.
72. A. V. Chadwick, E. J. Schofield, A. M. Jones, G. Cibin and J. F. W. Mosselmans, *Solid State Ion*, 2012, doi:10.1016/j.ssi.2012.02.035.
73. P. C. Arni, G. C. Cochrane and J. D. Gray, *J. Appl. Chem.*, 1965, **15**, 305.
74. P. C. Arni, G. C. Cochrane and J. D. Gray, *J. Appl. Chem.*, 1965, **15**, 363.
75. R. Giorgi, D. Chelazzi, E. Fratini, S. Langer, A. Niklasson, M. Rådemar, J.-E. Svensson and P. Baglioni, *J. Cult. Herit.*, 2009, **10**, 206.

Conclusions: Future Horizons and Perspectives for Material Science in Cultural Heritage Conservation

P. BAGLIONI

After journeying through the complex and exciting world of materials and colloid science applied to conservation, the reader will be aware that both research and applicative work in this field are far from being concluded. Indeed, despite the wide range of different applications and case studies described within these pages, conservation science is still a growing field, and the potential of nanosciences can be further developed to address a variety of issues concerning the preservation of classic cultural heritage. Every case study shows peculiar difficulties, and the definition of unique protocols for the cleaning, consolidation and pH control of artistic objects is a demanding task, whose completion should meet the needs of end-users who constantly turn to scientists in order to obtain affordable and long-lasting solutions. For instance, collagen-based substrates such as parchment and leather may exhibit severe degradation (such as the infamous leather "red rot") and, being extremely sensitive to several solvents and restoration materials, pose a serious challenge to both scientists and conservators.

In addition, a new threat to cultural heritage preservation is represented by the conservation and restoration of the latest Modern (1940s–1970s) and Contemporary works of art. Most of these objects exhibit complex compositions and are affected by degradation processes that can greatly differ

RSC Nanoscience & Nanotechnology No. 28
Nanoscience for the Conservation of Works of Art
Edited by Piero Baglioni and David Chelazzi

Published by the Royal Society of Chemistry, www.rsc.org

from those related to classic works of art. As a result, many of the conventional methods presently available are not effective in counteracting degradation processes that are often severe and can be dramatically fast. A typical example concerns the degradation of modern inks. The industrial synthesis of dyes replaced the tedious manual preparation of iron-gall inks, allowing mass production of writing fluids with optimal aesthetic and functional features; however, the practical and functional needs overcame concerns about the long-term preservation of modern materials. As a result, both documents and artefacts (such as artists' or writer's manuscripts, notes, *etc.*) commonly exhibit severe degradation, such as ink fading and corrosion, which hinders the readability of such artefacts or historical documents. Other representative problematic cases in modern conservation science include the cleaning of acrylic paintings, the preservation of movie films and the restoration of plastic sculptures, to mention only some of the most important issues.

The degradation processes that develop within artistic substrates often take place at the interface and are at the meso-scale domain. Therefore, materials with the same dimensional scale (*e.g.* nanomaterials) can prove highly effective in repressing such processes. However, the pathways that lead to the production of the nanomaterials tailored for specific degradation issues are, in most cases, not developed yet. Comparing works of art to patients demanding care, we can state that in many cases a cure has not yet been found.

We can only mention here some of the most exciting and open research topics in this field. Functionalized nanoparticles and nanocomposite materials exhibit promising potential for future applications. Functionalization may grant anchoring of nanoparticles selectively onto specific sites to direct any intervention and induced chemical process at desired areas of the artistic substrates. Moreover, tailored chemical functions might grant nanoparticles (such as alkaline buffers) specific characteristics such as plasticizing or mechanical properties. Nanocomposites, as already stated in Chapter 14, can provide the best qualities of two very different worlds, *i.e.* inorganic nanoparticles and organic polymers. Concerning cleaning and detergency, inert nanostructured fluids (water-in-oil or "waterless" oil-in-oil microemulsions) could be excellent candidates for the highly effective and non-invasive cleaning of sensitive surfaces that are swelled or leached by several common solvents and water-based systems.

Since colloids and material scientists took their first steps in the complex and fascinating world of conservation science, almost 30 years ago, a great door has been opened to us, revealing a long, intriguing and challenging path.

The hope of this book's authors and editors is that this treatise will contribute to making conservators aware of the possibilities of colloid and materials science (in a word, nanoscience) in the restoration of works of art, while familiarizing scientists and students with the issues involved in the preservation of cultural heritage and with some of the most innovative and advanced solutions that have been developed in the last decades. Our hope is this contribution will succeed in giving two fundamental concepts to readers: 1) the importance of cooperation between scientists and conservators, in triggering an iterative and interactive transfer of knowledge that leads to the production of reliable and long-term solutions; 2) the necessity of facing all the problematic aspects of cultural heritage preservation with a scientific interdisciplinary and creative approach.

In fact, creativity is the key to shaping and realizing new restoration nanomaterials. After a long period of time science and art merge again, as during the Italian renaissance when artists were also scientists; for example, just to mention one: Leonardo da Vinci.

Subject Index

Page numbers in *italics* refer to figures or tables.